Meat processing

Related titles from Woodhead's food science, technology and nutrition list:

Meat refrigeration (ISBN: 1 85573 442 7)

Based on the work of the internationally renowned Food Refrigeration and Process Engineering Research Centre (FRPERC) at the University of Bristol, this will be the standard work on meat refrigeration, covering both individual quality issues and the management of the cold chain from carcass to consumer.

Lawrie's meat science Sixth edition (ISBN: 1 85573 395 1)

This book remains a standard for both students and professionals in the meat industry. It provides a systematic account of meat science from the conception of the animal until human consumption, presenting the fundamentals of meat science. This sixth edition incorporates the significant advances in meat science which have taken place during the past decade including our increasingly precise understanding of the structure of the muscle, as well as the identification of the aberrations in DNA which lead to the development of BSE syndrome in meat.

HACCP in the meat industry (ISBN: 1 85573 448 6)

Following the crises involving BSE and *E.coli*, the meat industry has been left with an enormous consumer confidence problem. In order to regain the trust of the general public the industry must establish and adhere to strict hygiene and hazard control systems. HACCP is a systematic approach to the identification, evaluation and control of food safety hazards. It is being applied across the world, with countries such as the US, Australia, New Zealand and the UK leading the way. However, effective implementation in the meat industry remains difficult and controversial. This book is a survey of key principles and best practice, providing an authoritative guide to making HACCP systems work successfully in the meat industry.

Details of these books and a complete list of Woodhead's food science, technology and nutrition titles can be obtained by:

- visiting our web site at www.woodhead-publishing.com
- contacting Customer Services (email: sales@woodhead-publishing.com; fax: +44 (0) 1223 893694; tel.: +44 (0) 1223 891358 ext. 30; address: Woodhead Publishing Limited, Abington Hall, Abington, Cambridge CB1 6AH, England)

If you would like to receive information on forthcoming titles in this area, please send your address details to: Francis Dodds (address, tel. and fax as above; e-mail: francisd@woodhead-publishing.com). Please confirm which subject areas you are interested in.

Meat processing

Improving quality

Edited by
Joseph Kerry, John Kerry and David Ledward

CRC Press
Boca Raton Boston New York Washington, DC

WOODHEAD PUBLISHING LIMITED
Cambridge England

Published by Woodhead Publishing Limited
Abington Hall, Abington,
Cambridge CB1 6AH
England
www.woodhead-publishing.com

Published in North America by CRC Press LLC
2000 Corporate Blvd, NW
Boca Raton FL 33431
USA

First published 2002, Woodhead Publishing Limited and CRC Press LLC
© 2002, Woodhead Publishing Limited
The authors have asserted their moral rights.

This book contains information obtained from authentic and highly regarded sources. Reprinted material is quoted with permission, and sources are indicated. Reasonable efforts have been made to publish reliable data and information, but the authors and the publishers cannot assume responsibility for the validity of all materials. Neither the authors nor the publishers, nor anyone else associated with this publication, shall be liable for any loss, damage or liability directly or indirectly caused or alleged to be caused by this book.

Neither this book nor any part may be reproduced or transmitted in any form or by any means, electronic or mechanical, including photocopying, microfilming and recording, or by any information-storage or retrieval system, without permission in writing from the publishers.

The consent of Woodhead Publishing Limited and CRC Press LLC does not extend to copying for general distribution, for promotion, for creating new works, or for resale. Specific permission must be obtained in writing from Woodhead Publishing Limited or CRC Press LLC for such copying.

Trademark notice: Product or corporate names may be trademarks or registered trademarks, and are used only for identification and explanation, without intent to infringe.

British Library Cataloguing in Publication Data
A catalogue record for this book is available from the British Library.

Library of Congress Cataloging-in-Publication Data
A catalog record for this book is available from the Library of Congress.

Woodhead Publishing Limited ISBN 1 85573 583 0 (book); 1 85573 666 7 (e-book)
CRC Press ISBN 0-8493-1539-5
CRC Press order number: WP1539

Cover design by Martin Tacchi
Project managed by Macfarlane Production Services, Markyate, Hertfordshire
(macfarl@aol.com)
Typeset by MHL Typesetting Limited, Coventry, Warwickshire
Printed by TJ International Limited, Padstow, Cornwall, England

Contents

Contributors xi

1 Introduction 1
   D. Ledward, The University of Reading

2 Defining meat quality 3
   T. Becker, University of Honenheim, Stuttgart
   2.1   Introduction: what is quality? 3
   2.2   Consumer perceptions of quality 6
   2.3   Supplier perceptions of quality 14
   2.4   Combining consumer and supplier perceptions: the quality
          circle 16
   2.5   Regulatory definitions of quality 19
   2.6   Improving meat and meat product quality 21
   2.7   References 23

Part I Analysing meat quality 25

3 Factors affecting the quality of raw meat 27
   R. K. Miller, Texas A & M University, College Station
   3.1   Introduction 27
   3.2   Quality meat composition and structure 27
   3.3   Breed and genetic effects on meat quality 37
   3.4   Dietary influences on meat quality 49
   3.5   Rearing and meat quality 52
   3.6   Slaughtering and meat quality 52
   3.7   Other influences on meat quality 56

vi Contents

	3.8	Summary: ensuring consistency in raw meat quality	56
	3.9	Future trends	57
	3.10	References	57

4 The nutritional quality of meat 64
J. Higgs, Food to Fit, Towcester and B. Mulvihill, Republic of Ireland

	4.1	Introduction	64
	4.2	Meat and cancer	66
	4.3	Meat, fat content and disease	69
	4.4	Fatty acids in meat	71
	4.5	Protein in meat	78
	4.6	Meat as a 'functional' food	79
	4.7	Meat and micronutrients	82
	4.8	Future trends	88
	4.9	Conclusion	92
	4.10	References	92

5 Lipid-derived flavors in meat products 105
F. Shahidi, Memorial University of Newfoundland, St John's

	5.1	Introduction	105
	5.2	The role of lipids in generation of meaty flavors	106
	5.3	Lipid autoxidation and meat flavor deterioration	108
	5.4	The effect of ingredients on flavor quality of meat	110
	5.5	The evaluation of aroma compounds and flavor quality	116
	5.6	Summary	117
	5.7	References	117

6 Modelling colour stability in meat 122
M. Jakobsen and G. Bertelsen, Royal Veterinary and Agricultural University, Frederiksberg

	6.1	Introduction	122
	6.2	External factors affecting colour stability during packaging and storage	123
	6.3	Modelling dynamic changes in headspace composition	123
	6.4	Modelling in practice: fresh beef	124
	6.5	Modelling in practice: cured ham	128
	6.6	Internal factors affecting colour stability	131
	6.7	Validation of models	133
	6.8	Future trends	134
	6.9	References	135

7 The fat content of meat and meat products 137
A. P. Moloney, Teagasc, Dunsany

	7.1	Introduction	137
	7.2	Fat and the consumer	138

7.3	The fat content of meat	138
7.4	Animal effects on the fat content and composition of meat	141
7.5	Dietary effects on the fat content and composition of meat	144
7.6	Future trends	147
7.7	Sources of further information and advice	149
7.8	References	150

Part II Measuring quality 155

8 Quality indicators for raw meat 157
M. D. Aaslyng, Danish Meat Research Institute, Roskilde

8.1	Introduction	157
8.2	Technological quality	157
8.3	Eating quality	160
8.4	Determining eating quality	166
8.5	Sampling procedure	166
8.6	Future trends	168
8.7	References	168
8.8	Acknowledgemnts	174

9 Sensory analysis of meat 175
G. R. Nute, University of Bristol

9.1	Introduction	175
9.2	The sensory panel	176
9.3	Sensory tests	178
9.4	Category scales	185
9.5	Sensory profile methods and comparisons with instrumental measurements	186
9.6	Comparisons between countries	189
9.7	Conclusions	189
9.8	References	190

10 On-line monitoring of meat quality 193
H. J. Swatland, University of Guelph

10.1	Introduction	193
10.2	Measuring electrical impedance	195
10.3	Measuring pH	199
10.4	Analysing meat properties using NIR spectrophotometry	201
10.5	Measuring meat colour and other properties	201
10.6	Water-holding capacity	203
10.7	Sarcomere length	203
10.8	Connective tissue	204
10.9	Marbling and fat content	206
10.10	Meat flavour	207
10.11	Boar taint	207

10.12	Emulsions	207
10.13	Measuring changes during cooking	208
10.14	Conclusion	211
10.15	Sources of further information and advice	211
10.16	References	212

11 Microbiological hazard identification in the meat industry ... 217
P. J. McClure, Unilever Research, Sharnbrook

11.1	Introduction	217
11.2	The main hazards	218
11.3	Analytical methods	231
11.4	Future trends	233
11.5	Sources of further information and advice	234
11.6	References	234

Part III New techniques for improving quality ... 237

12 Modelling beef cattle production to improve quality ... 239
K. G. Rickert, University of Queensland, Gatton

12.1	Introduction	239
12.2	Elements of beef cattle production	240
12.3	Challenges for modellers	244
12.4	Simple model of herd structure	251
12.5	Future developments	254
12.6	References	255

13 New developments in decontaminating raw meat ... 259
C. James, Food Refrigeration and Process Engineering Research Centre (FRPERC), University of Bristol

13.1	Introduction	259
13.2	Current decontamination techniques and their limitations	260
13.3	Washing	262
13.4	The use of chemicals	263
13.5	New methods: steam	267
13.6	Other new methods	272
13.7	Future trends	273
13.8	Sources of further information and advice	276
13.9	References	277

14 Automated meat processing ... 283
K. B. Madsen and J. U. Nielsen, Danish Meat Research Institute, Roskilde

14.1	Introduction	283
14.2	Current developments in robotics in the meat industry	284
14.3	Automation in pig slaughtering	285

14.4	Case study: the evisceration process	287
14.5	Automation of secondary processes	290
14.6	Future trends	294
14.7	References and further reading	296

15   New developments in the chilling and freezing of meat   297
S. J James, Food Refrigeration and Process Engineering Research Centre (FRPERC), University of Bristol

15.1	Introduction	297
15.2	The impact of chilling and freezing on texture	299
15.3	The impact of chilling and freezing on colour	300
15.4	The impact of chilling and freezing on drip loss and evaporative weight loss	302
15.5	The cold chain	304
15.6	Temperature monitoring	306
15.7	Optimising the design and operation of meat refrigeration	308
15.8	Sources of further information and advice	310
15.9	References	310

16   High pressure processing of meat   313
M. de Lamballerie-Anton, ENITIAA, Nantes, R. Taylor and J. Culioli, INRA, Theix

16.1	Introduction: high pressure treatment and meat quality	313
16.2	General effect of high pressure on food components	314
16.3	Structural changes due to high pressure treatment of muscle	315
16.4	Influence on enzyme release and activity	318
16.5	High pressure effects on the sensory and functional properties of meat	318
16.6	Pressure assisted freezing and thawing	320
16.7	Effects on microflora	321
16.8	Current applications and future prospects	323
16.9	References	324

17   Processing and quality control of restructured meat   332
P. Sheard, University of Bristol

17.1	Introduction	332
17.2	Product manufacture	333
17.3	Factors affecting product quality: temperature, ice content, particle size and mechanical properties	338
17.4	Factors affecting product quality: protein solubility and related factors	343
17.5	Factors affecting product quality: cooking distortion	347
17.6	Sensory and consumer testing	349
17.7	Future trends	351

17.8	Sources of further information and advice	353
17.9	References	353

18 Quality control of fermented meat products 359
D. Demeyer, Ghent University and L. Stahnke, Chr. Hansen A/S, Hørsholm

18.1	Introduction: the product	359
18.2	The quality concept	360
18.3	Sensory quality and its measurement	361
18.4	Appearance and colour: measurement and development	363
18.5	Texture: measurement and development	365
18.6	Flavour: measurement and development	368
18.7	Taste and aroma: measurement and development	372
18.8	The control and improvement of quality	377
18.9	Future trends in quality development	381
18.10	References	382

19 New techniques for analysing raw meat 394
A. M. Mullen, The National Food Centre, Dublin

19.1	Introduction	394
19.2	Defining meat quality	394
19.3	Current state of art techniques	397
19.4	Emerging technologies	399
19.5	The genetics of meat quality	405
19.6	The future	407
19.7	Sources of further information and advice	408
19.8	References	408

20 Meat packaging 417
H. M. Walsh and J. P. Kerry, University College Cork

20.1	Introduction	417
20.2	Factors influencing the quality of fresh and processed meat products	419
20.3	Vacuum packaging	424
20.4	Modified atmosphere packaging	428
20.5	Bulk, master or mother packaging	435
20.6	Controlled atmosphere packaging and active packaging systems	437
20.7	Packaging materials used for meat products	439
20.8	Future trends	443
20.9	References	444

Index 452

Contributors

Chapter 1

Professor David Ledward
Department of Food Science and
    Technology
The University of Reading
Whiteknights
Reading RG6 6AP
England

Tel: +44 (0) 118 9316623
Fax: +44 (0) 118 9310080
E-mail:
D.A.Ledward@afnovell.reading.ac.uk

Dr Joseph Kerry and Dr John Kerry
Faculty of Food Science and
    Technology
University College Cork,
Cork,
Ireland

Fax: +35 32 12 70 213
E-mail: joe.kerry@ucc.ie
E-mail: Jfkerry@indigo.ie

Chapter 2

Professor Dr Tilman Becker
Institute for Agricultural Policy and
    Marketing
University of Hohenheim
D-70593 Stuttgart
Germany

Tel: +711 4592599
Fax: +711 4592601
E-mail: tbecker@uni-hohenheim.de

Chapter 3

Professor R. K. Miller
2471 TAMU
Meat Science Section
Department of Animal Science
Texas A & M University
College Station
TX 77843-2471
USA

Tel: 979 845 3935
Fax: 979 845 9454
E-mail: Rmiller@tamu.edu

Chapter 4

Jennette Higgs
Food To Fit
PO Box 6057
Greens Norton
Towcester NN12 8GG
Northamptonshire
England

E-mail: jennette@foodtofit.co.uk

Dr Breda Mulvihill
Glenalappa
Moyvane
County Kerry
Republic of Ireland

E-mail: bredamulvihill@hotmail.com

Chapter 5

Professor Feridoon Shahidi
University Research Professor
Department of Biochemistry
Memorial University of
    Newfoundland
St John's
Newfoundland A1B 3X9
Canada

Tel: (709) 737 8552
Fax: (709) 737 4000
E-mail: fshahidi@mun.ca

Chapter 6

Dr Marianne Jakobsen and Associate
    Professor Grete Bertelsen
Department of Dairy and Food
    Science
The Royal Veterinary and
    Agricultural University
Rolighedsvej 30
1958 Frederiksberg C
Denmark

Tel: +45 35283268
Tel: +45 35283212
Fax: +45 35283344
Fax: +45 35283190
E-mail: mj@kvl.dk
E-mail: grb@kvl.dk

Chapter 7

A.P. Moloney
Teagasc
Grange Research Centre
Dunsany
County Meath
Republic of Ireland

Tel: +353 46 25214
Fax: +353 46 26154
E-mail: amoloney@grange.teagasc.ie

Chapter 8

Dr Margit Dall Aaslyng
Danish Meat Research Institute
Maglegaardsvej 2
DK-4000 Roskilde
Denmark

Tel: +45 4630 3194
Fax: +45 4630 3132
E-mail: mas@danishmeat.dk

Chapter 9

Geoffrey R. Nute
Division of Food Animal Science
School of Veterinary Science
University of Bristol
Langford
Bristol BS40 5DU
England

Tel: +44 (0) 117 928 9305
Fax: +44 (0) 117 928 9324
E-mail: Geoff.Nute@bris.ac.uk

Chapter 10

Professor H. J. Swatland
Department of Food Science
Department of Animal and Poultry
    Science
University of Guelph
Canada

E-mail: HSwatland@uoguelph.ca

Chapter 11

Dr Peter McClure
Unilever R&D Colworth
Colworth House
Sharnbrook
Bedford MK44 ICQ
England

Tel: +44 (0) 1234 781781
Fax: +44 (0) 1234 222277
E-mail: peter.mcclure@unilever.com

Chapter 12

Dr K. G. Rickert
Director of Research
Faculty of Natural Resources,
    Agriculture and Veterinary Science
The University of Queensland
Gatton Campus
Gatton
Queensland 4343
Australia

Fax: +61 7 5460 1324
E-mail: krickert@uqg.uq.edu.au

Chapter 13

Dr Christian James
Food Refrigeration and Process
    Engineering Research Centre
    (FRPERC)
University of Bristol
Churchill Building
Langford
Bristol BS40 5DU
England

Tel: +44 (0) 117 928 9239
Fax: +44 (0) 117 928 9314
E-mail: chris.james@bristol.ac.uk

Chapter 14

Dr K. B. Madsen and Dr Jens Ulrich
    Nielsen
Danish Meat Research Institute
Maglegaardsvej 2
PO Box 57
DK-4000 Roskilde
Denmark

Tel: +45 4630 3030
Fax: +45 4630 3132
E-mail: madsenkb@adr.dk

Chapter 15

Dr Stephen J. James
Food Refrigeration and Process
  Engineering Research Centre
  (FRPERC)
University of Bristol
Churchill Building
Langford
Bristol BS40 5DU
England

Tel: +44 (0) 117 928 9239
Fax: +44 (0) 117 928 9314
E-mail: steve.james@bristol.ac.uk

Chapter 16

Dr Marie de Lamballerie-Anton
Genie des Procedes Alimentaires
ENITIAA
BP 82225
44322 Nantes Cedex 3
France

Tel: +33 (0) 2 51785465
Fax: +33 (0) 2 51785467
E-mail: anton@enitiaa-nantes.fr

Dr Joseph Culioli and Dr Richard G.
  Taylor
Station de Recherches sur la Viande
INRA
Theix 63122
Saint Genès-Champanelle
France

Tel: +33 (0) 4 73624183
Fax: +33 (0) 4 73624089
E-mail: Culioli@sancy.clermont.inra.fr

Chapter 17

Dr Peter Sheard
Division of Food Animal Science
University of Bristol
Langford
Bristol BS40 5DU
England

Tel: +44 (0) 117 928 9240
Fax: +44 (0) 117 928 9324
E-mail: peter.sheard@bristol.ac.uk

Chapter 18

Professor Dr Ir Daniel Demeyer
University of Ghent
Department of Animal Production
Proefhoevestraat 10
9090 Melle
Belgium

Tel: +32 9 264 9001
Fax: +32 9 264 9099
E-mail: daniel.demeyer@rug.ac.be

Professor Louise Stahnke
Meat and Food Safety
Chr. Hansen A/S
Boege Alli 10–12
PO Box 407
DK-2970 Hørsholm
Denmark

Tel: +45 45 74 8566
Fax: +45 45 74 8994
E-mail: louise.stahnke@dk.
  chr-hansen.com

Chapter 19

A. M. Mullen
The National Food Centre
Teagasc
Castleknock
Dublin 15
Republic of Ireland

Tel: +353 1 8059519
Fax: +353 1 8059550
E-mail: amullen@nfc.teagasc.ie

Chapter 20

H. M. Walsh
Faculty of Food Science and
 Technology
University College Cork
Cork
Ireland

Fax: +353 21 270213
E-mail: helenamwalsh@hotmail.com

Dr Joseph Kerry
Faculty of Food Science and
 Technology
University College Cork
Cork
Ireland

Fax: +353 21 270213
E-mail: joe.kerry@ucc.ie

1

Introduction

D. Ledward, The University of Reading, J. Kerry and J. Kerry, University College Cork

Meat has long been a central component of the human diet, both as a food in its own right and as an essential ingredient in many other food products. Its importance has also attracted controversy. Meat consumption has, for example, been associated with chronic diseases such as cancer and heart disease. These and other concerns, such as those over safety, have led to declining consumption of some types of red meat in regions such as the EU. As a result, the questions of what defines meat quality in the minds of consumers, and the ways these quality attributes can be maintained or enhanced during processing, are of particular importance to the food industry. This volume addresses these questions.

Chapter 2 provides the foundation for the rest of the book by discussing what defines meat quality. It explores changing consumer perceptions, the cues they use to measure quality attributes, and suggests ways in which the meat industry can meet consumer expectations more effectively. Part 1 considers individual aspects of quality, beginning with a discussion of the factors affecting the quality of raw meat. The nutritional role of meat has been a subject of concern to some consumers. Chapter 4 addresses such concerns and discusses recent research on the nutritional importance of meat in the modern diet. The following chapters consider other aspects of quality such as flavour, colour and the changing fat content of meat.

Following on from the discussion in Part 2 of individual quality attributes, Part 3 explores ways in which quality can be measured, beginning with a discussion of how to establish reliable and measurable indicators for quality attributes. Sensory analysis remains essential in both defining and measuring quality, and is reviewed in chapter 9. Whilst the use of trained sensory panels provides the foundation for measuring meat quality, instrumental techniques are essential for effective control during processing. Chapter 10 discusses the range

2   Meat processing

of on-line instrumentation available, whilst the following chapter considers the important topic of identifying microbiological hazards in ensuring meat safety.

The final part of the book looks at a range of new techniques that have been applied at the various stages in the supply chain to provide improved and more consistent quality. The use of computer models to understand and control processes more effectively is growing throughout the food industry. Chapter 12 looks at its application at the beginning of the supply chain to beef cattle production. The following two chapters then review new developments in the subsequent stages of production, discussing automation in slaughtering and carcass handling, and the key area of carcass decontamination after slaughter. If its safety and quality are to be preserved before it is either sold to the consumer or goes on for further processing, raw meat requires effective refrigeration. The collection therefore includes a review of the impact of chilling and freezing on meat quality and ways of optimising the design and operation of the meat cold chain. This chapter is complemented by a comprehensive review of current developments in meat packaging. Finally, the book concludes with chapters on the processing and quality control of such products as restructured meat and fermented meat products.

2
Defining meat quality

T. Becker, University of Hohenheim, Stuttgart

2.1 Introduction: what is quality?

Price and quality are key factors for success in food markets and, as such, are important both for the competitiveness and economic efficiency of firms and of the whole supply chain in meeting consumer demands. The price premium, which high quality products receive compared to low price products, is one measure (in this case financial) of the quality of a product. This price premium is the result of the interplay of the supply of and demand for quality. In terms of the demand side of the market, it represents the marginal willingness of consumers to pay a premium for quality. In terms of the supply side, if markets are competitive, it is equal to the marginal cost of producing a higher quality product. If the supplier is in a monopolistic quality position, prices will be higher than marginal production cost.

In general, food markets are rather competitive and price is the predominant parameter of success, but delivering a premium quality may lessen price competition and give the supplier the opportunity to increase revenue. In some cases a certain level of quality, defined for example by a farm assurance scheme, may be made a prerequisite by those customers with market power. Products produced according to a premium quality standard as requested by large retailers may gain no price premium but just the opportunity to stay in the market. Food retailers in Great Britain have significant market power (Northen, 2000a, b) compared to the food retailing sector in Germany for example. As a result, some of the price premium for quality paid by the consumer accrues to the large retailers.

The members of each stage of the food supply chain (Fig. 2.1) in general and the meat supply chain in particular have their own economic interests and goals. Consumers would like to pay low prices whilst retailers prefer high prices for

4 Meat processing

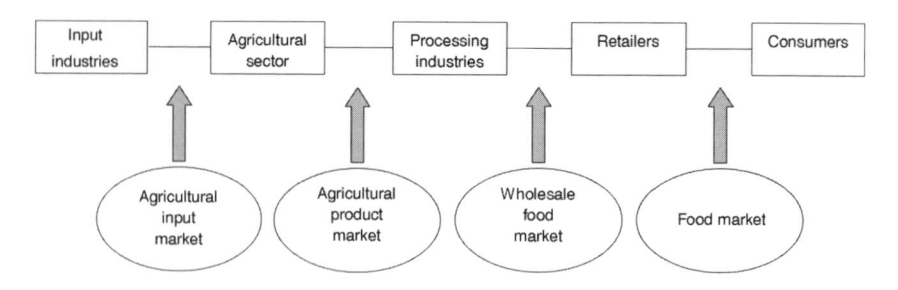

Fig. 2.1 The supply chain for food products.

food products. Retailers would like to purchase at low prices in the wholesale food market, while the processing industry tries to maximise its returns. In turn, the processor would like to purchase raw materials cheaply from the agricultural sector, while farmers try to get the best price for their produce. The strategic interests of each stage of the supply chain are in conflict both with respect to price and, therefore, potentially with respect to quality as well.

Each stage of the supply chain has its own definition of quality. Consumers ask both for sensory quality and products that are safe to consume. They may also demand a range of other potential quality attributes such as nutritional quality which may itself be variously defined to include a range of effects on health (such as level of fat content). They may also include in their definition of quality how a product is manufactured, ranging from animal welfare standards and environmental impacts to product composition and ingredients. Quality is defined by consumers according to their own personal preferences and goals.

Retailers are interested in a high margin and accordingly in products that are cheap to purchase yet can command a premium price, are easy to handle, have a long shelf-life and quick turn-over, and which contribute positively to their image. Price is of utmost importance and quality is defined according to the extent to which a product contributes to the economic goals of retailers. Food manufacturers are interested in a high margin and a good product which contributes to their brand image. The larger manufacturers in particular invest heavily in value-added products which can be used to create strong brands to gain a competitive advantage in the market. A strong industry brand is not in general in the interest of retailers who prefer to establish their own brands to improve their own market position at the expense of food manufacturers.

The processing stage itself may include more than one stage. In the meat chain, slaughterhouses are only the first step in the processing of the agricultural product. Furthermore we have to distinguish here at least between two chains, the fresh meat and the meat product chain. Food processors are interested in agricultural raw materials in large homogeneous batches produced to quality criteria geared to the demands of manufacturing. Producers usually source their raw materials from a number of suppliers. It is often not perceived to be in the interest of one farmer to co-ordinate on quality with other farmers if the cost of co-ordination for the individual farmer is higher than the benefit received. This will often be the case even if the total benefits of co-ordination among farmers would be much higher than the total co-ordination cost if the latter was shared between them.

Definitions of quality thus differ between the different stages of the supply chain and, as a result, consumer needs are not always met efficiently. In cases where the reputation of manufacturers depends decisively on the quality and safety of the agricultural products used as inputs, as in the case of baby food, the industry prescribes farming production methods or even reduces the role of the farmer to a supplier of land and labour. In creating their own brands, retailers may also impose their own quality standards on the manufacturers they contract to supply their products and on farmers producing fresh produce for the retail sector. Contracting of this kind or other forms of vertical integration may prevent the inefficient supply of quality through competing stages in the supply chain, and are a means of ensuring more uniform quality through the supply chain as a whole.

In order to facilitate an efficient supply chain response to the needs of the consumer, interests in the supply chain need to be aligned. There needs to be an understanding of and commitment to meeting the consumer definition of quality at all stages of the supply chain from retailer through to the agricultural sector. In the case of meat this consensus needs to extend even further through the supply chain to include, for example, the animal feed industry, and other sectors providing inputs into agricultural production. An efficient response of the supply chain to the consumer demand for quality implies a communication of quality through all the stages of the supply chain. This implies a definition of quality shared by all the stages of the supply chain and the willingness of all stages of the supply chain to work together to meet consumer quality demand. This might sound Utopian, but Utopian worlds may give us signs for the direction to go in the real world. Vertical integration is only one, and sometimes very costly, means to co-ordinate on quality. Other forms of co-ordination, like quality standards shared by all the stages of the supply chain, might be more efficient.

However, definitions of quality differ not only between the different stages of the supply chain, but also between the different scientific disciplines involved in meat quality management and policy. We will present here a framework for defining quality which is designed to take on board a range of different definitions and ways of defining quality. This framework requires information distributed among many scientific disciplines. It also needs to take into account gaps in knowledge and suggests where new research might be directed.

We will approach the definition of quality by distinguishing between two extremes. One view is that quality may be regarded as a construct in the mind of the customer which is highly subjective and which cannot be measured consistently and objectively. The other extreme view is that quality is objectively defined and exists only to the extent it is scientifically measurable. The subjective view is taken to the extreme in the following statement: 'Quality cannot be defined. It can only be recognised'. The objective view is taken to the extreme in this statement: 'Quality exists only to the extent it can be measured with laboratory methods'. We will not take either of these extreme views here. We will regard quality, on the one hand, as a subjective concept since it is dependent on the perceptions, needs and goals of the individual customer. In some parts of the literature following this approach the term 'perceived quality' is used to stress the view that quality is neither absolute nor objective. On the other hand, we take the view that quality can be described and defined even if it is more than what is measurable with laboratory methods. While the objective concept of quality is predominant in the supply chain and in the meat sciences, the subjective concept of quality drives consumer demand and is shared in the marketing and management literature. In our approach both the subjective and the objective quality concept are combined.

We will approach our definition of quality by presenting first the consumer perspective and then the producer perspective. These two sections are intended to give a short overview and to lay the ground for the integration of both approaches. The approach, presented in the following section, is basically a combination and, more importantly, an integration of different approaches already available in the literature, which either take the consumer or the producer perspective.

2.2 Consumer perceptions of quality

Sensory studies are frequently used to evaluate the quality of meat and meat products. According to these studies, preferences for meat seem to be strongly affected by colour/appearance and texture, and to a lesser extent by changes in flavour. Texture may be understood with juiciness and tenderness as different dimensions of textural quality (Risvik, 1994). Flavour may be regarded as consisting of taste and smell. However, eating or sensory quality is only one dimension of consumer perceived quality. Many consumer surveys in several countries of the European Union clearly demonstrate that consumers not only care about eating quality but also other quality attributes such as product safety (in the light of outbreaks such as BSE and foot-and-mouth disease), animal welfare, ecological production methods, or the presence of residues or additives such as hormones or antibiotics used in animal production.

There are two main approaches to investigating and modelling consumer behaviour: the consumer studies/marketing approach and the microeconomic approach. In the former, several models are available that seek to capture different aspects of consumer attitudes and behaviour. This approach regards the perception of quality as closely linked to the personal goals and ends of each consumer. The means-end chain theory is a good example of this approach. Consumers are assumed to choose products because they believe that specific attributes of the product can help them achieve desired ends. Audenaert and Steenkamp (1996) apply this approach to beef and explain the consequences for marketing beef to consumers in Belgium. Attributes like tender, succulent and the lack of visible fat are linked in the mind of the consumer with the consequence of an enjoyable eating experience, while attributes such as leanness and the absence of growth-promoting hormones used in animal rearing are linked to the consequence of a healthy life. Animal welfare, ecological production and other process attributes were excluded from this analysis. Even with this restricted design focusing exclusively on eating quality, it is clear that, aside from sensory properties, there exists at least one other quality dimension in the mind of the consumer, the contribution to a healthy lifestyle.

This subjective aspect of quality is further pursued in the perceived quality approach. Here quality may be defined as fitness for use, fitness for certain goals, or as the composite of all product attributes which yield consumer satisfaction. The distinction between quality cues and quality attributes becomes decisive. Quality cues are what the consumer observes, and quality attributes are what the consumer wants. Quality cues are important only to the extent that they act as consumer perceived indicators for attributes. Quality cues may be intrinsic or extrinsic to the product. Quality attributes may be experienced or have to be inferred (Steenkamp, 1990).

Consumer studies distinguish between quality expectations and quality performance (Steenkamp and van Trijp, 1996). At the point of purchase the consumer forms an impression about the expected product quality of alternative food products and accordingly decides which product to buy. It is generally acknowledged that consumers' expectations about quality are based on perceptions of quality cues. Quality cues are any informational stimuli that can be ascertained through the senses prior to consumption, and, according to the consumer, have predictive validity for the product's quality performance upon consumption. In the case of fresh meat, place of purchase and colour are among the more important quality cues, as confirmed in European consumer surveys (Glitsch, 2000a). These cues are essentially subjective.

While consumer studies literature stresses the subjective view of quality, economists analysing consumer behaviour in the context of markets, take a view more in accordance with the objective view of quality. Products are regarded as bundles of objectively measurable characteristics. Demand is assumed to depend on income and prices (Heien *et al.*, 1996). Chalfant and Alston (1991) stress the view that changes in meat demand can be explained using only relative prices and income variables without assuming a structural change in consumer preferences or tastes. Davis (1997) discusses the problems of identifying and measuring structural change in consumer preferences from a microeconomic perspective. However, economists have begun to stress the importance of non-price/income factors. The non-price/income factors which appear to be more important for consumption trends in the UK, for example, are associated with such issues as health and convenience (Bansback, 1995). Anderson and Shugan (1991) stress these factors as the reason for the increase in poultry consumption and the decline in beef sales. Von Alvensleben (1995) regards the comparatively low product diversity in the case of meat as influencing consumption patterns. Richardson *et al.*, (1993) regard health, taste and concerns over additives as determinants of changes in meat consumption.

In order to take account of the distinction between the subjectively perceived quality approach and the objective quality approach, we will use the following terms:

- **quality attributes** (QA) to denote those quality features of the product perceived as important by the consumer
- **quality characteristics** (QC) to denote those quality features which are scientifically measurable.

Food has to be prepared to become a dish. In order to take this on board, the consumption process itself may be regarded as a production process where food products with other inputs like time, skill, and other goods are transformed into the final good which is consumed. Consumer preferences also exist for these final goods. These final goods have no market prices but only subjective 'shadow prices' in the mind of the consumer (Stigler and Becker, 1977). These final goods become the foundation on which the consumer builds quality attributes. Both the perceived quality approach and the objective quality approach are linked together if we regard product attributes as the output of the home production of final goods. The willingness to pay a price premium for objectively measurable quality characteristics may be derived from the internal personal valuations of quality attributes as validated by the final goods prepared and consumed within the home.

The recent focus of microeconomic theory on asymmetric information adds other important insights to the consumer perspective on quality. According to Becker (1996), there are three decision frames for the consumer when assessing the quality attributes of a product:

1. The decision under certainty, made at the point of purchase.
2. The decision under risk, when the consumer assumes quality attributes will be realised later at the point of consumption.
3. The decision under uncertainty, where the consumer may not be able to establish the quality attribute independently.

An example of the first frame, the decision under certainty, is the size of a piece of meat. The consumer can be sure of the quality attribute by inspection. We will denote those attributes as **inspection quality attributes** (IQA). Consumers use other cues, such as the kind of shop or the colour, as inspection quality attributes. Colour, for example, is regarded by consumers as both a quality attribute in itself and even more important as an indicator for eating quality and meat safety (Glitsch, 2000b).

An example of the second frame, the decision under risk, is the attribute of meat 'tenderness'. The consumer cannot assess this attribute when buying the piece of meat, but experiences the tenderness only after preparing the product and consuming it. Accordingly, we will use the term **experience quality attribute** (EQA) or 'eating quality' to denote those quality attributes that are experienced only in consumption. Consumers look for certain quality cues that suggest the meat will be tender to eat after it is cooked.

Examples of the third frame, the decision under uncertainty, are 'animal welfare friendly' or 'organic' production methods. Here the consumer has in general no means to establish whether the product has these quality attributes but instead has to rely on third-party information. As an example, the consumer has no way of inferring from the product itself whether the animal had been reared and slaughtered humanely. As in the case of experience quality attributes, the consumer has to rely on cues but, in the case of **credence quality attributes** (CQA), trust has to substitute for personal experience. The consumer is not only interested in the sensory or eating quality, but also in issues like animal welfare, environmentally friendly production and, in particular in the case of beef, in the safety of the meat. The safety of a meat product may be seen either as an EQA, since the consumer may be immediately exposed to any risk after consumption, or as a CQA if potential health effects are long term. Within the group of credence quality attributes we will distinguish accordingly between ethical and safety/health credence quality attributes. The distinction is important because food safety and health issues are of importance for the well-being of the consumer, while ethical issues are more important for the well-feeling of the consumer.

This asymmetric or incomplete information approach sheds some light on the cue processing process internalised within the consumer. Any content of information, whether cues or attributes, can be categorised according to the three decision frames. Research shows (Glitsch 2000a, b) that in the case of meat and meat products, the place of purchase, whether butcher's shop or supermarket, is regarded by consumers (even in those countries where butcher's shops have hardly any importance) as a primary indicator, both of safety and eating quality. Price is regarded as a much less important indicator. Even in those countries like Sweden, where independent butcher's shops are comparatively rare, their importance as a cue for quality equals that of price. Colour and, to a lesser extent, country of origin are, together with place of purchase, also among the first-ranked indicators used by consumers to infer eating quality. In most countries in Europe, producer labels and brands have only minor importance as indicators for quality. Except for beef in Sweden and chicken in Germany, this quality cue is regarded as less important than the place of purchase.

Trust is decisive for the consumer in the case of credence quality attributes. The two most trusted sources of information on the safety of meat in six countries investigated in Europe (Table 2.1) are the independent retailer or butcher and the butcher in the supermarket (Glitsch, 2000a). German consumers rank consumer groups in third place, while British and Irish consumers rank their own opinion third. In the case of Sweden newspapers are ranked third and, in the case of Italy and Spain, the Department of Health. Italy and Spain, compared to the other countries investigated, are characterised by a direct involvement of national and regional ministries in quality policy.

Freshness, as indicated by attributes such as colour and a juicy texture, seems to be the most important indicator for the safety of meat. Among the more important experience-quality attributes related to safety are flavour, tenderness,

Table 2.1 Most trusted sources of information about meat (% of all answers)

	Germany	Ireland	Italy	Spain	Sweden	United Kingdom
1	Independent retailers/butchers (37.5%)	Butchers in the supermarket (36.7%)	Independent retailers/butchers (28.3%)	Independent retailers/butchers (25.6%)	Independent retailers/butchers (10.7%)	Butchers in the supermarket (23.1%)
2	Butchers in the supermarket (6.8%)	Independent retailers/butchers (9.8%)	Butchers in the supermarket (28.2%)	Butchers in the supermarket (15.1%)	Butchers in the supermarket (10.2%)	Independent retailers/butchers (9.3%)
3	Consumer groups (6.6%)	Own opinion (4.6%)	Department of Health (6.0%)	Department of Health (5.6%)	Newspapers (7.6%)	Own opinion (6.0%)
4	Magazines (3.8%)	Reports (2.8%)	Friends (3.8%)	Consumer groups (4.7%)	Own opinion (4.6%)	Newspapers (2.9%)
5	Reports (3.7%)	Farmer representatives (2.4%)	Consumer groups (3.7%)	Own opinion (4.0%)	Friends (4.0%)	Government (2.4%)
6	Friends (3.5%)	Newspapers (2.3%)	Reports (3.6%)	Government (2.8%)	Food safety board (3.5%)	Labelling (2.0%)

Source: Glitsch (2000a), p. 139.

in % of respondents	Tending to get worse	Tending to improve	Neither
Fresh meat:	45	32	23
Fresh fish:	32	38	31
Fresh vegetables:	28	44	28
Fresh fruit:	28	46	27
Pre-cooked meals:	25	43	32
Eggs:	24	39	37
Canned foods:	23	38	39
Fresh milk:	21	42	36
Bread and bakery products:	21	49	30
Frozen foods:	18	49	33
Cheese:	17	48	35

Some people think that the quality of food products sold in [our country] is improving, whilst others think it is getting worse. For each of the following products sold in [our country], please tell me if you think its quality is tending to improve or tending to get worse? (Show card.)

EU 15 average

Fig. 2.2 Food quality perception in the European Union (February 1997). Source: International Research Associates (INRA): Eurobarometer 47.0, 20 March 1997.

juiciness and smell. Where possible consumers look for cues for credence quality attributes. Place of purchase, country of origin and, in the case of chicken, free-range production have some importance as cues. Price and the name of the producer have comparatively little importance as indicators for the safety of meat. Among the more important safety concerns are the use of hormones and antibiotics in animal rearing, and the presence of salmonella in chicken and BSE in the case of beef. Fat and cholesterol seem to be comparatively minor concerns for European consumers (Fig. 2.2). This would certainly not hold for consumers in the United States of America.

From the perspective of the consumer, food should be safe, convenient to prepare, good for the health, tasty and produced in accordance with personal ethical values. The results of consumer interviews show that, though objective quality may not have changed, the perceived quality of meat is regarded as having worsened, though the situation varies between countries (Fig. 2.3). This may be accounted for by changing product attributes demanded by consumers but not sufficiently taken care of by the supply side. The consumption data shows that overall meat consumption has stayed relatively constant throughout the last decade. However, while pork consumption and, in particular, poultry consumption has increased further, beef consumption has decreased (Fig. 2.4). If quality is defined as fitness for consumption, both consumer interviews and consumption patterns show that the (perceived) quality of beef in particular has decreased over time. Per capita beef consumption in Europe has suffered a long-term decline, accelerated more recently by the BSE crisis. Consumption patterns after food scandals recover after some time, in most cases after six months or a year. The BSE crisis is no exception in this regard, but the quality image of beef

12   Meat processing

	Tending to get worse	Tending to improve	Neither
Germany	62	24	14
Greece	60	30	11
Belgium	53	24	23
France	51	29	21
Italy	45	28	27
Portugal	42	40	18
Luxemborg	39	32	29
Denmark	36	35	30
Ireland	35	43	22
Austria	34	37	29
Netherlands	34	38	27
United Kingdom	32	33	35
Finland	29	24	46
Spain	29	54	18
Sweden	25	27	48

Some people think that the quality of food products sold in [our country] is improving, whilst others think it is getting worse. For each of the following products sold in [our country], please tell me if you think its quality is tending to improve or tending to get worse? (Show card.)

in % of respondents

Fig. 2.3   Meat quality perception in the European Union (February 1997). Source: International Research Associates (INRA): Eurobarometer 47.0, 20 March 1997.

has suffered further damage. Current patterns thus suggest that an increase in overall meat consumption is very unlikely and that, at best, consumption will remain stable or even decline. Safety is clearly a credence quality attribute which, in the case of beef in particular, the industry has found it difficult to satisfy.

Food preparation should be convenient. The size of households has decreased whilst income has increased. As more women have entered the labour market, households have less time to prepare meals. Consumers now invest less time and skill in preparing food, and look for convenience foods that require little preparation. Meat is only one ingredient among many in most convenience foods, and has less importance than in a home-made meal preparation. Consumer surveys clearly show that the skills required to prepare meat dishes have decreased from generation to generation (Glitsch, 2000a). This loss of domestic culinary skills may also have reduced the perceived eating quality attributes of meat, consolidating a perceived decline in quality.

Food should be healthy. From the perspective of the consumer, meat does not contribute to health as much as other foods such as vegetables. On the contrary, meat is regarded as a food which may contribute to coronary heart disease and other diseases, and should therefore be consumed less to produce a healthy diet (Wildner, 2000). This perception has again contributed to a decrease in meat consumption.

Food should be produced in accordance with personal values. In order to consume meat, animals have to be raised and slaughtered. Consumer attitudes to

Fig. 2.4 Per capita meat consumption in the EU.

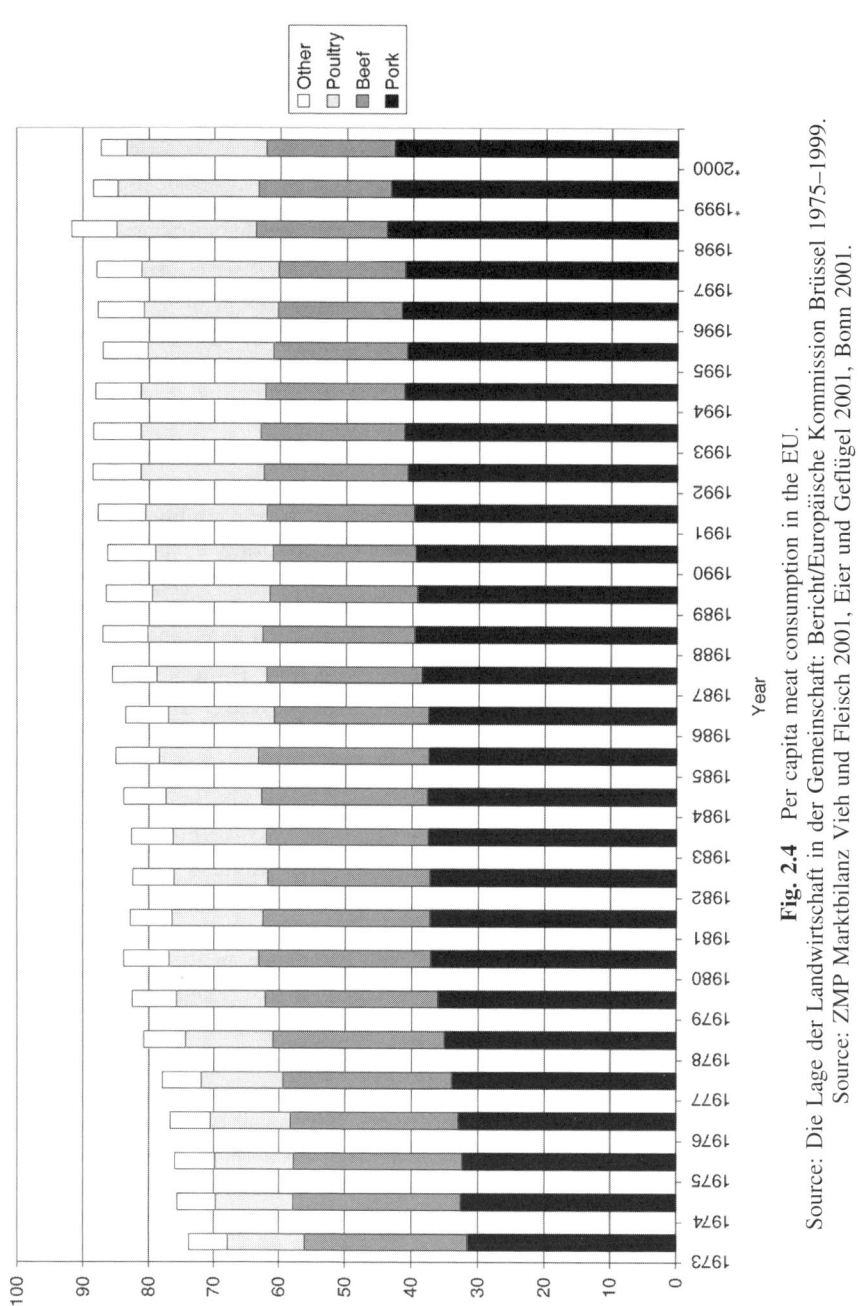

Source: Die Lage der Landwirtschaft in der Gemeinschaft: Bericht/Europäische Kommission Brüssel 1975–1999. Source: ZMP Marktbilanz Vieh und Fleisch 2001, Eier und Geflügel 2001, Bonn 2001.

methods of production have changed profoundly. There has been increasing criticism of the use of synthetic additives to enhance quality attributes such as colour and shelf-life. Consumers have increasingly demanded foods that are more 'minimally' processed and retain their original sensory and nutritional qualities. In the case of meat, there has been increasing concern about the use of antibiotics and growth-promoting hormones to make animal rearing more productive. At the same time consumers have become more concerned that animals are reared, transported and slaughtered in humane conditions. Finally, there is increasing consumer pressure to make farming practices and manufacturing processes more environmentally friendly. This has been reflected, for example, in the increasing demand for organic produce.

2.3 Supplier perceptions of quality

As in the case of studies of consumer attitudes, there are two distinct approaches to the topic of quality from the supplier perspective: that of industrial economics and that of quality management. In microeconomic theory, quality is regarded as a parameter for competition. Firms choose from a variety of possible quality attributes the bundle of quality characteristics that maximises profit. Quality is regarded as only one parameter among others for competition. The optimal choice of quality depends on the behaviour of the other firms and customer needs (Tirole, 1988). From an economic perspective, firms need not always meet customer quality needs efficiently, but only to the degree needed to maintain a profitable position in the market. In particular, in the case of experience quality attributes, there exist market equilibrium conditions which may result in an undersupply of quality. If consumers prefer high to low quality and are willing to cover the additional production cost for high quality, there may still be an undersupply of quality products if consumers are not perfectly informed of the quality attributes at the time of purchase or have no reliable cues for judging higher quality. This is well demonstrated by the so-called 'Lemons Problem' (Akerlof, 1970). Signals for quality may lessen the problem of an undersupply of quality. Any reputation mechanism like brands, advertising or warranties may act as such a signal.

The focus of the quality management literature has changed significantly over the last century, shifting first from reactive end-product testing to a more proactive emphasis on improved process control. It has also moved from a focus on how to produce more consistently and meet specifications more exactly to a more consumer-orientated approach geared to identifying consumer needs more effectively and then designing products around those needs (Dalen, 1996). This quality-by-design approach has been developed, for example, in the concept of quality function deployment that requires a product development team to find measurable characteristics describing customer needs (Akao, 1990). By using the customer's own rating of importance of needs and the specialist knowledge of the relationship between needs and characteristics, it is possible to find the most important characteristics valued by customers and to focus on these in product design. This is not an easy task: 'In some instances it is necessary to establish new expressions to describe the wanted quality. Sometimes new ways of measuring quality must be found. This is one of the challenges to meat research scientists as well as their colleagues in related fields: what to measure in the end product and how to measure it' (Dalen, 1996).

This approach has been developed for industrial design specifications but falls short in covering all the peculiarities of food, in particular meat product design. In the case of meat, both process and product attributes, experience and credence quality attributes, are of importance. Credence attributes seem to play a dominant role in consumer food demand in general compared to non-food products. The Quality Guidance Approach suggested by Steenkamp and van Trijp (1996) seeks to adapt a Quality Function Deployment approach to food product design and can be characterised as follows:

> The quality guidance philosophy consists of the following steps: (1) measurement of the quality judgements made by consumers in the target markets; (2) disentanglement of the quality judgements into its constituents, viz. perceptions of intrinsic quality cues and quality attributes; (3) linking consumer perceptions with respect to intrinsic quality cues and quality attributes to physical product characteristics (Steenkamp and van Trijp 1996).

This adds the important distinction between cues and attributes to the Quality Function Deployment approach above. Empirical research employing this framework is available for the case of blade steak. (Steenkamp and van Trijp, 1996). Characteristics measured were colour, fatness, pH, water-binding capacity, shear force and sarcomere length. These characteristics were linked to consumer perceptions of freshness, visible fat, appearance, tenderness, flavour, non-meat components and the quality expectations and quality performance of the steak. The general results show that visible fat in the raw steak and perceived tenderness at consumption have relatively accurate counterparts in physical product characteristics. For other consumer quality attributes such as perceived freshness, presence of non-meat components and flavour, the conventionally employed physical measures appear to be less effective predictors.

In most, if not all, countries of the European Union there are meat quality schemes. Only a few of these cover all the stages of the supply chain. Nevertheless, these schemes are efforts to co-ordinate quality between at least some stages of the supply chain. Schemes may be led by manufacturers, retailers, industry associations or by government agencies. In Ireland, for example, quality schemes are publicly administered, while those in Spain have a combination of farmer, industry and publicly administration involvement. In Germany, Sweden and Italy quality schemes are mainly run by farmer organisations/co-operatives. In the United Kingdom, schemes are run by industry-led organisations and retailers who are able to exercise considerable influence over the supply chain. In general, commercial quality schemes for the supply chain tend to be run by the channel 'captain', that is, the most powerful player in the supply chain. In the United Kingdom, Ireland and Sweden, quality schemes cover most of the supply chain and have a national coverage. In Germany, Italy and Spain quality schemes tend to be of regional nature and some of the schemes account for less than 1% of national supply. (Northen, 2000a)

2.4 Combining consumer and supplier perceptions: the quality circle

Quality in a market is the result of the interplay of demand and supply. Only if the cost of producing high-quality products is less than the price received will a firm produce high-quality products. If firms and consumers acting in a perfectly competitive market have complete information on quality characteristics, through consumer reliance on measurable inspection quality attributes, the supply of quality products will be efficient in the sense that, if the consumer is prepared to pay more than the production cost, products of the desired quality will be supplied to consumers by the market. No quality market failure occurs in this case.

However, if we assume more realistically that quality is based on experience or credence quality attributes, quality is less easily defined and measured by consumers and producers, and there may be a suboptimal supply of quality products. This risk may be lessened if there are reliable predictive indicators or signals for eating quality such as brands or warranties though, as we have seen, these are not significant indicators for consumers of meat products. Credence quality attributes depend largely on trust. In these cases, claims by the producers may have to be endorsed and regulated by some kind of reputable third party. Consumers need to be convinced both of the nature and reliability of producer claims and the standing and effectiveness of any third-party regulation (Caswell and Mojduszka, 1996).

The scientific literature has contributed to several aspects of quality design, production and consumer perception. This chapter is aimed at putting the different views together. The concept of a 'circle of quality' (Fig. 2.5) has been developed to define quality in a manner that takes into account both consumer and producer views on quality and ensures an effective dialogue and consensus in the market. Current research on quality has tended to neglect the problems of communication between consumers and producers in defining what each means by quality. Clearly, communication is very important in the case of experience quality attributes and even more so in the case of credence quality attributes.

While the circle of quality is intended to give a frame for consumer-orientated meat quality management, the frame has to be filled by the respective experts. Researchers respectively in meat science and quality management have to come together with those analysing consumer behaviour and the

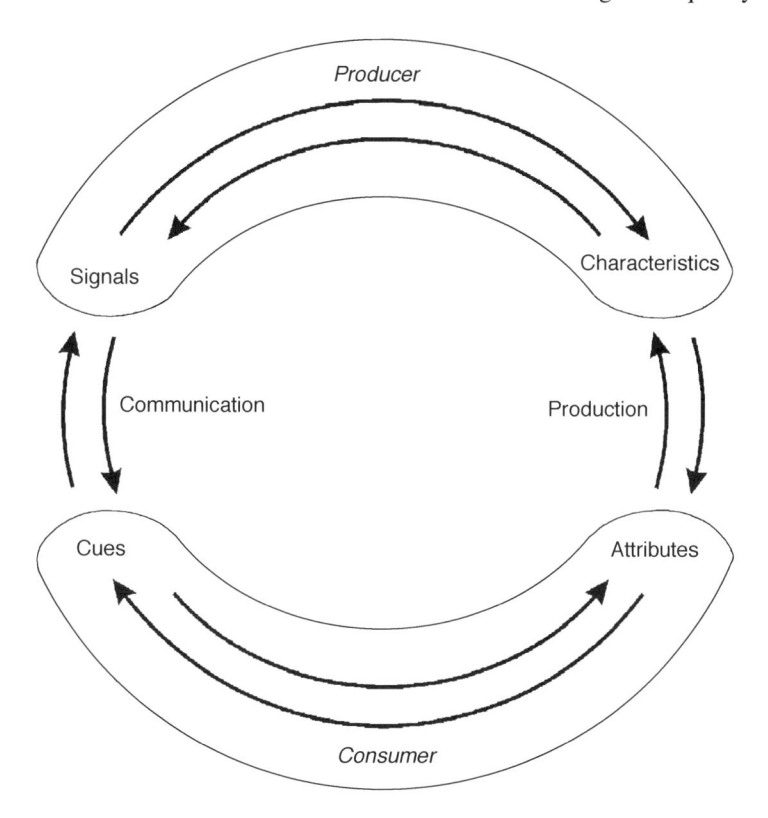

Fig. 2.5 The quality circle.

microeconomic behaviour of firms to develop robust quality management schemes. The circle of quality is intended to clarify the approaches used in the different disciplines and to contribute to a common understanding. It is clear that focusing on inspection quality attributes, such as the appearance of meat, and even experience quality attributes, that is the eating quality of meat, may be inadequate if this means neglecting consumer demands for credence quality attributes. It also becomes clear that quality production is only one part of quality management and needs to be matched by an equal emphasis on understanding how consumers assess quality and effective quality communication to the consumer.

The assumptions underlying the quality circle have been reviewed in the previous two sections. On the one side there is the consumer with her or his perception of quality based on both product and process. This perception is based on quality attributes, which are assessed by inspection, experience or credence. Cues are used by the consumer as indicators to infer quality, in particular for experience or credence quality attributes. The consumer side of the quality circle is only one side, the demand side. The other side is the production side of quality. The whole supply chain is regarded as one producer to keep the exposition simple. Problems accruing from the different interests of the different stages of the supply chain have already been discussed in the beginning of this chapter. Quality management includes both the design of the product and the process to meet consumers' needs. While quality characteristics may need to be expressed as technical specifications, these characteristics have to be translated into signals, like brands, labels, advertising and other information to be communicated to the consumer. The producer may be able to control the signals for product characteristics, like the brand image, but not all the cues used by the consumer to infer quality attributes. Here the communication process itself is decisive. Accordingly, we need to distinguish not only between attributes and characteristics, but also between signals from the producer and cues for the consumer as indicators for quality attributes. Attributes and characteristics, signals and cues are regarded here as the constituents of the universe of a consumer-orientated quality management. The full quality circle consists of two parts, the production and the communication part.

The production part of the circle needs to bridge the gap between characteristics and attributes. There is now a significant body of research, for example, on the effects of breed, feeding, rearing, transporting and slaughtering (Chapter 3) and chilling (Chapter 15) on raw meat quality. Research has also improved understanding in such areas as the control of colour and flavour (Chapters 5 and 6), and there are increasingly sophisticated ways of measuring these characteristics (Chapter 10). Meat and meat product producers therefore have increasingly sophisticated means of measuring inspection and experience quality characteristics. Improvements in consumer and sensory research (Chapter 9) have helped to show how these characteristics correlate with attributes perceived by the consumer.

However, eating or experience quality is only one category of quality characteristics. Hardly any information on the 'production' of credence quality characteristics or the link of these characteristics to attributes is available. As we have seen, consumer research has made important advances in identifying these attributes and the cues used by consumers to assess them. Consumer surveys clearly show that credence quality attributes have gained in importance for the consumer in judging the quality of meat and meat products. Though we know from consumer surveys which credence quality attributes are perceived as important, we hardly know how to define product and process characteristics to meet these consumer demands, or to match signals to the cues customers look for in assessing these attributes.

Communication is of particular importance for experience and credence quality attributes. In the case of credence quality attributes, trust has to substitute for personal experience. Though there may be product and process characteristics available to meet consumer demands for credence quality attributes, unless they are defined, enshrined in technical regulations and efficiently communicated to the consumer, a quality problem occurs. The whole circle consists of both the management of quality production and of quality communication to the consumer. Both the production and communication parts have to be added to create the full quality circle. Without both parts, the circle is incomplete. The quality production circle starts from the characteristics and links these to attributes. The quality communication circle starts with these characteristics, identifies cues and corresponding signals to turn them into the quality attributes that the consumer is looking for.

Quality communication is of particular importance in the case of process characteristics. Technical regulations may define product and process quality characteristics in detail, but have to be communicated to the consumer using the right signals, particularly in the case of process characteristics extrinsic to the product. The signals are intended by the sender of the signal, the producer, to act as indicators for product and process quality characteristics but have to be received and interpreted by the respective consumer as cues. Some of these signals may act as cues for the consumer, others may be ineffective. Some of these signals may communicate to the consumer what the producer originally intended. Other signals may even be interpreted in a way contrary to the meaning intended. Only those cues which are trusted and accepted as reliable indicators by the consumer have some influence on consumer attitudes and behaviour. Those that are not may even deepen consumer mistrust. The perception of credence quality attributes, like animal welfare friendly production, is particularly sensitive to the degree of trust in the information source. The claim of a producer that production is welfare friendly, is not a direct verifiable signal unless, for example, it is validated by a reputable third party in which the consumer has trust.

Our approach has demonstrated that, for credence quality attributes in particular, unresolved issues in communicating quality hinder an efficient supply of quality in the market. Ethical and safety/health issues have to be produced and communicated. Without trusted indicators and signals consumer needs are not fulfilled. Both public and private quality management schemes in the meat sector could be improved to meet consumer needs by taking more into account the communication element in quality management.

2.5 Regulatory definitions of quality

Quality standards for meat, as laid down in mandatory public quality schemes are predominantly targeted towards food safety and hygiene, though they do cover eating quality, animal welfare and other ethical issues.

2.5.1 Standards for food hygiene and safety

The general rules for food hygiene in the European Union are laid down in Directive 93/43/EEC. This directive lays down general rules for hygiene control, covering meat processing though not primary production. Food hygiene is defined as 'all measures necessary to ensure the safety and wholesomeness of foodstuffs'. The HACCP system is made mandatory. Member States are able to maintain or introduce national hygiene provisions that are more specific than those laid down in the Directive, providing that these are not less stringent, and that they do not constitute a barrier to trade in foodstuffs produced in accordance with the Directive. While the exclusion of primary production from product liability has subsequently been abolished, the introduction of HACCP systems in the agricultural sector is still limited, although many microbiological problems have their source in the agricultural sector, such as salmonella in chicken. The White Paper on Food Safety (Commission of the European Communities, 2000) has announced that:

> A new comprehensive Regulation will be proposed recasting the existing legal requirements to introduce consistency and clarity throughout the food production chain. The guiding principle throughout will be that food operators bear full responsibility for the safety of the food they produce. The implementation of hazard analysis and control principles and the observance of hygiene rules, to be applied at all levels of the food chain, must ensure this safety.

Within the overall framework of general hygiene regulation, more than 20 Directives cover different aspects of meat hygiene. As a result of the BSE crisis in particular, these Directives have been supplemented by more than 30 Decisions of the European Commission. As an example, Regulation 2377/90/EEC lays down procedures for establishing maximum residue limits for veterinary medicinal products in foods of animal origin. Maximum allowable residue-levels of veterinary medical products are defined in Regulation 675/92/EEC and subsequent regulations. Regulation 315/93/EEC defines contaminants, how they should be handled (through, for example Good Manufacturing Practice (GMP)) and, where appropriate, maximum allowable levels. Directive 86/363/EEC specifies maximum levels for pesticide residues in foods of animal origin. The BSE crisis has resulted in a significant extension in food safety regulation. The establishing of a system for the identification and registration of bovine animals (Regulation 1760/2000/EU) would probably not have been politically feasible without the BSE crisis. Though a system for beef traceability was introduced mainly to meet safety considerations, it also underpins labelling of meat products, an area covered by other legislation governing quality.

2.5.2 Standards for animal welfare and other ethical issues

The first European Union legislation on animal welfare was introduced in 1974 (Directive 74/577/EEC), laying down requirements for the stunning of animals before slaughter. Since that time a wide body of animal welfare legislation has been introduced. The Treaty of Maastricht included a declaration on the protection of animals which called for EU member states to take proper account of the welfare requirements of animals when drafting and implementing legislation. Specific welfare standards have also been laid down for individual species. Regulation 1804/99 extends the Regulation for organic production

(2092/91) to cover animal production. These Regulations define the requirements for the production process of products to be labelled as organic. Standards for organic production methods can be regarded as combining both environmental and animal welfare standards. These regulations also cover other ethical and safety issues, like restrictions on the use of genetically modified organisms in agricultural production. These standards can be categorised as regulating a wide range of ethical credence quality attributes.

2.5.3 Standards for eating quality

Compulsory carcass classification for cattle and pigs was implemented under Regulation 1208/81 and Regulation 3220/84 respectively. The main aim of this system was to reference the processing quality of carcasses for standardising price reporting systems across member states. This carcass classification can be regarded as providing an internal quality standard for the meat trade and, because it is not communicated to the consumer, has only an indirect impact on eating quality. The European Quality Beef Scheme is an effort to establish a quality label for fresh beef directed specifically at the consumer. So far, the success of this scheme seems to be rather limited (Becker, 2000)

2.6 Improving meat and meat product quality

A prerequisite for the improvement of (perceived) meat and meat product quality is a common understanding of quality between the consumer and the producer. In particular, communication to the consumer of quality attributes seems to be the weakest part of quality management in the meat sector. Communication is a two-way concept and implies listening to the needs and perceptions of the consumer and informing the consumer effectively about quality attributes in a way that is meaningful to the consumer. The supply chain has to listen and to speak to the consumer more carefully than it has done so far.

Quality attributes perceived as important by the consumer have been divided into three categories: inspection, experience and credence attributes. Of particular importance for meat and meat products are credence attributes. Within this category we have distinguished between ethical and safety/health attributes. The management of inspection quality attributes and the appearance of meat in the shop needs still further improvement, but we will focus on experience quality attributes and on credence quality attributes.

2.6.1 Experience quality attributes

Eating quality could be improved further by integrating sensory research with research in meat science. In an ideal world from the quality management perspective, research should start with consumer perceptions of important attributes and investigate profitable ways of producing the product and process characteristics linked to these quality attributes. The eating quality of meat is the result of the home production process of the consumer. An efficient quality management programme has to include this stage of processing. Consumer skills in preparing meat and knowledge of meat recipes have decreased over time, contributing to the decline in meat consumption. Product development and innovation in the meat sector should take more into account the growing demand of consumers for convenience in preparation.

Consumers use relatively few predictive cues to judge the eating quality of meat. The place of purchase, whether butcher's shop or supermarket, and the colour provide the main cues indicating eating quality. This clearly demonstrates that more predictive indicators, from an objective point of view, are needed. There is nothing like a quality grade or quality classification scheme available for consumers in the case of fresh meat. Even more important, predictive cues already available, like age and sex of the animals, and what the animal is fed on, are not communicated to the consumer. In the case of fresh meat, brands and labels play hardly any role as cues to indicate eating quality. Efforts should be made to give consumers better and fuller predictive cues to indicate eating quality, for example by developing and establishing a sensory quality index and communicating this index to the consumer. In the fish sector, efforts have already been made to establish such a quality index.

2.6.2 Credence quality attributes

In comparison with the US, meat in the EU seems to have a better health image, in particular with respect to cholesterol. There needs to be further improvement in the health image of meat (see Chapter 4). Improvements in communication seem to be more promising in increasing perceived quality than improvements in production which are already significant. The BSE crisis has undermined the safety image of meat. Public policy has taken over the task of meeting consumer demands for food safety by, for example, establishing traceability systems for meat. The recent stabilisation of beef consumption and the lack of media interest in further BSE cases seem to indicate that public policy has been quite successful in this respect. However, other safety concerns such as the use of antibiotics in animal rearing are still significant and need to be taken account of by both the industry and government. The focus of public mandatory standards has traditionally been on the management and control of safety. Private standards, as laid down in quality schemes, have been targeted more towards eating quality and ethical credence quality attributes. However, progress has sometimes been piecemeal with price competition, for example, putting pressure on the implementation of more expensive animal welfare standards in production. Public standards are increasingly taking over the task of defining technical regulations for ethical credence attributes as well.

2.7 References

AKAO, Y. (1990): *Quality Function Deployment.* Productivity Press: Portland, 1990.

AKERLOF, G. (1970): The Market for 'Lemons': Quality Uncertainty and the Market Mechanism. In: *Quarterly Journal of Economics*, Vol. 84, pp. 488–500.

ALVENSLEBEN, R. VON (1995): Die Imageprobleme bei Fleisch. Ursachen und Konsequenzen. In: *Berichte über Landwirtschaft*, Vol. 73, pp. 65–82.

ANDERSON, E. and S. SHUGAN (1991): Repositioning for Changing Preferences: The Case of Beef versus Poultry. In: *Journal of Consumer Research*, Vol. 18, pp. 219–232.

AUDENAERT, A. and J.-B. STEENKAMP (1996): Exploring the Nature of Consumer's Cognitive Structures for Beef. In: *Agricultural Marketing and Consumer Behavior in a Changing World.* Wageningen, 1996, pp. 81–89.

BANSBACK, B. (1995): Towards a Broader Understanding of Meat Demand Presidential Address. In: *Journal of Agricultural Economics*, Vol. 46, pp. 287–308.

BECKER, T. (1996): Quality Policy and Consumer Behaviour. In: Schiefer, G. and R. Helbig (eds): *Quality Management and Process Improvement for Competitive Advantage in Agriculture and Food.* Volume I. Proceedings of the 49th Seminar of the European Association of Agricultural Economists (EAAE), February 19–21, 1997, Bonn, Germany. Bonn 1997, pp. 7–28.

BECKER, T. (2000): EU Policy Regulating Meat Quality. In: Becker, T. (ed.) *Quality Policy and Consumer Behaviour in the European Union.* Kiel: Wissenschaftsverlag Vauk, 2000, pp. 53–72.

CASWELL, J. and E. MOJDUSZKA (1996): Using Informational Labeling to Influence the Market for Quality in Food Products. In: *American Journal of Agricultural Economics*, Vol. 78, No. 5, pp. 1248–1253.

CHALFANT, J. and J. ALSTON (1991): Accounting for Changes in Tastes. In: *Journal of Political Economy*, Vol. 96, No. 2, pp. 391–410.

COMMISSION OF THE EUROPEAN COMMUNITIES (2000): *White Paper on Food Safety.* Brussels, 12. January 2000.

DALEN, A. (1996): Assuring Eating Quality of Meat. In: *Meat Science*, Vol. 43, No. S, S21–S33.

DAVIS, G. (1997): The Logic of Testing Structural Change in Meat Demand: A Methodological Analysis and Appraisal. In: *American Journal of Agricultural Economics*, Vol. 79, pp. 1186–1192.

GLITSCH, K. (2000a): Consumer Requirements for Fresh Meat: Results of the Survey. In: Becker, T. (ed.) *Quality Policy and Consumer Behaviour in the European Union.* Kiel: Wissenschaftsverlag Vauk, 2000, pp. 113–156.

GLITSCH, K. (2000b): Consumer Perceptions of Fresh Meat Quality: Cross-National Comparison. In: *British Food Journal*, Vol. 102, No. 3, pp. 177–194.

HEIEN, D., T.-N. CHEN, X.-L. CHIEN and A. GARRIDO (1996): Empirical Models of Meat Demand: How Do They Fit Out of Sample? In: *Agribusiness*, Vol. 12, No. 1, pp. 51–66.

NORTHEN, J. (2000a): Quality Attributes and Quality Cues. Effective Communication in the UK Meat Supply Chain. In: *British Food Journal*, Vol. 102, No. 3, pp. 230–245.

NORTHEN, J. (2000b): Private Initiatives to Manage Safety and Quality of Meat in Selected Member States of the EU. In: Becker, T. (ed.) *Quality Policy and Consumer Behaviour in the European Union*. Kiel: Wissenschaftsverlag Vauk, 2000, pp. 193–220.

RICHARDSON, N., H. MACFIE and R. SHEPHERD (1993): Consumer Attitudes to Meat Eating. In: *Meat Science*, Vol. 36, pp. 57–65.

RISVIK, E. (1994): Sensory Properties and Preferences. In: *Meat Science*, Vol. 36, pp. 67–77.

STEENKAMP, J.-B. (1990): Conceptual Model of the Quality Perception Process. In: *Journal of Business Research*, Vol. 21, pp. 309–333.

STEENKAMP, J.-B. and H. VAN TRIJP (1996): Quality guidance: A Consumer-Based Approach to Food Quality Improvement using Partial Least Squares. In: *European Review of Agricultural Economics*, Vol. 23, No. 2, pp. 195–215.

STIGLER, G. and G. BECKER (1977): De Gustibus Non Est Disputandum. In: *The American Economic Review*, Vol. 67, No. 2, pp. 77–90.

TIROLE, J. (1988): *The theory of industrial organization*. MIT Press: Cambridge 1988.

WILDNER, S. (2000): Die Nachfrage nach Nahrungsmitteln in Deutschland unter besonderer Berücksichtigung von Gesundheitsinformationen. In: *Agrarwirtschaft, Zeitschrift für Betriebswirtschaft, Marktforschung und Agrarpolitik*, special issue no. 169.

Part I

Analysing meat quality

3

Factors affecting the quality of raw meat

R. K. Miller, Texas A & M University, College Station

3.1 Introduction

Establishing an understanding of raw meat eating quality and consistency is an important component of meat production systems. It is generally understood that production of meat must be tied to the production of a product that consumers find visually appealing, that they will continually purchase and that consistently delivers an acceptable eating experience. Therefore, meat quality encompasses the visual appearance and eating quality. Both of these quality factors can be influenced by ante-mortem and post-mortem production factors. This chapter will concentrate on ante-mortem production factors of breed and genetic effects, dietary influences, and rearing effects on meat quality and the post-mortem factor of the slaughter effect will be discussed as a post-mortem production factor. This information will provide a basis of understanding for subsequent discussions on meat quality in ensuing chapters.

3.2 Quality, meat composition and structure

Meat is composed of lean tissue or muscle fiber cells, fat and connective tissue. Fat or adipose cells can be found in up to three depots or locations in meat. Fat can be deposited intramuscularly as marbling or contained between muscles (defined as seam fat) or it can be found as external fat or subcutaneous fat. Additionally, meat may include bone, but the trend has moved toward boneless meat cuts and therefore bone will not be discussed in this chapter. Nervous tissue and components of the blood system are contained within meat but their total weight or proportional contribution to meat is small and so will not be discussed. These three major components of meat, fat, lean or the myofibrillar component, and connective tissue, affect meat quality in different ways.

3.2.1 Fat component

Intramuscular fat content has been shown to affect flavor, juiciness, tenderness and visual characteristics of meat. Savell and Cross (1988) developed the Window of Acceptability to demonstrate the general relationship between the role of increased intramuscular fat on meat pork, lamb and beef palatability (Fig. 3.1). In general, as fat content increases, palatability increases; however, improvements in palatability with increasing fat percentage are not equal across all fatness levels. If fat content is less than 3%, palatability decreases markedly with each decrease in fat percentage. In fact, this is the steepest slope on the curve. As fat increases from 3% to about 6%, meat palatability improves, but not as dramatically as reported at the lower levels. As fat content exceeds 7.3%, fat is highly visible and has been identified as too fatty by health-conscious consumers. Too much visible fat has raised questions about consumption of fat in meat products and increased incidence of coronary heart disease, obesity or some forms of cancer in humans; these issues can affect consumers' perception of acceptability. Therefore, meat with fat content between 3 and 7.3% is generally considered acceptable. Diet/health-conscious consumers may be willing to sacrifice palatability for lower fat content.

How does intramuscular fat affect palatability? One way is through the relationship of intramuscular fat with meat juiciness. As intramuscular fat increases, humans perceive that the meat is juicier. During mastication or during the first bites, if fat is present, some of it is released and the salivary glands are stimulated. This results in a perception of juiciness, additionally, meat with a

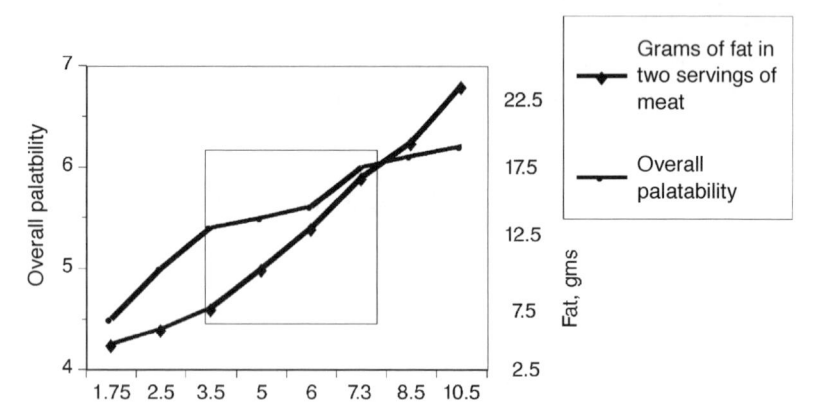

Fig. 3.1 The Window of acceptability. Adapted with permission from *Designing Foods: Animal Product Options in the Marketplace.* Copyright 1988 by the National Academy of Sciences. Courtesy of the National Academy Press, Washington, DC.

higher fat content may give a longer sustained perception of juiciness. Savell and Cross (1988) stated that 'fat may affect juiciness by enhancing the water-holding capacity of meat, by lubricating the muscle fibers during cooking, by increasing the tenderness of meat and thus the apparent sensation of juiciness, or by stimulating salivary flow during mastication'.

A second way that intramuscular fat affects palatability is through the relationship between fat content and tenderness. Interestingly, there is conflicting evidence as to the meat tenderness and fat relationship. Savell and Cross (1988) supported the relationship between increased intramuscular fat and meat tenderness by proposing four hypotheses. The first hypothesis, the Bulk Density Theory, states that as fat is lower in density than heat-denatured protein in cooked meat, as the fat percentage increases, the overall density of the meat decreases. As bulk density decreases within a given bite of meat, the meat is more tender. The second hypothesis is defined as the Lubrication Effect. Intramuscular fat is mainly triglycerides stored in adipose cells embedded in the perimysial connective tissue wall of the muscle. As meat is cooked, triglycerides melt and bathe the muscle fibers. As the meat is chewed, fat is released, salivation increases and the meat is perceived as juicy. Additionally, the muscle fibers give or slide more easily resulting in an increased perception of tenderness. The third hypothesis, the Insurance Theory, states that fat provides protection against the negative effects of over-cooking or high heat on protein denaturation. Meat proteins are involved in binding water in the muscle fiber. As meat is cooked, proteins denature and lose some of their ability to bind water. Fat can act to insulate the transfer of heat or slow down the heat transfer so that protein denaturation is less severe and less moisture is lost during cooking. The fourth theory or the Strain Theory relates to the weakening of the perimysial connective tissue surrounding muscle bundles. As marbling is deposited as adipose cells dispersed in perimysial connective tissue, development and an increased number of adipose cells weaken the connective tissue structure resulting in more tender meat.

To understand if marbling or intramuscular fat affected consumer acceptance and the subsequent relationship with trained sensory responses, the Beef Customer Satisfaction study was conducted (Lorenzen *et al.*, 1999; Neely *et al.*, 1998; Savell *et al.*, 1999) in the United States. Beef top loin steaks from four USDA Quality Grade classifications were selected to represent four Quality Grade classifications where Low Select would contain beef top loin steaks with Slight^{00} to Slight^{50} degrees of marbling that would equate to about 3 to 3.5% chemical lipid; High Select steaks had Slight^{51} to Slight^{100} degrees of marbling or about 3.5 to 4.0% chemical lipid; Low Choice steaks had a small degree of marbling or about 4 to 5% chemical lipid; and Top Choice consisted of steaks with modest and moderate degrees of marbling or about 6 to 7% chemical lipid. Chemical lipid approximations were projected from Savell and Cross (1988). Steaks were evaluated by 300 households in four cities where each household contained two adult consumers who ate beef three or more times per week. Four top loin steaks from each carcass was served to four consumers in each city and one steak was evaluated by a trained meat descriptive attribute panel and Warner-Braztler shear force was conducted as a mechanical measurement of tenderness as described by AMSA (1995). Consumers rated Top Choice steaks highest for overall like and juiciness (Table 3.1). They liked the tenderness and flavor of Choice (Top Choice and Low Choice) steaks compared to Select steaks and they indicated that the Choice steaks had a higher intensity of flavor than Select steaks. Trained sensory panels also indicated that as marbling score increased, cooked beef top loin steaks were juicier, more tender, more intense in flavor and they had higher levels of beef flavor and beef fat flavor (Table 3.1). Warner-Bratzler shear force values decreased as marbling score increased (Table 3.1). In this same study, top sirloin and top round steaks also were evaluated. These steaks had slightly lower fat content than top loin steaks and the marbling to palatability relationship was not as strong.

In pork, a similar study was conducted in three cities with pork consumers in the United States. Pork loin chops were selected to vary in pH, lipid content and tenderness as determined by Warner-Bratzler shear force value (Table 3.2). Pork consumers in the US did not rate pork loin chops differently based on lipid content. However, when a similar study was conducted with Japanese consumers (Table 3.3), Japanese consumers rated pork loin chops with higher National Pork Producer Council (NPPC) marbling score (NPPC marbling scores are a visual assessment of intramuscular fat and they are related to a chemical lipid value) as juicier, they liked the flavor and taste, they liked the color and they tended to like the amount of fat and visual appearance. Pork loin chops with the highest level of lipid tended not to be preferred by Japanese consumers most likely due to too much visible fat. In summary, there is a marbling to meat palatability relationship, but this relationship may vary across meat species and across consumer populations. While this relationship is not strong across all meat species, increased marbling or intramuscular fat assists in improving the eating quality of meat.

Intramuscular fat also has an indirect relationship to meat tenderness. As animals grow and develop, fat is deposited sequentially and marbling is the last fat depot to fill. Marbling therefore is an indication of growth and nutritional status of animals. If animals are fed high-energy-based diets they grow rapidly or they have high rates of protein and lipid accretion. The end result is heavier animals with higher levels of subcutaneous, seam and intramuscular fat and greater muscle mass. These heavier, fatter and more muscular carcasses chill slower and are less susceptible to cold-induced toughening (see discussion in 3.2.2). Additionally, animals fed energy-based diets, that grow rapidly have higher collagen solubility (see discussion in 3.2.3) that improves meat tenderness. It becomes apparent that interrelationships between the connective tissue, muscle fiber and fat component are involved in understanding meat palatability.

Marbling has been shown to affect consumer and trained sensory panel meat flavor attributes (Tables 3.1, 3.2, 3.3). As fat level increases, consumers tend to like the flavor of beef and pork. Fat has a characteristic flavor and has been

Table 3.1 Least squares means of top loin steaks from US Beef Customer Satisfaction Study for consumer sensory attributes,[a] trained meat descriptive sensory attributes and Warner-Bratzler shear force (kg) as effected by USDA quality grade

Quality attribute	USDA quality grade				Root mean square error	P-value
	Top choice	Low choice	High select	Low select		
Consumer sensory attributes[a]						
Overall like/dislike	19.2[c]	19.1[c]	18.8[c]	18.7[c]	3.06	0.0004
Juiciness	18.5[c]	18.5[c]	18.3[d]	18.0[c]	3.57	0.0006
Tenderness like/dislike	19.0[cd]	19.2[d]	18.6[cd]	18.6[c]	3.28	0.0001
Flavor intensity	19.1[c]	19.2[d]	18.9[cd]	18.9[c]	2.87	0.0009
Flavor like/dislike	19.3[cd]	19.3[d]	19.0[cd]	18.9[c]	2.88	0.0002
Trained meat descriptive sensory attribute[b]						
Juiciness	5.8[d]	5.6[c]	5.5[c]	5.4[c]	0.58	0.0001
Muscle fiber tenderness	6.7[d]	6.6[cd]	6.5[c]	6.5[c]	0.58	0.01
Connective tissue amount	6.8[c]	6.9[d]	6.9[c]	6.9[c]	0.45	0.55
Overall tenderness	6.6[d]	6.6[cd]	6.5[c]	6.5[c]	0.56	0.06
Flavor intensity	5.7[d]	5.7[d]	5.6[c]	5.6[c]	0.31	0.002
Beef flavor intensity	3.5[d]	3.5[d]	3.3[c]	3.3[c]	0.32	0.0001
Beef fat flavor intensity	2.1[e]	2.0[d]	1.8[c]	1.8[c]	0.23	0.000a
Mechanical tenderness measurement[b]						
Warner-Bratzler shear force, kg	2.70[d]	2.75[d]	3.00[c]	2.95[c]	0.71	0.0002

[a] Values from Neely et al. (1998) and Lorenzen et al. (1999). Values differ from those reported as models differed slightly in order to generate these least squares means. Consumers' sensory attributes were rated as 1 = dislike extremely, not at all juicy, not at all tender, dislike extremely, and no flavor at all, respectively and 23 = like extremely, extremely tender, extremely juicy, like extremely, and an extreme amount of flavor, respectively.
[b] Values are unpublished data, but they were derived from the same data set as published by Neely et al. (1998) and Lorenzen et al. (1999).
[cde] Least squares means within a row and a cut lacking a common superscript differ ($P < 0.05$).

Table 3.2 Least squares means for pork consumer sensory traits[a] as affected by predetermined categories of lipid, Warner-Bratzler shear force, and pH from loin chops from the US Pork Consumer Sensory Study. Adapted from Miller et al. (2000).

Trait	n	Juiciness	Tenderness	Flavor	Overall like
pH category		0.04	0.0165	0.06	0.03
Low	648	3.3[d]	3.3[d]	3.2	3.2[d]
Medium	620	3.3[d]	3.3[d]	3.2	3.2[d]
High	498	3.5[e]	3.4[e]	3.4	3.4[e]
RSD[c]		1.13	1.08	1.10	1.03
Lipid category		0.20	0.19	0.09	0.18
Low	427	3.4	3.3	3.3	3.2
Medium	857	3.3	3.3	3.2	3.2
High	482	3.4	3.4	3.4	3.3
RSD[c]		1.3	1.08	1.05	1.03
Shear category		0.0004	0.0001	0.0004	0.0001
High	379	3.2[d]	3.1[d]	3.1[d]	3.0[d]
Medium	844	3.4[d]	3.3[e]	3.3[e]	3.3[e]
Low	520	3.5[e]	3.5[f]	3.4[e]	3.4[e]
RSD[c]		1.12	1.07	1.05	1.03

[a] Consumer attributes were evaluated using a 5-point hedonic, end-anchored sensory scale where 1 = dislike extremely and 5 = like extremely.
[b] P-value from the Analysis of Variance table.
[c] RSD = Residual Standard Deviation from the Analysis of Variance table.
[def] Least squares means within a column and a trait lacking a common superscript differ ($P < 0.05$).

identified as one of the major components of the meat flavor lexicon (Johnsen and Civille, 1986). Whereas fat is not the predominant flavor in meat, it provides a balance between lean and fat flavors. When meat contains very low levels of fat, the predominant flavors are those associated with the lean such as cooked beef lean, serumy, bloody, grainy, metallic, livery/organy, and brothy (Johnsen and Civille, 1986; Lyon, 1987). As the level of fat or marbling increases, the cooked fat aromatic or flavor increases in meat and this aromatic can assist in decreasing or masking flavor attributes associated with lean, providing a balance of flavors.

3.2.2 Lean or muscle fiber component

The major component of meat is lean and lean is mainly composed of muscle fibers. Muscle fibers from the cellular structure that possesses the contractile apparatus of the muscle. Muscle proteins also are the components in the muscle fiber that binds water or interacts with water to hold it in the muscle fiber. The structural integrity and the ability of the muscle proteins to bind water affect meat tenderness and juiciness.

There are two components of the muscle fiber structure, the contractile state and the degradative state, that influence meat tenderness. In living tissue, muscle

Table 3.3 Least squares means for consumer sensory scores of pork loin chops from the Japanese Pork Consumer Study that vary by NPPC marbling scores determined at the 10th rib in the Longissimus muscle. Adapted from Miller et al. (2000)

Consumer attribute	Marbling score[c]						P Value
	1	2	3	4	5	6	
Aroma like/dislike[a]	3.20	3.11	3.16	3.27	3.87	3.00	0.13
Juiciness like/dislike[a]	3.09[de]	3.00[d]	3.01[de]	3.13[de]	4.12[e]	3.36[de]	0.048
Tenderness like/dislike[a]	3.34	3.29	3.25	3.39	4.25	3.82	0.07
Flavor like/dislike[a]	3.15[d]	3.19[d]	3.14[d]	3.29[d]	4.12[e]	3.64[de]	0.04
Overall taste like/dislike[a]	3.15[d]	3.16[d]	3.12[d]	3.34[de]	4.25[f]	3.82[ef]	0.006
Appearance like/dislike[a]	3.01[d]	3.11[de]	3.19[de]	3.32[de]	3.75[e]	2.82[d]	0.02
Color like/dislike[a]	3.07[d]	3.17[d]	3.23[de]	3.28[de]	3.87[e]	2.82[d]	0.04
Color intensity[b]	3.16[d]	3.36[de]	3.15[d]	3.25[a]	3.87[e]	2.91[d]	0.02
Amount of fat like/dislike[a]	3.06[d]	3.19[de]	3.26[e]	3.36[e]	3.75[e]	3.09[de]	0.02
Overall visual like/dislike[a]	3.00[d]	3.13[d]	3.23[d]	3.34[d]	3.50[d]	2.82[d]	0.009

[a] Consumer attributes were evaluated using a 5-point scale where 1 = dislike extremely and 5 = like extremely.
[b] Consumer attributes were evaluated using a 5-point scale where 1 = light and 5 = dark.
[c] National Pork Producers Council new fresh meat marbling scores where 1 ≤ 1% lipid, 2 = 2% lipid; 3 = 3% lipid, 4 = 4% lipid, 5 = 5% lipid and 6 ≥ 6% lipid.
[def] Least squares means within a row lacking a common superscript differ ($P < 0.05$).

fibers are elastic and have the ability to contract and relax. Through the conversion of muscle to meat, muscle proceeds through rigor mortis where muscle fibers lose their ability to relax and that results in loss of much of their elasticity. Biochemical and physical conditions present when a muscle proceeds through rigor mortis affect the final contractile state or sarcomere length and tenderness of the muscle. One physiological phenomenon that can occur during rigor mortis is called cold-induced toughening or cold-shortening. It occurs during the onset of rigor mortis when the muscle is chilled rapidly, the sarcoplasmic reticulum loses its ability to bind calcium, so calcium concentrations in the cytosol of the cell increases. The end result is that energy stores in the form of ATP are still available when calcium concentration increases and the contractile apparatus of the muscle still has the ability to contract. The muscle fibers then contract more rigorously than normal and upon the inability to relax, the contractile state of the muscle is shorter than normal. The result is tougher meat. Sarcomere length is a measure of cold-induced toughening and it is the distance between the two Z lines within a sarcomere. A sarcomere is the smallest contractile apparatus of a muscle fiber and the Z lines are rigid structures that compose the exterior of a sarcomere. Z lines are very strong structures that have to withstand the forces applied during contraction. In cold-shortened meat, Z line density increases within a given quantity of meat. As long as the Z line has not been disrupted by either degradation or fragmentation from contractile proteins in super-contracted meat, increased density of Z lines has been related to increased meat toughness (Locker, 1960; Marsh and Leet, 1966; Marsh et al., 1968). Additionally, muscle fibers that are shorter have been shown not to degrade as rapidly post-mortem as there is not sufficient room for degradative enzymes to work.

Strength of the structural components within the muscle fiber also has been related to meat tenderness. The basic premise is that as the structural apparatus of the muscle is degraded and weakened, meat tenderness improves (Koohmaraie, 1988, 1992; Koohmaraie et al., 1988). The structural apparatus of muscle fibers is composed of Z lines and multiple structural proteins that hold the myofilaments of muscle fiber in an organized, structural array. Also, the major contractile proteins, myosin and actin, are part of the structural array in that they are the predominant proteins in the muscle fiber. Degradative enzymes work to break apart the muscle fiber structural apparatus. In living tissue, these enzymes are responsible for protein degradation and repair of protein structure. In meat, these enzymes degrade large structural proteins such as titan and nebulin and loosen the strength of the muscle fiber component. The major enzyme system shown to affect post-mortem muscle fiber degradation is the calpain proteolytic system (Koohmaraie, 1988, 1992, Koohmaraie et al., 1988; Goll et al., 1995). This system is composed of β-calpain, m-calpain and calpastatin. Calpastatin regulates the activities of the calpains that have been shown to degrade structural proteins post-mortem (Koohmaraie, 1988, 1992, Koohmaraie et al., 1988). Increased myofibrillar degradation post-mortem has been related to improvements in muscle fiber tenderness. Goll et al. (1995)

proposed the theory that actomyosin interactions also may play a role in improved meat tenderness post-mortem, even though actomyosin does not degrade post-mortem. Goll *et al.* (1995) proposed that weakening or a decrease in the strength of the actomyosin interaction post-mortem may contribute to improved tenderness.

The ability of myofibrillar proteins to bind water within the muscle fiber is also related to meat tenderness and juiciness. Myofibrillar proteins have charged side-groups that contain ionic charges and these ionic charges bind water. Actin and myosin, the most abundant proteins in the muscle fiber, bind the majority of water within the muscle fiber. The charge on proteins can be either positive or negative and changes in charge can be altered by pH. As pH increases, there is a net increase of negative charges and as pH decreases, protein side-groups become more positively charged. As the net charge of proteins become either more positively or negatively charged, ionic forces increase and water is bound or held more tightly to the proteins. Change in net charge of proteins is accomplished by changing meat pH. An increase or decrease in meat pH will change the ratio of positive and negative charges on protein side-chains and will alter the ability of muscle proteins to bind water. The isoelectric point of a protein is the pH where there is a balance of positive and negative charges on the protein side-groups. The isoelectric point is where muscle proteins have the least ability to bind water and it is where water-holding capacity is lowest. As meat pH reaches the isoelectric point, meat loses more water as drip loss during storage and upon cooking or a lower cook yield. The resultant meat is drier and tougher. Therefore, meat pH is an important component of meat quality as it relates to the ability of muscle proteins to bind water and the subsequent juiciness and tenderness of the meat.

3.2.3 Meat color

Meat color, the major visual factor affecting meat quality, is imbedded within the muscle fiber component as meat color is a result of pigment-containing proteins that can either absorb or reflect light. In meat, myoglobin is the major pigment-containing compound. The level of myoglobin, the oxidative state of the heme-ring within myoglobin and what is bound to the myoglobin ligand affects meat color. The level of myoglobin within a muscle is influenced by species, muscle function within the animal, and age of the animal. The state of iron within the phorforin ring of myoglobin (Fe^{+2} or ferrous; Fe^{+3} or ferric) and what compound is bound to the myoglobin ligand is mainly affected by storage conditions of the meat.

Meat color from different species of animals and the corresponding muscle myoglobin content are presented in Table 3.4. As myoglobin content increases, color intensity of the meat increases from white or pink to very dark red. The higher myoglobin content in beef differentiates it from the lighter color of pork or poultry meat that has a lower myoglobin content. However, muscles within a species and a carcass can also vary in color. Muscles vary in myoglobin content

36 Meat processing

Table 3.4 The relationship between species of origination of muscle foods, raw meat color, muscle foods myoglobin content, and major factors influencing quality of meat

Species of origin of muscle food	Animal age	Myoglobin content, mg/g	Visual color	Major factors influencing quality listed in decreasing order of importance within a species
Beef	12 days 3 years > 10 years	0.70 4.60 16 to 20	Brownish pink Bright, cherry red to dark red	Tenderness Juiciness and flavor
Lamb	Young	2.50	Light red to red	Flavor Juiciness and tenderness
Poultry dark meat	8 weeks 26 weeks (females) 26 weeks (males)	0.40 1.12 1.50	Dull red	Flavor Juiciness Tenderness
Fish dark meat species		5.3 to 24.4	Dull red to dark red	Flavor Juiciness Texture
Turkey dark meat	14 weeks (female) 14 weeks (male) 24 weeks (female) 24 weeks (male)	0.37 0.37 1.00 1.50	Dull red	Flavor Juiciness Tenderness
Pork	5 months	0.30	Grayish pink	Juiciness Flavor Tenderness
Poultry white meat	8 weeks 26 weeks (females) 26 weeks (males)	0.01 0.08 0.10	Grayish white	Flavor Juiciness Tenderness
Turkey white meat	14 weeks (female) 14 weeks (male) 24 weeks (female) 24 weeks (male)	0.12 0.12 0.25 0.37	Dull red	Flavor Juiciness Tenderness
Fish white meat species		0.3 to 1.0	Grayish white	Flavor Juiciness Texture

Adapted from Miller (1994).

based on the physiological role of the muscle. High use muscles, such as the leg muscle in chicken and other species, have higher myoglobin content due to the need for myoglobin to store and deliver oxygen in the muscle. Myoglobin content also increases as animals increase in age so that meat from older animals is darker than meat from younger animals. For example, veal is brownish pink versus beef from three-year-old steers that is bright, cherry red. The increased redness in color within beef is due to higher myoglobin content (Table 3.4).

Therefore, muscle color has been used as an indication of maturity and quality within meat species.

While myoglobin is the major pigment in meat, accounting for 50 to 80% of the total pigment, hemoglobin, the major color pigment in blood, can also contribute to meat color. Conditions during slaughter that influence proper blood removal can influence hemoglobin content. A higher hemoglobin content results in darker lean. Other meat pigments, cytochromes, catalase, and flavins, exist within muscle and influence meat color, but only to a very minor extent.

3.2.4 Connective tissue component

Perimysium, connective tissue surrounding muscle bundles, and endomysium, connective tissue surrounding muscle fibers, provide structural support to muscles. High-use muscles used for work or major movements have higher connective tissue content than low-use muscles or muscles that provide structural support. Muscles with higher amounts of connective tissue are tougher. This phenomenon is why muscles from the hindquarter of animals that are used for locomotion, such as the *Biceps femorus*, *Semimembranosus*, and *Semitendinosus,* are inherently tougher than support muscles, such as the *Longissimus lumborium* in the loin region. Another aspect of connective tissue is the type of crosslinking within the connective tissue matrix. There are two classifications of bonds within connective tissue, heat-soluble bonds and heat-insoluble bonds. Collagen is the main fiber in the perimysium and endomysium connective tissue matrix. During heating or cooking, a proportion of the bonds can be solubilized or broken. As animals age, the percentage of insoluble bonds increases. Increased toughness due to increases in animal age is mainly attributed to the increase in heat-insoluble collagen bonds. Therefore, connective tissue contributes to meat quality mainly by its influence on meat tenderness. In young animals, connective tissue affects meat tenderness mainly through the total amount of connective tissue between muscles within the same animal. As animals increase in age, meat becomes tougher mainly by increasing the percentage of heat-insoluble collagen cross-links.

In summary, the three major components of meat, fat, lean and connective tissue, contribute to meat quality with each uniquely contributing to meat juiciness, tenderness and flavor. While each of these components has been discussed separately, they are not independent components, but they are interconnected and interact biologically within the muscle or meat system. Therefore, ante-mortem and post-mortem factors that affect meat quality may affect any of the three components and subsequently affect meat quality.

3.3 Breed and genetic effects on meat quality

As meat quality is affected by the lipid, muscle fiber and connective tissue components within an animal, it is not surprising that animal genetics can play a major role in meat quality. It has long been understood that the unique genetic code for each animal regulates the production of proteins and that genetic variation exists within meat animal species for important meat quality attributes. Meat quality traits are generally recognized as being moderate to highly heritable.

3.3.1 Beef
In beef cattle, variation in quality in the US has been well documented through the National Beef Quality Audits in 1991, 1995, and 2000 (Lorenzen *et al.*, 1993; Boleman *et al.*, 1998; McKenna *et al.*, 2002) and the National Beef Tenderness Surveys (Morgan *et al.*, 1991; Brooks *et al.*, 2002). Extensive research on factors that contribute to this variation has been conducted. One source of variation implicated as contributing to this variation has been breed type.

Biological type within *Bos taurus* cattle, British (Hereford, Angus and Shorthorn) and Exotic or Continental (Charolais, Chianina, Gelbvieh, Limousin, Maine Anjou, Pinzgauer, Simmental, and Tarentais, for example) and dairy breeds (Holstein, Jersey and Brown Swiss) has been shown to influence tenderness, but mainly through differences in growth rate, weight at the time of slaughter and fatness at slaughter. Continental-influenced cattle tend to be taller, have heavier carcasses at a constant fatness, and require a longer time on high-energy diets to reach a constant fat endpoint when compared to British-based cattle. As biological type influences growth rate, fatness, weight, and body mass, these factors have a direct or indirect influence on meat tenderness, especially as carcass fatness and weight can influence cold-induced toughness and marbling levels. As long as cattle are managed similarly and slaughtered at the same fatness endpoint, differences in meat quality are minimal. However, dairy-based breeds tend to have higher marbling levels at a constant subcutaneous fatness level as selection for milking ability appears to have resulted in selection for higher marbling.

Researchers at the United States Department of Agriculture (USDA), Agricultural Research Service (ARS), Roman L. Hruska US Meat Animal Research Center in Clay Center, NE, have examined genetic differences between many breed types. The Germ Plasm Evaluation (GPE) program has completed multiple cycles. In Cycles I, II and III, F_1 crosses out of Hereford and Angus dams and sired by Angus, Brahman, Brown Swiss, Charolais, Chianina, Gelbvieh, Hereford, Jersey, Limousin, Maine Anjou, Pinzgauer, Red Poll, Sahiwal, Simmental, South Devon and Tarentais bulls were evaluated for multiple carcass and beef quality characteristics (Koch *et al.*, 1982a; Koch *et al.*, 1976; Koch *et al.*, 1982b; Koch *et al.*, 1979). While differences in marbling score and Warner-Bratzler tenderness existed between Continental- and British-based cattle, differences were slight (Table 3.5) as long as cattle had been fed to a similar fat thickness or days-on-feed endpoint. If cattle are fed to contain varying levels of fatness or they are at different physiological points in their

Table 3.5 Summary of marbling and Warner-Bratzler shear force (kg) (WBS) differences between beef cattle breed-types evaluated in the Germ Plasm Evaluation program at the USDA, ARS Roman L. Hruska US Meat Animal Research Center in Clay Center, NE

Breed group	Cycle I		Cycle II		Cycle III	
	Marbling	WBS	Marbling	WBS[a]	Marbling	WBS[a]
Hereford	10.0	3.1	10.6	3.2		
Angus	13.2	3.2	14.2	3.0		
Hereford × Angus	11.4	3.4	10.8	3.4	11.4	3.4
Angus × Hereford	12.1	3.1	11.9	3.1	11.9	3.2
Jersey ×	13.7	3.0				
South Devon ×	11.7	3.0				
Limousin ×	9.2	3.4				
Simmental ×	10.3	3.4				
Charolais	10.9	3.2				
Red Poll ×			11.3	3.3		
Brown Swiss ×			11.7	3.4		
Gelbvieh ×			9.6	3.4		
Maine Anjou ×			11.1	3.1		
Chianina ×			9.2	3.4		
Brahman ×					9.2	3.9
Sahiwal ×					9.6	4.2
Pinzgauer ×					10.8	3.3
Tarentaise ×					10.0	3.7

Adapted from GPEP, 1974; GPEP, 1975; GPEP, 1978.

growth curve, breed differences may exist. These breed differences are then a factor of not comparing cattle at the same endpoint and they may differ in body mass, fatness level and marbling level. Differences in tenderness then are most likely due to the effects of marbling on meat palatability, cold-shortening effects and connective tissue differences.

The main breed effect for meat tenderness has been between *Bos indicus* versus *Bos Taurus* cattle. It has been well documented that *Bos indicus*-influenced cattle have higher shear force values and greater variation. Research has documented that as the percentage of *Bos indicus* breeding increases, beef tenderness tends to decrease and the variability in tenderness increases (Damon *et al.*, 1960; Ramsey *et al.*, 1963; Koch *et al.*, 1982b; Crouse *et al.*, 1989; Wheeler *et al.*, 1990; Whipple *et al.*, 1990; Shackelford *et al.*, 1991) (Table 3.6 as adapted from Shackelford (1992)). Early research hypothesized that *Bos indicus* cattle were tougher due to lower levels of intramuscular fat and higher connective tissue content when compared to *Bos taurus* cattle. Wheeler *et al.* (1990) showed that *Bos indicus* cattle had lower levels of μ-calpain and higher levels of calpastatin. They concluded that calpain activity, as modulated by calpastatin, seemed to play a major role in the inherent tenderness differences between Hereford and American Gray Brahman steers.

Table 3.6 Warner-Bratzler shear force (kg) means from the *Longissimus* muscle of cattle differing in *Bos indicus* versus *Bos taurus* inheritance

Reference	Breed[a]	\multicolumn{6}{c}{Percentage *Bos indicus* breeding}					
		0	25	38	50	62	75	100
Damon et al., 1960	B	6.22	6.72	6.68	7.14	–	7.79	9.27
Carpenter et al., 1961	B	–	3.93	–	5.02	–	4.65	5.29
Ramsey et al., 1963	B	2.31	–	3.03	2.46	–	–	3.23
Luckett et al., 1975	B	3.94	4.37	–	–	–	–	6.29
Koch et al., 1982b	B	3.44	–	–	3.92	–	–	–
Koch et al., 1982b	S	3.44	–	–	4.27	–	–	–
McKeith et al., 1985	B	4.79	–	–	5.77	–	–	7.18
Bidner et al., 1986	B	3.90	4.30	–	–	–	–	–
Riley et al., 1986	B	4.30	–	–	6.00	–	–	–
Crouse et al., 1987	B	4.00	–	–	7.50	–	–	–
Crouse et al., 1987	S	4.00	–	–	8.00	–	–	–
Crouse et al., 1989	B	4.40	5.16	–	5.80	–	6.68	–
Crouse et al., 1989	S	4.40	5.64	–	6.64	–	8.41	–
Cundiff et al., 1990	N	5.50	–	–	7.00	–	–	–
Wheeler et al., 1990	B	4.75	–	–	4.75	–	–	6.40
Whipple et al., 1990	S	4.70	–	6.40	–	7.70	–	–
Shackelford et al., 1991	B	4.50	–	–	–	5.40	–	–

[a] B = Brahman, S = Sahiwal and N = Nelore.
Adapted from Shackelford (1992).

As with other breed types, variation in beef quality within the *Bos indicus* breed exists. To understand the effect of major sire lines within *Bos indicus* breeds on beef quality in the US, a five-year research study was conducted in the 1990s. This research evaluated steers ($n = 252$) from 15 Brahman sires and one Nelore sire and born from Hereford ($n = 44$) or Angus ($n = 208$) cows under standard environmental conditions to understand if difference in tenderness existed (Hager, 2000). Sixty pure-bred Angus steers were included in the last three years. The overall goal of this research was to identify *Bos indicus* sires that produced progeny that had positive carcass traits and that were tender and less variable in tenderness. Quality grade characteristics were obtained and Warner–Bratzler shear force values (kg) were determined after 1, 7, 14, 21, 28, and 35 days of ageing at 4°C. Sire influenced ($P < 0.05$) lean maturity, overall maturity, marbling, quality grade and shear force values (Tables 3.7 and 3.8) and sire affected ($P < 0.05$) shear force values after ageing for 1, 7, and 21 days (Table 3.8). The F_1 *Bos indicus*-influenced steers were less tender at 1 day and 7 days post-mortem than Angus steers (Fig. 3.2). Angus steers reached their maximum tenderness after 7 days post-mortem. However, F_1 steers showed a faster rate of ageing and were not different in tenderness after 14 days post-mortem ageing than meat from Angus steers. The rate of ageing was faster for F_1 steers than the Angus steers, and shear force values did not differ between the two breeds after 21 days ageing. It can be hypothesized that sufficient post-mortem ageing can remove variation in tenderness between *Bos*

Table 3.7 Least squares means and standard errors for quality grade carcass traits of F_1 *Bos indicus* × Angus or Hereford steers as influenced by sire from Hager (2000).

Sire	Lean maturity[a]	Skeletal maturity[a]	Overall maturity[a]	Marbling[b]	Quality grade[c]
1 ($n = 21$)	167.7±3.54[d,e]	149.5±2.07	157.3±1.92[d,e]	433.6±12.29[h]	697.0±8.32[l]
2 ($n = 14$)	177.3±4.15[e,f]	157.6±2.42	166.1±2.25[h]	320.7±14.40[d,e]	622.5±9.75[d,e,f]
3 ($n = 18$)	171.4±3.76[d,e]	154.8±2.19	161.9±2.03[d,e,f,g,h]	341.3±13.03[d,e,f,g]	637.4±8.82[e,f,g,h]
4 ($n = 16$)	172.8±3.87[e]	154.2±2.25	162.2±2.09[e,f,g]	365.2±13.41[f,g]	660.2±9.08[h,i,j]
5 ($n = 16$)	178.6±4.35[e,f]	146.0±2.54	158.9±2.36[d,e,f,g]	310.8±15.11[d]	611.1±10.23[d]
6 ($n = 13$)	180.5±4.33[e,f]	154.1±2.52	165.3±2.34[g,h]	362.8±15.01[f,g]	652.1±10.16[g,h,i,j]
7 ($n = 17$)	172.0±3.87[e]	153.4±2.26	161.4±2.10[d,e,f,g,h]	348.8±13.45[e,f,g]	646.6±9.10[f,g,h,i]
8 ($n = 17$)	170.4±3.63[d,e]	153.8±2.11	161.0±1.96[d,e,f,g,h]	367.3±12.60[f,g]	660.9±8.52[h,i,j]
9 ($n = 17$)	173.3±3.75[e,f]	152.6±2.18	161.1±2.03[d,e,f,g,h]	343.1±13.00[d,e,f,g]	638.2±8.80[e,f,g,h]
10 ($n = 15$)	168.3±3.98[d,e]	147.6±2.32	156.3±2.15[d]	348.2±13.80[e,f,g]	648.4±9.34[g,h,i,j]
11 ($n = 16$)	171.7±3.97[d,e]	151.9±2.31	161.4±2.15[d,e,f,g,h]	376.1±13.76[g]	671.7±9.31[j]
12 ($n = 15$)	164.0±4.07[d,e]	152.4±2.37	157.3±2.20[d,e]	372.7±14.11[g]	665.2±9.55[i,j]
13 ($n = 14$)	168.6±4.23[d,e]	150.0±2.47	157.9±2.29[d,e]	332.2±14.68[d,e,f,g]	632.0±9.94[d,e,f,g]
14 ($n = 16$)	183.8±3.89[f]	151.0±2.27	165.1±2.11[e,f,g,h]	339.5±13.50[d,e,f,g]	639.1±9.14[e,f,g,h,i]
15 ($n = 13$)	159.8±4.36[d]	154.3±2.55	156.6±2.36[d,e]	451.2±15.14[h]	704.1±10.25[l]
16 ($n = 14$)	170.4±4.71[e,f]	149.7±2.75	158.6±2.55[d,e]	303.3±16.34[d]	618.1±11.06[d,e]
P-value	0.003	0.05	0.01	0.0001	0.0001
RSD[k]	14.6	8.5	7.9	50.5	34.2

[a] $100 = A^{00}$ and $500 = E^{00}$.
[b] $100 =$ Practically devoid^{00} and $900 =$ Abundant^{00}.
[c] $100 =$ Canner^{00} and $800 =$ Prime^{00}.
[d,e,f,g,h,i,j] Means with different superscripts within a column are different ($P < 0.05$).
[k] RSD = residual standard deviation.

Table 3.8 Least squares means and standard errors for Warner-Bratzler shear force values (kg) of F_1 *Bos indicus* × Angus or Hereford steers as influenced by sire for post-mortem ageing period of 1, 7, 14, 21, 28, and 35 days from Hager (2000)

Sire	Length of storage, day					
	1	7	14	21	28	35
1 ($n=21$)	4.30±.23[c]	3.45±.20[b,c,d]	3.03±.17	2.87±.18[a,b,c]	3.10±.17	2.79±.18
2 ($n=14$)	4.41±.23[c]	3.57±.23[b,c,d]	3.17±.20	2.90±.21[a,b,c]	3.21±.20	3.39±.21
3 ($n=18$)	3.89±.24[a,b,c]	3.87±.21[d]	3.01±.18	2.99±.19[b,c,d]	3.06±.18	3.21±.19
4 ($n=16$)	3.85±.25[a,b,c]	3.20±.21[a,b,c]	2.76±.18	2.59±.19[a,b]	2.80±.19	2.75±.20
5 ($n=16$)	3.32±.27[a]	3.18±.24[a,b,c]	2.85±.20	2.54±.21[a,b]	2.87±.20	2.59±.21
6 ($n=13$)	4.03±.29[b,c]	3.60±.25[c,d]	2.79±.21	3.12±.23[c,d]	3.27±.22	3.03±.23
7 ($n=17$)	4.34±.27[c]	3.68±.24[c,d]	2.91±.20	2.89±.21[a,b,c]	3.40±.20	3.19±.21
8 ($n=17$)	3.59±.23[a,b]	3.00±.20[a,b]	2.76±.17	2.77±.18[a,b,c]	3.05±.17	2.79±.18
9 ($n=17$)	4.21±.25[c]	3.50±.22[b,c,d]	2.88±.17	3.43±.20[d]	3.57±.19	2.82±.20
10 ($n=15$)	3.44±.24[a,b]	3.32±.21[a,b,c]	2.66±.18	2.82±.19[a,b,c]	2.98±.18	2.69±.19
11 ($n=16$)	4.09±.25[b,c]	3.33±.22[a,b,c]	3.27±.19	3.02±.20[b,c,d]	2.95±.19	2.91±.20
12 ($n=15$)	3.56±.26[a,b]	3.00±.23[a,b]	2.59±.19	2.42±.21[a]	2.99±.20	2.82±.21
13 ($n=14$)	3.96±.27[a,b,c]	3.27±.23[a,b,c]	2.66±.19	2.61±.23[a,b,c]	2.97±.20	2.81±.21
14 ($n=16$)	3.56±.25[a,b]	3.16±.22[a,b,c]	2.68±.18	2.75±.20[a,b,c]	3.18±.19	2.86±.20
15 ($n=13$)	3.28±.27[a]	3.00±.24[a,b]	2.57±.20	2.66±.22[a,b,c]	2.74±.21	2.53±.22
16 ($n=14$)	3.44±.29[a,b]	2.70±.26[a]	2.65±.22	2.61±.23[a,b,c]	2.98±.22	2.81±.23
P-value	0.0001	0.03	0.16	0.01	0.09	0.15
RSD[e]	0.85	0.74	0.85	0.67	0.65	0.68

[a,b,c,d] Means with different superscripts within a column are different ($P < 0.05$).
[e] RSD = residual standard deviation.

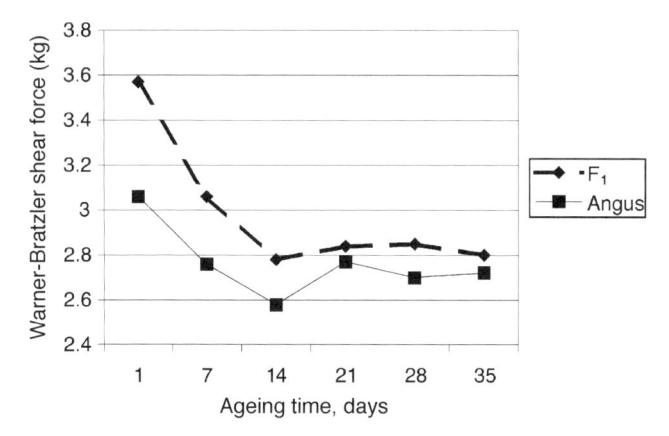

Fig. 3.2 Warner-Bratzler shear force values (kg) for top loin steaks from Angus and F_1 Angus or Hereford × Bos indicus steers from Hager (2000).

indicus and *Bos taurus* cattle. Also, differences in tenderness between *Bos indicus* and *Bos taurus* cattle can be partially attributed to differences in post-mortem muscle fiber ageing effects attributed to differences in calpastain and/or calpain levels. Interestingly, the relationship between marbling and Warner-Bratzler shear force is very low in *Bos indicus*-influenced cattle (Hager, 2000). To understand what chemical factors (Table 3.9) were related to Warner-Bratzler shear force over 35 days of post-mortem ageing, simple correlation coefficients were calculated (Table 3.10).

Components related to the myofibrillar component, sarcomere length and calpastatin activity, and the connective tissue component, collagen amount and solubility, were not highly related to Warner-Bratzler shear force values. However, fat was significantly, but only slightly, correlated to Warner-Bratzler shear force after 14 days of ageing. While it would be expected that calpastatin would have a higher relationship based on the previous discussion, it should be noted that calpain levels were not measured. While the evidence is strong that *Bos indicus* cattle differ in tenderness from *Bos taurus* cattle, there is not one factor that contributes to this effect. The lean, fat and connective tissue components are interrelated. Obviously, differences in rate of ageing occurred and marbling differences may be contributing to differences, but the data do not support singling out one component as the contributing factor.

Extensive research in the 1990s has been directed at development of beef genetic markers. Genetic markers for marbling developed in Australia, a marker for marbling that was developed out of the Angelton Project at Texas A&M University, and seven markers for tenderness that were developed out of the Angelton Project, are being examined for commercial production. However, genetic markers identify the genetic propensity of an animal and they do not guarantee that high-quality beef will result. Production and management factors that can influence beef quality need to be carefully controlled to ensure an animal has the opportunity to express its genetic potential. When commercial

Table 3.9 Least squares means and standard errors for chemical components of F_1 steers as influenced by sire including Angus steers adapted from Hager (2000)

Sire	Sarcomere length, μm	Calpastatin, activity/g	Moisture, %	Fat, %	Total collagen, mg/g	Collagen solubility, %
1 ($n=21$)	1.73	2.73	72.28^{b,c}	4.93^{e,f}	2.42	7.55^b
2 ($n=14$)	1.69	2.51	73.15^{c,d}	3.51^{b,c,d}	2.79	7.12^b
3 ($n=18$)	1.68	2.53	72.97^c	3.75^{b,c,d}	2.62	6.98^b
4 ($n=16$)	1.73	2.36	73.07^{c,d}	3.88^{b,c,d}	2.57	7.24^b
5 ($n=16$)	1.68	2.61	73.38^{c,d}	3.15^{b,c}	2.28	9.28^{b,c}
6 ($n=13$)	1.66	2.68	72.44^{b,c}	4.14^{b,c,d,e}	2.57	7.40^b
7 ($n=17$)	1.66	2.16	72.71^{b,c}	4.171^{c,d,e}	2.38	8.34^b
8 ($n=17$)	1.70	2.53	72.91^c	3.89^{b,c,d}	2.69	7.37^b
9 ($n=17$)	1.69	3.02	72.53^{b,c}	3.82^{b,c,d}	2.49	7.23^b
10 ($n=15$)	1.68	2.12	72.70^{b,c}	4.00^{b,c,d,e}	2.56	6.97^b
11 ($n=16$)	1.69	1.90	72.26^{b,c}	4.05^{b,c,d,e}	2.52	7.80^b
12 ($n=15$)	1.70	2.44	72.71^{b,c}	4.44^{d,e,f}	2.44	8.02^b
13 ($n=14$)	1.74	3.56	74.27^d	2.94^b	2.46	8.23^b
14 ($n=16$)	1.69	2.56	72.47^{b,c}	4.04^{b,c,d,e}	2.37	11.32^c
15 ($n=13$)	1.75	2.63	72.62^{b,c}	4.54^{d,e,f}	2.73	7.49^b
16 ($n=14$)	1.67	2.64	73.42^{c,d}	3.56^{b,c,d}	2.61	8.26^b
Angus ($n=60$)	1.75	2.43	71.61^b	5.39^f	2.65	7.36^b
P-value	0.27	0.13	0.005	0.0001	0.95	0.03
RSD^a	0.11	0.74	1.8	1.5	0.75	3.2

^{a} RSD = residual standard deviation.
^{b,c,d,e,f} Means with different superscripts within a column are different ($P < 0.05$).

Table 3.10 Simple correlations between shear force values and chemical components adapted from Hager (2000)

	Shear force at different length post-mortem ageing time, days					
Chemical components	1	7	14	21	28	35
Sarcomere length, μm	−0.01	−0.02	0.15	0.09	−0.004	0.0003
Calpastatin, activity/g	−0.01	0.04	0.03	−0.03	0.03	−0.03
Moisture, %	−0.04	−0.17^a	−0.01	−0.12	−0.08	−0.11
Fat, %	−0.02	−0.10	−0.20^a	−0.27^a	−0.26^a	−0.30^a
Total collagen, mg/g	−0.01	−0.01	−0.05	−0.06	−0.04	−0.06
Collagen solubility, %	−0.08	−0.16^a	0.03	−0.01	0.06	−0.03

^{a} Correlations with superscript are significant ($P < 0.05$).

use of genetic markers is viable, these markers will assist in removing variability associated with breed differences.

3.3.2 Pork

Differences in ultimate muscle pH, lean color, water-holding capacity and marbling are the major pork quality issues as color and the ability of muscle proteins to bind water affect pork quality. Pork is inherently more tender than beef as pork is much less susceptible to cold-shortening effects and post-mortem ageing occurs at a much more rapid rate than in beef. Also, pork is slaughtered at physiologically younger ages so connective tissue plays a very minor role in meat quality. Pork is also more susceptible to pre-slaughter stress that induces more quality defects related to the color and water-holding capacity of the lean. As a large proportion of pork meat goes into further processed products, water-holding capacity or the ability to hold brines or non-meat ingredients becomes a much more important issue. Marbling, while not as important an issue as in beef, has economic value in some international markets and so marbling differences can be an important trait.

Berkshire pigs have darker colored lean, higher marbling scores, higher ultimate pH and more tender meat (Table 3.11) (Goodwin, 1994; NPPC, 1995; Goodwin, 1997); however, high levels of overall carcass fat and low lean production are issues related to the practical production of these animals except for specialty markets. Hampshire hogs have been shown to have moderate pink to grayish-pink color, intermediate levels of marbling, but low ultimate pH and low water-holding capacity (Goodwin, 1997). Most of this effect has been contributed to the Napole gene effect (see discussion below). Durocs have been shown to have a higher lipid content and Landrace have a pale meat color and low pH (NPPC, 1995; Goodwin, 1997). Therefore differences in pork quality are related to breed group, however, most commercial pork operations use composite genetic types. Comparison of differences in pork quality between these genetic types is not available and, therefore, direct comparisons can not be made. However, most major breeding companies provide carcass and meat quality data for comparative purposes. The pork industry has genetic markers, the Halothane gene and the Napole gene, for pork quality commercially available. Use of these genetic markers assists in removing quality variation and improving overall quality of pork.

3.3.3 Halothane gene effects on pork quality

The halothane gene has been associated with the Pork Stress Syndrome in pigs. Fujii *et al.* (1991) reported a point mutation in the ryanodine receptor regulatory region of chromosome 6 and this mutation resulted in pigs that were susceptible to malignant hypothermia when exposed to halothane gas. The ryanodine receptor is involved in calcium release and regulation from the sarcoplasmic reticulum into the cytosol of the cell. When mutant or homozygote (nn) pigs are exposed to stress, calcium concentrations in the cytosol of the cell increase abnormally and pigs do not have the ability to adequately decrease calcium concentration in the cytosol or re-establish calcium concentrations for muscle relaxation. The result can be either death or near death due to malignant hyperthermia. If these animals are stressed immediately prior to slaughter, one of two conditions may exist. Either the animals prematurely die or due to increased metabolism, pH declines very rapidly post-mortem and results in pale,

Table 3.11 Ultimate pH as influenced by breed type

Reference	Berkshire	Cinta Senese	Duroc	Hampshire	Belgian Landrace	Italian Landrace	Swedish Landrace	US Landrace	Poland Pietrain	China	Spot	Yorkshire
Lawrie and Gatherum, 1962								5.44				5.49
Jensen et al., 1967			5.46	5.33						5.42	5.38	4.57
Hedrick et al., 1968			5.59	5.43								
Monin and Sellier, 1985				5.40					5.45			5.53
Dazzi et al., 1987		6.03	5.72		5.72	5.92			5.68			5.53
Sellier et al., 1988					5.86				5.78			5.86
Barton-Gade, 1990			5.56	5.48				5.46				5.47
Goodwin, 1994	5.92		5.72	5.53				5.65		5.72	5.69	5.72
NPPC, 1995	5.91		5.85	5.70							5.83	5.84
Lindahl et al., 2001				5.33			5.42					5.44[a]

[a] Swedish Yorkshire pigs

soft and exudative (PSE) meat. Selection of pigs with one copy of the halothane gene, or heterozygotes (Nn), has been implemented by some genetic companies due to the relationship between heterozygotes and decreased carcass fatness and increased carcass meat yields; however, a decrease in pork quality has been associated with heterozygote pigs.

Halothane heterozygote pigs have been shown to be paler in color (Christian and Rothschild, 1981; Lundstrom *et al.*, 1989; Wilson, 1993; Louis *et al.*, 1994; Goodwin, 1994), have more drip loss (Lundstrom *et al.*, 1989), have softer meat (Louis *et al.*, 1994) with less marbling (Louis *et al.*, 1994; Goodwin, 1994), and they were tougher (Goodwin, 1994) than normal or non-carrier pigs. Therefore, Halothane gene status, either Nn or nn, affects pork quality through increased susceptibility to short-term pre-slaughter stress that results in lower than normal ultimate pH and paler, softer meat that has a higher than normal drip and cook loss. These effects result in drier, tougher and less flavorful meat. Goodwin (2002) found that the overall frequency of Halothane heterozygote in eight US breeds from the National Barrow Show in 1999, 2000 and 2001 was 6.2% with Berkshire, Chester White, Duroc, Hampshire, Landrace, Poland China, Spot and Yorkshire having 4.2, 0.6, 1.6, 0, 1.2 39, 12, and 3.3%, respectively, frequencies. He found that Halothane heterozygote pigs had less 10th rib backfat, larger ribeye areas, lower pH, lighter color, lower intramuscular fat, lower water-holding capacity, higher cook yield loss, and the cooked loin chops were drier and tougher.

The Halothane gene obviously affects pork quality. While pre-slaughter handling systems to reduce stress decrease the impact of this gene on meat quality, the use of homozygote and heterozygote animals for meat production negatively affects pork quality.

3.3.4 Napole gene effect on pork quality

The Rendement Napole (RN-) gene, commonly called the Napole gene, in pork is believed to be the cause of red, soft, exudative (RSE) pork (Warner *et al.*, 1997). RSE pork has a red color that consumers' desire, but is soft and exudative indicating that the muscle proteins have low water-holding capacity. The RN allele was first suggested to be responsible for the RSE condition by LeRoy *et al.* (1990) where the RN gene was identified in two French composite lines including Hampshire lines. The Hampshire hogs examined had lower processing yield and higher drip loss in cooked cured products. Warner *et al.* (1997) later proposed that lower processing yields in RSE pork were due to lower pH and high amounts of glycogen in the muscle and they suggested that RSE pork was a result of the presence of the RN gene. Research has shown that the RN allele increases muscle glycogen content by 70% of homozygous and heterozygous RN carriers (Estrade *et al.*, 1993). The RN homozygous and heterozygous RN animals have a modified adenosine monophosphate kinase. Adenosine monophosphate regulates glycogen synthase and the altered enzyme cannot effectively inhibit glycogen production as in normal animals. As glycogen is converted to lactic acid post-mortem, the higher levels of glycogen in meat from Napole gene carriers results in higher levels of lactic acid being produced post-mortem. Increased production of lactic acid results in a lower ultimate meat pH (Lundstrom et al., 1996; Enfalt et al., 1997a). A low muscle pH results in higher drip loss or as lower water-holding capacity (LeRoy et al., 1996) decreases due to the meat pH approaching the meat isoelectric point. When compared to non-carriers, drip loss and cooking loss increased by 21% and 12%, respectively, in meat from carriers (Lundstrom et al., 1996).

Studies have shown that the RN gene has other detrimental effects on pork quality besides lower ultimate pH, water-holding capacity and cook yields. Carriers of the Napole gene also have lower protein extractability (Lundstrom et al., 1996). This is due to the lower protein content of RN carriers. Estrade et al. (1993) found that carriers of the RN allele had a 10% lower protein content of all protein fractions compared to non-carriers. The extraction of salt soluble proteins is essential to the manufacturing of hams and other processed pork products. A decrease in protein extractability results in a lower-quality product. The lower protein content also has an effect on water-holding capacity of meat with the RN gene. The decrease in protein content of RN carriers leads to a decrease in the water content in the myofibrils after curing (Lundstrom et al., 1996).

Other negative effects of meat from RN carriers include higher surface and internal reflectance and ash values (Lundstrom et al., 1996). Also, meat from Hampshire pigs carrying the RN allele has been found to have lower intramuscular marbling scores than non-carriers or Yorkshires (Miller K.D. et al., 2000). This can affect the flavor and tenderness of meat.

The effect of the RN gene on pork palatability has not been consistent. Some research when comparing non-carriers to RN carriers, found lower Warner-Bratzler shear force values and higher trained descriptive attribute sensory taste intensity and flavor (LeRoy et al., 1996; Lundstrom et al., 1996); whereas other studies failed to find the same taste and tenderness differences between carriers and non-carriers (Lundstrom et al., 1998). Other positive attributes found in RN carriers have been higher daily gains, fewer days on test, and carcasses with higher lean meat content and a larger proportion of ham (Enfalt et al., 1997a). Another study from Enfalt et al. (1997b) found that carriers had less sidefat than non-carriers, larger proportions of whole ham and whole back compared to non-carriers and Landrace and Yorkshire pigs. Goodwin (2002) showed that the Napole gene did not have an effect on carcass composition in eight breeds evaluated in 1999, 2000, and 2001, in the US.

A genetic marker test is commercially available to test pork for the RN gene. Miller R.K. et al. (2000) found that the RN allele exists at high frequencies in the American Hampshire breed in the Unites States and they reported a frequency of 0.630 for the dominant RN allele using the Hardy-Weinberg equilibrium, in the American Hampshire breed. Goodwin (2002) reported the frequency of the RN gene in a representative sample of the US pure-bred pork population from the 1999, 2000 and 2001 National Barrow Show was 5.6%. The percentage of pigs within breed-type having one copy of the RN gene were 6.3, 1.3, 0, 66, 0, 16, 25 and 1.3 for Berkshire, Chester White, Duroc, Hampshire, Landrace, Poland China, Spot and Yorkshire, respectively. It is apparent that the highest frequency is within the Hampshire breed, but other breeds, except Duroc and Landrace, has a small incidence of the RN gene. Goodwin (2002) found that heterozygote Napole pigs did not differ from normal pigs in 10th rib carcass backfat, loineye area, intramuscular fat and cooked loin juiciness. However, heterozygote Napole pigs had lower pH, slightly lighter color and marbling scores, lower water-holding capacity and higher cook loss, but the cooked loin chops were more tender than normal cooked pork loin chops.

The Napole gene affects meat quality and the incidence of it is at a high enough frequency that pork quality is affected by its presence. Selection to remove this gene from the pork population would improve overall pork quality without significantly affecting lean composition.

3.4 Dietary influences on meat quality

Dietary influences on meat quality have been extensively studied in a number of meat species. In general, as the energy density of the diet increases, either through the use of high-quality grains that replace forages or by adding fat, the growth rate of the animals increase, animals reach slaughter weight at younger ages, the resultant carcass is heavier and higher in overall fatness and marbling, the meat is juicier and species specific-flavors are somewhat diluted by an increase in fat flavor. When animals are fed forages, growth rate is slower, animals are older at slaughter, the carcass has less fat and the meat is leaner (a positive attribute for diet/health-conscious consumers), the meat is darker in color and has more species specific lean flavors. Forage fed animals also may retain β caratene derived from forages in their fat that results in more yellow fat. Additionally, some forages contain compounds that can be stored in the fat portion of meat that results in meat off-flavors. Therefore, animal diet can negatively or positively affect meat quality.

3.4.1 Feeding high concentrate diets to beef

Extensive research has shown that intensive feeding of high concentrate diets prior to slaughter positively affects beef sensory properties (Meyer *et al.*, 1960; Hawrysh *et al.*, 1975; Kropf *et al.*, 1975; Bowling *et al.*, 1977; Schroeder *et al.*, 1980; Tatum, 1981). This improvement in beef palatability has been associated with multiple factors. First, by feeding cattle on high concentrate diets, animal overall fatness, muscle mass and carcass weight increase. Therefore, with increased time on high concentrate diets, marbling scores increase and sensory panel palatability ratings increase in beef (Greene *et al.*, 1989; Williams *et al.*, 1992; May *et al.*, 1992). When cattle are fed high concentrate diets, they grow more rapidly and they reach slaughter weight in a shorter period of time.

Secondly, cattle fed high concentrate diets are usually slaughtered at younger ages and therefore, the negative effects of increased age on meat palatability are diminished. Thirdly, as cattle fed high concentrate diets are heavier at slaughter with carcasses containing higher amounts of subcutaneous fat and greater muscle mass, beef carcasses from fed cattle are not as susceptible to cold shortening as these fatter, heavier, more muscular carcasses chill slower. Fourthly, cattle fed high concentrate diets that experience rapid rates of growth have been shown to have higher amounts of collagen solubility in young cattle (Aberle *et al.*, 1981; Wu *et al.*, 1981). Therefore, feeding high concentrate diets to cattle prior to slaughter has a positive effect on the structural components of the muscle and results in improved meat palatability. Feeding cattle high concentrate diets prior to slaughter also has been associated with removing variation associated with nutritional effects prior to the high concentrate feeding period. The lower the energy density of the diet, the more restricted cattle growth is. Cattle entering the feedlot that have been fed varying energy-based diets from high to low energy would be expected to differ in live weight and composition and therefore their quality also would vary. By feeding high concentrate diets, this variation is reduced (Harrison *et al.*, 1978; Skelley *et al.*, 1978; Schroeder *et al.*, 1980; Miller *et al.*, 1987).

Feeding cattle high concentrate diets is also related to improving beef flavor and juiciness. The meat from animals fed high concentrate diets is brighter, cherry red and the fat is whiter. It is generally recommended that the effect of forages high in β carotene that results in yellow fat can be decreased or eliminated by feeding high concentrate diets for up to 90 days prior to slaughter.

The diet fed to cattle prior to slaughter can affect beef flavor. As the diet can affect overall fatness level, the affect of flavor may be due to changes in fat content. Meat with lower fat content is often described as being more beefy or brothy, higher in serumy, bloody, livery and grainy/cowy flavors and is more metallic than beef with higher amounts of fat. Fat most likely either masks the other flavor attributes or by slightly coating the mouth, the ability to detect other flavor attributes may be diminished. Other off-flavors can also be derived from dietary forage sources. Dietary flavor compounds can be deposited in adipose cells and result in off-flavors. Melton (1983) summarized that the corn in beef high-concentrate finishing diets can be partially or totally replaced by corn silage, a combination corn silage and alfalfa, alfalfa hay, or a combination of alfalfa hay and timothy and beef flavor was not affected. Changes in the grain sources within a high concentrate diet most likely will not affect beef flavor. Miller *et al.* (1997) fed beef steers either a corn, corn/barley or barley based high-concentrate finishing diet 102 to 103 days prior to slaughter to a final live weight of approximately 495 kg. Cooked top loin steaks did not differ in cooked beef flavor intensity or in any beef flavor attribute due to grain source of the diet.

The meat from beef fed corn-based diets can differ in flavor from pasture-fed beef. Most of the flavor difference is due to fatness differences in the beef (Melton, 1983). When pasture- and grain-fed cattle are slaughtered at similar fatness, Melton (1983) found that beef from pasture-fed cattle was still less desirable. These differences were most likely due to deposition of feed-derived compounds deposited in the fat. Cattle fed either bromegrass and bluestem; bluegrass and clover; fescue, orchardgrass, and clover; fescue alone; flint hills grass; native range grass; forage sorghum; orchard grass and clover; oats, rye, and ryegrass; millet or coastal bermuda grass; and bermuda grass-clover and sudan grass had lower flavor ratings. Supplementing cattle with grain during pasture feeding will dilute out these effects or feeding cattle for 90 to 100 days on grain-based diets prior to slaughter will reduce the negative flavor effects of these grasses.

The effects of nutrition on lamb quality are very similar as discussed for beef. As lambs are ruminants, feeding grain-based diets has similar effects as those discussed for beef. However, feeding high-concentrate diets to lambs prior to slaughter is not as common a practice as in beef. Lamb can be slaughtered directly after being fed forage or grass-based diets. As long as lambs are fed to a fat-constant endpoint, differences in tenderness are not generally reported. The most significant impact of feeding lamb or beef on forage-based diets is the potential for off-flavors derived from forage-based compounds.

Feeding of dietary supplements such as vitamins has been shown to affect meat quality. Feeding of vitamin E has been shown to improve color stability and extend shelf-life of beef. Vitamin E is fat-soluble and is deposited in cell membranes and adipose cells. It is a strong antioxidant and most likely works to control lipid oxidation and color deterioration through its antioxidant function (Faustman and Wang, 2000).

3.4.2 Dietary effects in pork

In pork production, the use of high energy grain diets (soy bean and corn) is standard, the main effects of diet on pork quality are related to the lysine level in the diet or the level and/or quality of the fat (fatty acid composition) in the diet. The NPPC conducted a study with six pork genotypes that were fed one of four diets differing in lysine level (1.25, 1.1, 0.95 and 0.8% lysine). Hogs fed the diet containing the lowest lysine level had higher overall carcass fatness and higher intramuscular fat content. So diet could affect marbling level, but at the detriment of decreasing carcass leanness. The increased level of marbling would most likely not offset the increased production and carcass yield cost associated with decreasing carcass leanness.

Altering the fatty acid composition of the diet in non-ruminants influences the final fatty acid composition of the animal's fat. In ruminants, the microflora biohydrogenate unsaturated fatty acids and it is more difficult to modify beef and lamb fatty acid composition by dietary means unless rumen-protected fats are fed. St. John *et al.* (1987) fed growing pigs a high-oleate diet and found higher muscle and adipose tissue oleate levels. However, meat flavor and palatability were altered. High-oleic cooked Longissimus chops were juicier, had higher tenderness scores, and flavor was similar to chops from traditionally fed hogs (St. John *et al.*, 1987). The higher unsaturated fatty acid composition found in the meat from animals fed the high-oleic sunflower containing diet was softer, oilier and would be considered a visual quality defect. While fat flavor is dependent on the composition or fatty acid profile of the fat, slight alterations in the fatty acid profile may not significantly affect flavor. The source of the fat may also affect cooked pork quality. When pigs were fed canola oil, the subsequent meat had more off-flavor, lower quality scores, and lower overall palatability ratings than meat from pigs fed either a normal swine ration, animal-fat-, safflower oil-, or sunflower oil-based diets. While altering the fat source in the diet can affect the fatty acid profile in the subsequent meat, the decrease in the fatty acids associated with increasing human serum cholesterol levels is minimal and most likely would not affect the overall health of consumers.

3.5 Rearing and meat quality

The rearing or housing of animals prior to slaughter can affect meat quality. These effects are mainly due to the lack of stress or the level of stress inflicted on the animal due to the rearing environment. If animals are housed in conditions that result in lower rates of gain then animals may be slightly older and may not have the same level of fatness as their counterparts reared in more desirable conditions. Free-range animals also have the potential to have access to a higher variety of feedstuffs prior to slaughter that may affect the flavor of the subsequent meat. For example, range fed hogs would have access to forages during some parts of the year that may result in off-flavors in their meat (see previous discussion), but this effect would be seasonal and based on what type of forages were available. In general, confinement feeding has minimal to no effects on meat quality. If animals are over-crowded, there may be limited access to feed and water and animals exhibit undesirable social behaviors such as fighting, chewing and inability to rest properly. In these situations animal growth will be affected and the subsequent meat may be lower in overall fatness and meat from these animals may have a higher incidence of quality problems related to stress during slaughter.

3.6 Slaughtering and meat quality

The conversion of muscle to meat or the live animal to a carcass can impact meat quality. Rigor mortis, Latin for 'stiffening after death', is the process that the muscle proceeds through in order to become meat. During this process, stress induced on the animal, either long- or short-term, will affect how rapidly the process of rigor mortis will proceed.

3.6.1 Long-term stress on meat quality
Animals exposed to long-term pre-slaughter stress have reduced glycogen supplies at slaughter. Upon onset of rigor mortis, pH decline does not proceed at

a normal rate. The substrate glucose, that is derived from glycogen, is converted to lactic acid. The build up of lactic acid is responsible for post-mortem pH decline in muscle. If post-mortem pH decline does not proceed normally and the ultimate pH is higher than normal (greater than 6.0), the resultant meat is darker in color, has a firm texture, has a high water-holding capacity and has less drip loss and free moisture on the meat surface. The meat is defined as dark, firm and dry or DFD. When DFD meat is cooked, it is often described as being juicy, tender and very intense in flavor. Some describe the flavor as serumy, musty or old. Additionally, due to the high pH, DFD meat will spoil more rapidly. Conditions that induce long-term stress are long transit times, exposure to extremes in temperature (hot or cold), extended periods without food, bulls expressing sexual behavior, or improper handling prior to slaughter. It is obvious that conditions imposing long-term stress should be avoided in order to improve quality.

3.6.2 Short-term stress effects on meat quality

Short-term stress results in pale, soft and exudative meat (PSE). This meat has a lower than normal pH that results in meat that is pale in color and does not have the ability to hold water. During cooking, PSE meat will lose a high amount of moisture and the resultant meat will be drier, tougher and not as flavorful. Improper handling, rough handling, mixing of pens during transportation or at the slaughter plant, poorly designed holding and handling facilities at the slaughter plant, and other conditions that induce stress just immediately prior to slaughter can result in PSE meat. When animals become excited pre-slaughter, metabolism associated with the flight or fight mechanisms is increased. Body temperature increases and glycolytic metabolism is stimulated. During exsanguination, the blood supply is no longer able to help regulate body temperature and remove the products of anaerobic metabolism. As a result increased body temperature in combination with rapid metabolism results in a faster than normal pH decline and some protein denaturation due to higher body temperature. The combined effect of a lower pH and protein denaturation contribute to the lower water-holding capacity of the meat. The higher amount of free-water provides a higher reflective surface for light so that the meat is paler in color. The weaker protein interactions result in softer, less firm meat that also provides greater reflectance surface for light contributing to the paler color. The NPPC defined visual standards of PSE meat in the ham and loin chop because final meat pH is a continuous variable that is difficult to categorize. Ultimate pH should be considered an important quality variable as it accounts for short-term pre-slaughter stress effects and the subsequent visual and eating quality of the meat.

The RSE condition discussed previously in 3.3.4 is a result of higher glycogen levels in meat prior to slaughter. During the conversion of muscle to meat, as there is a higher amount of substrate for the conversion of glucose to lactic acid, the resultant meat has a lower than normal pH, but not as low as in

54   Meat processing

PSE meat. The RSE meat does not have the pale color associated with PSE meat and the RSE condition, in general, is not associated with pre-slaughter stress. Elimination or minimization of short-term pre-slaughter stress obviously affects pork quality. To improve overall pork quality and consistency, management and handling practices must not induce short-term pre-slaughter stress.

3.6.3   Stunning method effects on meat quality

The type of stunning method used to immobilize animals during the slaughter process can affect meat quality either through inducing short-term pre-slaughter stress or it can affect blood removal upon exsanguination. The most common methods of stunning include an apparatus that induces a concussion, referred to as concussive methods, the use of electricity to immobilize and concuss the animal, or exposure to CO_2 that results in an immobilized state. Cattle are more commonly immobilized using concussive stunning methods and hogs are immobilized generally with electricity, but also CO_2 stunning is used in some countries. Electrical stunning can be placed on the head only, on the head to the back or on the head to the brisket. With electrical stunning an epileptiform seizure is induced so that the animal is insensible to pain. There are two phases of these seizures, tonic and clonic (Gregory, 1985). The induction and strength of the seizures is dependent on the amount of current and the area of the brain that the current affects. Hoenderken (1978) defined a minimum requirement of a current of 1.25 to 1.3A that is maintained for three seconds using a voltage of at least 240V for electrical stunning. During electrical stunning, kicking can occur during the clonic phase as the brain's inhibitory influence on the spinal cord is reduced. This can result in increased time for shackling, increased worker risks and less effective exsanguination.

Pigs immobilized with CO_2 are more relaxed (Channon et al., 2002). Use of head-to-back or head-to-brisket stunning electrodes has been considered the most humane compared to head-to-head stunning as these methods induce cardiac fibrillation that results in cardiac arrest and these pigs show less kicking (Wotton et al., 1992). With cardiac arrest, pigs will not regain consciousness. With head-to-head stunning, there are some incidences where pigs regain consciousness. A disadvantage of head-to-back electrical stunning is that there can be some incidence of broken vertebrae if too high a voltage is applied. Broken vertebrae are reduced by applying head-to-brisket stunning.

Carbon dioxide stunning has been shown to reduce the incidence of ecchymosis (Gregory, 1985) and as animals can remain motionless for up to 60 seconds, kicking is reduced during shackling (Larsen, 1982). Worker safety therefore is reduced with CO_2 stunning. However, CO_2 stunning in itself does not reduce the incidence of PSE, but as animals stunned with CO_2 have reduced stress, meat quality is better than with electrical stunning (Barton-Gade, 1993). Channon et al. (2002) found that hogs stunned with CO_2 had higher pH than head-to-brisket and head-only stunned hogs in the 5th–6th thoracic after 40 minutes, 90 minutes, 3 hours and 6 hours, but after 24 hours, pH did not differ in meat from animals stunned by the three methods. Drip loss (%) was higher in meat from animals stunned with head-to-brisket stunning and ecchymosis was lower in animals stunned using CO_2.

Hemoglobin content of meat is strongly influenced by conditions immediately prior to and during exsanguination at slaughter. Pre-slaughter stress, inadequate severing of the artery or vein used for exsanguination, extended time between stunning and exsanguination, and improper suspension of the carcass during exsanguination can restrict the volume of blood removed. When hemoglobin content is higher in muscle tissue due to improper bleeding, ecchymosis can be increased and meat can be darker red and taste more metallic. Proper application of stunning and proper blood removal during exsanguination can affect meat quality. Care in properly applying these methods is needed to reduce variation in quality.

3.6.4 Electrical stimulation effects on meat quality

The use of electrical pulses to use up energy reserves in meat is called electrical stimulation. Savell *et al.* (1978) showed that by applying electrical stimulation to beef carcasses, cold-induced toughening was reduced. They showed that electrically stimulated beef carcasses had accelerated post-mortem pH decline and longer sarcomeres that resulted in more tender meat. High or low voltage electrical stimulation can be used to reduce variation in beef quality. The major differences between low-voltage and high-voltage electrical stimulation is that low-voltage electrical stimulation must be applied early in the post-mortem process and results in more gentle muscle contractions when compared to high-voltage electrical stimulation. Additionally, with high-voltage electrical stimulation there can be some tearing at the molecular level in muscles that are rigorously worked that provides additional tenderness improvements. Electrically stimulated carcasses also have brighter cherry red color at shorter chilling times post-mortem. As rigor proceeds at a more rapid rate in electrically stimulated beef carcasses, ultimate pH is obtained more rapidly and post-mortem physiological changes stabilize sooner. Some research has shown that electrically stimulated beef carcasses have brighter cherry-red color and higher amounts of marbling. Carcasses with low levels of external fat, usually less than 0.64 cm, chill rapidly and the resultant Longissimus muscle may appear darker red along the exterior rim of the meat and be lighter in color in the center. This condition is called heat ring, but it is actually a result of pH differences in the muscle due to a more rapid chilling of the exterior surface of the cut compared to the center.

The exterior surface will have a higher pH that is a result of rapid chilling in lean carcasses. At cold temperatures, glycolysis proceeds at a reduced rate until it eventually is halted. In rapidly chilled muscle, glycolysis halts when there is still substrate, glucose, available for further pH decline, but the system to convert glucose to lactic acid is not functioning. Therefore, ultimate pH is higher. In the center of the muscle, rigor mortis continues at a more normal rate and glycolysis is not limited due to cold temperatures. The ultimate pH is lower.

56   Meat processing

Electrical stimulation reduces or eliminates this effect by forcing the muscles to work and use up ATP reserves, rigor mortis proceeds at a more rapid rate and ultimate pH is more closely reached before chilling can inhibit glycolysis and rigor mortis development.

Electrical stimulation has not commonly been applied to pork. As pork has more problems with rapid rates of post-mortem pH decline due to short-term excitement, electrical stimulation traditionally has induced higher levels of PSE. Research is continuing on modifying electrical stimulation to address quality problems in pork.

3.7   Other influences on meat quality

Storage of meat can strongly affect quality positively and negatively. The positive effect of meat storage influences meat tenderness, also referred to as meat ageing. During refrigerated post-mortem storage, meat tenderness improves. The major factor responsible for post-mortem improvement in meat tenderness is degradation or proteolysis of muscle proteins. Proteolysis of muscle post-mortem has mainly contributed to sarcoplasmic Ca^{2+}-dependent proteases, the calpains, and the level of their inhibitor, calpastatin. The physiological changes in post-mortem muscle have been associated with degradation of Z-lines, troponin-T, titin, nebulin and desmin and the appearance of a 95,000 dalton and 30,000 dalton components. During post-mortem storage, the major improvements in tenderness occur with the first 7 to 14 days (Fig. 3.2), but degradation continues with increased storage, but at a slower rate. The negative effect of meat storage on meat quality is due to microbial growth and/or lipid oxidation. Both of these processes result in ending the shelf-life of meat.

3.8   Summary: ensuring consistency in raw meat quality

Meat quality and consistency are important in ensuring consumer satisfaction. Quality of meat is affected by the genetic propensity of the animal, how the animal is reared, and the nutritional status during production. These factors affect the fat, lean and connective tissue component of meat and therefore influence meat quality. Genetic differences are being understood as genetic markers are being developed for many major quality characteristics within species. As the production segment selects animals to maximize quality, reduction in meat quality can be obtained. However, these animals must be fed and reared to maximize quality. Quality also is strongly influenced by conditions at the slaughter plant. How animals are handled pre-slaughter affects the rate of rigor mortis. The application of stunning and exsanguination methods that ensure reduced animal stress are important to meat quality. The application of electrical stimulation and how the carcass is chilled influence the rate of rigor mortis and subsequent meat quality.

3.9 Future trends

Quality as a meat industry issue will continue. Providing consumers with high quality, consistent product is a key to the success of the meat industry as with any food entity. Today's consumers demand consistency and quality and their demands are met by other segments of the food industry. Those livestock/meat producers who can ensure consistent quality will be the viable players of the future. To ensure consistency and quality, links between the production segments of the livestock industry that have genetic verification of animals and that then manage these animals to maximize their genetic propensity will be producers of the future. The seedstock and commercial production segments will either be vertically integrated or there will be alliances between producers to form joint ventures to produce animals of consistent quality, much like the large poultry companies in the United States. The slaughter and manufacturing segments of the meat industry will have control points within their production segments to assure meat quality and will control the end product from slaughter to the final package for the consumer.

Technological advances to improve meat quality will be viewed as interventions to help control consistency and quality. A fully integrated meat production system that assures quality will have economic benefit and returns. Products will be brand-identified and carry a quality reputation as a component of marketing the product.

3.10 References

ABERLE E D, REEVES E S, JUDGE M D, HUNSLEY R E and PERRY T W (1981), 'Palatability and muscle characteristics of cattle with controlled weight gain. Time on a high energy diet', *J Anim Sci*, 52, 757–763.

AMSA (1995), 'Research Guidelines for Cookery, Sensory Evaluation and Instrumental Measurements of Fresh Meat', *American Meat Science Association and National Livestock and Meat Board*, Chicago, IL.

BARTON-GADE P A (1990) Pork quality in generic improvement programmes – the Danish experience. *Proc. of the National Swine Improvement Federation Annual Meeting*. Des Moines, IA.

BARTON-GADE P A (1993), 'Effect of stunning on pork quality and welfare: Danish experience', *Allen D. Leman Swine Conference*, 20, 173–178.

BIDNER T D, SCHUPP A R, MOHAMAD A B, RUMORE N C, MONTGOMERY R E, BAGLEY C P and MCMILLIN K W (1986), 'Acceptabiltiy of beef from Angus-Hereford or Angus-Hereford-Brahman steers finished on all-forage or a high-energy diet', *J Anim Sci*, 62, 381–387.

BOLEMAN S L, BOLEMAN S J, MORGAN W W, HALE D S, GRIFFIN D B, SAVELL J W, AMES R P, SMITH M T, TATUM J D, FIELD T G, SMITH G C, BARDNER B A, MORGAN J B, NORTHCUTT S L, DOLEZAL H G, GILL D R and RAY F K (1998), 'National Beef Quality Audit-1995: Survey of producer-related defects

and carcass quality and quantity attributes', *J Anim Sci*, 76, 96–103.
BOWLING R A, SMITH G C, CARPENTER Z L, DUTSON T R and OLIVER W M (1977), 'Comparison of forage-finished and grain-finished beef carcasses', *J Anim Sci*, 45, 209–215.
BROOKS J C, BELEW J B, GRIFFIN D B, GWARTNEY B L, HALE D S, HENNING W R, JOHNSON D D, MORGAN J B, PARRISH JR F C, REAGAN J O and SAVELL J W (2002), 'National Beef Tenderness Survey 1998', *J Anim Sci*, (Accepted).
CARPENTER J W, PALMER A Z, KIRK W G, PEACOCK F M and KOGER M (1961), 'Slaughter and carcass characteristics of Brahman and Brahman crossbred steers', *J Anim Sci*, 20, 336–340.
CHANNON H A, PAYNE A M and WARNER R D (2002), 'Comparison of CO_2 stunning with manual electrical stunning (50 Hz) of pigs on carcass and meat quality', *Meat Sci*, 60, 63–68.
CHRISTIAN L L and ROTHSCHILD M F (1981), 'Performance characteristics of normal, stress carrier and stress susceptible swine', *Anim Res Rep*, AS-528-F.
CROUSE J D, SEIDEMAN S C and CUNDIFF L V (1987), 'The effect of carcass electrical stimulation on meat obtained from *Bos indicus* and *Bos taurus* cattle', *J Food Qual*, 10, 407.
CROUSE J D, CUNDIFF L V, KOCH R M, KOOHMARAIE M and SEIDEMAN S C (1989), 'Comparisons of *Bos indicus* and *Bos taurus* inheritance for carcass beef characteristics and meat palatability', *J Anim Sci*, 67, 2661–2668.
CUNDIFF L V, KOCH R M, GREGORY K E, CROUSE J D and DIKEMAN M E (1990), 'Germ plasm evaluation program', Progress Report No. 12, pp. 1–6.
DAMON R A JR, CROWN R M, SINGLETARY C B and MCCRAINE S F (1960), 'Carcass characteristics of purebred and crossbred beef steers in the Gulf coast region', *J Anim Sci*, 19, 820–844.
DAZZI G, MADARENA G, CAMPANINI G, CAMPESATO E, CHIZZOLINI R and BADIANI A (1987), 'Interrelationships between various compositional and quality parameters of pork from pure breeds', *Proc International Congress of Meat Science and Technology*, 33, 6–9.
ENFALT A, LUNDSTROM K, HANSSON I, JOHANSEN S and NYSTROM P (1997a), 'Comparison of non-carriers for carcass composition, muscle distribution and technological meat quality in Hampshire-sired pigs', *Livestock Production Science*, 47, 221–229.
ENFALT A, LUNDSTROM K, KARLSSON A and HANSSON I (1997b), 'Estimated frequency of the RN allele in Swedish Hampshire pigs and comparison of glycolytic potential, carcass composition, and technological meat quality among Swedish Hampshire, Landrace and Yorkshire pigs', *J Anim Sci*, 75, 2924–2935.
ESTRADE M, VIGNON X and MONIN G (1993), 'Effect of the RN gene on ultrastructure and protein fractions in pig muscle', *Meat Sci*, 35, 313–319.
FAUSTMAN C and KE-WE WANG X (2000), In E A Decker, C Faustman and C J Lopez-Botez, 'Antioxidants in muscle foods', OCED, John Wiley Publishing, pp. 135–152.

FUJII J, OTSU K, ZORZATO F, DELEON S, KHANNA V K, WEILER J E, O'BRIEN P J and MACLENNON D H (1991), 'Identification of a mutation in the porcine ryanodine receptor associated with malignant hyperthermia', *Science*, 25, 448–451.

GOLL D E, GEESINK G H, TAYLOR R G and THOMPSON V F (1995), 'Does proteolysis cause all postmortem tenderization, or are changes in the actin/myosin interaction involved?', *Proc International Congress of Meat Science and Technology*, 41, 537–544.

GOODWIN R N (1994), 'Genetic parameters of pork quality traits', Ph.D. dissertation. Iowa State University, Ames.

GOODWIN R N (1997), 'Genetic effects on pork quality', *The Pork Quality Summit*, National Pork Producers Council, Des Moines, IA, p. 25.

GOODWIN R N (2002), 'Effects of Rendement Napole gene and HAL 1843 gene on fresh and cooked pork loin quality', *J Anim Sci* (Suppl. 1), (Accepted).

GPEP (1974), 'Progress Report: Germ Plasm Evaluation Program Report No. 1'. *US Meat Animal Research Center*, Clay Center, NE.

GPEP (1975), 'Progress Report: Germ Plasm Evaluation Program Report No. 2', *US Meat Animal Research Center*, Clay Center, NE.

GPEP (1978), 'Progress Report: Germ Plasm Evaluation Program Report No. 6', *US Meat Animal Research Center*, Clay Center, NE.

GREENE B B, BACUS W R and RIEMANN M J (1989), 'Changes in lipid content of ground beef from yearling steers serially slaughtered after varying lengths of grain finishing', *J Anim Sci*, 67, 711–715.

GREGORY N G (1985), 'Stunning and slaughter of pigs', *Pig News and Information*, 6(4), 407–413.

HAGER L B (2000), 'Evaluation of carcass traits, connective tissue, and myofibrillar proteins characteristics on tenderness of F_1 steers sired by *Bos indicus* bulls', M.S. Thesis, Texas A&M University, College Station, TX.

HARRISON A R, SMITH M E, ALLEN D M, HUNT M C, KASTNER C L and KROPF D H (1978), 'Nutritional regime effects on quality and yield characteristics of beef', *J Anim Sci*, 47, 383–388.

HAWRYSH Z J, BERG R T and HOWES A D (1975), 'Eating quality of beef from steers fed full or restricted levels of moisture treated barley', *Can J Anim Sci*, 55, 179–185.

HEDRICK H B, LEAVITT R K and ALEXANDER M A (1968), 'Variation in porcine muscle quality of Duroc and Hampshire barrows', *J Anim Sci*, 27, 48–52.

HOENDERKEN R (1978), 'Electrical stunning of pigs', In S. Fabiansson (ed.), *Hearing on pre-slaughter stunning* (Report No. 52), pp 29–38, Kavlinge: Swedish Meat Research Centre.

JENSEN P H, CRAIG H B and ROBISON O W (1967) 'Phenotypic and genetic associations among carcass traits of swine', *J Anim Sci*, 26, 1252–1260.

JOHNSEN P B and CIVILLE G V (1986) 'A standardized lexicon of meat WOF descriptors', *J Sensory Studies*, 1, 99–104.

KOCH R M, DIKEMAN M E, ALLEN D M, MAY M, CROUSE J D and CAMPION D R (1976),

'Characterization of biological types of cattle. III. Carcass composition, quality and palatability', *J Anim Sci*, 43, 48–62.

KOCH R M, DIKEMAN M E, LIPSEY R J, ALLEN D M and CROUSE J D (1979), 'Characterization of biological types of cattle cycle II:III. Carcass composition, quality and palatability', *J Anim Sci*, 4, 448–460.

KOCH R M, CUNDIFF L V and GREGORY K E (1982a), 'Heritabilities and genetic, environmental and phenotypic correlations of carcass traits in a population of diverse biological types and their implications in selection programs', *J Anim Sci*, 55, 1319–1329.

KOCH R M, DIKEMAN M E and CROUSE J D (1982b), 'Characterizations of biological types of cattle (Cycle III). III. Carcass composition, quality and palatability', *J Anim Sci*, 54, 35–45.

KOOHMARAIE M (1988), 'The role of endogenous proteases in meat tenderness', *Proc Recip Meat Conf*, 41, 89–100.

KOOHMARAIE M (1992), 'The role of Ca^{2+}-dependent proteases (calpains) in post mortem proteolysis and meat tenderness', *Biochimie*, 74, 239–245.

KOOHMARAIE M, BABIKER A S, MERKEL R A and DUTSON T R (1988). 'Role of Ca-dependent proteases and lysosomal enzymes in postmortem changes in bovine skeletal muscle', *J Food Sci*, 53, 1253–1257.

KROPF D H, ALLEN D M and THOUVENELLE G J (1975), 'Short-fed, grass-fed and long-fed beef compared', *Kansas Agr Exp Sta Rep 230*.

LARSEN H K (1982), 'Comparison of 300 volt manual stunning, 700 volt automatic stunning and CO_2 compact stunning with respect to quality parameter, blood splashing, fractures and meat quality', In G. Eikelenboom (ed.), *Stunning of animals for slaughter*, pp. 73–81, The Hague: Martinus Nijhoff Publishers.

LAWRIE R A and GATHERUM D P (1962), 'Studies on the muscle of meat animals. II. Differences in the ultimate pH and pigmentation of longissimus dorsi muscles from two breeds of pigs', *J Agric Sci*, 58, 97.

LEROY P, NAVEAU J, LOOFT C and KALM E (1990), 'Evidence for a new major gene influencing meat quality in pigs', *Genet Res Camb*, 55, 33–40.

LEROY P, JUIN H, CARITEZ J C, BILLON Y, LAGANT H, ELSEN J M and SELLIER P (1996), 'Effet du genotype RN sur les qualities sensorielles de la viande de porc', *Journees Rech Porcine en France*, 28, 53–56.

LINDAHL G, LUNDSTROM K and TORNBERG E (2001), 'Contribution of pigment content, myoglobin forms and internal reflectance to the colour of pork loin and ham from pure breed pigs', *Meat Sci*, 59, 141–151.

LOCKER R H (1960). 'Degree of muscular contraction as a factor in tenderness of beef', *J Food Sci*, 25, 304–307.

LORENZEN C L, HALE D S, GRIFFIN D B, SAVELL J W, BELK K E, FREDERICK T L, MILLER M F, MONTGOMERY T H and SMITH G C (1993), 'National Beef Quality Audit: Survey of producer-related defects and carcass quality and quantity attributes', *J Anim Sci*, 71, 1495–1502.

LORENZEN C L, NEELY T R, MILLER R K, TATUM J D, WISE J W, TAYLOR J F, BUYCK M J, REAGAN J O and SAVELL J W (1999), 'Beef Customer Satisfaction:

Cooking method and degree of doneness effects on the top loin steak', *J Anim Sci*, 77, 637–644.
LOUIS C F, MICKELSON J R and REMPEL W E (1994), 'The effect of skeletal muscle ryanodine receptor (ryrl) genotype on swine performance and carcass traits', *J Anim Sci*, 72(Suppl. 1), 248.
LUCKETT R L, BIDNER T D, ICAZA E A and TURNER J W (1975), 'Tenderness studies in straightbred and crossbred steers', *J Anim Sci*, 40, 468–475.
LUNDSTROM K, ESSEN-GUSTAVSSON B, RUNDGREN M, EDFORES-LILJA I and MALMFORS G (1989), 'Effect of halothane genotype on muscle metabolism at slaughter and its relationship with meat quality: a with-in-litter comparison', *Meat Sci*, 25, 251–263.
LUNDSTROM K, ANDERSSON A and HANSSON I (1996), 'Effect of the RN Gene on technological and sensory meat quality in crossbred pigs with Hampshire as terminal sire', *Meat Sci*, 42, 145–153.
LUNDSTROM K, ENFALT A, TORNBERG E and AGERHEM H (1998), 'Sensory and technological meat quality in carriers and noncarriers of the RN allele in Hampshire crosses and in purebred Yorkshire pigs', *Meat Sci*, 48, 115–124.
LYON B G (1987), 'Development of chicken flavor descriptive attribute terms aided by multivariate statistical procedures', *J Sensory Stud*, 2, 55–67.
MCKEITH F K, SAVELL J W, SMITH G C, DUTSON T R and CARPENTER Z L (1985), 'Physical, chemical, histological and palatability characteristics of muscle from three breed-types of cattle at different times-on-feed', *Meat Sci*, 15, 37–50.
MCKENNA D R, ROEBER D L, BATES P K, SCHMIDT T B, HALE D S, GRIFFIN D B, SAVELL J W, BROOKS J C, MORGAN J B, MONTGOMERY T H, BLEK K E and SMITH G C (2002), 'National Beef Quality Audit-2000: Survey of targeted cattle and carcass characteristics related to quality, quantity, and value of feed steers and heifers', *J Anim Sci*, 80, 1212–1222.
MARSH B B and LEET N G (1966), 'Studies in Meat Tenderness: 3. The effect of cold shortening on tenderness', *J Food Sci*, 31, 450.
MARSH, B B, WOODHAMS P R and LEET N G (1968), 'Studies in meat tenderness. 5. The effect on tenderness of carcass cooling and freezing before the completion of rigor mortis', *J Food Sci*, 33, 12–18.
MAY S G, MIES W L, EDWARDS J W, WILLIAMS F L, WISE J W, MORGAN J B, SAVELL J W and CROSS H R (1992), 'Beef carcass composition of slaughter cattle differing in frame size, muscle score, and external fatness', *J Anim Sci*, 70, 2431–2445.
MELTON S L (1983), 'Effect of forage feeding on beef flavor', *Food Technol*, 37(5), 239–248.
MEYER B, THOMAS J, BUCKLEY R and COLE J W (1960), 'The quality of grain-finished and grass-finished beef as affected by ripening', *Food Technol*, 14, 4–7.
MILLER K D, ELLIS M, MCKEITH F K, BIDNER B S and MEISINGER D J (2000), 'Frequency of the rendement napole RN allele in a population of American Hampshire pigs', *J Anim Sci*, 78, 1811–1815.

MILLER R K (1994), 'Quality characteristics', In D M Kinsman, A W Kotula and B C Breidenstein (eds) *Muscle Foods*, pp. 296–332, Chapman and Hall, New York.

MILLER R K, CROSS H R, CROUSE J D and TATUM J D (1987), 'The influence of diet and time on feed on carcass traits and quality', *Meat Sci*, 19, 303–313.

MILLER R K, ROCKWELL L C, LUNT D K and CARSTENS G E (1997), 'Determination of the flavor attributes of cooked beef from steers fed corn or barley based diets', *Meat Sci*, 44, 235–243.

MILLER R.K., MOELLER S J, GOODWIN R N, LORENZEN C L AND SAVELL J W (2000), 'Consistency in meat quality', *Inter Congress of Meat Sci and Tech*, 46, 566–580.

MONIN G and SELLIER P (1985), 'Pork of low technological quality with a normal rate of muscle pH fall in the immediate post-mortem period: the case of the Hampshire breed', *Meat Sci*, 13, 49–63.

MORGAN J B, SAVELL J W, HALE D S, MILLER R K, GRIFFIN D B, CROSS H R and SHACKELFORD S D (1991), 'National Beef Tenderness Survey', *J Anim Sci*, 69, 3274–3283.

NEELY T R, LORENZEN C L, MILLER R K, TATUM J D, WISE J W, TAYLOR J F, BUYCK M J, REAGAN J O and SAVELL J W (1998), 'Beef Customer Satisfaction: Role of cut, USDA Quality Grade, and city on in-home consumer ratings', *J Anim Sci*, 76, 1027–1032.

NPPC (1995), 'Genetic Evaluation: Terminal Line Program', *National Pork Producers Council*, Des Moines, IA.

RAMSEY C B, COLE J W, MEYER B H and TEMPLE R S (1963), 'Effects of type and breed of British, Zebu and dairy cattle on production, palatability and composition. II. Palatability differences and cooking losses as determined by laboratory and family panels', *J Anim Sci*, 22, 1001–1008.

RILEY R R, SMITH G C, CROSS H R, SAVELL J W, LONG C R and CARTWRIGHT T R (1986), 'Chronological age and breed-type effects on carcass characteristics and palatability of bull beef', *Meat Sci*, 17, 187–198.

SAVELL J W and CROSS H R (1988), 'The role of fat in the palatability of beef, pork, and lamb'. In: *Designing Foods: Animal Product Options in the Marketplace*. National Academy Press, Washington, D.C.

SAVELL J W, DUTSON T R, SMITH G C and CARPENTER Z L (1978), 'Structural changes in electrically-stimulated beef muscle', *J Food Sci*, 43, 1606–1607.

SAVELL J W, LORENZEN C L, NEELY T R, MILLER R K, TATUM J D, WISE J W, TAYLOR J F, BUYCK M J and REAGAN J O (1999), 'Beef Customer Satisfaction: cooking method and degree of doneness effects on the top sirloin steak', *J Anim Sci*, 77, 645–652.

SCHROEDER J W, CRAMER D A, BOWLING R A and COOK C W (1980), 'Palatability, shelflife and chemical differences between forage- and grain-finished beef', *J Anim Sci*, 50, 852–859.

SELLIER P, MEJENES-QUIJANO A, MARINOVA P, TALMONT A, JACQUET B and MONIN G (1988), 'Meat quality as influenced by halothane sensitivity and ultimate

pH in three porcine breeds', *Livestock Prod Sci*, 18, 171–186.
SHACKELFORD S D (1992), 'Heritabilities and phenotypic and genetic correlations for bovine postrigor calpastatin activity and measures of meat tenderness and muscle growth', Ph.D. Dissertation, Texas A&M University, College Station, TX.
SHACKELFORD S D, KOOHMARAIE M, MILLER M F, CROUSE J D and REAGAN J O (1991), 'An evaluation of tenderness of the longissimus muscle of Angus by Hereford versus Brahman crossbred heifers', *J Anim Sci*, 69, 171–177.
SKELLEY G C, EDWARD R L, WARDLAW F B and TORRENCE A K (1978), 'Selected high forage rations and their relationship to beef quality, fatty acids and amino acids' *J Anim Sci*, 47, 1102–1108.
ST. JOHN L C, YOUNG C R, KNABE D A, THOMPSON L D, SCHELLING G T, GRUNDY S M and SMITH S B (1987), 'Fatty acid profiles and sensory and carcass traits of tissues form steers and swine fed an elevated monosaturated fat diet', *J Anim Sci*, 64, 1441–1447.
TATUM D J (1981), 'Is tenderness nutritionally controlled?', *Proc Rec Meat Conf*, 34, 65–67.
WARNER R D, KAUFFMAN R G and GREASER M L (1997), 'Muscle protein changes post mortem in relation to pork quality traits', *Meat Sci*, 45, 339–352.
WHEELER T L, SAVELL J W, CROSS H R, LUNT D K and SMITH S B (1990), 'Mechanisms associated with the variation in tenderness of meat from Brahman and Hereford cattle', *J Anim Sci*, 68, 4206–4220.
WHIPPLE G, KOOHMARAIE M, DIKEMAN M E, CROUSE J D, HUNT M C and KLEMM R D (1990), 'Evaluation of attributes that affect longissimus muscle tenderness in *Bos taurus* and *Bos indicus* cattle', *J Anim Sci*, 68, 2716–2728.
WILLIAMS C B, KEELE J W and WALDO D R L (1992), 'A computer model to predict empty body weight in cattle from diet and animal characteristics', *J Anim Sci*, 70, 3215–3222.
WILSON E R (1993), 'Comparison of mutant ryanodine receptor gene carrier (Nn) and normal (NN) pigs for growth and carcass traits', *J Anim Sci*, 71(Suppl 1), 37.
WOTTON S B, ANIL M H, WHITTINGTON P E and MCKINSTRY J L (1992), 'Pig slaughtering procedures: head-to-back stunning', *Meat Sci*, 32, 245–255.
WU J J, ALLEN D M, HUNT M C, KASTNER C L and KROPF D H (1981), 'Nutritional effects on beef palatability and collagen characteristics', *J Anim Sci*, 51(suppl. 1), 71.

4
The nutritional quality of meat
J. Higgs, Food To Fit, Towcester and B. Mulvihill, Republic of Ireland

4.1 Introduction

The most common dietary problems in developed countries are due mainly to over-nutrition. The incidences of overweight, obesity and adult onset-diabetes are increasing steadily. Cancer is now the most common cause of death in many developed countries. The most common cancers are breast, lung, bowel and prostate, which are virtually absent in some developing countries. However, even in our affluent society, we also see signs of nutritional inadequacies. For instance, in the UK nearly a half of females aged between 11 and 14 are not getting enough iron in their diet, while more than a third are not getting enough zinc (Gregory et al., 2000). We are living in a society where both signs of over- and under-nutrition occur side by side. To correct for these nutritional paradoxes we as consumers have to get the balance of nutrients, energy and physical activity right. The objective of this chapter is to highlight the nutritional role that meat can play in modern society.

The National Food Survey for 1999 (Ministry of Agriculture Fisheries and Food, 1999), included a special analysis on meat and meat products consumption in the UK. It stated that 'meat, meat products ... are important contributors to the intakes of many nutrients in the British diet'. Data from this survey showed that meat and meat products supply: energy 15%, protein 30%, fat 22% (saturated fatty acids (SFA) 22%, monounsaturated fatty acids (MUFA) 27%, polyunsaturated fatty acids (PUFA) 15%), vitamin D 19%, B_2 14%, B_6 21%, B_{12} 22%, vitamin A equivalents 20%, niacin 37%, zinc 30%, iron 14%.

Meat has been a major part of the human diet for at least 2 million years. Human genetic make-up and physical features have been adapted over 4.5 million years for a diet containing meat. An example of this adaptation is our present teeth and jaw structure, which have developed to become efficient at chewing and swallowing meat. Meat is a highly nutritious and versatile food. The primary importance of meat as a food lies in the fact that when digested its protein is broken down releasing amino acids, these are assimilated and ultimately used for the repair and growth of cells. Meat is a nutrient dense food, providing valuable amounts of many essential micronutrients. Meat supplies fatty acids, vitamins, minerals, energy and water and is involved in the synthesis of protein, fat and membranes in the body.

Traditionally, meat was considered a highly nutritious food, highly valued and associated with good health and prosperity. As such, western societies gradually increased consumption with increasing affluence. The healthy image of red meat gradually became eroded during the 1980s, when the lipid hypothesis focused attention on the fat contributed from meat. The British Government's Committee on Medical Aspects of Food and Nutrition (COMA) report on coronary heart disease (CHD) in 1984 identified meat as a major source of saturated fat, contributing a quarter of UK intakes (COMA, 1984). Although the multifactorial nature of CHD risk is now widely acknowledged (British Nutrition Foundation,1996; COMA, 1994), the health image of red meat remains tarnished due to this negative association. More recently, we have seen the publication of two reports on diet and cancer (World Cancer Research Fund, 1997; COMA, 1998). These reports associated red meat consumption with increased incidence of certain cancers, in particular, colorectal cancer (CRC), despite the existence of conflicting evidence. Both of these reports issued guidelines on the limits of red meat one should consume to reduce the risk of developing CRC, thereby negatively influencing the image of red meat.

The 1990s also saw major publicity on non-nutritional issues including animal health concerns such as bovine spongiform encephalopathy (BSE) and more recently the return of foot and mouth disease (FMD) to Britain. The last 25 years have been the most turbulent regarding issues surrounding meat consumption with much of the publicity being negative thus playing down meat's nutritional value.

The negative nutritional image that surrounds red meat is in some way responsible for the decrease in expenditure. In 1999, 25.8% of expenditure on home food in Great Britain was spent on meat and meat products (Ministry of Agriculture Fisheries and Food, 1999). This is a significant drop compared with 32.1% in 1979. During this time period there have been major changes in the type of meat that people are buying in the UK. Expenditures on beef, lamb, pork, bacon and ham each fell, whilst expenditure shares on poultry and on other meats has risen. The major growth area in processed meats and meat products has been frozen convenience meat products, meat-based ready meals and other meat products such as Chinese and Italian meals containing meat (Ministry of Agriculture Fisheries and Food, 1999). There are many factors responsible for these changes, the tarnished image of red meat being one such factor. Other influencing factors include changes in lifestyle trends which saw the drive for convenience foods and the resultant responsiveness of the industry to this has greatly influenced the changing meat buying habits of consumers.

4.2 Meat and cancer

Meat consumption has been implicated in many cancers, as being either protective or causative, depending on the type of cancer. Meat consumption has been shown to protect against cancers of the stomach (Hirayama, 1990; Tuyns *et al.*, 1992; Azevedo *et al.*, 1999), liver and the oesophagus (Zeigler *et al.*, 1981; Tuyns *et al.*, 1987, Nakachi *et al.*, 1988). These are three of the top five cancers globally. On the other hand meat consumption has been implicated as a cause of colorectal (colon and rectal), breast and prostate cancer, with the main emphasis being on (CRC). CRC is the fourth most common cancer in the world, but in Europe and other Western countries it is second in terms of incidence and mortality (after lung cancer in men and breast cancer in women) with 190,000 new cases per year in Europe (Black *et al.*, 1997; Bingham, 1996). There is strong evidence from epidemiological studies showing that diet plays an important role in most large bowel cancers, implying that it is a potentially preventable disease (Higginson, 1966; COMA, 1998) The precise dietary components that influence CRC risk have not been fully elucidated. However, epidemiological studies suggest that high intakes of fat, meat and alcohol increase risk, whereas vegetables, cereals and non-starch polysaccharides, found in fruit and many other foods, decrease the risk (Bingham, 1996). For many of these dietary factors the evidence is equivocal. In the case of meat, the evidence is conflicting, early cross-sectional comparisons attributed much of the worldwide variation in CRC incidence to fat and animal protein consumption (Armstrong and Doll, 1975). In contrast, subsequent case-control and cohort studies are much less consistent (Hill, 1999a).

Meat consumption and CRC became a high-profile issue during 1997–1998 with the global launch of the World Cancer Research Fund (1997) report, timed to coincide with the publication of the British COMA report, both on diet and cancer. The WCRF report was particularly negative towards red meat, which fuelled the launch publicity. This stimulated several critical appraisals of the report, all challenging the conclusions regarding meat (Hill, 1999b). The scientific evidence is not sufficiently robust to recommend a maximum of 80g/day red meat as pronounced by the WCRF and the initial announcement by COMA for a similar recommendation was subsequently revised. Most of the data showing an association between meat consumption and CRC are American, whereas several studies conducted outside the US (many in Europe) have shown no such relationship (Hill, 1999a). On final publication, COMA (1998) reassured UK consumers that average consumption levels (90g/day of cooked red meat) were acceptable. COMA suggests that high consumers, less than 15% of the UK population, eating above 140g/day might benefit from a reduction. Equally important, this report acknowledged that meat and meat products remain a valuable source of a number of nutrients including iron and that for many a moderate intake makes an important contribution to micronutrient status. The potential effect on iron status of further reductions to red meat intakes was subsequently investigated, as recommended within the COMA report. Given that a 50% reduction in intake would result in a third of women having low iron intakes (below 8 mg/d), the appropriateness of public health messages concerning meat consumption should be carefully considered prior to reaching the media (Gibson and Ashwell, 2001).

Various components of meat (protein, iron, and heterocyclic amines) have been suspected of contributing to the development of CRC. Dietary protein is broken down in the body to amino acids, which are further degraded to ammonia, which may have cancer-initiating effects. The human colon is also rich in amides and amines that are substrates for bacterial nitrosation by nitric oxide (NO) to N-nitroso compounds that are found in human faeces. There is no conclusive evidence that protein derived compounds can increase cancer risk in humans. It is hypothesised, but not yet established, that the intake of iron from meat and other iron-rich foods may increase the risk of cancer via the production of free radicals in the body. Heterocyclic amines are formed by the Maillard reactions that involve amino acids, sugars and creatine, during cooking. They are usually produced during cooking at very high temperatures on the surface of meat, such as frying, grilling or barbecuing but they are minimal when meat is steamed, microwaved or marinated. The heterocyclic amines are known mutagens *in vitro* and carcinogens in rodents. The most abundant heterocyclic amines produced in meat is phenylimadazo pyridine (PhIP), which is a relatively weak carcinogen compared to other heterocyclic amines such as IQ and MeIQ. The role of heterocyclic amines in causing CRC is not fully elucidated in humans.

Truswell summarised the evidence in Hill (2000) and showed that 20 out of 30 case-control studies and 10 out of 14 prospective studies showed no relationship between meat intake and CRC with some of the results of the remaining studies being confused and one prospective study showing an inverse correlation between meat consumption and CRC risk (Hill, 2000). If meat consumption were associated with increased risk for cancer, one would expect mortality from cancer to be much lower among vegetarians. In a recent meta-analysis of five cohort studies, results have shown no significant differences in mortality from cancer in general, and more specifically mortality in stomach, breast, lung, prostate and colorectal cancer between vegetarians and omnivores (Key *et al.*, 1998, 1999). If red meat consumption were associated with increased risk for CRC, one would expect a decrease in the incidence of CRC to occur over time as a result of decreasing meat consumption trends. During the past 30 years, red meat consumption in the UK has decreased by approximately 25%, while during the same time the incidence of CRC has increased by about 50% (Hill, 1999b). Similarly, if meat consumption were associated with increased risk for CRC, one would expect the rates of CRC to be higher in countries with high meat consumption and lower in countries with low meat consumption. The Mediterranean countries eat more red meat than for instance the UK yet these countries have lower CRC rates (Hill, 2000). Such paradoxical evidence is further evidence that, at current levels, meat consumption is not a risk factor for CRC incidence.

Epidemiological associations between dietary components, specific foods or food groups and chronic disease, such as cancer, can identify risk factors, but are generally not sufficient to establish cause and effect relationships. Findings from epidemiological studies must be combined with other types of evidence (e.g. animal experiments, human clinical trials) before a persuasive causal relationship can be established. CRC is multi-factorial; it is confounded by diet, smoking, alcohol, physical activity, obesity, aspirin use, age and family history. There are known protective and causative factors. It is well known that daily consumption of vegetables and meat reduces the risk of cancer at many sites, whereas daily meat consumption with less frequent vegetable consumption increases risk (Hirayama, 1986; Kohlmeier et al., 1995; Cox and Whichelow, 1997). Evidence suggests that it is the reduced intakes of the protective factors such as vegetables and cereals that are the main determinants of CRC risk with meat being coincidentally related.

There is a need to assess the role of meat when consumed in normal quantities, by normal cooking methods, and within the context of a mixed, balanced, diet. The method of cooking meat and the degree of browning is of particular importance to this whole issue. A major effort by International Meat Industry partners has attempted to raise awareness of the complexities of meat preparation and cooking habits and how these differ between countries. Dietary assessment techniques adopted by nutrition scientists currently do not take full account of the diverse differences between meat products world wide and the consequent influences these may have on the body. For example, it is well recognized that meat is often cooked more evenly through the muscle within Europe, whereas it tends to be 'blackened' on the outside whilst remaining rare on the inside in North America. This may be one reason for the greater negative findings in American studies of the role of meat in CRC, compared to European studies. This hitherto unexplored facet of meat consumption may have far-reaching implications for interpretation of epidemiological data and ultimately for public health recommendations. Certain marinades applied to meat before cooking will reduce the quantity of potential carcinogenic materials present. The application of knowledge in this area to the production of processed meat products with all the nutritional benefits and none of the potentially harmful components would be progressive indeed.

In summary, it is important not only to examine the relationship between meat consumption and CRC alone, but also meat preparation and cooking differences in conjunction with protective factors, such as vegetables and cereals. At a meat and diet workshop, it was stated: 'It is time that the meat CRC story was laid to rest, so that we can get back to recommending that young women of childbearing age eat meat as a ready source of available iron' (Hill, 2000). Nevertheless, it is sensible to consider that there must be an optimal range for meat intakes in order to ensure a balanced diet is achieved, whilst optimal weight is maintained. From this practical perspective COMA's (1998) suggested intake range of 90–140g cooked meat per day, is sensible as a public health message. The overemphasis on reducing meat however, rather than encouraging greater accompanying plant food intake has served only to confuse the public (Hill, 1999b). Evidence suggests that the risk of cancer will be reduced to a greater extent by increasing intakes of fruit and vegetables than by lowering meat intakes. Once again, the move towards pre-prepared meal solutions provides an opportunity for manufacturers to develop recipes with a healthy balance of meat and vegetable ingredients such that the nutritional profile of the dish is optimised.

4.3 Meat, fat content and disease

Regular consumption of red meat is associated, epidemiologically with increased risk of coronary heart disease, due to its fat composition. Conversely a growing bank of evidence is showing that a healthy diet that includes lean red meat can produce positive blood lipid changes (Watts *et al.*, 1988; Scott *et al.*, 1990; Davidson *et al.*, 1999; Beauchesne-Rondeau *et al.*, 1999). Blood cholesterol levels are increased by inclusion of beef fat, not lean beef in an otherwise low-fat diet. Equal amounts of lean beef, chicken, and fish added to low-fat, low-saturated-fat diets, similarly reduce plasma cholesterol and LDL-cholesterol levels in hypercholesterolaemic and normocholesterolaemic men and women.

Meat is a source of arachidonic acid (20:4n-6), both in the lean and visible fat components (Duo *et al.*, 1998). Assumptions that the 20:4n-6 content of meat was responsible for increasing thrombotic tendencies in Western societies are too simplistic. The presence of large amounts of linoleic acid (18:2n-6) in current diets results in plasma increases of linoleic and arachidonic acids only. However, in the absence of linoleic acid, the long chain n-6 and n-3 PUFAs present in lean meat can influence the plasma pool, increasing plasma eicosatrienoic acid (20:3n-6), 20:4n-6, and eicosapentanoic acid (20:5n-3), and probably reducing thrombotic tendencies. It is the imbalance of n-6: n-3 PUFAs in the diet, brought about by excessive 18:2n-6 that causes high tissue 20:4n-6 levels, so encouraging metabolism to eicosanoids, (Sinclair *et al.*, 1994; Mann *et al.*, 1997).

Meat contributes between one-third to half of the UK daily cholesterol intake, (Chizzolini *et al.*, 1999; British Nutrition Foundation, 1999). Meat's cholesterol content is, for consumers, another negative influence on meat's health image, although it is now accepted that dietary intake of cholesterol has little bearing on plasma cholesterol. A review of the cholesterol content of meat indicates surprisingly that levels of cholesterol are generally not higher in fatty meat or meat products. The cholesterol content of a meat is related to the number of muscle fibres so tends to be higher the more red the muscle.

Twenty years ago red meat and meat products were identified as major contributors to fat intake in the UK. Most of the visible (subcutaneous) fat in the meat was consumed. In the early 1980s the red meat industry began to shift production systems to favour less fat, reflecting more energy efficient animal husbandry. For many years now there has been emphasis on reducing the fat content of our diets and this continued consumer demand for less fat, further prompted the meat industry to consider ways to reduce the fat content of meat. The fat content of the carcass has reduced in Britain by over 30% for pork, making British pork virtually the leanest in the world, 15% for beef, and 10% for lamb, with further reductions anticipated for beef and lamb over the next 5–10 years. The fat content of fully trimmed lamb, beef and pork is now 8%, 5% and 4% respectively (Chan *et al.*, 1995).

These achievements are due to three factors: selective breeding and feeding practices designed to increase the carcass lean to fat ratio; official carcass classification systems designed to favour leaner production; and modern butchery techniques (seaming out whole muscles, and trimming away all intermuscular fat. It is easier to appreciate the process and extent of fat reduction by looking at the changes over time for a single cut of meat such as a pork chop (Fig. 4.1). The reduction in fat for pig meat is well illustrated by the trend downwards in P_2 fat depth from the 1970s to the 1990s (P_2 is fat depth at the position of the last rib) (Fig. 4.2). Since 1992 it has remained stable at around 11mm.

Although updated compositional figures for British meat were published from 1986 onwards (Royal Society of Chemistry, 1986; 1993; 1996; Meat and

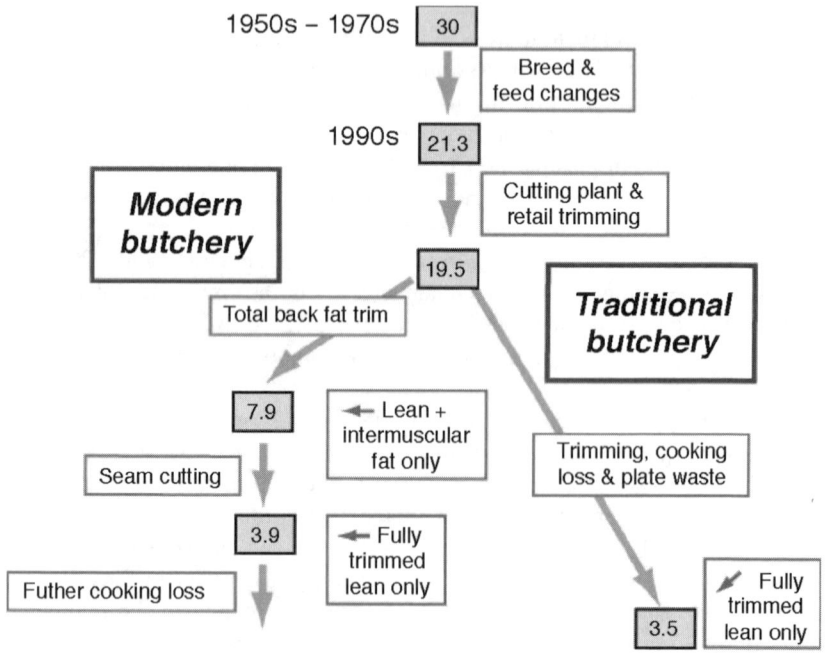

Sources of data: McCance and Widdowson (1940, 1960, 1978); Royal Society of Chemistry (1995); MLC/RSC report to MAFF (1990).

Fig. 4.1 Pork loin – change in fat content of pork loin for 100g of raw edible tissue. Adapted from Higgs and Pratt, 1998.

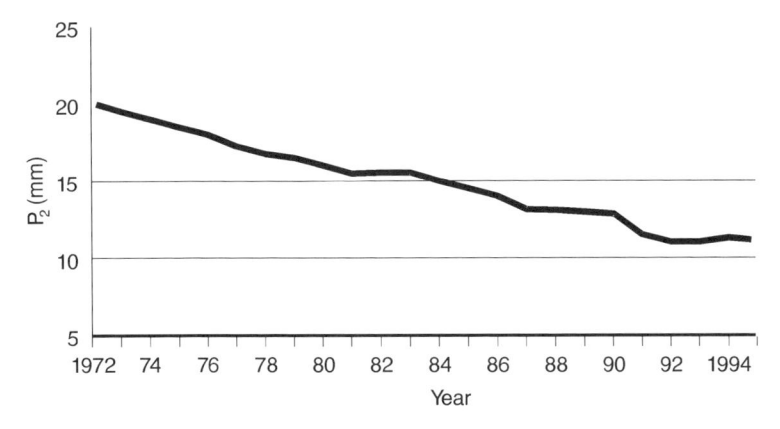

Fig. 4.2 Average P_2 fat depth of British slaughter pigs 1972–95. Source: MLC (1990).

Livestock Commission and Royal Society of Chemistry, 1990), it is only since updated supplements to the McCance and Widdowson tables were published in 1995 (Chan *et al.*, 1995, 1996), that the achievement of the meat industry in reducing the fat content of meat has been more widely acknowledged (Department Of Health, 1994a; Scottish Office, 1996; Higgs, 2000).

A fat audit for the UK, commissioned by the Government's Ministry of Agriculture, Fisheries and Food to trace all fat in the human food chain provides a more accurate picture than National Food Survey (NFS) (Ministry of Agriculture, Fisheries and Food, 1981–1999) data for identifying principal sources of fat in the diet, between 1982–1992 (Ulbricht, 1995). It illustrates that whereas the fat contributed by red meat reduced by nearly a third, that from fats and oils as a group increased by a third to contribute nearly half of our fat intakes (Fig. 4.3). This striking picture is lost in NFS data since vegetable fats (in particular) are consumed within a broad range of end products – from chips (so hidden within the vegetables section) to meat products (so artificially inflating the apparent fat contributed by meat).

The fat content of meat products can vary considerably, dependent on the proportion of lean and fat present and the amount of added non-meat fat (Higgs and Pratt, 1998). Traditional types such as sausages, pastry-covered pies and salami are high in fat (up to 50%) but modern products include ready meals and prepared meats that can be low in fat (5%). The trend downwards in fat for red meat is reflected in the reduced fat content of a number of meat products, such as hams and sausages. Some reduced-fat meat products are now available although the potential for product development in this area has not been fully exploited.

4.4 Fatty acids in meat

The fatty acid composition of food, including meat, has become increasingly important in recent years, because of concerns with the effects they have on

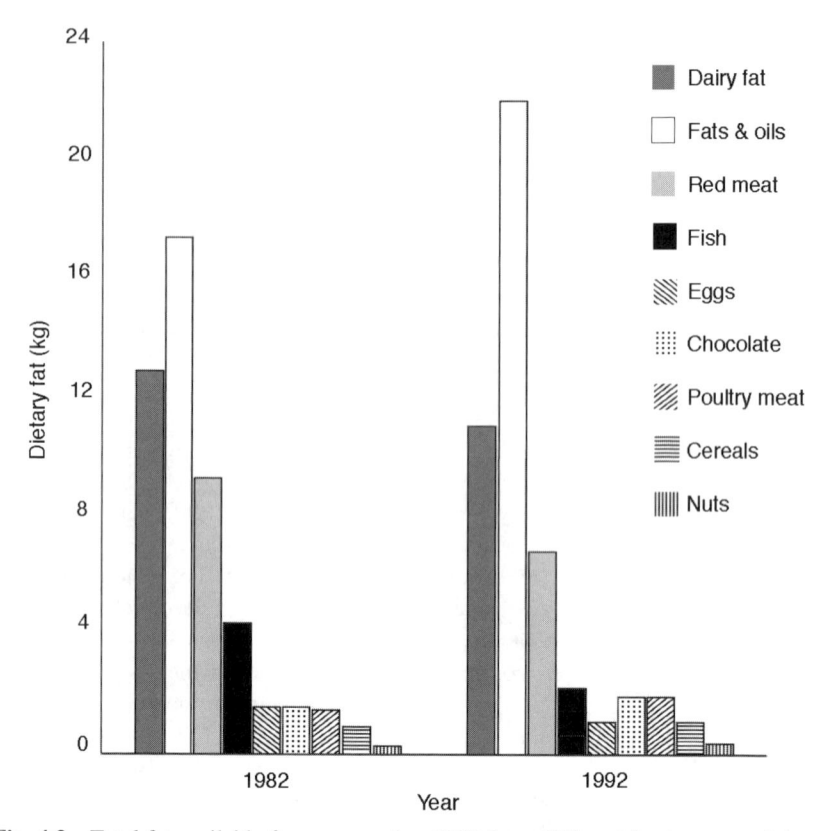

Fig. 4.3 Total fat available for consumption (UK) from different food sources. Adapted from: Ulbricht 1995 fat in the food chain.

human health. Fatty acids play a role in many conditions such as CHD, cancer, obesity, diabetes and arthritis. These roles can be protective, causative, or relatively neutral, depending on the disease, the fatty acid, and the opposing effects of other dietary components. Current dietary advice emphasises balancing the intake of the different fatty acids. The Department of Health (COMA, 1994) has recommended a reduction in the intake of saturated fat and an increase in the intake of unsaturated fat. Within the unsaturated fatty acids it is recommended to increase the omega-3 (n-3) PUFAs relative to the omega-6 (n-6) PUFAs.

4.4.1 Saturated fatty acids

Probably the main misconception about meat fat is that it is assumed to be totally saturated. Meat contains a mixture of fatty acids both saturated and unsaturated and the amount of saturated fat in meat has been reduced in recent years. Nowadays, less than half the fat in pork and beef and 51% of the fat in lamb is saturated. The saturated fat contributed to the diet from red meat and meat products has gradually fallen from 24% in 1979 to 19.6% in 1999. Carcass meats now provide 6.7% of total saturated fat intake (Ministry Of Agriculture Fisheries And Food, 1981). In reality even this figure is an overestimate, since there is a disproportionate wastage in terms of trimming, cooking losses and plate waste (Leeds *et al.*, 1997).

The predominant saturated fatty acids in meat are stearic acid (C18:0) and palmitic acid (C16:0). In general terms, saturated fats are known as the 'bad' fats as they tend to raise blood cholesterol and cause atherosclerosis. However, not all saturated fats are equal in their effects on blood cholesterol. For instance, stearic acid does not appear to raise blood cholesterol (Bonanome and Grundy, 1988) or other thrombotic risk factors (Kelly *et al.*, 1999, 2001). Stearic acid is a prominent saturated fat in meat, for example, it accounts for approximately one-third of the saturated fat in beef. Similarly, palmitic acid, another major saturated fat in meat does not consistently raise blood lipids. On the other hand, myristic acid (C14:0) is the most atherogenic fatty acid, it has four times the cholesterol-raising potential of palmitic acid (Ulbricht and Southgate, 1991). Myristic acid is found only in minor quantities in meat.

4.4.2 Monounsaturated fatty acids

Meat contains a mixture of unsaturated fatty acids, polyunsaturated fatty acids and monounsaturated fatty acids (MUFAs). MUFAs are the dominant unsaturated fatty acid in meat and they account for approximately 40% of the total fat in meat. It is a neglected fact that meat and meat products are the main contributors to MUFAs in the British diet, supplying 27% of total MUFA intake (Ministry Of Agriculture Fisheries And Food, 1999). MUFAs are considered to be neutral with respect to blood cholesterol levels. The principal MUFA in meat is oleic acid (*cis* C18:1n-9), which is also found in olive oil and is associated with the healthy Mediterranean diet.

4.4.3 Polyunsaturated fatty acids

The PUFAs have a structural role as they are found in the membrane phospholipids and they are also involved in eicosanoid synthesis. There are two types of polyunsaturated fatty acids, the omega-3 (n-3) and the omega-6 (n-6). Meat and meat products, supply 17% n-6 and 19% n-3 PUFA intake (Gregory *et al.*, 1990). Linoleic acid (C18:2 n-6) and α-linolenic acid (C18:3n-3) are essential fatty acids as we cannot synthesise them ourselves, so we are dependent on diet to provide them. In the body these are further elongated and desaturated to longer chain derivatives, arachidonic acid (C20:4n-6), docosapentaenoic acid (C22:5n-6), eicosapentaenoic acid (C20:5n-3) and docosahexaenoic acid (C22:6n-3). These are found in useful quantities in meat. Over the past 30 years there has been a major shift in the intakes of the different fatty acids, the saturated fats being replaced by the unsaturated fats. The increase in the unsaturated fatty acids was mainly due to an increase in n-6 fatty acids as a consequence of replacing vegetable oils for animal fat. Today, the usual

Western diet contains 10 to 20 times more n-6 than n-3. For instance, in the UK, the n-6 PUFA intake is now responsible for 87.5% of total PUFA intake, the remainder being the n-3 PUFAs. However, evidence now indicates that it is the n-3 PUFAs which are cardioprotective, in particular, the very long chain n-3 PUFAs, eicosapentaenoic acid (C20:5n-3) and docosahexaenoic acid (C22:6n-3). The GISSI trial showed that 1g of eicosapentaenoic acid (C20:5n-3) and docosahexaenoic acid (C22:6n-3) daily reduced coronary heart disease deaths by 20% (GISSI, 1999). The exact mechanism for this effect is not clear but they may reduce blood cholesterol. Other beneficial effects of the very long chain n-3 PUFAs include anti-inflammatory and anti-tumourigenic properties. Docosahexaenoic acid (C22:6n-3) also plays a role in neuronal development, cognitive function and visual acuity. It appears that newborn babies have a reduced ability to make the longer chain derivatives and docosahexaenoic acid (C22:6n-3) is an essential fatty acid for the newborn. Meat and fish are the only significant sources of preformed very long chain n-3 PUFAs in the diet. The chief sources of n-3 PUFAs are oily fish and fish oils, however, only one-third of the UK population consume oily fish weekly. Not so surprising then in the UK, meat and meat products supply more n-3 PUFAs (19%) than fish and fish dishes, (14%) (Gregory *et al.*, 1990). In a report on n-3 fatty acids the British Nutrition Foundation summarised this fact with the following statement 'red meat is likely to rival fish as a source of n-3 PUFAs in many peoples diet' (British Nutrition Foundation, 1999).

Animals can convert α-linolenic acid to 20- and 22-carbon n-3 PUFAs but plants cannot, hence, there are no long chain PUFAs in vegan diets. Diets that exclude meat and fish, such as vegetarian diets, are practically devoid of very long chain n-3 PUFAs. Vegans rely solely on the endogenous synthesis of very long chain n-3 PUFA from α-linolenic acid. This fact is verified by studies that have shown that vegetarians have lower n-3 PUFA intake than their omnivore counterparts. This imbalance may have nutritional consequences for vegans and vegetarians. For instance, results from a recent observation study showed that the n-3:n-6 ratio in plasma phospholipids was significantly lower among ovo-lactovegetarians and vegans compared with meat eaters and this may be responsible for an increased platelet aggregation tendency among vegetarians, which is a risk factor for cardiovascular disease, (Li *et al.*, 1999).

Meat is already a valuable source of n-3 PUFAs among omnivores, thus any further increase in the n-3 PUFA content of meat will make useful contributions to their overall intakes. Nowadays, researchers are looking at ways to enhance the n-3 PUFA content of meat. Feeding trials of cattle, pigs and sheep have shown dietary modification to be successful in raising n-3 PUFA content of their meats. The n-3 PUFA content of meat can be enhanced by increasing the amount of n-3 PUFAs in the diet of the animal. For instance, grass is rich in α-linolenic acid (C18:3n-3) and grass-fed meat has a higher n-3 fatty acid content than grain-fed meat (Enser *et al.*, 1998). Similarly, experiments have shown that including fish oil, marine algae, oils and oilseeds, such as linseed, which are rich sources on n-3 PUFAs, in the animals' diet can favourably enhance the n-3 content of the resultant meat. Enhancing the n-3 PUFA content of meat is much easier to achieve in monogastrics, such as pigs and poultry, than in ruminants. In the rumen, the dietary unsaturated fatty acids are susceptible to biohydrogenation. Biohydrogenation is a process that occurs in the rumen where the dietary unsaturated fatty acids are hydrogenated by ruminant microorganisms to more saturated end products. Evidence indicates that some unsaturated fatty acids appear to be more resistant to biohydrogenation than others. Examples include the very long chain n-3 PUFAs. However more research is required to clarify this issue. Researchers are looking at ways to overcome biohydrogenation in ruminants by protecting the n-3 PUFA. Altering the fatty acid composition of meat can have negative impacts on the meat quality, its shelf-life, colour and flavour. Therefore animal scientists, food technologists and nutritionists are looking at ways to improve the nutritional quality of meat by enhancing its n-3 PUFA content without causing any adverse sensory qualities or negatively affecting its shelf-life.

The Department of Health (1994a) has issued guidelines regarding the recommended intake of saturated and polyunsaturated fats. The current recommendation for the polyunsaturated:saturated ratio (P:S ratio) is about 0.4. Pork has a positive P:S ratio whereas the P:S ratio of lamb and beef is lower (Table 4.1), as a consequence of biohydrogenation. The Department of Health (1994) has also issued an index regarding the ratio of n-6:n-3 PUFAs. The recommended value for this ratio (n-6:n-3) is less than 4. The n-6:n-3 ratios of trimmed beef, lamb and pork are approximately 2.2, 1.3 and 7.5, respectively (Table 4.1). Therefore, both beef and lamb have acceptable n-6:n-3 ratios whereas that for pork needs to be reduced, to reach acceptable values. The reason for the high n-6:n-3 ratio in pork, is due to significant amounts of linoleic acid (C18:2 n-6) present in its adipose tissue (Enser *et al.*, 1996). In summary, researchers are focusing on ways of enhancing the n-3 PUFA content of meat and meat products. However, when increasing the n-3 fatty acid composition of ruminant meats such as beef and lamb, they are focusing on ways to increase the P:S ratio whilst retaining the positive n-6:n-3 ratio. On the other hand, for monogastric meat, such as pork, the n-3 PUFA content should be increased, whilst maintaining its positive P:S ratio. Many of the results to date are promising, for instance, beef and lamb liver from animals raised on grass are particularly good sources of n-3 PUFAs with the n-6:n-3 being 0.46 (Enser *et al.*, 1998). Such data highlights the potential for carcass meat with improved fatty acid composition as a highly acceptable and effective vehicle for providing optimal fatty acid intake for the consumer.

4.4.4 Conjugated linoleic acid (CLA)
Another emerging dietary benefit for meat, in particular, ruminant meat, is conjugated linoleic acid (CLA). CLA is a fatty acid that occurs naturally in ruminant meats such as beef and lamb. The acronym CLA is a collective term used to describe a mixture of positional (7,9-; 8,10-; 9,11-; 10,12- or 11,13-) and

76 Meat processing

Table 4.1 Fatty acid ratios related to healthy nutrition

Source of meat	Sample	P:S	n-6:n-3
Beef	Muscle	0.11	2.11
Beef	Adipose tissue	0.05	2.30
Beef	Steak	0.07	2.22
Lamb	Muscle	0.15	1.32
Lamb	Adipose tissue	0.09	1.37
Lamb	Chop	0.09	1.28
Pork	Muscle	0.58	7.22
Pork	Adipose tissue	0.61	7.64
Pork	Chop	0.61	7.57

Values for steaks and chops calculated for whole cut as purchased.
Adapted from Enser et al. (1996).

geometrical (c,c-; c,t-; t,t- or t,c-) isomers of linoleic acid ($9c,12c$-18:2). CLA has the same chain length as linoleic acid (18C), but in CLA the double bonds are conjugated. Conjugated double bonds are separated by only one single carbon bond. The $c9$-$t11$-18:2 isomer (rumenic acid) is the predominant isomer of CLA (Kramer et al., 1998). This isomer has been shown to account for at least 60% of total CLA in beef (Shantha et al., 1994; O'Shea et al., 1998). Factors influencing the CLA content of meat include the breed, age and diet of the animal (O'Shea et al, 1998; Mulvihill, 2001). As well as having a high n-3 PUFA content, grass fed meat also has higher CLA content (Shantha et al., 1994). Since, CLA is formed predominately in the rumen, the CLA content of ruminant meat, beef and lamb, is much higher than non-ruminant meat such as pork, chicken and game (Chin et al., 1992). The best natural dietary sources of CLA are ruminant products such as beef and lamb (Ma et al., 1999). Meat and meat products supply approximately a quarter of dietary CLA in Germany (Fritsche and Steinhart, 1998).

CLA appears to have a variety of potential health benefits. It has been shown to have tumour-reducing (Belury, 1995; Ip et al., 1991, 1994, 1999; Ip and Scimeca, 1997) and atherosclerotic-reducing properties (Lee et al., 1994; Nicolosi et al., 1997; Gavino et al., 2000). CLA may also reduce adiposity (Park et al., 1997; West et al., 1998) and delay the onset of diabetes (Houseknecht et al., 1998). The different isomers of CLA appear to be responsible for its differing biological effects. For instance, the c-9,t-11 isomer may play an anti-carcinogenic role, while the t-10,c-12 isomer appears to play a role in reducing adiposity. So far, most of the research work demonstrating the health benefits of CLA has been conducted in experimental animals or cell culture models. The jury is still out for its effect on human health. The American Dietetic Association has endorsed beef and lamb as functional foods because of the anti-tumourigenic properties of the CLA they contain (ADA, 1999). We are just beginning to fully understand the effect(s) that CLA has on human health and the role that meat plays in its dietary provision. In a review, Mulvihill (2001) raised a number a questions that need to be answered to improve our knowledge about CLA in meat. They include: how is CLA formed in the rumen? Can this be regulated? What CLA isomers are in meat? Can meat consumption influence CLA levels in the human body?

4.4.5 *Trans* fatty acids
Trans fatty acids raise LDL cholesterol and decrease HDL cholesterol. It is recommended by the Department of Health (1991) that *trans* fatty acids contribute less than 2% of total energy. Ruminant meats are a source of *trans* fatty acids, contributing around 18% of total intakes. These are formed during biohydrogenation in the rumen. In the British diet the main source of *trans* fatty acids are cereals and cereal products and fat spreads which use partially hydrogenated vegetable and fish oils in their products. Other significant sources include ruminant meat and milk (Gregory *et al.*, 1990). It appears from the analysis of 14 European countries that the fat content of meat does not correlate with the percentage of *trans* fatty acid content (Hulshof *et al.*, 1999). *Trans* fats have been highlighted as contributing to atherogenesis, although the hydrogenated fats from vegetable sources used in bakery goods and other processed foods appear to be more of a concern than the natural *trans* fats found in ruminant meats and milk fat (British Nutrition Foundation's Task Force, 1995). After assessing the intake of *trans* fatty acids in 14 European countries (TRANSFAIR study), the conclusion was that the current intake of TFA in most Western European countries including the United Kingdom does not appear to be a reason for major concern (Hulshof *et al.*, 1999; van de Vijver *et al.*, 2000). In fact, the TRANSFAIR study, showed that intakes of *trans* fatty acids did not influence LDL and HDL cholesterol and a weak inverse association was found in total serum cholesterol (van de Vijver *et al.*, 2000). In the USA, there is a much greater reliance on processed foods, the consequent higher intakes (6% dietary energy) of non-ruminant *trans* fatty acids are causing some concern.

4.4.7 Cholesterol
Much research has looked at the effect that individual fatty acids have on blood cholesterol rather than the mixture that we digest. It is now obvious that we should be looking at the effect that diet as a whole has on blood cholesterol. In the United States, the National Cholesterol Education Program (NCEP) recommends dietary guidelines for people with hypercholesterolaemia (raised blood cholesterol). The NCEP dietary guidelines are a first-line therapy for the management of high blood cholesterol. A recent study compared the effect of including lean red meat (beef, veal and pork) and lean white meat (poultry and fish) in the NCEP diet, on blood cholesterol of people with hypercholesterolaemia (Davidson *et al.*, 1999). This study showed that the inclusion of approximately 170g lean red meat per day, five to seven times per week in the

NCEP diet was as effective as lean white meat in reducing both total and LDL cholesterol while simultaneously raising HDL cholesterol. Thus the inclusion of lean red meat in such a diet had a positive impact on blood cholesterol levels. The authors also indicated that the study participants who consumed the lean red meat were more likely to follow their dietary regimen as they had a wider food choice than those on the white meat diet. This study not only highlights the nutritional value of red meat in such a diet but also the practical value, as no diet can possibly work unless it is adhered to!

An earlier study conducted in the United Kingdom, showed similar results, where mildly hypercholesterolaemic men ate 180g of lean meat every day, a quantity we would consider high today. This diet was low fat, low saturated fat and high in PUFA and it proved to be effective in lowering total and LDL cholesterol (Watts et al., 1988). In Canada, a study was conducted comparing the effects of lipid-lowering diets containing lean beef, poultry (without skin) and lean fish on plasma cholesterol levels in men with raised blood cholesterol. The results indicated that when compared to the usual diet, the lean beef and poultry diets significantly reduced both total cholesterol and LDL ('bad') cholesterol in men with raised blood cholesterol. Whereas, in the fish-containing diet, only total cholesterol levels fell significantly when compared to the usual diet (Beauchesne-Rondeau et al., 1999). There is now a wealth of studies showing similar results (Scott et al, 1990; Mann et al, 1997; Davidson et al., 1999), which are not that surprising, as lean red meat is low in fat, low in SFA and contains a mixture of beneficial unsaturated fatty acids, such as linoleic acid, n-3 PUFAs, MUFAs and CLA.

4.5 Protein in meat

Protein is the basic building material for making cells and its adequate intake can be of particular benefit for those growing or in adults where muscle tissue is being rebuilt, such as athletes or those recuperating post surgery. Meat is a good source of protein and it contains all the essential amino acids. In the United Kingdom, meat and meat products supply 30% of dietary protein intakes (Ministry Of Agriculture Fisheries And Food, 1999). Emphasis on a prudent diet for health that recommended just 11E% (National Advisory Committee On Nutrition Education, 1983) from protein has led us to underplay the potential role of high protein foods in the diet. Recent interest in the use of high protein diets (25E%) for weight reduction have utilised the higher satiating properties of protein, important for dietary compliance, and achieved significantly more weight loss over a six months dietary intervention compared to lower (12E%) protein. These results were achieved without adverse effects on renal function (Skov et al., 1999a, b).

Meat protein has a higher biological value than plant protein as some of the amino acids are limiting in plant protein. For example, lysine is the limiting amino acid in wheat, tryptophan is the limiting amino acid in maize and sulphur-containing amino acids are limiting in soyabean. It is necessary for vegans and vegetarians to eat a wide variety of vegetable protein foods to provide the necessary amounts of each amino acid. Meat is a rich source of taurine. Taurine is considered to be an essential amino acid for newborns, as they seem to have a limited ability to synthesise it. Taurine concentrations in the breast milk of vegans were shown to be considerably lower than in omnivores (Rana and Saunders, 1986). The significance of this finding is unknown.

4.6 Meat as a 'functional' food

Typical Western omnivorous diets over the last 40 years have been relatively high in protein and fat with insufficient dietary fibre, fruit and vegetables. Meat intake is by definition the key difference between vegetarian and omnivorous diets, thus comparative studies have tended to exaggerate the health benefits of a vegetarian diet so reinforcing a negative health image for meat. It has long been recognised (Burr, 1988) that although vegetarianism seems to confer some protection against heart disease, it is not clear if this is due to abstinence from meat or high consumption of vegetables. Meat intake has provided a marker for a generally 'unhealthy' diet, in the past (American Dietetic Association, 1993; COMA, 1991; Sanders and Reddy, 1994; Thorogood *et al.*, 1994). Furthermore, vegetarians have tended to be more health conscious, they traditionally smoke less, consume less alcohol, tea, and coffee, and tend to exercise more, thus their good health could be attributed to any or a combination of these habits. CHD and cancer are multifactorial, diet is one factor playing a role in these conditions, but diet alone is a very broad term, because within diet there are protective and causative factors. Comparing current omnivorous and vegetarian diets shows that the meat content of the former is not responsible for its higher fat content. Australian research has shown that when the meat component was removed from an omnivore diet, the remaining part of the diet was still significantly higher in total fat, saturated fat and cholesterol than a vegetarian diet (Li *et al.*, 1999). This suggests that the overall diet rather than the meat is responsible for these diet characteristics.

The significance of meat to nutrient intake depends on the importance given to meat in an individual's, or society's diet and culture. With a limited range of foods available in primitive societies throughout history, meat provided a concentrated source of a wide range of nutrients (Davidson and Passmore, 1969; Sanders, 1999). Considering the diet of modern man, where meat is excluded within traditional vegetarian cultures, the nutrients it provides can be supplied from a combination of other foods and this appears at least adequate, provided the diet is not too restrictive and dependent on nutritionally inferior staples such as maize or cassava (Sanders, 1999). With the range and abundance of foods available to developed societies today, the nutritional significance of any one food is reduced.

Traditionally, the vegetarian was likely to consume a wider range of foods than the meat eater. Consequently vegetarians in Europe and North America historically had similar energy intakes to meat eaters and greater intakes of vitamins B1, C, E, folic acid, β-carotene, potassium and fibre (Sanders, 1999). Nowadays vegetarianism cannot be assumed to provide a favourable fatty acid intake. Comparative studies of vegetarian and omnivorous children surveyed from 9 to 17 years found that saturated fat intakes were no lower in the vegetarian children (Nathan *et al.*, 1994, 1997; Burgess *et al*, 2001). There was no significant difference between energy intakes and the percentage energy from fat, or saturated fat intakes between vegetarian and omnivore adolescents in North West England (Burgess *et al.*, 2001). Vegetarian women have lower zinc intakes and status than their omnivore counterparts (Ball and Ackland, 2000). A recent study in Australia showed vegetarians had a lower intake of beneficial very long chain n-3 PUFAs (Li *et al.*, 1999). A study comparing meat eaters with vegetarians has shown that plasma homocysteine, an independent risk factor for heart disease, among vegetarians was significantly higher than their omnivore counterparts, and this was correlated with a lower intake of vitamin B_{12} among the vegetarians (Mann *et al.*, 1999; Krajcovicova-Kudlackova *et al.*, 2000; Mann 2001b). Vegans have significantly lower intakes of protein, vitamin D, calcium, and selenium but no difference in energy and iron intakes to omnivores and the vegans have significantly lower vitamin B_{12} blood concentration (Larsson and Johansson, 2001).

Modern eating habits are contributing to erosion of the traditional vegetarian diet in developed countries as there is now a greater dependence on vegetarian convenience foods, coinciding with increased availability and choice. Whilst vegetarian convenience foods may appear attractive in terms of health as well as for ease and speed of preparation, they are not necessarily of superior nutritional value compared to meat containing equivalents. There is wide variation in the fat content of vegetarian products, ranging from 2% to 58%, with nearly a third supplying more than 50% of their energy from fat (Reid and Hackett, 2001).

Excluding meat whilst paying little attention to selecting appropriate alternative food combinations to ensure adequate nutrients are supplied is cause for concern, especially in children and adolescents. Today's busy lifestyles give rise to more erratic dietary practices making it easier to obtain all nutrients required for health by including meat as a component of the diet. The time spent planning and preparing meals is minimal and an increasing proportion of our daily food intake is consumed outside the home, as snacks and quick meals. NFS data suggest that in 1998, 28% of total expenditure on food and drink was outside the home (MAFF, 1999). Data on the dietary intakes and nutritional status of young people aged 4–18 years in the UK show that energy intakes of young people are now approximately 20% below estimated average requirements (EAR) for age. Growth patterns suggest such intakes are adequate and merely reflect the corresponding lower activity levels of youngsters today, which in itself is a concern. Reduced energy intakes must increase the emphasis on a more nutrient dense diet, particularly in growing children. The survey has recorded intakes of iron, zinc and copper below the RNI particularly in older girls (Gregory *et al.*, 2000). It is possible that the recorded lower meat intakes are partly responsible for this. The decision to become vegetarian should be accompanied by adequate nutritional information and education. Despite popular opinion, vegetarianism *per se* does not guarantee a nutritionally adequate diet. Conversely, using meat as a significant protein source in the diet provides a concentrated nutrient supplement, thus ensuring the diet is nutritionally adequate (Department Of Health, 1994; Millward, 1999). The potential for producing nutritionally superior, convenience products, that include meat as a functional ingredient, is enormous and deserves more thorough exploitation.

4.6.1 Meat and paleolithic diets

Humans are omnivores. Evidence such as dentition, gut structure and ecosystem, enzymic range and adaptability and our dependence on both plant and animal sources for our essential nutrients are all supportive of this issue. We begin life as omnivores, because as babies *in utero*, all the nutrients we receive are of animal origin. During the Ice Age, plants could not grow thus man had to depend on meat as his main source of nutrition. There is much historical evidence and data from carbon isotopes, gut morphology, brain size, cranio-dental features, tools, weapons and rock art depiction of hunting all trace the evolution of man as an omnivore (Mann, 2001a). There is considerable weight to the argument that our brains evolved because we could eat a variety of foods including meat.

As we begin the new Millennium, some experts are looking at the diet of Paleolithic (stone-age) man in a search for ways to reduce the incidences of 'modern' diseases such as obesity, cancer and coronary heart disease. Research from hunter-gatherer societies has indicated that these people were relatively free of many of the chronic and degenerative diseases that plague us today, this is in part, attributable to the different dietary practices. Investigation of the dietary habits of modern hunter-gatherer societies, as an approximation of Paleolithic practices, has shown a high reliance on animal foods compared with plant foods for basic energy requirements (Cordain *et al.*, 2000). It has been estimated that the hunter-gatherers obtained approximately 45–65% of their total energy intake from meat, which was either hunted or fished (Cordain *et al.*, 2000). It is only with the relatively recent rise in agriculture that humans have begun to consume high levels of carbohydrates. This is now recognised as a major contributor of 'Western lifestyle' diseases. We have changed from a diet high in meat to a diet where grains and refined foods dominate. The hunter-gatherer diet was high in protein (19–35%E) and low in carbohydrate (22–40%E) whereas nowadays the opposite prevails – lower in protein (15%E) and much higher in carbohydrates (55%E) (Cordain *et al.*, 2000). The fatty acid profiles of such diets may have differed with higher levels of unsaturated fatty acids in wild animals, compared to domesticated farm animals.

Studies have shown that Australian Aborigines have shown significant health improvements, including a reduction in blood cholesterol levels, after returning to their natural diets – where there is a high reliance on animal foods (O'Dea,

1991). Research of macronutrient proportions in the diet of hunter-gatherer populations show a clear relationship between high protein content and the evolution of insulin resistance, which offered a survival and reproductive advantage (Brand-Miller and Colagiuri, 1994). However, the advent of agriculture and the rise of a diet higher in carbohydrate has meant that people were unprepared for the high glycaemic load and this is responsible for the current incidence of non-insulin dependent diabetes mellitus (Brand-Miller and Colagiuri, 1994). However, we must also remember that humans are not carnivores and thus we cannot exist on protein intakes above 35% energy for extended periods of time. 'A clear role for lean red meat in a healthy balanced diet becomes evident as the diet history of our species is uncovered' (Mann, 2001a).

4.6.2 Meat and satiety
The prevalence of obesity has increased dramatically in recent years (National Audit Office, 2001). Satiety influences the frequency of meals and snacks, whereas, satiation influences the size of meals and snacks. Macronutrients have differing effects on satiety, protein is more satiating than carbohydrates which are more satiating than fat (Hill and Blundell, 1986; Barkeling et al., 1990; Stubbs, 1995). The exact mechanism by which protein exerts its satiating effect is not known, but it may involve changes in the levels and patterns of metabolites and hormones (e.g. amino acids, glucose and insulin), cholecystokinin and amino acid precursors of the neurotransmitters serotonin, noreadenaline and dopamine. A meat containing meal was shown to have more sustained satiety than a vegetarian meal (Barkeling et al., 1990). Other studies have shown that different meats have different satiating powers (Uhe et al., 1992). These differences may be related to differences in amino acid profiles or digestibilities. More research on the effects that different meats have on satiety will prove invaluable in assessing whether or not meat can, in the future, be promoted as a food that can negatively curb the growing levels of obesity.

4.7 Meat and micronutrients

4.7.1 Iron in meat
Iron deficiency (Schrimshaw, 1991) and iron deficiency anaemia (Walker, 1998) remain the most common nutritional disorders in the world today. Iron deficiency is the only widespread nutrient deficiency occurring in both developed and developing countries. Iron deficiency affects between 20–50% of the world's population (Beard and Stoltzfus, 2001). There are many causes of iron deficiency, including hook worm infestation, low iron intakes, low bioavailability of dietary iron and increased demand due to physiological requirements. The most common result of iron deficiency is anaemia. Some of the liabilities associated with iron deficiency and anaemia are defective

psychomotor development in infants, impaired education performance in schoolchildren, adverse perinatal outcome in pregnancy and diminished work capacity (Cook, 1999). All of the iron in our body comes from our diet, and meat is a rich dietary source. Concern about iron deficiency is one nutritional reason for recommending eating at least some meat (WHO, 1990; COMA, 1998).

Food iron can be classified as haem iron or non-haem iron. Haem iron is derived from haemoglobin and myoglobin and its chief food source is meat, whereas non-haem iron is derived mainly from cereals, fruits and vegetables. Meat is distinctive as it contains both types of iron, haem (50–60%) and non-haem. Our bodies readily absorb haem iron (20–30%) as it is not affected by other dietary factors. Meat positively influences the bioavailability of non-haem iron. Bioavailability of iron refers to the proportion of ingested iron that is absorbed and utilised by the body (O'Dell, 1989). Only two dietary factors enhance non-haem iron bioavailability, they are vitamin C (Hallberg et al., 1989) and meat (Cook and Monsen, 1999; Taylor et al., 1986; Hazell et al., 1978; Kapsokefalou and Miller, 1991, 1993, 1995; Mulvihill and Morrissey, 1998a, b; Mulvihill et al., 1998). Absorption of non-haem iron from meat is typically 15–25%, compared with 1–7% from plant sources (Fairweather-Tait, 1989). The presence of meat in a meal enhances the bioavailability of non-haem iron contained in the other foods present such as cereals, fruits and vegetables.

The enhancing effect of meat on non-haem iron bioavailability is commonly referred to as the 'meat factor'. The exact mechanism by which the 'meat factor' works still remains unknown despite the fact that numerous efforts have concentrated on this topic. Research indicates that the mechanism of the 'meat factor' may not be due solely to a single factor but due to a number of contributing factors which work together promoting non-haem iron bioavailability. These factors include the release of cysteine-rich small molecular weight peptides during the proteolysis of meat; the ability of these peptides to reduce ferric iron to the more soluble ferrous iron; the chelation of soluble non-haem iron by these peptides and the ability of meat to promote gastric acid secretion and gastrin release better than other food components (Mulvihill, 1996).

Glutathione is a tri-peptide containing cysteine, and this is considered to play a role in the 'meat factor'. However, reduced glutathione represents only 3% of total cysteine in meat and this is considered too low to have such a profound positive influence on non-haem iron bioavailability (Taylor et al., 1986). Elucidation of the mechanism(s) of the 'meat factor' is extremely important in the search for more effective ways to improve iron nutrition. Isolation of the 'meat factor' will allow the potential to produce stable non-haem iron absorption enhancers which can be added to other foods, thus improving iron bioavailability.

Meat and meat products provide 14% of iron intake (MAFF, 1999) within this, carcass meat and meat products supply 12.5% of total iron intakes. This figure grossly underestimates the value of meat for influencing iron status. Meat has an important influence on iron bioavailability and thus iron status due to its enhancing properties and overall greater absorption capacity. Low iron intakes and status are common among certain subgroups of the population; toddlers (Gregory *et al.*, 1995; Edmond *et al.*, 1996), adolescents (Nelson *et al.*, 1993; Nelson, 1996), pregnant women (Allen, 1997) and the elderly (Finch *et al.*, 1998). Data from the National Diet and Nutrition Survey of children shows that 20% have low iron stores and 8% have iron deficiency anaemia (Gregory *et al.*, 1995). Iron deficiency anaemia among toddlers is often associated with late weaning practices. A Spanish study showed that children who first ate meat before eight months of age showed a better iron status than those who were introduced to meat later than eight months (Requejo *et al.*, 1999). Another study showed that low iron stores in one- and two-year old children is related to a low meat iron intake (Mira *et al.*, 1996). The COMA report *Weaning and the Weaning Diet* recommends that foods containing haem iron should be incorporated into the diets of infants by 6–8 months of age. Soft-cooked puréed meat can be introduced. This goes against the modern trend to delay introduction, the basis for which appears to be non-scientific.

Adolescents have high demands for iron to allow for muscle development and increased blood volume while the onset of menstruation in females makes them vulnerable to iron deficiency. Half the female population living in the UK, aged between 15–18 years, have iron intakes below the recommended level. This is reflected by the fact that 27% of that age group have low iron stores (Gregory *et al.*, 2000). The prevalence of low iron stores among adolescent girls in the UK has been cited to be as high as 43% (Nelson *et al.*, 1993). During pregnancy, more lactovegetarians (26%) reported suffering from iron deficiency than omnivores (11%) (Drake *et al.*, 1999b). Lyle *et al.*, (1992) has demonstrated that meat supplements were more effective than iron tablets in maintaining iron status during exercise in previously sedentary young women. Among the elderly, both low iron intakes and low iron status has been shown to increase with age (Finch *et al.*, 1998).

Serum ferritin, the body's iron store, is strongly correlated with haem iron (Reddy and Sanders, 1990). Bioavailability of iron plays an important role in determining iron status. Studies have shown that despite the fact that vegetarians have either a similar or a higher iron intake than their omnivore counterparts, their iron status is lower (Nathan *et al.*, 1996; Ball and Bartlett, 1999; Wilson and Ball, 1999). Vegetarians should consume iron-rich foods to compensate for the low bioavailability of non-haem iron from the foods they eat. The importance of meat in iron nutrition cannot be overemphasised. The effects of meat and meat products on iron nutrition are threefold. Firstly, they are a rich source of iron. Secondly, they contain haem iron, which is readily absorbed. Thirdly, they promote the absorption of non-haem in the diet.

4.7.2 Zinc in meat
All meats, but in particular beef, are excellent sources of dietary zinc. It takes 41oz. milk, 15oz. tuna or 6½ eggs to equal the amount of zinc in an average 4oz. portion of beef (Hammock, 1987). On average, meat and meat products account

for a third of total zinc intakes (MAFF, 1999). Zinc absorption is suppressed by inhibitors such as oxalate and phytate which are found in plant foods (Johnson and Walker, 1992; Zheng *et al.*, 1993; Hunt *et al.*, 1995). On the contrary, meat facilitates the absorption of zinc – 20–40% of zinc is absorbed from meat. For instance, one study showed that female omnivores who had a significantly lower zinc intake than their vegetarian counterparts had a higher zinc status (Ball and Ackland, 2000), such data highlights the role that meat plays in providing an assured source of dietary zinc. Because of the low bioavailability of zinc from plant foods, vegetarians should strive to meet or exceed their RDA for zinc, to ensure adequate zinc intakes.

Zinc is necessary for growth, healing, the immune system, reproduction (Aggett and Comerford, 1995) and cognitive development (Sandstead, 2000). Low zinc intakes are becoming more prevalent, especially among adolescents. An NDNS survey showed that one in ten 7–10-year-old girls and one in three of 11–14-year-old girls have intakes of zinc below the recommended level (Gregory *et al.*, 2000). Long-term, low zinc intakes lead to zinc deficiencies, which may become a public health problem in the future (Sandstead, 1995). Iron and zinc deficiencies can often occur simultaneously, particularly among adolescents (Sandstead, 2000). Adolescents often avoid eating meat; in some incidences meat is providing up to just 25% of total zinc intakes compared to 40% of adult intakes (Gregory *et al.*, 1995; Mills and Tyler, 1992; Gregory *et al.*, 2000). Therefore including meat in the diet of adolescents can aid in averting both iron and zinc deficiencies in concert, as these minerals in meat are in easily absorbable forms. Similarly, concern over low zinc status among infants prompted the DoH, in its COMA weaning report, to recommend increasing meat portion sizes for infant's at the weaning stage (Department Of Health, 1994b).

4.7.3 Selenium in meat

Selenium acts as an antioxidant and is considered to protect against coronary heart disease and certain cancers, such as prostate. Meat contains about 10mg selenium per 100g, which is approximately 25% of our daily requirement. Beef and pork contain more selenium than lamb, which may be due to the age of the animal as selenium may collect in the meat over time. Bioavailability of selenium from plant foods was thought to be greater than that from animal foods, but recent data demonstrate that meat, raw and cooked, provides a highly bioavailable source (Shi and Spallholz, 1994).

4.7.4 Other minerals in meat

Meat also contains phosphorus, a typical serving provides roughly 20–25% of an adult's requirement. Phosphorus has important biochemical functions in carbohydrates, fat and protein metabolism. Meat also provides useful amounts of copper, magnesium, potassium, iodine and chloride.

4.7.5 B-vitamins in meat

Meat is a significant and an important source of many B vitamins. The B-vitamins in meat are thiamin (vitamin B_1), riboflavin (vitamin B_2), niacin, pantothenic acid, vitamin B_6 and vitamin B_{12}. B-vitamins are water soluble hence, lean meat contains more of these vitamins than fattier meat. Some losses of B-vitamins occur during cooking, the amount lost depends upon the duration and the temperature of the cooking method.

Thiamin and riboflavin are found in useful amounts in meats. Pork and its products including bacon and ham are one of the richest sources of thiamin. Pork contains approximately 5–10 times as much thiamin as beef or lamb. Thiamin aids the supply of energy to the body by working as part of a co-enzyme that converts fat and carbohydrates into fuel. It also helps to promote a normal appetite and contributes to normal nervous system function. Typical servings of pork provide all the daily requirement of thiamin. Offal meats are good sources of riboflavin, for example, a single portion (100g) of kidney or liver provides more than its daily requirement. Riboflavin, like thiamin, also aids in supplying energy and also promotes healthy skin, eyes and vision.

Meat is the richest source of niacin. Half the niacin provided by meat is derived from tryptophan, which is more readily absorbed by the body than that bound to glucose in plant sources. Niacin helps to supply energy to the body as it plays a role in converting carbohydrates and fats into fuel. Meat and meat products supply more than a third of total niacin intakes in Britain (MAFF, 1999). Liver and kidney are rich sources of pantothenic acid. Although most of this vitamin is leached into the drip loss associated with frozen meat, this is unlikely to be of any nutritional consequence as pantothenic acid is universal in all living matter.

A 100g portion of veal liver provides half our daily vitamin B_6 needs and other meats provide around a third. Vitamin B_6 is a necessary co-factor for more than 100 different cellular enzyme reactions, including those related to amino acid metabolism and inter-conversion. Vitamin B_{12} is exclusively of animal origin as it is a product of bacterial fermentation, which occurs in the intestine of ruminant animals such as cattle, sheep and goats. Vitamin B_{12} is required to produce red blood cells and acts as a cofactor for many enzyme reactions. Deficiency of vitamin B_{12} causes megaloblastic anaemia, neuropathy and gastrointestinal symptoms. Groups at risk of vitamin B_{12} deficiency include vegans and strict vegetarians because vitamin B_{12} is exclusively of animal origin and the elderly, because their ability to absorb this vitamin from the diet diminishes with age (Allen and Casterline, 1994; Swain, 1995; Baik and Russell, 1999; Drake et al., 1999a). In the past some vitamin B_{12} was provided from the soil of poorly cleaned foods. This may in part explain the apparent absence of deficiency in some vegan groups. Today with the emphasis on good food hygiene practices, this source can no longer protect against deficiency in vulnerable individuals. Vegans are recommended to take vitamin B_{12} supplements since the quantity consumed from foods fortified with the vitamin is too low (Jones, 1995; Draper, 1991: Sanders and Reddy, 1994). The RNI for vitamin B_{12} among the elderly is 1.5µg/day (Department of Health, 1991). A 100g portion of lean trimmed beef contains 2µg vitamin B_{12}, thus supplying all their daily needs for this vitamin. In Britain, meat and meat products supply more than a fifth of both vitamin B_6 and B_{12} intakes (MAFF, 1999). The need for vitamin B_{12} has been a part of the rationale for recommending the consumption of animal foods among all age groups (WHO, 1990).

Raised homocysteine, an amino acid metabolite, is an independent risk factor for cardiovascular disease. It is estimated that 67% of the cases of hyperhomocysteinemia are attributable to inadequate plasma concentrations of one or more of the B-vitamins namely; folate, vitamin B_6 and vitamin B_{12}. Some enzymes that reduce homocysteine levels require vitamins B_6 and B_{12} as co-factors. Vitamin B_6 is a co-factor for two enzyme reactions that catabolise homocysteine to cysteine via a transulfuration pathway, they are cystathionine β-synthase and cystathionase. Meanwhile, vitamin B_{12} is a co-factor for the remethylation enzyme, methionine synthase, which converts homocysteine to methionine. Research has shown that low levels of both vitamins B_6 and B_{12} independently correlate with raised homocysteine. For instance, ovo-lactovegetarians or vegans who had significantly lower serum vitamin B_{12} levels than meat eaters had significantly higher levels of plasma homocysteine (Mann et al., 1999, Krajcovicova-Kudlackova et al., 2000, Mann, 2001b). Similarly, low doses of vitamin B_6 can effectively lower fasting plasma homocysteine levels (McKinley et al., 2001). The role of meat in regulating homocysteine is intriguing and needs to be addressed further.

4.7.6 Meat and vitamin D

In the body vitamin D acts as a hormone, essential for the absorption of dietary calcium. Therefore, vitamin D is essential for skeletal development and severe deficiency is associated with defective mineralisation of the bone resulting in rickets in children or its adult equivalent, osteomalacia (Fraser, 1995; Dunnigan and Henderson, 1997; DeLuca and Zierold, 1998; Department of Health, 1998b). More subtle degrees of insufficiency lead to increased bone loss and osteoporotic fractures. Other functions of vitamin D includes its role in the immune system, and may be protective against tuberculosis, muscle weakness, diabetes, certain cancers and coronary heart disease (Department of Health, 1998b).

It is well established that sunlight exposure on the skin is the main source of vitamin D. However, there are certain subgroups in the population who are more at risk of vitamin D deficiency, and these depend on diet in addition to sunlight to obtain adequate vitamin D. Such subgroups include infants, toddlers, pregnant and lactating women, the elderly and those who have low sunlight exposure, such as certain ethnic minorities and those housebound (Department of Health, 1998a). The prevalence of vitamin D inadequacies among these groups is widespread. For instance, 27% of two-year-old Asian children living in England have low vitamin D status (Lawson and Thomas, 1999) and, 99% of elderly people living in institutions are not receiving enough of dietary vitamin D (Finch *et al.*, 1998). Vitamin D deficiency among the elderly will become much more apparent and a greater public health problem when we consider that we are living in an increasingly ageing population.

Liver aside, meat and meat products were considered poor sources of vitamin D. However, new analytical data for the composition of meat indicates that this is not true (Chan *et al.*, 1995). Meat and meat products contain significant amounts of 25-hydroxycholecalciferol, assumed to have a biological activity five times that of cholecalciferol. In fact, the meat group is now recognised as the richest natural dietary source of vitamin D, supplying approximately 21% (Gibson and Ashwell, 1997). Vitamin D is present in both the lean and the fat of meat although its exact function in the animal is not yet known. Since interest in the role of meat in supplying vitamin D is a relatively new subject, there are certain areas that need to be researched such as the effect of cooking meat on vitamin D levels, the bioavailability of vitamin D from meat and the influence of seasonal variation on the vitamin D content of meat and meat products.

Low intake of meat and meat products, has emerged as an independent risk factor for Asian rickets and absent intake of meat and meat products emerged as an independent risk factor for Asian osteomalacia (Dunnigan and Henderson, 1997). It has been hypothesised by this research group that there may be a 'magic factor' in meat which is protective against rickets and osteomalacia. In Glasgow, at the beginning of the century, the incidence of rickets was high, whereas, between 1987 and 1991, only one case of rickets was reported. This may be explained by the fact that nowadays infants are weaned onto an omnivorous diet from four months of age and this meat inclusion is offering protection against rickets (Dunnigan and Henderson, 1997). Obviously much more research is required to improve our knowledge on this subject. It is also of interest to note that signs of both iron and vitamin D deficiency can occur simultaneously among toddlers (Lawson and Thomas, 1999). For instance, during the winter half of the toddlers had both low vitamin D and low iron levels (Lawson and Thomas, 1999). Such evidence highlights the potential protective role that meat inclusion can play in a toddler's diet. It is important for toddlers and children to eat foods rich in both iron and vitamin D such as meat and meat products as well as playing outdoors to get sunlight.

4.8 Future trends

As we begin the 21st Century, we look to the future to predict the likely nutritional problems we will need to tackle. The four major nutritional problems today are heart disease, hypertension, obesity and diabetes. These are likely to remain significant public health problems in the future. The demographic structure of the population is changing. Throughout Europe both birth and death dates are falling, people are living longer and it is estimated that by the year 2030, more than half the population living in the UK will be over 50 years old. With this knowledge we shall try to ascertain the likely future nutritional role of meat.

This chapter clearly outlines ways to reduce the fat content of meat and manipulate its fatty acid composition. The meat that is on sale today has never been leaner. Fortunately, most of the valuable nutrients of meat are located in the lean component, so reducing the visible fat of meat has little bearing on its micronutrient status. Researchers are focusing on ways of further improving the fatty acid composition of meat, using the knowledge that grass feeding results in high levels of both n-3 PUFAs and CLA content. Both n-3 PUFAs and CLA may have many possible benefits for human health, and in particular may offer protection against predicted future health problems (Cordain *et al.*, 2002). N-3 PUFAs, in particular, have a very long chain, are cardioprotective and have anti-inflammatory and anti-tumourigenic properties. CLA can prevent formation and slow the growth for tumour development (Ip *et al.*, 1994), reduce atherosclerosis development (Lee *et al.*, 1994) and can help normalise blood glucose levels, which may be shown to prevent adult-onset diabetes (Houseknecht *et al.*, 1998). Studies in human subjects are needed before we fully realise the benefits of CLA on human health. The fat and fatty acid story for meat so far is positive and only research and time will tell whether this story will be further improved.

The prevalence of overweight and obesity is increasing steadily in many developed countries. In the UK, over a quarter of the population are either overweight or obese. Obesity is a risk factor for many conditions. During a ten-year follow-up study, the incidence of colon cancer, diabetes, heart disease, hypertension, stroke (men only) and gallstones increased in line with the degree of overweight among adults (Field *et al.*, 2001). Thus, reducing the incidence of overweight and obesity is a major public health priority. A positive energy balance is the cause of practically all cases of overweight and obesity. Factors regulating food intake are hunger, appetite, satiation and satiety.

Meat-containing meals have higher satiety values than vegetable-containing meals (Barkeling *et al.*, 1990). Research needs to be undertaken to determine whether meat can play a role in curtailing obesity, as a result of its high satiety value. Media hype about CLA has concentrated on its ability to reduce body fat and increase lean body mass. Studies have noted that CLA induces a relative decrease in body fat and an increase in lean muscle (Park *et al.*, 1997; West *et al.*, 1998). Trials are currently under way to confirm whether or not these benefits occur in humans. Lean meat already is low in fat, but other attributes such as its high satiety value and the CLA it contains may be used in the future to market meat as a food than can help to reduce overweight and obesity. Furthermore, the capacity of meat to encourage greater vegetable and salad consumption, due to the way it is eaten should not be overlooked in this regard.

An increase in the incidence of hip fractures is an inevitable consequence of people living longer. Research has shown that an increase in meat protein consumption among elderly women correlates with a decrease in the risk of hip fracture (French *et al.*, 1997). Decreasing the risk of hip fracture is a public health priority. Vegetarian women tended to have lower spinal bone mineral density than non-vegetarians (Barr et al., 1998). Dunnigan and Henderson (1997) suggested that there may be a 'magic factor' in meat protecting against rickets and osteomalacia. To suggest that meat plays a role in bone health is relatively new and exciting and warrants further investigation.

Another emerging benefit for meat is selenium. Up to the middle part of the last century the main source of selenium in the diet was from wheat-containing products. Wheat, which was imported mainly from the United States, was high in selenium. Nowadays, there is a much greater reliance on European wheat, which is much lower in selenium. This has resulted in the fact that our intake of selenium has decreased steadily during the past fifty years, but the proportion of selenium we get from meat has increased. Recent studies have found that selenium may reduce the risk of heart disease and certain types of cancer such as prostate and enhance the body's ability to fight infections.

Meat provides a wide range of valuable nutrients, for example, one study has shown that young women consuming a high meat diet have greater intakes of thiamine, niacin, zinc and iron than those consuming a low meat diet (Ortega et al., 1998). In a review on optimal iron intakes, iron contained in animal foods is far better assimilated than in vegetarian foods (Cook, 1999). Meat is one of the richest natural sources of glutathione, an important reducing agent providing a major cellular defence against a variety of toxicological and pathological processes. Moderate levels of glutathione are found in fruit and vegetables and low levels are present in dairy and cereal products. Glutathione inhibits formation of mutagens in model systems (Trompeta and O'Brien, 1998). It also maintains ascorbate in a reduced and functional form. Glutathione importance in the defence against chronic disease provides positive potential for meat and merits further research (Bronzetti, 1994; Trompeta and O'Brien, 1998).

There has never before been such a wide variety and choice of food on sale in Western societies and in the recent past we have seen the development of functional foods. A functional food can be loosely described as a food that provides a health benefit beyond its basic nutritional content. In the United States beef and lamb are now described as functional foods (ADA, 1999), because of the CLA they contain. At a Meat Marketing/Communication Workshop, Dr Lynne Cobiac (CSIRO) (Cobiac, 2000) described some nutritive and non-nutritive meat components that may have potential health promoting properties. They are summarised as follows:

- *Lipoic acid* has antioxidant properties and has been shown to be beneficial in diabetics and in the prevention of cataract development in animal models and cell lines. Organ meats contain higher quantities of lipoic acid than muscle meats.
- *Carnosine* is a dipeptide composed of alanine and histidine. Carnosine is found in meats and its antioxidant properties may confer some protection against oxidative stress. Its an anti-inflammatory agent and has anti-tumourigenic properties in rats and it also plays a role in cellular homeostasis.

- *Biogenic amines* are naturally formed from bacterial decarboxylation of amino acids or natural decarboxylase activity. They have been linked with improving gut health and cognitive performance.
- *Nucleotides* are added to enteral feeds to enhance the general immune function. Organ meats are good sources of nucleotides.
- *Glutathione* is a tripeptide containing the sulphur amino acid cysteine. Glutathione may be the 'meat factor' which enhances non-haem iron absorption.
- *Choline* is now termed a nutrient. In the United States, it is an essential nutrient and the estimated adequate intake is 550mg/day for men and 425mg/day for women. Choline is a precursor of the neurotransmitter acetylcholine, it is necessary for central nervous system development, folate/homocysteine metabolism, it plays a role in the immune system, fat metabolism and improves athletic performance. Beef and in particular liver is one of the richest sources of choline.
- *Carnitine* is composed of lysine and methionine. Seventy-five percent of carnitine comes from the diet, mainly from red meat, lamb being a particularly good source. Carnitine carries the long chain fatty acids to the mitochondria for oxidation to give energy and thus can be used to improve athletic performance. It also has antioxidant capabilities and it may be critical for normal brain development by providing acetyl groups to synthesise acetylcholine, a neurotransmitter.

This range of meat components may have the ability to fight against certain cancers, CHD, anaemia and cataracts, enhance immunity and cognition, improve gut and bone health, regulate body weight and may be used in sports nutrition. However, a lot of the evidence indicating beneficial effects of these components comes from animal or cell culture models. Research will have to be conducted in humans to demonstrate their effect on human health. But even glancing at the amount of 'potential' components present in meat indicates a positive and competitive future for meat.

However, when looking to the future we must also try to visualise what changes are likely to occur that may influence meat consumption. Traditionally, food purchase was mainly influenced by price and sustenance. Current and future food choice depends on these values but alongside other factors such as health, food safety, convenience and welfare concerns. Changes in our social patterns, such as moving away from the formal family meat-eating patterns to a 'grazing' or 'snacking' habit will become much more apparent. Increasing loss of culinary skills is already evident and is likely to rise. The market will demand more convenience and processed meat products in place of traditional cuts of meats. Eating outside the home will place a greater emphasis on the catering sector as food providers. Availability of 'exotic' meats will escalate. Demand for organic meat is expected to rise. Competition from other foods will intensify. The emergence of more functional foods is likely to occur. These are some of the factors that will sculpture the future demand for meat and meat products.

Meat must adapt to the changing environment. However, the emphasis between food choice and health was never as great and is likely to become even more important. In the past, meat responded to consumer demands by decreasing its fat content. Meat is a versatile food, however, it is time to banish the misinformation that surrounds the nutritional value of meat. Meat is a relatively low fat nutrient dense food. Meat and meat products are an integral part of the UK diet and for those who choose to consume meat, it makes a valuable contribution to nutritional intakes (BNF, 1999).

4.9 Conclusion

There has been considerable emotive and public health debate over the last two decades on the relative importance of meat in the diet of modern man. Early dismissive arguments have more recently been revisited and challenged as a result of the continual progress and review of nutritional science. The early focus on fat as the route cause of Western-style diseases of affluence led, naively, to meat being blamed for diet-related problems. More recently, the focus on the diets of our ancestors has effectively reversed this thinking and lean red meat has been rediscovered as a mainstay of human diet evolution. The serious health concerns resulting from the epidemic rise in CHD, obesity, diabetes and cancers require more carefully guided public health advice, based on an holistic approach to diet and lifestyle.

Lean meat can be seen as the ultimate natural functional food. Eaten in moderate quantities as part of a meal along with sufficient plant foods, it provides a valuable, arguably essential nutrient-dense supplement to the diet with beneficial effects for health, both in the short and long term. As a key ingredient of modern processed pre-prepared meals, meat, when added as a quality ingredient, can enhance the nutritional benefits of the food product and make a significant, positive contribution to our health. It would be naïve to ignore this potential.

4.10 References

ADA REPORT (1999). Position of The American Dietetic Association: Functional Foods. *Journal of The American Dietetic Association,* **99**: 1278–1285.

AGGETT PJ and COMERFORD JG (1995). Zinc and Human Health. *Nutrition Reviews,* **53**: S16–S22.

ALLEN, L (1997). Pregnancy and iron deficiency: unresolved issues. *Nutrition Reviews,* **55**(4): 91–101.

ALLEN LH and CASTERLINE J (1994). Vitamin B_{12} deficiency in elderly individuals: diagnosis and requirements. *American Journal of Clinical Nutrition,* **60**: 12–14.

AMERICAN DIETETIC ASSOCIATION (1993). 'Position Of The American Dietetic Association: Vegetarian Diets'. *Journal of The American Dietetic Association,* 1317–1319.

ARMSTRONG B and DOLL R (1975). Environmental factors and the incidence and mortality from cancer in different countries with special reference to dietary practices. *International Journal of Cancer*, **15**: 617–631.

AZEVEDO LF, SALGUEIRO LF, CLARO R, TEIXEIRA-PINTO A and COSTA-PEREIRA A (1999). Diet and gastric cancer in Portugal – a multivariate model. *European Journal of Cancer Prevention*, **8**: 41–48.

BAIK HW and RUSSELL RM (1999). Vitamin B12 deficiency in the elderly. *Annual Reviews of Nutrition*, **19**: 357–377.

BALL M and ACKLAND M (2000). Zinc intake and status in Australian vegetarians. *British Journal of Nutrition*, **83**: 27–33.

BALL M and BARTLETT M. (1999). Dietary intake and iron status of Australian vegetarian women. *American Journal of Clinical Nutrition*, **70**: 353–358.

BARKELING B, ROSSNER S and BJORVELL H (1990). Effects of a high protein meal (meat) and a high carbohydrate meal (vegetarian) on satiety measured by automated computerized monitoring of subsequent food intake, motivation to eat and food preferences. *International Journal of Obesity*, **14**: 743–751.

BARR SI, PRIOR JC, JANELLE KC and LENTLE BC (1998). Spinal bone mineral density in premenopausal vegetarian and nonvegetarian women: cross-sectional and prospective comparisons. *Journal of the American Dietetic Association*, **98**(7): 760–5.

BEARD J and STOLTZFUS R (2001). Foreword – Iron deficiency Anaemia: Reexamining the Nature and Magnitude of the Public Health Problem. *Journal of Nutrition*, **131**: 563S.

BEAUCHESNE-RONDEAU E, GASCON A, BERGERON J and JACQUES H (1999). Lean Beef In Lipid Lowering Diet: Effects On Plasma Cholesterol And Lipoprotein B In Hypercholesterolaemic Men. *Canadian Journal of Dietetic Practice and Research*, **60** June Supplement.

BELURY MA (1995). Conjugated dienoic linoleate: A polyunsaturated fatty acid with unique chemoprotective properties. *Nutrition Reviews*, **53**: 83–89.

BINGHAM SA (1996). Epidemiology and mechanisms relating to risk of colorectal cancer. *Nutrition Research Reviews*, **9**: 197–239.

BLACK *et al.* (1997). Cancer incidence and mortality in the European Union. Cancer registry data and cancer incidence for 1990. *European Journal of Cancer*, **33**: 1075–1107.

BONANOME A and GRUNDY SM (1988). Effect of dietary stearic acid on plasma cholesterol and lipoprotein levels. *New England Journal of Medicine*, **318**: 1244–1248.

BRAND-MILLER J and COLAGIURI S (1994). The carnivore connection: dietary carbohydrate in the evolution of NIDDM *Diabetologia*, **37:** 1280–1286.

BRITISH NUTRITION FOUNDATION (1996). *Diet And Heart Disease: A Round Table Of Facts.* Second edition, November 1996. Editor: M. Ashwell.

BRITISH NUTRITION FOUNDATION (1999). *Meat In The Diet*, Briefing Paper BNF, London.

BRITISH NUTRITION FOUNDATION'S TASK FORCE. (1995). *Trans Fatty Acids – The*

Report Of The BNF Task Force. BNF. London.
BRONZETTI G (1994). Antimutagens In Food. *Trends in Food Science and Technology.* **5:** December, 390–395.
BURGESS L, HACKETT AF, MAXWELL S and ROUNCEFIELD M (2001). The nutrient intakes of vegetarian and omnivorous adolescents in North-West England. *Proceedings of the Nutrition Society,* **60**(4): 69A.
BURR ML (1988). Heart Disease In British Vegetarians. *American Journal of Clinical Nutrition,* **48**: 30–32.
CHAN W, BROWN J, LEE SM and BUSS DH (1995). *Meat, poultry and game. Supplement to McCance and Widdowson's The Composition of Foods.* The Royal Society of Chemistry and Ministry of Agriculture, Fisheries and Food.
CHAN W, BROWN J, CHURCH SM and BUSS DH (1996). Meat Products and Dishes. *Supplement To McCance and Widdowson's The Composition Of Foods.* Royal Society Of Chemistry, Ministry Of Agriculture Fisheries And Foods. HMSO London.
CHIN SF, LIU W, STORKSON JM, HA YL and PARIZA MW (1992). Dietary sources of conjugated dienoic isomers of linoleic acid, a newly recognized class of anticarcinogens. *Journal of Food Composition and Analysis,* **5**: 185–97.
CHIZZOLINI R, ZANARDI E, DORIGONI V and GHIDINI S (1999). Calorific Value And Cholesterol Content Of Normal And Low Fat Meat And Meat Products *Trends in Food Science and Technology,* **10**: 119–128.
COBIAC L (2000). *Could red meat be a functional food of the future?* Marketing/ Communication Workshop, 6–7 July, 2000, organised by the International Meat Secretariat and hosted by The Meat and Livestock Commission, PO Box 44, Winterhill House, Winterhill, Milton Keynes MK6 1AX, UK.
COMMITTEE ON MEDICAL ASPECTS OF FOOD POLICY (1984). Diet And Cardiovascular Disease. Report On Health And Social Subjects. *Department Of Health And Social Security.* No. 28. HMSO London.
COMMITTEE ON MEDICAL ASPECTS OF FOOD POLICY (1991). *Dietary Reference Values For Food, Energy And Nutrients For The United Kingdom,* Report of the Panel on Dietary Reference Values. no. 41 HMSO London.
COMMITTEE ON MEDICAL ASPECTS OF FOOD POLICY (1994). *Nutritional Aspects Of Cardiovascular Disease,* Report of the Working Group on Diet and Cancer. no. 46 HMSO London.
COMMITTEE ON MEDICAL ASPECTS OF FOOD POLICY (1998). *Nutritional aspects of the development of cancer.* Report of the working group on diet and cancer of the Committee on Medical aspects of Food and Nutrition Policy, no. 48. HMSO London.
COOK JD (1999). Defining optimal body iron. *Proceedings of the Nutrition Society,* **58**: 489–495.
COOK JD and MONSEN ER (1999). Food iron absorption in human subjects III. Comparison of the effect of animal proteins on non-haem iron absorption. *American Journal of Clinical Nutrition,* **29**: 859–867.
CORDAIN L, BRAND-MILLER J, EATON SB, MANN NJ, HOLT SH and SPETH JD (2000).

Plant-animal subsistence ratios and macronutrient energy estimations in worldwide hunter-gatherer diets. *American Journal of Clinical Nutrition*, **71**: 682–692.

CORDAIN L, WATKINS BA, FLORANT GL, KELHER M, ROGERS L and LI Y (2002). Fatty acid analysis of wild ruminant tissues: evolutionary implications for reducing diet-related chronic disease: *European Journal of Clinical Nutrition,* March, **56**(3).

COX BD and WHICHELOW MJ (1997). Frequent consumption of red meat is not a risk factor for cancer. *British Medical Journal,* **315**: 1018.

DAVIDSON MH, HUNNINGHAKE D, MAKI KC, KWITEROVICH PO and KAFONEK S. (1999). Comparison of the effects of lean red meat vs lean white meat on serum lipid levels among free-living persons with hypercholesterolaemia *Archives of Internal Medicine*, **159**: 1331–1338.

DAVIDSON S and PASSMORE, R (1969). *Human Nutrition And Dietetics.* Churchill Livingstone. London.

DE LUCA HF and ZIEROLD C (1998). Mechanisms and functions of vitamin D. *Nutrition Reviews,* **56**(2): S4–S10.

DEPARTMENT OF HEALTH (1991). *Dietary reference Values for Food Energy and Nutrients for the United Kingdom.* Report on Health and social subjects: 41, HMSO, London.

DEPARTMENT OF HEALTH (1994a). *Eat Well! An Action Plan From The Nutrition Task Force To Achieve The Health Of The Nation Targets On Diet And Nutrition.* Department Of Health: HMSO London.

DEPARTMENT OF HEALTH (1994b). *Weaning And The Weaning Diet.* Report On Health And Social Subjects: 45. HMSO, London.

DEPARTMENT OF HEALTH (1998a) *Nutrition And Bone Health. Report On Health And Social Subjects*: 49 The Stationery Office; London.

DEPARTMENT OF HEALTH (1998b). *Nutrition and bone health: with particular reference to calcium and vitamin D.* Committee on the Medical Aspects of food and Nutrition Policy. Working Group on the Nutritional Status of the Population Subgroup on Bone Health. London: The Stationery Office.

DEPARTMENT OF HEALTH (1998c). *Nutritional Aspects Of The Development Of Cancer.* Report On Health And Social Subjects 48: The Stationery Office London.

DRAKE R, REDDY S and DAVIES GJ (1999a). Dietary and supplement intake of vegetarians during pregnancy. *American Journal of Clinical Nutrition,* **70**(suppl): 627S.

DRAKE R, REDDY S and DAVIES GJ (1999b). Health of vegetarians during pregnancy and pregnancy outcome. *American Journal of Clinical Nutrition*, **70**(suppl): 628S.

DRAPER A (1991). The Energy And Nutrient Intakes Of Different Types Of Vegetarians A Case For Supplements. *British Journal of Nutrition,* **69**: 3–19.

DUNNIGAN MG and HENDERSON JB (1997). An epidemiological model of privational rickets and osteomalacia. *Proceedings of the Nutrition Society,*

56: 939–956.

DUO LI, NG A, MANN NJ and SINCLAIR AJ (1998). Contribution Of Meat Fat To Dietary Arachidonic Acid. *Lipids*, **33**(4): 437–440.

EDMOND AM, HAWKINS N, PENNICK C, GOLDING J and the ALSPAC Children in Focus Team (1996). Haemoglobin and ferritin concentrations in infants at 8 months of age. *Archives of Disease in Childhood*, **74**: 36–39.

ENSER M, HALLETT K, HEWITT B, FURSEY GAJ and WOOD JD (1996). Fatty acid content and composition of English beef, lamb and pork at retail. *Meat Science*, **42**(4): 443–456.

ENSER M, HALLETT KG, HEWITT B, FURSEY GAJ, WOOD JD and HARRINGTON G (1998). Fatty acid content and composition of UK beef and lamb muscle in relation to production system and implications for human nutrition. *Meat Science*, **49**: 329–341.

FAIRWEATHER-TAIT SJ (1989). Iron in foods and its availability. *Acta Paediatrica Scand*, Suppl. **361**: 12–20.

FIELD AE, COAKLEY EH, MUST A, SPADANO JL, LAIRD N, DIETZ WH, RIMM, E and COLDITZ GA (2001). Impact of overweight on the risk of developing common chronic diseases during a 10-year period. *Archives of Internal Medicine*, **161**(13): 1581–1586.

FINCH S, DOYLE W, LOWE C, BATES CJ, PRENTICE A, SMITHERS G and CLARKE PC (1998). *National Diet and Nutrition Survey: people aged 60 years and over*. HMSO. London.

FRASER DR (1995). Vitamin D. *Lancet,* **345**: 104–107.

FRENCH SA, FOLSOM AR, JEFFERY RW, ZHENG W, MINK PJ and BAXTER JE (1997). Weight variability and incident disease in older women: the Iowa Women's Health Study. *International Journal of Obesity Related Metabolic Disorders*, **21**(3): 217–223.

FRITSCHE J and STEINHART H (1998). Amounts of conjugated linoleic acid (CLA) in German foods and evaluation of daily intake. *Z Lebensm Unters Forsch A,* **206**: 77–82.

GAVINO VC, GAVINO G, LEBLANC M and TUCHWEBER B (2000). An isomeric mixture of conjugated linoleic acids but not pure cis-9, trans-11-octadecadienoic acid affects body weight gain and plasma lipids in hamsters. *Journal of Nutrition,* **130**: 27–29.

GIBSON SA and ASHWELL M (1997). New Vitamin D values for meat and their implication for vitamin D intake in British adults. *Proceedings of the Nutrition Society,* **56**: 116A.

GIBSON S and ASHWELL M (2001). Implications for reduced red and processed meat consumption for iron intakes among British women. *Proceedings of the Nutrition Society*, **60**(4): 60A.

GISSI-PREVENZIONE INVESTIGATORS (1999). Dietary supplementation with n-3 PUFAs and vitamin E after myocardial infarction: results of the GISSI-Prevenzione trial. *Lancet*, **354**: 447–455.

GREGORY J, FOSTER K, TYLER H and WISEMAN M (1990). *The Dietary And Nutritional Survey Of British Adults*. HMSO London.

GREGORY JR, CLARKE PC, COLLINS DL, DAVIES PSW and HUGHES JM (1995). *National Diet And Nutrition Survey; Children Aged 1½ To 4½ Years.* HMSO. London.

GREGORY J, LOWE S, BATES CJ, PRENTICE A, JACKSON LV, SMITHERS G, WENLOCK R and FARRON M (2000). 'National Diet And Nutrition Survey: Young People Aged 4 To 18 Years'. *The Stationery Office*, London.

HALLBERG L, BRUNE M and ROSSANDER L (1989). Iron absorption in man ascorbic acid and dose dependent inhibition by phytate. *American Journal of Clinical Nutrition*, **49**: 140–144.

HAMMOCK DA (1987). The red meat in our diet-good or bad? In: Cook Fuller CC, ed. *Nutrition 87/88.* Guildford, Conn: Sushkin Publishing group: 16–18.

HAZELL T, LEDWARD DA and NEALE RJ (1978). Iron availability from meat. *British Journal of Nutrition*, **39**: 631–638.

HIGGINSON J (1966). Etiological factors in gastrointestinal cancer in man. *Journal – National Cancer Institute*, **37**: 527–545.

HIGGS JD (2000). An overview of the compositional changes in red meat over the last 20 years and how these have been achieved. *Food Science and Technology Today*, **14**(1): 22–26.

HIGGS JD and PRATT J (1998). Meat Poultry And Meat Products: Nutritional Value Volume 2, 1272–1282. In: Sadler MJ, Strain JJ and Cabalerro B (eds) *The Encyclopaedia of Human Nutrition.* Academic Press Limited: London.

HILL AJ and BLUNDELL JE (1986). Macronutrients and satiety: the effects of a high protein or high carbohydrates meal on subjective motivation to eat and food preferences. *Nutrition and Behavior*, **3**: 133–134.

HILL MJ (1999a). Meat and colorectal cancer. *Proceedings of the Nutrition Society,* **58**: 261–264.

HILL MJ (1999b). Meat and colorectal cancer. A European perspective *European Journal of Cancer Prevention,* **8**: 183–185.

HILL, MJ (2000) Meeting Report: Meat and nutrition, Hamburg: 17–18 October 2000. *European Journal of Cancer Prevention* **9**: 465–470.

HIRAYAMA T (1986). A Large Scale Cohort Study On Cancer Risks By Diet – With Special Reference To The Risk Reducing Effects Of Green-Yellow Vegetable Consumption. In: Hayashi T *et al. Diet Nutrition and Cancer.* Japan Sci Soc Press Tokyo/VNU SCI Press Utrecht, 41–53.

HIRAYAMA T (1990). *Lifestyle and mortality: a large scale census based study in Japan.* Karger, Basle.

HOUSEKNECHT KL, VANDEN HEUVEL JP, MOYA-CAMARENA SY, PORTOCARRERO CP, PECK LW, NICKEL KP and BELURY MA (1998). Dietary conjugated linoleic acid acid normalises impaired glucose tolerance in the Zucker diabetic fatty *fa/fa* rat. *Biochemical and Biophysical Research Communications*, **244**: 678–682.

HULSHOF KF, VAN ERP-BAART MA, ANTTOLAINEN M, BECKER W, CHURCH SM, COUET C, HERMANN-KUNZ E, KESTELOOT H, LETH T, MARTINS I, MOREIRAS O, MOSCHANDREAS J, PIZZOFERRATO L, RIMESTAD AH, THORGEIRSDOTTIR H, VAN AMELSVOORT JM, ARO A, KAFATOS AG, LANZMANN-PETITHORY D and

VAN POPPEL G (1999). Intake of fatty acids in western Europe with emphasis on *trans* fatty acids: the TRANSFAIR Study. *European Journal of Clinical Nutrition*, **53**(2): 143–157.

HUNT JR, GALLAGHER SK, JOHNSON LK and LYKKEN GI (1995). High- versus low-meat diets: effects on zinc absorption, iron status, and calcium, copper, iron, magnesium, manganese, nitrogen, phosphorus, and zinc balance in postmenopausal women. *American Journal of Clinical Nutrition*, **62**: 621–632.

IP C and SCIMECA JA (1997). Conjugated linoleic acid and linoleic acid are distinctive modulators of mammary carconogenesis. *Nutrition and Cancer*, **27**(2): 131–135.

IP C, CHIN SF, SCIMECA JA and PARIZA MW (1991). Mammary cancer prevention by conjugated dienoic derivative of linoleic acid. *Cancer Research*, **51**: 6118–6124.

IP C, SINGH M, THOMPSON HJ and SCIMECA JA (1994). Conjugated linoleic acid suppresses mammary carcinogenesis and proliferative activity of the mammary gland in the rat. *Cancer Research*, **54**: 1212–1215.

IP MM, MASSO-WELCH PA, SHOEMAKER SF, SHEA-EATON WK and IP C (1999). Conjugated linoleic acid inhibits proliferation and induces apoptosis of normal rat mammary epithelial cells in primary culture. *Experimental Cell Research*, **250**: 22–34.

JOHNSON JM and WALKER PM (1992). Zinc and iron utilization in young women consuming a beef-based diet. *Journal of the American Dietetic Association*, **12**, 1474–1478.

JONES DP (1995). Glutathione Distribution In Natural Products. *Methods in Enzymology*, **252** pp 3–13.

KAPSOKEFALOU M and MILLER DD (1991). Effects of meat and selected food components on the valence of non-haem iron during *in vitro* digestion. *Journal of Food Science*, **56**: 352–358.

KAPSOKEFALOU M and MILLER DD (1993). Lean beef and beef fat interact to enhance non-haem iron absorption in rats. *Journal of Nutrition*, **123**: 1429–1434.

KAPSOKEFALOU M and MILLER DD (1995). Iron speciation in intestinal contents of rats fed meals composed of meat and non-meat sources of protein and fat. *Food Chemistry*, **52**: 47–56.

KELLY FD, MANN NJ, TURNER AH and SINCLAIR AJ (1999). Stearic acid-rich diets do not increase thrombotic risk factors in healthy males. *Lipids*, **34**: S199.

KELLY FD, SINCLAIR AJ, MANN NJ, TURNER AH, ABEDIN L and LI D (2001). A stearic acid-rich diet improves thrombogenic and atherogenic risk factor profiles in healthy males. *European Journal of Clinical Nutrition*, **55**: 88–96.

KEY TJ, DAVEY GK and APPLEBY PN (1999). Health Benefits Of A Vegetarian Diet. In *Meat Or Wheat For The Next Millennium?* Proceedings of the Nutrition Society, **58**(2): 271–275.

KEY TJ, FRASER GE, DAVEY GK, THOROGOOD M, APPLEBY PN, BERAL V, REEVES G, BURR ML, CHANG-CLAUDE J, FRENTZEL-BEYNE R, KUZMA JW, MANN J and

MCPHERSON K (1998). Mortality in vegetarians and non-vegetarians: a collaborative analysis of 8,300 deaths among 76,000 men and women in five prospective studies. *Public Health Nutrition*, **1**(1): 33–41.

KEY TJ, FRASER GE, THOROGOOD M, APPLEBY PN, BERAL V, REEVES G, BURR ML, CHANG-CLAUDE J, FRENTZEL-BEYME R, KUZMA JW, MANN J and MCPHERSON K (1998). Mortality in vegetarians and non-vegetarians: detailed findings from a collaborative analysis of 5 prospective studies. *American Journal of Clinical Nutrition*, **70** (suppl): 516S–524S.

KOHLMEIER L, SIMMANSEN N and MOTTINS K (1995). Dietary Modifiers Of Carcinogenesis. *Environmental Health Perspectives*, **103**: Supplement 8 P180–184.

KRAJCOVICOVA-KUDLACKOVA M, BLAZICEK P, KOPCOVA J, BEDEROVA A and BABINSKA K (2000). Homocysteine levels in vegetarians versus omnivores. *Annual Nutrition Metabolism*, **44**: 135–138.

KRAMER JKG, PARODI PW, JENSEN RG, MOSSOBA MM, YURAWECZ MP and ADLOF RO (1998). Rumenic acid: a proposed common name for the major conjugated linoleic acid isomer found in natural products. *Lipids*, **33**: 835.

LARSSON CL and JOHANSSON, G (2001). Dietary intake and nutritional status of young vegans and omnivores in Sweden. *Proceedings of the Nutrition Society*, **60**(4): 69A.

LAWSON M and THOMAS M (1999). Vitamin D concentrations in Asian children aged 2 years living in England: population survey. *British Medical Journal*, **318**: 28–29.

LEE KN, KRITCHEVSKY D and PARIZA MW (1994). Conjugated linoleic acid and atherosclerosis in rabbits. *Atherosclerosis*, **108**: 19–25.

LEEDS AR, RANDLE A and MATTHEWS KR (1997). A Study Into The Practice Of Trimming Fat From Meat At The Table, And The Development Of New Study Methods. *Journal of Human Nutrition and Dietetics*, **10**: 245–251.

LI D, SINCLAIR A, MANN N, TURNER A, BALL M, KELLY F, ABEDIN, L and WILSON A (1999). The association of diet and thrombotic risk factors in healthy male vegetarians and meat-eaters. *European Journal of Clinical Nutrition*, **53**: 612–619.

LYLE RM, WEAVER CM, SEDLOCK DA, RAJARAM S, MARTIN B and MELBY CL (1992). Iron Status In Exercising Women: The Effect Of Oral Iron Therapy vs Increased Consumption Of Muscle Foods. *American Journal of Clinical Nutrition*, **56**(6): 1049–1055.

MA DWL, WIERZBICKI AA, FIELD CJ and CLANDININ MT (1999). Conjugated linoleic acid in Canadian diary and beef products. *Journal of Agricultural and Food Chemistry*, **47**(5): 1956–1960.

MCKINLEY MC, MCNULTY H, MCPARTLIN J, STRAIN JJ, PENTIEVA K, WARD M, WEIR DG and SCOTT JM (2001). Low dose vitamin B_6 effectively lowers fasting plasma homocysteine in healthy elderly persons who are folate and riboflavin replete. *American Journal of Clinical Nutrition*, **73**: 759–764.

MANN NJ (2001a). The evidence for high meat intake during the evolution of hominids. *Proceedings of the Nutrition Society*, **60**: 61A.

MANN NJ (2001b). Effect of vitamin B12 status on homocysteine levels in healthy male subjects. *Proceedings of the Nutrition Society*, **60**: 58A.

MANN NJ, SINCLAIR AJ, PILLE M, JOHNSON L, WARRICK G, REDER E and LORENZ R (1997). The Effect Of Short Term Diets Rich In Fish, Red Meat Or White Meat On Thromboxane And Prostacyclin Synthesis In Humans. *Lipids* **32**(6): 635–643.

MANN NJ, LI D, SINCLAIR AJ, DUDMAN NP, GUO XW, ELSWORTH GR, WILSON AK and KELLY FD (1999). The effect of diet on plasma homocysteine concentrations in healthy male subjects. *European Journal of Clinical Nutrition,* **53**(11): 895–899.

MEAT AND LIVESTOCK COMMISSION AND ROYAL SOCIETY OF CHEMISTRY (1990). *The Chemical Composition Of Pig Meat.* Report To The Ministry Of Agriculture, Fisheries And Food On The Trial To Determine The Chemical Composition Of Fresh And Cured Pork From British Pigs Of Different Breed Types, Sexes And Origins. Meat And Livestock Commission, Milton Keynes.

MILLS A and TYLER H (1992). *Food And Nutrient Intakes Of British Infants Aged 6–12 Months.* Ministry Of Agriculture, Fisheries And Food. London: HMSO.

MILLWARD DJ (1999). Meat Or Wheat For The Next Millennium? *Proceedings of the Nutrition Society*, **58**(2): 209–210.

MINISTRY OF AGRICULTURE FISHERIES AND FOOD (1981) *Household Food Consumption And Expenditure: 1979.* National Food Survey Committee. HMSO: London.

MINISTRY OF AGRICULTURE FISHERIES AND FOOD (1999). *National Food Survey 1999.* The Stationery Office, London.

MINISTRY OF AGRICULTURE FISHERIES AND FOOD, FOOD SAFETY DIRECTORATE (1992, 1993, 1995, 1998) *National Food Survey-Household Consumption For 1991, 1992, 1994, 1997* MAFF London.

MIRA M, ALPERSTEIN G, KARR M, RANMUTHUGALA G, CAUSER J, NIEC A and LILBURNE AM (1996). Haem iron intake in 12–36 month old children depleted in iron: case-control study. *British Medical Journal*, **312**: 881–883.

MULVIHILL B (1996). The effect of meat systems on the bioavailability of non-haem iron. PhD Thesis, University College Cork, Republic of Ireland.

MULVIHILL B (2001). Ruminant meat as a source of conjugated linoleic acid (CLA). *Nutrition Bulletin*, **26**(4): 295–300.

MULVIHILL B and MORRISSEY PA (1998a). Influence of the sulphydryl content of animal proteins on *in vitro* bioavailability of non-haem iron. *Food Chemistry*, **61**(No. 1–2): 1–7.

MULVIHILL B and MORRISSEY PA (1998b). An investigation of factors influencing the bioavailability of non-haem iron from meat systems. *Irish Journal of Agricultural and Food Research*, **37**(2): 219–226.

MULVIHILL B, KIRWAN FM, MORRISSEY PA and FLYNN A (1998). Effect of myofibrillar proteins on the *in vitro* bioavailability of non-haem iron. *International Journal of Food Science and Technology,* **49**: 187–192.

NAKACHI K, IMAI K, HOSHIYAMA Y and SASABA T (1988). The joint effects of two factors in the aetiology of oesphageal cancer in Japan, *Journal of Epidemiology and Community Health*, **42**: 755–761.
NATHAN I, HACKETT IF and KIRBY S (1994). Vegetarianism And Health: Is A Vegetarian Diet Adequate For The Growing Child? *Food Science And Technology Today*, **8**(1): 13–15.
NATHAN I, HACKETT AF and KIRBY S (1996). The dietary intake of a group of vegetarian children aged 7–11 years compared with matched omnivores. *British Journal of Nutrition*, **75**(4): 533–544.
NATHAN I, HACKETT IF and KIRBY SP (1997). A Longitudinal Study Of The Growth Of Matched Pairs Of Vegetarian And Omnivorous Children Aged 7–11 Years In The North-West Of England. *European Journal of Clinical Nutrition*, **51**: 20–25.
NATIONAL ADVISORY COMMITTEE ON NUTRITION EDUCATION (1983). *Proposals For Nutritional Guidelines For Health Education In Britain.* September. Health Education Council, London.
NATIONAL AUDIT OFFICE REPORT (2001). *Tackling Obesity in England*, The Stationery Office, London, February.
NELSON M (1996). Anaemia in adolescent girls: effects on cognitive function and activity. *Proceedings of the Nutrition Society*, **55**: 359–367.
NELSON M, WHITE J and RHODES C (1993). Haemoglobin, ferritin, and iron intakes in British children aged 12–14 years: a preliminary investigation. *British Journal of Nutrition*, **70**: 147–155.
NICOLOSI RJ, ROGERS EJ, KRITCHEVSKY D, SCIMECA JA and HITH PJ (1997). Dietary conjugated linoleic acid reduces plasma lipoproteins and early aortic atherosclerosis in hypercholesterolemic hamsters. *Artery*, **22**(5): 266–277.
O'DEA K (1991). Traditional diet and food preferences of Australian aboriginal hunter-gatherers. *Philos Trans R Soc Lond B Biol Sci,* **334**: 233–241.
O'DELL BL (1989). Bioavailability of trace elements. *Nutrition Reviews*, **42**: 301–308.
ORTEGA R, LOPEZ-SOBALER A, REQUEJO A, QUINTAS M, GASPAR M, ANDRES P and NAVIA (1998). The influence of meat consumption on dietary data, iron status and serum lipid parameters in young women. *International Journal for Vitamin and Nutrition Research*, **68**: 255–262.
O'SHEA M, LAWLESS F, STANTON C and DEVERY R (1998). Conjugated linoleic acid in bovine milk fat: a food-based approach to cancer chemoprevention. *Trends in Food Science and Technology*, **9**: 192–196.
PARK Y, ALBRIGHT KJ, LIU W, STORKSON JM, COOK ME and PARIZA MW (1997). Effect of conjugated linoleic acid on body composition in mice. *Lipids*, **32**: 853–858.
RANA SK and SANDERS TAB (1986). Taurine Concentrations In The Diet, Plasma, Urine And Breast Milk Of Vegans Compared With Omnivores. *British Journal of Nutrition*, **56**: 17–27.
REDDY S and SANDERS TAB (1990). Haemotological Studies On Pre-Menstrual Indian And Caucasian Omnivores. In *British Journal of Nutrition*, **64**:

331–338.
REID RL and HACKETT AF (2001). A database of vegetarian convenience foods. *Proceedings of the Nutrition Society*, **60**(4): 4A.
REQUEJO AM, NAVIA B, ORTEGA RM, LOPEZ-SOBALER AM, QUINTAS E, GASPAR MJ and OSORIO O (1999). The age at which meat is first included in the diet affects the incidence of iron deficiency and ferropenic anaemia in a group of pre-school children from Madrid. *International Journal for Vitamin and Nutrition Research*, **69**(2): 127–131.
ROYAL SOCIETY OF CHEMISTRY (1986). Nitrogen Factors For Pork *Analyst*, **111**(8): 969–973.
ROYAL SOCIETY OF CHEMISTRY (1993). Nitrogen Factors For Beef; A Reassessment. *Analyst*, **118**(9): 1217–1226.
ROYAL SOCIETY OF CHEMISTRY (1996). Nitrogen Factors For Sheep Meat. *Analyst*, **121**(7): 889–896.
SANDERS TAB (1999). The Nutritional Adequacy Of Plant Based Diets. In: *Meat Or Wheat For The Next Millennium?* Proceedings of The Nutrition Society, **58**(2): 265–269.
SANDERS TAB and REDDY S (1994). Vegetarian diets and children. *American Journal of Clinical Nutrition*, Supplement 59 **120**(4): 1176S–1181S.
SANDSTEAD HH (1995). Is zinc deficiency a public health problem? *Nutrition*, **11**: 87–92.
SANDSTEAD HH (2000). Causes of iron and zinc deficiencies and their effects on brain. *Journal of Nutrition,* **130**(2S Suppl): 347S–349S.
SCOTT LE, KIMBALL KT, WITTELS EH, DUNN JK, BRAUCHI DJ, POWNALL HJ, HERD JA, SAVELL JW and PAPADOPOULOUS LS (1990). The Effect Of Lean Beef, Chicken And Fish On Lipoprotein Profile. In *63rd Scientific Sessions Of The American Heart Association*. 12–15 Nov. Circulation 82. Dallas, Texas, USA.
SCOTTISH OFFICE DEPARTMENT OF HEALTH (1996). *Eating For Health – A Diet Action Plan For Scotland*. HMSO. Edinburgh.
SCRIMSHAW NS (1991). Iron Deficiency. *Scientific American*, 24–30 October.
SHANTHA NC, CRUM, AD and DECKER EA (1994). Evaluation of conjugated linoleic acid concentrations in cooked beef. *Journal of Agricultural and Food Chemistry*, **42**: 1757–1760.
SHI B and SPALLHOLZ JE (1994). Selenium Peroxidase Is Highly Available As Assessed By Liver Glutathione Peroxidase Activity And Tissue Selenium. *British Journal of Nutrition*, **72**(6): 873–881.
SINCLAIR AJ, JOHNSON L, O'DEA K and HOLMAN T (1994). 'Diets Rich In Lean Beef Increase The Arachidonic Acid And Long Chain N-3 PUFA Levels In Plasma Phospholipids. *Lipids*, **29**(5): 337–343.
SKOV AR, TOUBRO S, RONN B, HOLM L and ASTRUP A (1999a). Randomised Trial On Protein Vs Carbohydrate In *Ad Libitum* Fat Reduced Diet For The Treatment Of Obesity. *International Journal of Obesity*, **23**: 528–536.
SKOV AR, TOUBRO S, BULOW J, KRABBE K, PARVING H-H and ASTRUP A (1999b). Changes In Renal Function During Weight Loss Induced By High Vs

Low-Protein Diets In Overweight Subjects. *International Journal of Obesity*, **23**: 1170–1177.
STUBBS RJ (1995). Macronutrient effects on appetite. *International Journal of Obesity,* **19**: S11–S19.
SWAIN R (1995). An update on vitamin B12 metabolism and deficiency states. *The Journal of Family Practice,* **41**(6): 595–600.
TAYLOR PG, MARTINEZ-TORRES C, ROMANO EL and LAYRISSE M (1986). The effect of cysteine containing peptides released during meat digestion on iron absorption in humans. *American Journal of Clinical Nutrition,* **43**: 68–71.
THOROGOOD M, MANN J, APPLEBY P and MCPHERSON, K (1994). 'Risk Of Death From Cancer And Ischaemic Heart Disease In Meat And Non-Meat Eaters'. *British Medical Journal*, **308**: 1667–1671.
TROMPETA V and O'BRIEN J (1998). 'Inhibition Of Mutagen Formation By Organosulphur Compounds. *Journal of Agriculture and Food Chemistry*, **46**: 4318–4323.
TUYNS AJ, RIBOLI E, DOORNBOS G and PEGUINOT G (1987). Diet and esophageal cancer in Calvados (France). *Nutrition and Cancer,* **9**: 81–92.
TUYNS AJ, *et al.* (1992) Diet and gastric cancer. A case-control study in Belgium. *International Journal of Cancer*, Apr 22; **51**(1): 1–6.
UHE AM, COLLIER GR and O'DEA K (1992). A comparison of the effects of beef, chicken and fish protein on satiety and amino acid profiles. *Journal of Nutrition*, **122**: 467–472.
ULBRICHT TLV (1995). Fat In The Food Chain. *A Report to The Ministry of Agriculture Fisheries and Food*. April. MAFF. London.
ULBRICHT TLV and SOUTHGATE DAT (1991) Coronary Heart Disease: Seven Dietary Factors. *Lancet*, **338**: 985–992.
VAN DE VIJVER LPL, KARDINAAL AFM, COUET C, ARO A, KAFATOS A, STEINGRIMSDOITTIR L, AMORIM CRUZ JA, MOREIRAS O, BECKER W, VAN AMELSVOORT JMM, VIDAL-JESSEL S, SALMINEN I, MOSCHANDREAS J, SIGFUSSON N, MARTINS I, CARBAJAL A, YTTERFORS A and VAN POPPEL G (2000). Association between *trans* fatty acid intake and cardiovascular risk factors in Europe: the TRANSFAIR study. *European Journal of Clinical Nutrition,* **54**: 126–135.
WALKER ARP (1998). The Remedying Of Iron Deficiency: What Priority Should It Have? *British Journal of Nutrition,* **79**: 227–235.
WATTS G, AHMED W, QUINEY J, HOULSTON R, JACKSON P, ILES C and LEWIS B (1988). Effective Lipid Lowering Diets Including Lean Meat. *British Medical Journal (Clin Res Ed)*, **296**: (6617) 235–237.
WEST DB, DELANY JP, CAMET PM, BLOHM F, TRUETT AA and SCIMECA J (1998). Effect of conjugated linoleic acid on body fat and energy metabolism in the mouse. *American Journal of Physiology*, **44**: R667–R672.
WHO (1990). *Diet, Nutrition and the prevention of chronic diseases.* Report of a WHO Study Group. WHO Technical Series 797. Geneva, Switzerland: World Health Organisation, 1990.
WILSON A and BALL M (1999). Nutrient intake and iron status of Australian male

vegetarians. *European Journal of Clinical Nutrition,* **53**: 189–194.
WORLD CANCER RESEARCH FUND (1997). Food Nutrition And The Prevention Of Cancer: A Global Perspective. *American Institute Of Cancer Research*: Washington DC.
ZEIGLER RG, MORRIS LE, BLOT WJ, POTTERN LM, HOOVER R and FRAUMENI JFJ (1981). Esophageal cancer among black men in Washington DC. II Role of Nutrition. *Journal – National Cancer Institute*, **67**: 1199–1206.
ZHENG JJ, MASON JB, ROSENBERG IH and WOOD RJ (1993). Measurement of zinc bioavailability from beef and a ready-to-eat high-fiber breakfast cereal in humans: application of a whole-gut lavage technique. *American Journal of Clinical Nutrition*, **58**: 902–907.

5
Lipid-derived flavors in meat products

F. Shahidi, Memorial University of Newfoundland, St John's

5.1 Introduction

Flavor is an important quality attribute of muscle foods and comprises mainly the two sensations of taste and aroma or smell. Although both of these factors affect the overall acceptability of foods, the aroma or flavor volatiles are of utmost importance because they influence the judgement of the consumer even before the food is eaten. While muscle foods, namely red meat, poultry and, to a lesser extent, seafoods in the fresh, raw state have little flavor of their own, upon heat processing their specific meaty aroma develops (Shahidi, 1989; Farmer, 1992). Nonetheless, it should be recognized that the fishy aroma of raw, stored seafoods may arise from the presence of amines, derived from trimethylamine oxide which is present in gadoid fish species as an osmoregulator, or via lipoxygenase-assisted oxidation and generation of aldehydes and alcohols, especially 2,4,7-dicatrienals and cis-4-heptenal (Josephson, 1991; Lindsay, 1994).

Upon heat processing, meat constituents undergo a series of thermally induced reactions to afford a large number of volatile compounds that contribute to its aroma. Some of the non-volatile precursors of volatile flavor compounds are also known to contribute to the taste of cooked meat. The most important taste-active components of meats are amino acids, peptides, organic acids, nucleotides and other flavor enhancers, among others (Shahidi, 1989). It has further been concluded that high-molecular-weight fibrillar and sarcoplasmic proteins have little effect on the development of meat flavor volatiles. Since the free amino acids and carbohydrates in meat from different species are similar, their flavor upon heat processing is also expected to be similar. However, the lipid components of meat, mainly the intramuscular lipids, are known to modify the flavor quality of thermally-processed muscle foods. Both desirable and undesirable aromas are formed. Of course, the species of muscle food being examined, method of heat processing employed, dietary regime and sex of the slaughtered animal, as well as freshness of the product and storage conditions affect the flavor quality of products (e.g. Ladikos and Lougovois, 1990; Buckley et al., 1996).

More than 1000 chemicals have so far been identified in the volatiles of different muscle foods (Shahidi et al., 1985). While many of these have little influence on flavor of meat, no single compound has been identified as being primarily responsible for the aroma of cooked muscle foods. The aroma-impact compounds have been found to comprise a myriad of compounds belonging to different classes of heterocyclic and acyclic compounds containing N, O and S atoms. In addition, lipid breakdown products have been found to participate in the development of meaty aroma as well as meat flavor deterioration. Some aspects of aroma of muscle foods, both desirable and undesirable are discussed.

5.2 The role of lipids in generation of meaty flavors

The role of lipids in meat flavor generation has been the subject of extensive studies. It has been suggested that the basic meaty aroma of beef, pork and mutton is the same and is derived from the water-soluble fraction of the muscle which is a reservoir of low-molecular-weight compounds (Hornstein and Crowe, 1960; 1963). Meanwhile, species-specific flavors in meats originate from the involvement of their lipid components in Maillard reaction during heat processing. Lipids may contribute both to desirable and undesirable flavors of meat from different species. Their effect on generation of desirable aromas in cooked meats may arise from mild thermal oxidative changes which produce important flavor compounds; they may also react with components of lean tissues to afford other flavor compounds and may act as a carrier for aroma compounds, thus affecting their sensible threshold values.

The effect of both triacylglycerols (TAG) and phospholipids (PL) on the development of meaty aroma in meat and model systems has been studied (Mottram, 1983, 1991; Mottram and Edwards, 1983). In these studies, a meat sample was heat processed as such or was extracted with hexane or methanol-chloroform prior to cooking. While there was little difference in the development of meaty aroma in the untreated and hexane-extracted samples, meat extracted with methanol-chloroform had little meaty aroma, but possessed a sharp roast and biscuit-like odor. In particular, the concentration of dimethylpyrazine in the headspace volatiles was significantly increased with a concurrent decrease in the content of lipid oxidation products. Thus, it was concluded that phospholipids present in the intramuscular lipids of meat were primarily responsible for the development of meaty aromas.

Based on these experiments, other model systems were devised in order to unravel the role of phospholipids in the formation of Maillard reaction products (MRP) (Farmer and Mottram, 1990). Cysteine and ribose, with or without phospholipids, were used to assess the involvement of lipids in the formation of

Table 5.1 Relative concentration of selected acyclic and heterocyclic volatiles from the reaction of cysteine and ribose in the absence or presence of beef triacylglycerols (BTAG) and beef phospholipids (BPL)

Compound	No lipid	BTAG	BPL
3-Pentanone, 2-mercapto	1	0.72	0.49
2-Pentanone, 3-mercapto	1	0.77	0.47
2-Furylmethanethiol	1	0.67	0.63
3-Furanethiol, 2-methyl	1	0.40	0.15
2-Thiophenethiol	1	0.32	0.03
3-Thiophenethiol, 2-methyl	1	0.08	0.01
Pyridine, 2-pentyl	0	0.09	1
Thiophene, 2-pentyl	0	0.00	1
Thiophene, 2-hexyl	0	0.15	1
2-H-Thiapyran, 2-pentyl	0	0.10	1

aroma compounds via Maillard reaction following heating in buffered solutions.

There was a marked reduction in the amount of thiols when phospholipids, and to a lesser extent triacylglycerols, were present (Table 5.1). The compounds that were formed only in the presence of lipids were 2-pentylpyridine, 2-alkylthiophenes, alkenylthiophenes, pentylthiapyran and alkanethiols (Table 5.1). Furthermore, the impact of PL was much greater than that of TAG in affecting the flavor of systems under investigation. Formation of 2-alkylheterocyles was generally due to the reaction of 2,4-decadienal with ammonia or hydrogen sulfide formed from cysteine or other precursors (Fig. 5.1). Direct reaction of hexanal with amino acids would also lead to the formation of 2- hexylpyridine. Formation of other alkyl substituted heterocyclic compounds from participation of lipid-derived aldehydes is also contemplated. Reaction of 2,4-decadienal directly with amino acids has also been reported. The compound 2,4-dacadienal is a major breakdown product of the omega-6 fatty acids. Furthermore, 1-heptanethiol and 1-octanethiol were present only in the systems containing phospholipids and their formation is presumed to be due to the interaction of alcohols with hydrogen sulfide. Furthermore, MRPs may also interact with meat lipids by acting as antioxidants in order to stabilize them. Bailey (1988), and Bailey et al. (1997) have demonstrated that MRPs act as important antioxidants in meat model systems. It has also been shown that furanthiols and thiophenethiols exert antioxidative activity in lipids (Eiserich and Shibamoto, 1994), as well as in aqueous solutions as evidenced by their tyrosyl radical scavenging effect (Eiserich et al., 1995). The antioxidant activity of these compounds was similar to that of ascorbic acid. Figures 5.1 and 5.2 show examples of involvement of lipids in Maillard reaction and formation of volatile flavor compounds.

Other potentially desirable flavor components that might be formed in processed meats are free fatty acids and related compounds which are prevalent in dry-cured-ham. These compounds are formed in such products via

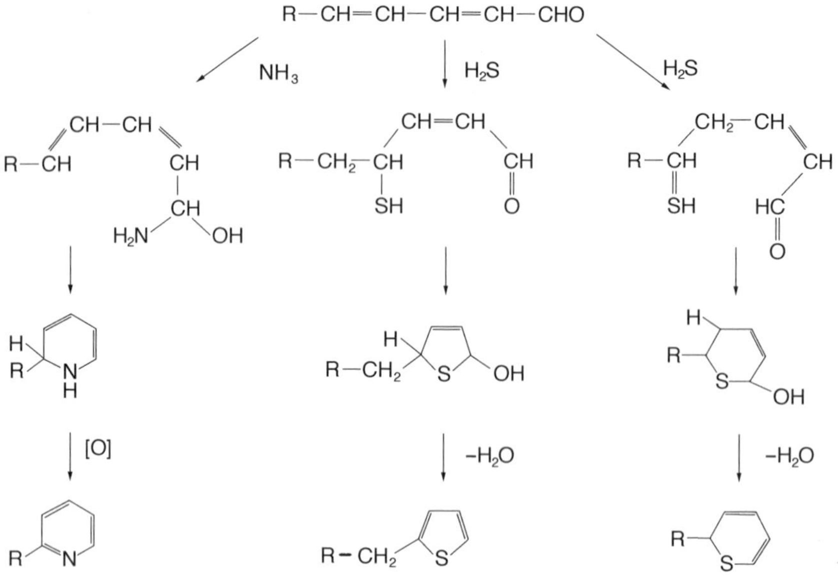

Fig. 5.1 Production of long-chain alkyl heterocycles from the reaction of 2,4-decadienal with ammonia and hydrogen sulfide.

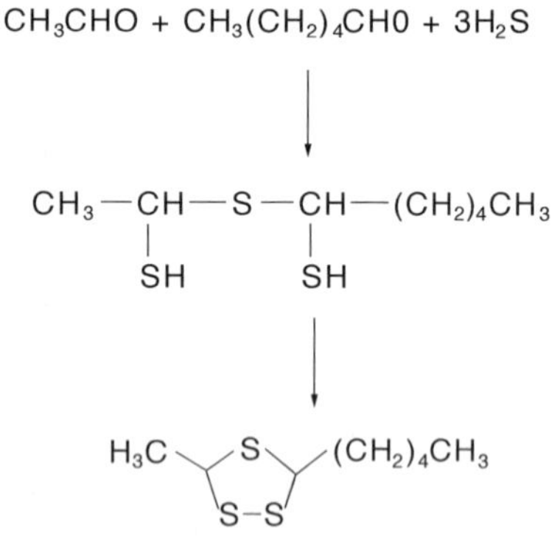

Fig. 5.2 Involvement of hexanal, an oxidation product of linoleic acid, in production of a trisulfide heterocyclic compound.

fermentation reactions. The flavor characteristics of dry-cured ham has recently been reported (Toldra et al., 1997).

5.3 Lipid autoxidation and meat flavor deterioration

The primary mechanism for the degradation of desirable flavor in stored meats is lipid autoxidation. Lipids in muscle foods, particularly their phospholipid

Table 5.2 Flavor thresholds of some lipid oxidation products.[1]

Compound	Threshold, ppm
Hydrocarbons, saturated	90–2150
Alkenes	0.02–9
Furans	1–27
Alcohols, saturated	0.3–2.5
Alcohols, unsaturated	0.001–3
Aldehydes, saturated	0.014–1
Aldehydes, monounsaturated	0.04–2.5
Aldehydes, diunsaturated	0.002–0.6
Ketones, methyl	0.16–5.5
Ketones, vinyl	0.0002–0.007

[1] Adapted from Drumm and Spanier (1991).

components, undergo degradation to produce a large number of volatile compounds. While hydroperoxides, the primary products of lipid oxidation, are odorless and tasteless, their degradation leads to the formation of an array of secondary products such as aldehydes, hydrocarbons, alcohols, ketones, acids, esters, furans, lactones and epoxy compounds as well as polymers. These latter classes of compounds are flavor-active, particularly aldehydes, and possess low threshold values in the parts per million or even parts per billion levels, thus they are responsible for the development of warmed-over flavor (WOF), as coined by Tims and Watts (1958), and meat flavor deterioration (MFD) (e.g. Drumm and Spanier, 1991). Table 5.2 summarizes the threshold values of selected classes of volatile compounds. In addition, lipid oxidation products lead to the loss of nutritional value and safety, color, texture and other functional properties and wholesomeness of foods.

The degree of unsaturation of the acyl constituents of meat lipids primarily dictates the rate at which MFD proceeds; unsaturated lipids being more susceptible. Autoxidation of meat lipids gives rise to a number of hydroperoxides which, in conjunction with the many different pathways possible, decompose to a large number of volatile compounds. However, other factors might also affect the oxidation of meat lipids and formation of WOF as well as shelf-life of products (Spanier et al., 1988; Gray et al., 1996). In chicken meat, lack of α-tocopherol is the main reason for MFD and formation of undesirable WOF in products. However, cooked turkey meat, despite its higher content of unsaturated lipids, may not readily develop WOF because it contains endogenous α-tocopherol. In addition, presence of heme compounds and metal ions may also hasten the oxidation of meat lipids. Furthermore, presence of salt and other ingredients used in cooking, such as onion, may also affect progression of MFD. It has been reported that aldehydes, generated from oxidation of lipids, react with thiols, a class of compounds in onion, such as propenethiol to produce 1,1-bis (propylthio)-hexane and 1,2-bis (propylthio)-

hexane. These compounds will definitely modify flavor profile of muscle foods (Ho *et al.*, 1994). Formation of volatile aldehydes and other lipid degradation products results in the masking of desirable meaty aroma of products. Thus, the role of both endogenous and exogenous factors on the oxidative status of muscle foods requires attention (see the following).

5.4 The effect of ingredients on the flavor quality of meat

In addition to the effects of low-molecular-weight water-soluble components and lipid constituents on quality attributes of meat, other factors must also be carefully considered. Thus, endogenous components of meat as well as ingredients added to it prior to heat processing are of major importance in the development of flavor and its quality characteristics. An attempt will be made to provide a cursory account of these factors.

5.4.1 Lipid classes and fatty acid composition
While adipose tissues generally contain over 98% triacylglycerols, phospholipids constitute a major portion of intramuscular lipids of muscle foods. The unsaturated fatty acids present in TAGs of red meat and poultry contain mainly oleic and linoleic acids, however, phospholipids contain a relatively higher proportion of linolenic and arachidonic acids. In seafoods, long chain omega-3 fatty acids such as eicosapentaenoic and docosahexaenoic acids are prevalent. Existing differences in the fatty acid constituents of phospholipids and triacylglycerols of muscle foods are primarily responsible for species differentiation in cooked samples of meat, as discussed earlier. Furthermore, depending on the type and proportion of unsaturated fatty acids in muscle foods, lipid autoxidation and flavor deterioration may proceed at different rates. In this respect, seafoods deteriorate much faster than chicken which is oxidized quicker than red meats.

5.4.2 Effect of transition metal ions
Transition metals such as iron, copper and cobalt may catalyze the initiation and enhance the propagation steps involved in lipid autoxidation. For example, Fe^{2+} will reductively cleave hydroperoxides to highly reactive alkoxy radicals which in turn abstract a hydrogen atom from other lipid molecules to form new lipid radicals. This reaction is known as hydroperoxide-dependent lipid peroxidation (Svingen *et al.*, 1979). Morrissey and Tichivangana (1985) and Tichivangana and Morrissey (1985) have reported that ferrous ion at 1–10 ppm levels acts as a strong prooxidant in cooked fish muscles. Similarly, copper (II) and cobalt (II) were effective prooxidants. These observations are in agreement with the findings of Igene *et al.* (1979) who reported that iron ions were the major catalysts responsible for enhancement of autoxidation in muscle foods. Furthermore, Shahidi and Hong (1991) demonstrated that the prooxidant effect

Table 5.3 Effect of chelators on TBA numbers of ground pork catalyzed by iron ions and Heme compounds[1]

Treatment	NO chelator	Na$_2$ EDTA, 500 ppm	STPP, 3000 ppm
Control	3.03	0.07	0.22
Fe^{2+}	4.80	0.09	0.27
Fe^{3+}	4.60	0.08	0.37
Mb	4.20	0.07	0.38
Hb	4.33	0.10	0.30
Hm	4.35	0.07	0.33
CCMP	0.42	0.30	0.30

[1] From Shahidi and Hong (1991).

of metal ions was more pronounced at their lower oxidation state and found that in the presence of chelators such as disodium salt of ethylenediaminentetraacetic acid (Na$_2$EDTA) and sodium tripolyphosphate (STPP), the prooxidant effect of metal ions was circumvented (Table 5.3). Furthermore, Cassens et al. (1979) have shown that 1–3% of the total amount of nitrite added to meat during curing was recovered in the lipid extracts using the Folch method (Folch et al., 1957). Thus, stabilization of meat lipids by nitrite may also be influenced by direct coupling of nitric oxide with lipid radicals, but most of the nitrite was in the protein-bound form as nitrosothiol, nitrite/nitrate and nitrosylheme complex, among others (e.g. Kanner et al., 1984).

Hemoproteins in meats are generally known for their prooxidant activity (Robinson, 1924; Younathan and Watts, 1959; Pearson et al., 1977, Igene et al., 1979; Rhee, 1988; Shahidi et al., 1988; Johns et al., 1989; Shahidi and Hong, 1991; Wettasinghe and Shahidi, 1997); some have also been reported to possess antioxidant properties (Ben Aziz et al., 1971; Kanner et al., 1984; Shahidi et al., 1987; Shahidi, 1989; Shahidi and Hong, 1991; Wettasinghe and Shahidi, 1997). The prooxidant activity of heme compounds arises, at least in part, from their decomposition upon cooking of meat and liberation of free iron. Meanwhile, nitric oxide derivatives of heme pigments, namely nitrosyl myoglobin and nitrosyl ferrohemochrome (or cooked cured-meat pigment, CCMP), are reported to have antioxidant effect (e.g. Wettasinghe and Shahidi, 1997). This topic is further discussed under the effect of nitrite and nitrite alternatives.

5.4.3 Effect of salt

Sodium chloride, or table salt, is an important ingredient in the meat industry. It acts generally as a prooxidant, but sometimes also as an antioxidant (Kanner and Kinsella, 1983). In comminuted meat samples, under different processing conditions, sodium chloride did not act as an antioxidant, but its neutral or prooxidant effects were clearly demonstrated (St. Angelo et al., 1992; Wettasinghe and Shahidi, 1996). Takiguchi (1989) and Kanner et al. (1991)

112 Meat processing

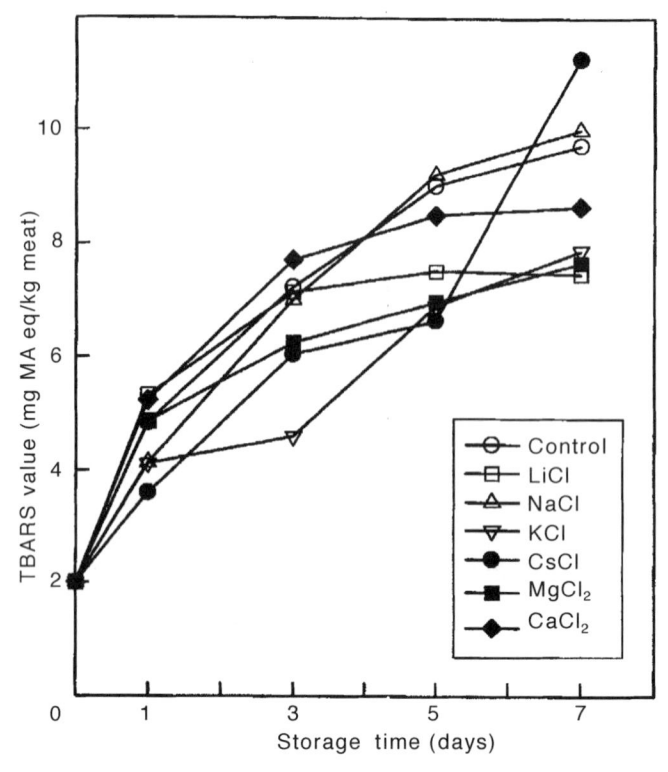

Fig. 5.3 Effect of chloride salts on oxidation of meat lipids, as reflected in 2-thiobarbituric acid reactive substances (TBARS) values, over a seven-day storage period.

have demonstrated the prooxidant effect of NaCl in a comminuted muscle system and suggested that it may promote the displacement of iron from binding sites of heme compounds by interfering with iron-protein interactions. The free iron ions so formed may catalyze lipid peroxidation. Recently, Wettasinghe and Shahidi (1996) reported that LiCl, KCl, CsCl, $MgCl_2$ and $CaCl_2$ exhibited prooxidant activities in a cooked meat model system (Fig. 5.3). Thus, the overriding prooxidant activity was thought to be due to the chloride ion of salts. The results of Rhee *et al.* (1983 a,b) and Cho and Rhee (1995) for the effect of NaCl, KCl and $MgCl_2$ in ground pork samples are in agreement with our findings. Further studies on the effect of replacing chloride with their fluoride and iodide counterparts showed inhibition of lipid oxidation of meats, but bromide salts behaved very similarly to their chloride counterparts (Fig. 5.4). However, the situation was somewhat different when alkali earth halides were used. Thus, it was concluded that pro- or antioxidative effects of salts are primarily dictated by their anions, but mediated by their cations because of existing differences in their ability to participate in iron-pairing interactions with the anion counterparts.

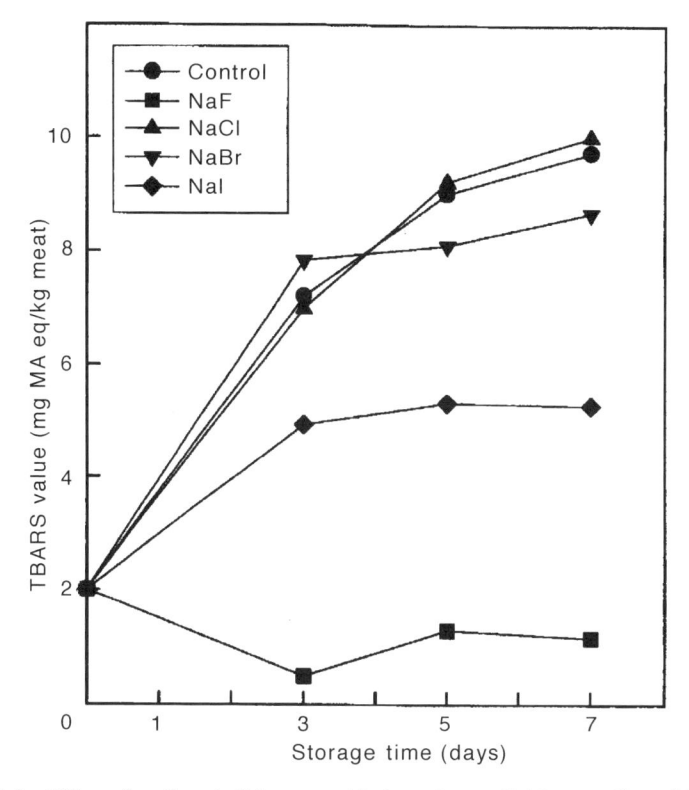

Fig. 5.4 Effect of sodium halides on oxidation of meat lipids, as reflected in 2-thiobarbituric acid reactive substance (TBARS) values and measured as malonaldehyde (MA) equivalents/kg sample, over a seven-day storage period.

5.4.4 Effect of nitrite and nitrite alternatives

Nitrite is a key ingredient of the cure and is responsible for producing the characteristic pink color in cooked-cured products and contributes to the typical flavor associated with cured meats and prevents the formation of warmed-over flavor as well as rendering microbial stability to products (Fox, 1966; Hadden *et al.*, 1975; Pearson *et al.*, 1977; Shahidi, 1992). The relationship of nitrite to cured meat flavor was first described by Brooks *et al.* (1940). These authors presumed that an adequate cured flavor could be obtained with a relatively low, 10 mg/kg, concentration of nitrite, but Simon *et al.* (1973) and MacDougall *et al.* (1975) showed higher taste panel scores when the addition level of nitrite was increased.

The role of nitrite in modifying flavor of cooked meat (Hadden *et al.*, 1975; MacDonald *et al.* 1980) and suppressing lipid oxidation and MFD in cooked meats (Fox, 1966; Pearson *et al.*, 1977; Fooladi *et al.*, 1979) is well documented. Sato and Hegarty (1971) reported that nitrite at 50 ppm was capable of suppressing oxidation of meat lipids. However, Shahidi *et al.* (1987) demonstrated that presence of a reductant such as sodium ascorbate was essential to eliminate lipid oxidation in meats when less than 150 ppm of nitrite was used. Meat flavor deterioration, therefore does not develop in cured meat. This observation might be attributed to any or a combination of the effects related to (a) stabilization of heme pigments (Zipser et al., 1964), (b) stabilization of membrane lipids (Zubillaga et al., 1984), (c) chelation of free metal ions and prooxidant catalysts (Shahidi and Hong, 1991), and (d) formation of nitrosylheme derivatives possessing antioxidant effects (Morrissey and Tichivangana, 1985; Kanner and Juven, 1980; Shahidi et al., 1988; Shahidi and Hong, 1991).

Heme proteins and their related products as well as transition metal ions have been implicated in meat lipid oxidation (e.g. Shahidi and Hong, 1991), as discussed earlier. As a result of heat processing, heme compounds in untreated meats are rapidly oxidized and produce ferrous and ferric ions. In cured meats nitric oxide, produced from nitrite, reacts with myoglobin and also combines with Fe^{2+} ions and thus suppresses MFD (see Table 5.3). The chelating properties of nitrite may be duplicated by the action of sequesterants such as Na_2EDTA and STPP. The antioxidant activity of some nitrosylheme compounds has been demonstrated by Ben Aziz et al. (1971) and Wettasinghe and Shahidi (1997). The exact mechanism by which this antioxidative effect is exerted remains elusive. Nonetheless, stabilization of iron in the protoheme compounds may be, at least in part, responsible.

Shahidi et al. (1987) and Wettasinghe and Shahidi (1997) have clearly shown that the preformed cooked cured-meat pigment (CCMP) (Fig. 5.5) acts as an antioxidant in a meat and in a β-carotene-linoleate model system, respectively. However, CCMP in the presence of ascorbates may form potent antioxidant combinations with evidence of strong synergism between the components (Shahidi, 1992). Furthermore, this synergistic effect was noted for both nitrosyl myoglobin and CCMP in which iron atoms are in the ferrous form and their coordination sites are occupied. Suggestions have been made that nitrosated iron porphyrin compounds act in the early stages of the reaction to neutralize substrate free radicals and thus inhibit oxidation of meat lipids (Kanner et al., 1984). During the curing process, S-nitrosocysteine (RSNO) may also be formed. This compound acts as a strong antioxidant as it has been reported to inhibit oxidation of turkey meat (Kanner and Juven, 1980). Effectiveness of nitrite in inhibition of MFD through other mechanisms has been further demonstrated in the literature (Ohshima et al., 1988).

Inhibition of oxidation of unsaturated fatty acids and formation of secondary carbonyl compounds as well as other degradation products of lipid hydroperoxides in cured products results in a drastic reduction in the number and concentration of volatiles in the cured meat products as compared with their uncured counterparts (e.g. see Table 5.4). In particular, formation of higher aldehydes is effectively suppressed (Cross and Ziegler, 1965; Shahidi, 1992). Cross and Ziegler (1965) have also reported that passage of volatiles of uncured beef and chicken through an acidic solution of 2,4-dinitrophenylhydrazine, which stripped their carbonyl compounds, resulted in the revelation of a flavor

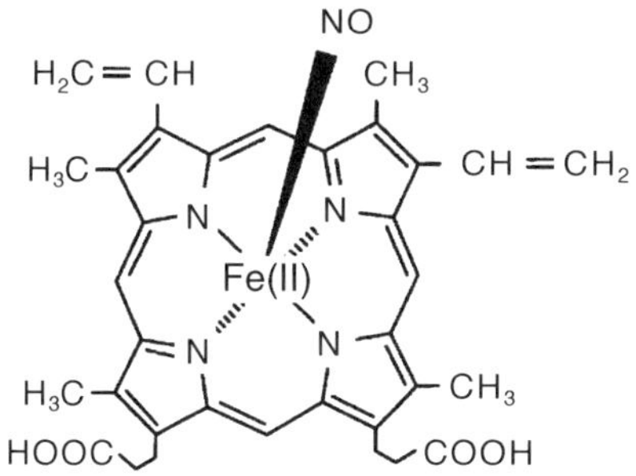

Fig. 5.5 The chemical structure of cooked cured-meat pigment (CCMP).

Table 5.4 Effect of curing on the relative concentration of major aldehydes in pork flavor volatiles

Aldehyde	Relative concentration	
	Uncured	Cured
Hexanal	100	7.0
Pentanal	31.3	0.5
Heptanal	3.8	<0.5
Octanal	3.6	<0.5
2-Octenal	2.6	–
Nonanal	8.8	0.5
2-Nonenal	1.0	–
Decanal	1.1	–
2-Undecenal	1.4	0.5
2,4-Decadienal	1.1	–

which was indistinguishable from that of cured ham. However, curing of sheep meat did not remove the specific 'sheep meat' odor of the product (Young *et al.*, 1994). This effect is thought to be due to the presence of 4-methylnonanoic and, to a lesser extent, 4-methyloctanoic acids in cooked samples of meat from mutton. Therefore, it appears that the basic flavor of cooked uncured meats without having been affected much by their lipid components is in most cases similar to their cured counterparts, but becomes masked with carbonyl products, formed via autoxidation. Lindsay (1997) has also thought that nitrophenols may have an influence on the characteristic aroma of cured meat products. However, this suggestion has not yet been supported by relevant experimental results.

Based on these findings, we have proposed that any other agent or combinations that could prevent lipid oxidation would, in principle and in general, duplicate the antioxidant role of nitrite in the curing process for species other than mutton, thus preventing MFD (Shahidi, 1991). To verify this simplistic view, we examined the effect of commonly used food antioxidants and chelators as well as other curing adjuncts such as salt, as discussed earlier, as

Table 5.5 Inhibition of formation of selected volatile oxidation products (%) in nitrite- and nitrite-free cured meats

Additive(s)	TBARS	Hexanal	2-pentyl-furan	Nonanal	2,4-deca-dienal
SA	48	50	80	21	63
STPP	70	62	40	34	57
SA+STPP	92	88	88	61	64
SA+STPP+CCMP	99	98	97	91	90
SA+STPP+CCMP +TBHQ	99	99	98	92	96
SA + NaNO$_2$	98	99	98	95	95

Abbreviations are: TBARS, 2-thiobarbituric acid reactive substances; SA, sodium ascorbate; STPP, sodium tripolyphosphate; CCMP, cooked cured-meat pigment; TBHQ, t-butylhydro quinone; and NaNO$_2$, sodium nitrite.

well as ascorbates and polyphosphates in prevention of MFD and formation of off flavors (Shahidi and Pegg, 1992). Surprisingly, curing adjuncts had a marked effect in controlling MFD as in cured meats. Furthermore, addition of CCMP to the mixture, with or without an auxiliary antioxidant such as TBHQ, resulted in a near-duplication of the action of nitrite in preventing lipid oxidation in cooked, stored meats (Table 5.5).

Having recognized the importance of curing adjuncts on the flavor of meats, we developed nitrite-free curing systems in which ascorbates, polyphosphates, CCMP and possibly an antimicrobial agent, with or without any additional agents were used to duplicate all properties of nitrite-cured meats. For further information on this topic, the reader is referred to publications by Shahidi and co-workers.

5.5 The evaluation of aroma compounds and flavor quality

Flavor quality of muscle foods may be evaluated by employing sensory techniques. Both desirable and undesirable aroma characteristics of products are easily recognized. A thorough discussion of sensory analysis of meat products as well as application of instrumental methods and chemical procedures is provided in the literature. While many indicators may be used for such evolutions, it is most appropriate to select the ones that show maximum variation under experimental conditions of heat processing, packaging and storage. In this connection, the use of the 2-thiobarbituric acid reactive substances (TBARS) test is commonplace. Although there are many shortcomings associated with this method of evaluation, nonetheless, it provides a realistic assessment of the relative oxidative stability of products (e.g. Shahidi and Wanasundara, 1996). This topic has been thoroughly discussed in the literature. Furthermore, hexanal, a prominant oxidation product of linoleic and other omega-6 fatty acids may be used effectively to evaluate the oxidative state and off flavor formation in cooked red meats and poultry. However, when fatty acids in the food are dominated by omega-3 fatty acids, such as eicosapentaenoic and docasahexaenoic acids, use of hexanal as an indicator might not be appropriate. Thus, propanal may be used as an indicator for evaluation of aroma deterioration of seafoods and marine oils.

5.6 Summary

The formation of desirable aromas in muscle foods has been found to arise, primarily, from the reaction of low-molecular-weight, water-soluble compounds and involvement of lipids in Maillard reaction. While lipids and their breakdown products contribute positively to the flavor quality of freshly prepared muscle foods, their further oxidation is recognized to cause MFD. Breakdown products of lipids such as aldehydes, ketones and related compounds have low threshold values and, even at relatively low concentration, may mask the desirable aroma of heterocyclic and acyclic heteroatomic compounds. Thus, control of oxidation in muscle foods is necessary in order to protect them against generation of undesirable aromas and flavor deterioration. Furthermore it is desirable to devise formulations which could accentuate the desirable meaty aroma in products which remain effective during prolonged storage.

5.7 References

BAILEY, M.E. 1988. Inhibition of warmed-over flavor with emphasis on Maillard reaction products. *Food Technol.* **42**(6): 123–132.

BAILEY, M.E., CLARKE, A.D., KIM, Y.S. and FERNANDO, L. 1997. Antioxidant properties of Maillard reaction products as meat flavor compounds. In *Natural Antioxidants: Chemistry, Health Effects and Application*. Edited by F. Shahidi, AOCS Press, Champaign, pp. 296–310.

BEN-AZIZ, A., GROSSMAN, S., ASCARELLI, I. and BUDOWSKI, P. 1971. Linoleate oxiation induced by lipoxygenase and heme proteins: a direct spectro photometric assay. *Anal. Biochem.* **34**: 88–100.

BROOKS, J., HAINER, R.B., MORAN, T. and PACE, J. 1940. The function of nitrate, nitrite and bacteria in the curing of bacon and hams. Department of Scientific and Industrial Research, Food Investigation Board. Special Report 49. His Majesty's Stationery Office, London. pp. 2–4.

BUCKLEY, D.J., MORRISSEY, P.A. and GRAY, J.I. 1996. Influence of dietary vitamin E on the oxidative stability and quality of pigment. *J. Anim. Sci.* **73**: 3122–3130.

CASSENS, R.G., GREASER, M.L., ITO, T. and LEE, M. 1979. Reactions of nitrite in meat. *Food Technol.* **33**(7): 46–57.

CHO, S.H. and RHEE, K.S. 1995. Calcium chloride effects on TBA values of cooked meat. *J. Food Lipids* **2**: 135–143.

CROSS, C.K. and ZIEGLER, P. 1965. A comparison of the volatile fractions from cured and uncured meat. *J. Food Sci.* **30**: 610–614.

DRUMM, T.D. and SPANIER, A.M. 1991. Changes in the content of lipid autoxidation and sulfur-containing compounds in cooked beef during storage. *J. Agric. Food Chem.* **39**: 336–343.

EISERICH, J.P. and SHIBAMOTO, T. 1994. Antioxidant activity and Maillard-derived sulfur-containing heterocyclic compounds. *J. Agric. Food Chem.* **42**: 1060–1063.

EISERICH, J.P., WONG, J.W. and SHIBAMOTO, T. 1995. Antioxidative activities of furan- and thiophenethiols measured in lipid peroxidation systems and by tyrosyl radical scavenging assay. *J. Agric. Food Chem.* **43**: 647–650.

FARMER, L.J. 1992. Meat flavor in *The Chemistry of Muscle-Based Foods*. Edited by D.E. Johnston, M.K. Knight and D.A. Ledward. Royal Society of Chemistry, London. pp. 169–182.

FARMER, L.J. and MOTTRAM, D.S. 1990. Interaction of lipid in the Maillard reaction between cysteine and ribose: the effect of a triglyceride and three phosphlipids on the volatile products. *J. Sci. Food Agric.* **53**: 505–525.

FOLCH, J., LEES, M. and SLOANE-STANLEY, G.H. 1957. A simple method for the isolation and purification of total lipids from animal tissues. *J. Biol. Chem.* 226: 497–509.

FOOLADI, M.H., PEARSON, A.M., COLEMAN, T.H. and MERKEL, R.A. 1979. The role of nitrite in preventing development of warmed-over flavour. *Food Chem.* **4**: 283–292.

FOX, J.B., JR. 1966. The chemistry of meat pigments. *J. Agric. Food Chem.* **14**: 207–210.

GRAY, J.I., GOMAA, E.A. and BUCKLEY, D.J. 1996. Oxidative quality and shelf-life of meats. *Meat Sci.* **41**: 8111–8123.

HADDEN, J.P., OCKERMAN, H.W., CAHILL, V.R., PARRETT, N.A. and BORTON, R.J. 1975. Influence of sodium nitrite on the chemical and organoleptic properties of comminuted pork. *J. Food Sci.* **40**: 626–630.

HO, C-T., OH, Y-C. and BAE-LEE, M. 1994. The flavour of pork. In *Flavor of Meat and Meat Products*. Edited by F. Shahidi, Blackie Academic and Professional, Glasgow, pp. 38–51.

HORNSTEIN, I. and CROWE, P.F. 1960. Flavour studies on beef and pork. *J. Agric. Food Chem.* **8**: 494–498.

HORNSTEIN, I. and CROWE, P.F. 1963. Meat Flavor: Lamb. *J. Agric. Food Chem.* **11**: 147–149.

IGENE, J.O., KING, J.A., PEARSON, A.M. and GRAY, J.I. 1979. Influence of haem pigments, nitrite, non-haem iron on development of warmed-over flavor in cooked meat. *J. Agric. Food Chem.* **27**: 838–842.

JOHNS, A.M., BIRKINSHAWA, L.H. and LEDWARD, D.A. 1989. Catalysts of lipid oxidation in meat products. *Meat Sci.* **25**: 209–220.

JOSEPHSON, D.B. 1991. Seafood. in *Volatile Compounds in Foods and Beverages*. Edited by H. Maarse, Marcel Dekker, Inc., New York, pp. 179–202.

KANNER, J. and JUVEN, B.J. 1980. S-nitrosocysteine as an antioxidant, colour developing and anticlostridial agent in comminuted turkey meat. *J. Food Sci.* **45**: 1105–1112.

KANNER, J. and KINSELLA, J.E. 1983. Lipid deterioration initiated by phagocytic cells in muscle foods: β-carotene destruction of a myoloperoxidase-hydrogen peroxide-liahide system. *J. Agric. Food Chem.* **31**: 370–376.

KANNER, J., HAREL, S. and JAFFE, R. 1991. Lipid peroxidation of muscle food as affected by NaCl. *J. Agric. Food Chem.* **39**: 1017–1021.

KANNER, J., HAREL, S., SHAGALOVICH, J. and BERMAN, S. 1984. Antioxidative effect of nitrite in cured meat products: nitric oxide-iron complexes of low molecular weight. *J. Agric. Food Chem.* **32**: 512–515.

LADIKOS, D. and LOUGOVOIS, V. 1990. Lipid oxidation in muscle foods: a review. *Food Chem.* **35**: 295–314.

LINDSAY, R.C. 1994. Flavor of fish. In: *Seafoods: Chemistry, Processing Technology and Quality*. Edited by F. Shahidi and J.R. Botta. Blackie Academic and Professional. Glasgow, pp. 75–84.

LINDSAY, R.C. 1997. *Off-flavors in muscle foods*. Abstract 7-3, Institute of Food Technologists Annual Meeting and Food Expo, 14–18 June.

MACDONALD, B., GRAY, J.I. and KAKUDA, Y. 1980. The role of nitrite in cured meat flavor. II. Chemical Analysis. *J. Food Sci.* **45**: 889–892.

MACDOUGALL, D.B., MOTTRAM, D.S. and RHODES, D.N. 1975. Contribution of nitrite and nitrate to the colour and flavour of cured meats. *J. Sci. Food Agric.* **26**: 1743–1754.

MORRISSEY, P.A. and TICHIVANGANA, J.Z. 1985. The antioxidant activities of nitrite and nitrosylmyoglobin in cooked meats. *Meat Sci.* **14**: 175–190.

MOTTRAM, D.S. 1983. The role of triglycerides and phospholipids in the aroma of cooked beef. *J. Sci. Food Agric.* **34**: 517-522.

MOTTRAM, D.S. 1991. Meat. In *Volatile compounds in Foods and Beverages*. Edited by Henk Maarse, Marcel Decker, Inc., New York, pp. 107–177.

MOTTRAM, D.S. and EDWARDS, E.A. 1983. The role of triglycerides and phospholipids in the aroma of cooked beef. *J. Sci. Food Agric.* **34**: 517–522.

OHSHIMA, T., WADA, S. and KOIZUMI, C. 1988. Influences of heme pigment, non-heme iron and nitrite on lipid oxidation in cooked mackerel meat. *Nippon Suisan Gakkaishi* **54**: 2165–2169.

PEARSON, A.M., LOWE, J.D. and SHORLAND, F.B. 1977. 'Warmed-over' flavor in meat, poultry and fish. *Adv. Food Res.* **23**: 1–74.

RHEE, K.S. 1988. Enzymatic and nonenzymatic catalysis of lipid oxidation in muscle foods. *Food Technol.* **42**(6): 127–138.

RHEE, K.S., SMITH, G.C. and TORRELL, R.N. 1983a. Effect of reduction and replacement of sodium chloride on rancidity development in raw and cooked ground pork. *J. Food Protec.* **46:** 578–581.

RHEE, K.S., TORRELL, R.N., QUINTANILLA, M. and VANDERZANT, C. 1983b. Effect of addition of chloride salts on rancidity of ground pork inoculated with a *Moraxalla* or *Lactobacillus* species. *J. Food Sci.* **48**: 302–303.

ROBINSON, M.E. 1924. Hemoglobin and methemoglobin as oxidative catalysts. *Biochem J.* **18**: 255–264.

SATO, K. and HEGARTY, G.R. 1971. Warmed-over flavor in cooked meats. *J. Food Sci.* **36**: 1098–1102.

SHAHIDI, F. 1989. Flavor of cooked meats. In *Flavor Chemistry: Trends and Developments*. Edited by R. Teranishi, R.G. Buttery and F. Shahidi. ACS Symposium Series 388. American Chemical Society, Washington, DC. pp. 188–201.

SHAHIDI, F. 1991. Developing alternative meat-curing systems. *Trends Food Sci. Technol.* 2219–222.

SHAHIDI, F. 1992. Prevention of lipid oxidation in muscle foods by nitrite and nitrite-free compositions. In *Lipid Oxidation in Food*. Edited by A.G. St. Angelo. ACS Symposium Series 222. American Chemical Society, Washington, D.C. pp. 161–182.

SHAHIDI, F. and HONG, C. 1991. Role of metal ions and heme pigments in autoxidation of heat-processed meat products. *Food Chem.* **42**: 339–346.

SHAHIDI, F. and PEGG, R.B. 1992. Nitrite-free meat curing systems: update and review; *Food Chem.* **43**: 185–191.

SHAHIDI, F. and WANASUNDARA, U. 1996. Methods for evaluations of the oxidative stability of lipid-containing foods. *Food Sci. Technol. Int.* **2**: 73–81.

SHAHIDI, F., RUBIN, L.J. and D'SOUZA, L.A. 1985. Meat flavor volatiles: a review of the composition, techniques and analysis, and sensory evaluation. *Crit. Rev. Food Sci. Nutr.* **24**: 141–243.

SHAHIDI, F., RUBIN, L.J. and WOOD, D.F. 1987. Control of lipid oxidation in cooked ground pork with antioxidants and dinitrosyl ferrohemochrome. *J. Food Sci.* **52**: 564–567.

SHAHIDI, F., RUBIN, L.J. and WOOD, D.F. 1988. Stabilization of meat lipids with nitrite-free curing mixtures. *Meat Sci.* **22**: 73–80.

SIMON, S., ELLIS, D.E., MACDONALD, B.D., MILLER, D.G., WALDMAN, R.C. and WESTERBERG, D.O. 1973. Influence of nitrite and nitriate curing ingredients on quality of packaged frankfurters. *J. Food Sci.* **38**: 919–923.

SPANIER, A.M., EDWARDS, J.V. and DUPUY, H.P. 1988. The warmed-over flavor process in beef: a study of meat proteins and peptides. *Food Technol.* **42**(6): 110–118.

ST. ANGELO, A.J., SPANIER, A.M. and BETT, K.L. 1992. Chemical and sensory evaluation of flavor of untreated and antioxidant-treated meat. In *Lipid Oxidation in Food*. Edited by A.J. St. Angelo. ACS Symposium Series 500. American Chemical Society, Washington, DC. pp. 140–160.

SVINGEN, B.A., BUEGE, J.A., O'NEIL, F.O. and AUST, S.O. 1979. The mechanism of NADPH-dependent lipid peroxidation: propagation of lipid peroxidation. *J. Biol. Chem.* **254**: 5892–5899.

TAKIGUCHI, A. 1989. Effect of NaCl on the oxidation and hydrolysis of lipids in salted sardine fillets during storage. *Nippon Suisan Gakkaishi*. **55**: 1649–1654.

TICHIVANGANA, J.Z. and MORRISEY, P.A. 1985. Metmyoglobin and inorganic metals as prooxidants in raw and cooked muscle systems. *Meat Sci.* **15**: 107–116.

TIMS, M.J. and WATTS, B.M. 1958. Protection of cooked meats with phosphates. *Food Technol.* **12**(5): 240–243.

TOLDRA, F., FLORES, M. and SANZ, Y. 1997. Dry-cured ham flavour: enzymatic generation and process influence. *Food Chem.* **59**: 523–530.

WETTASINGHE, M. and SHAHIDI, F. 1996. Oxidative stability of cooked comminuted lean pork as affected by alkali and alkali-earth halides. *J. Food Sci.* **61**: 1160–1164.

WETTASINGHE, M. and SHAHIDI, F. 1997. Antioxidant activity of preformed cooked cured-meat pigment in a β-carotene/linoleate model system. *Food Chem.* **58**: 203–207.

YOUNATHAN, M.T. and WATTS, B.M. 1959. Relationships of meat pigments to lipid oxidation. *Food Res.* **24**: 728–734.

YOUNG, O.A., REID, D.H., SMITH, M.E. and BRAGGINS, T.J. 1994. Sheep meat odour and flavour in *Flavour of Meat and Meat Products*. Edited by F. Shahidi. Blackie Academic & Professional, Glasgow, pp. 71–97.

ZIPSER, M.W., KWON, T.-W. and WATTS, B.M. 1964. Oxidative changes in cured and uncured frozen cooked pork. *J. Agric. Food Chem.* **12**: 105–109.

ZUBILLAGA, M.P., MAERKER, G. and FOGLIA, T.A. 1984. Antioxidant activity of sodium nitrite and meat. *J. Am. Oil Chem. Soc.* **61**: 772–776.

6
Modelling colour stability in meat

M. Jakobsen and G. Bertelsen, Royal Veterinary and Agricultural University, Frederiksberg

6.1 Introduction

Colour stability of meat products is influenced by a large number of factors some being of biochemical nature, some due to handling during the slaughter process and some due to packaging and storage conditions. This chapter focuses on modelling the effect of the external factors applied during packaging and storage. However, meat from different sources show different tendencies to undergo colour deterioration and this variation in internal factors influences the developed models. Therefore some consideration will also be given to discussing how internal factors like, e.g., muscle type and addition of nitrite in cured meat affect the models. Modelling can be used to identify the most important factors/interaction of factors affecting quality loss and to define critical levels of these factors, thereby forming the basis for proposing the optimal packaging and storage conditions or the best compromise, if several deteriorative reactions need to be considered. Caution in choosing the optimal packaging and storage conditions can largely improve the colour shelf life of meat products.

When packaging fresh meat products an elevated oxygen (O_2) partial pressure needs to be maintained to keep the meat pigment myoglobin in its oxygenated bright red state. Through modelling of a MAP system for fresh beef, the most critical external factors have been identified to be storage temperature and gas composition (Jakobsen and Bertelsen, 2000). Through modelling of a MAP system for cured meat products the most critical external factors have been identified to be a low availability of oxygen combined with exclusion of light (storage temperature was kept constant at 5°C) (Møller et al., 2002). However, low availability of oxygen is not solely ensured by reducing the residual oxygen level in the headspace during the packaging process. Other equally critical factors are a high product to headspace ratio and a packaging film of low oxygen transmission rate (OTR) of the packaging film (Møller *et al.*, 2002).

6.2 External factors affecting colour stability during packaging and storage

Modified atmosphere packed meat is a complex and dynamic system where several factors interact (Zhao *et al.*, 1994). Models can be used to describe how the initial package atmosphere changes over time and how these changes affect product quality and shelf-life. The dynamic changes in headspace gas composition during storage can be modelled as a function of gas transmission rates of the packaging material, initial gas composition, product and package geometry, gas absorption in the meat etc. Combined with the knowledge from models on quality changes in the meat as a function of packaging and storage conditions such as storage temperature, gas composition and light exposure, predictions of product shelf life can be made. Pfeiffer *et al.* (1999) developed simulations of how product shelf life changes with different packaging and storage conditions for a wide range of food products (primarily dry products). However, at present sufficient models for many quality deteriorative reactions are lacking and only a few attempts have been made to model chemical quality changes in meat products, in contrast to modelling of microbial shelf-life, where extensive work has been performed (McDonald and Sun, 1999).

6.3 Modelling dynamic changes in headspace composition

6.3.1 Permeability of the packaging film

Headspace gas composition changes dynamically due to several factors. Gas exchange with the environment occurs over the packaging film, if the partial pressure of a gas differs on the two sides of the film. The amount of gas that permeates over the film can be calculated from equation 6.1 (Robertson, 1993):

$$Q = P \cdot \Delta p \cdot t \cdot A \qquad \qquad 6.1$$

Q = the amount of gas that permeates over the film (cm^3)
P = the permeability of the packaging film ($cm^3/m^2/24h/atm$)
Δp = the difference in gas partial pressure on the two sides of the film (atm)
t = the storage time (24h)
A = the area of the package (m^2)

Different gases have different permeability through the same film. For conventional films, the permeability of CO_2 is generally 4–6 times greater than that of O_2 and 12–18 times larger than that of N_2. The permeability of a plastic film is roughly proportional to the thickness of the film. Doubling the film thickness approximately halves the permeability of the film.

124 Meat processing

Permeability is also influenced by storage temperature and relative humidity. Pfeiffer et al., (1999) found that the empirical equation 6.2 fitted well with literature data for oxygen permeability.

$$P(T, RH) = \exp(c_0 + c_1/T + c_2 \cdot RH + c_3 \cdot RH^2) \qquad 6.2$$

P = the permeability of the packaging film
T = storage temperature
RH = storage relative humidity
c = experimentally derived coefficient.

Gas exchange over the packaging film is of particular importance when the film needs to maintain a narrowly defined gas concentration as shown in the example in section 6.5, where the permeability of even small amounts of O_2 into a package containing a cured meat product is considered a critical packaging parameter.

6.3.2 Gas absorption in the meat

Headspace gas composition can also change due to gas absorption in the meat. Packaging in elevated levels of CO_2 can result in large amounts of CO_2 absorbed in the meat (Jakobsen and Bertelsen, 2002; Zhao et al., 1994) and thereby large changes compared to the initially applied gas composition. Absorption of O_2 and N_2 is negligible compared to the absorption of CO_2 (Jakobsen and Bertelsen, 2002). Models for CO_2 solubility as a function of packaging and storage parameters such as product to headspace volume ratio, temperature and initial CO_2 level were developed by Zhao et al., (1995) and Devlieghere et al., (1998). Fava and Piergiovanni (1992) developed models of CO_2 solubility as a function of different compositional parameters, a_w, pH, protein, fat and moisture content. As regards gas absorption, equilibrium is obtained during the first one or two days. Microbial or meat metabolism can also cause slight changes in gas composition by using O_2 and producing CO_2.

When it is understood how the gas atmosphere can change from the initially applied atmosphere under different packaging and storage conditions, this knowledge can be used to evaluate the effect on quality deteriorating reactions. Besides microbial growth, the primary concern when packaging both fresh and cured meat products is colour stability. The mechanisms of colour changes in fresh meat products and cured meat products are completely different as can be seen from the examples on modelling given in the following two sections.

6.4 Modelling in practice: fresh beef

Jakobsen and Bertelsen (2000) and Bro and Jakobsen (2002) modelled colour stability of fresh beef under different packaging and storage conditions. In all cases colour measurements were performed using a Minolta Colorimeter CR-

Modelling colour stability in meat   125

Table 6.1 Packaging and storage conditions used in the models developed in Jakobsen and Bertelsen (2000)

Modelling factor	Abbreviation	No. of levels	Setting of levels
Storage time (days)	Day	5	2, 4, 6, 8, 10
Temperature (°C)	Temp	3	2, 5, 8
O_2 level (%)	O2	5	20, 35, 50, 65, 80

300 (Minolta, Osaka, Japan) using the L, a, b coordinates (CIELAB colour system). Red colour was expressed as the a-value, the higher the a-value the redder the sample. When packaging fresh red meats elevated O_2 partial pressures are used to stabilise myoglobin in its bright red oxygenated form (oxymyoglobin). However, elevated O_2 levels may increase some deteriorative reactions, e.g., lipid oxidation. Consequently, it is interesting to investigate if a level of O_2 exists, which is acceptable when considering both colour stability and lipid oxidation. Jakobsen and Bertelsen (2000) investigated different packaging and storage conditions (Table 6.1) and developed a regression model/response surface model predicting the colour a-value as a function of storage time, storage temperature and O_2 level based on steaks of *Longissimus dorsi* muscles from four different animals. The resulting model (equation 6.3) contains main effects of the three factors plus two-way interactions and two squared effects. Interpretation of the model is best done by exploring the response surface plot (Fig. 6.1).

$$a\text{-value} = \beta_0 + \beta_1 \cdot \text{Day} + \beta_2 \cdot \text{Temp} + \beta_3 \cdot \text{O2} + \beta_4 \cdot \text{Day} \cdot \text{Temp} + \beta_5 \cdot \text{Day} \cdot \text{O2} + \beta_6 \cdot \text{Temp} \cdot \text{O2} + \beta_7 \cdot \text{Day} \cdot \text{Day} + \beta_8 \cdot \text{Temp} \cdot \text{Temp} \quad 6.3$$

Where β is a regression coefficient.

Figure 6.1 shows a response surface plot varying the two factors, temperature and O_2 level, while keeping the third factor, storage time, constant at day 6. Figure 6.1 reveals an interval of approximately 55–80% O_2, where the O_2 level does not affect the colour a-value significantly (the near horizontal lines in this interval mean that only temperature influences the a-value). The borders of this interval change a little depending on the setting of the day. But it is evident that the O_2 level can be reduced from the normally used 70–80% without adverse effect on colour shelf-life.

The complexity of the interactions/squared terms in equation 6.3 called for further search for adequate models. A novel approach called GEMANOVA (Generalized Multiplicative ANOVA) was therefore used in Bro and Jakobsen (2002). In this study the effect of different packaging and storage conditions (Table 6.2) on colour stability and lipid oxidation of steaks of *Longissimus dorsi* muscles from three different animals was investigated. The effect of light was evaluated as the time of exposure to a fluorescent tube commonly used for retail display (Philips Fluotone TLD 18W/830 yielding 1000 lux at the package surface for 0, 50 or 100% of the storage time). Even when considering only two

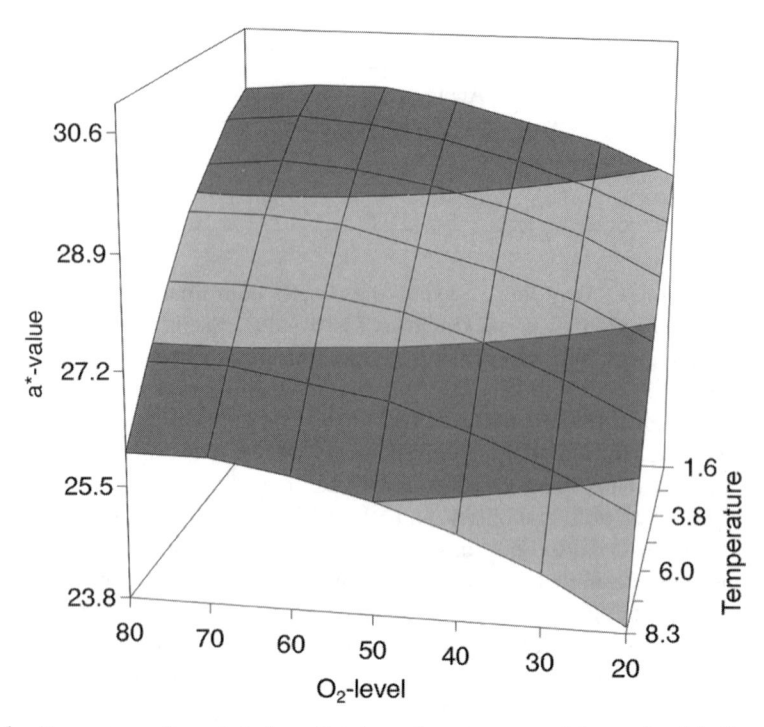

Fig. 6.1 Response surface plot of predicted a-values (average of four animals) after six days storage at different temperatures and different oxygen levels (Adapted from Jakobsen and Bertelsen, 2000).

Table 6.2 Packaging and storage conditions used in the models developed in Bro and Jakobsen (2002)

Modelling factor	Abbreviation	No. of levels	Setting of levels
Storage time (days)	Day	4	3, 7, 8, 10
Temperature (C)	Temp	3	2, 5, 8
Light exposure (%)	Light	3	0, 50, 100
O_2 level (%)	O2	3	40, 60, 80

factor interactions a traditional ANOVA model for the experiment in Table 6.2 would look like equation 6.4 (before removal of any insignificant effects).

$$a\text{-value} = \beta_0 + \beta_1 \cdot \text{Day} + \beta_2 \cdot \text{Temp} + \beta_3 \cdot \text{Light} + \beta_4 \cdot \text{O2}$$
$$+ \beta_5 \cdot \text{Day} \cdot \text{Temp} + \beta_6 \cdot \text{Day} \cdot \text{Light} + \beta_7 \cdot \text{Day} \cdot \text{O2} + \beta_8 \cdot \text{Temp} \cdot \text{Light}$$
$$+ \beta_9 \cdot \text{Temp} \cdot \text{O2} + \beta_{10} \cdot \text{Light} \cdot \text{O2} \quad\quad 6.4$$

Where β is a regression coefficient.

On the contrary, when applying the GEMANOVA model the interactions are modelled as one higher-order multiplicative effect, resulting in equation 6.5

(before removal of any insignificant effects). The interpretation of the GEMANOVA model is much more simple than the ANOVA model as is discussed in detail in Bro (1997) and Bro and Jakobsen (2002).

$$a\text{-value} = \text{Day} \cdot \text{Temp} \cdot \text{Light} \cdot \text{O2} \qquad 6.5$$

The resulting GEMANOVA model for the data in Table 6.2 can be written as equation 6.6, since the effect of the O_2 level is insignificant in the interval between 40–80% O_2 (Bro and Jakobsen 2002). The interaction term Day · Temp ·Light · c_{O2} describes deviations from the a-value on day 0 in a very simple way, and interpretation of the model parameters can be performed from Fig. 6.2.

$$a\text{-value} = a\text{-value}_0 + \text{Day} \cdot \text{Temp} \cdot \text{Light} \cdot c_{O2} \qquad 6.6$$

Where a-value$_0$ is the a-value at day 0 and c_{O2} is a constant.

For all settings of the factors the estimated response is simply the starting level of the a-value plus the product of the four effects seen from the ordinates in Fig. 6.2. The multiplicative term in equation 6.6 is 0 on day 0. Furthermore it is easily seen that:

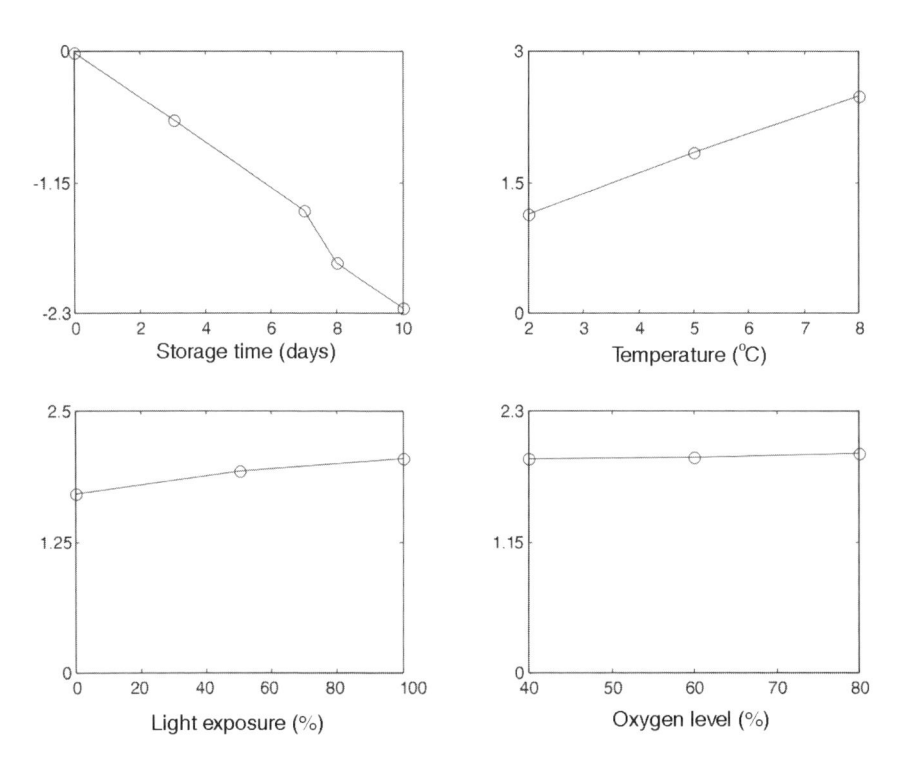

Fig. 6.2 Parameter levels for the interaction term (Day · Temp · Light · c_{O2}) in equation 6.6 (Adapted from Bro and Jakobsen, 2002).

128 Meat processing

- All changes in colour a-value are negative (colour becomes less red) compared to the starting colour. The change is calculated as the product of the four parameters Day, Temp, Light and O_2, which consist of one negative number (Day) and three positive numbers.
- The changes are relative and the effect of the individual factors can be interpreted individually. For example when going from 2°C to 8°C the Temp loading increases from 1.2 to 2.4, meaning that regardless of all other factors the decrease in a-value at 8°C is twice the decrease at 2°C.
- The effects of storage time and temperature are most important.
- The effect of light is minor, although an increase in time of exposure to light seems to result in a decreased colour a-value.
- The effect of the O_2 level is insignificant in the interval from 40–80% and is therefore contained in equation 6.6 as a constant.

The GEMANOVA model confirms the results from Jakobsen and Bertelsen (2000) by emphasising the importance of keeping a low storage temperature and showing no effect of O_2 level in the interval between approximately 40–80%. However, the interpretation of the model is much more simple, since the effect of each factor can be interpreted individually.

Likewise applying the GEMANOVA model on the data set in Table 6.1 results in equation 6.7 which is much more simple to interpret than equation 6.3.

$$a\text{-value} = a\text{-value}_0 + \text{Day} \cdot \text{Temp} \cdot O2 \qquad 6.7$$

Where a-value$_0$ is the a-value at day 0.

From Fig. 6.3 the effect of the individual factors can be interpreted, and the stable interval between 40–80% O_2 is evident.

It is rather surprising that 40% O_2 is sufficient to ensure the stability of the bright red meat colour, as an O_2 level of 70–80% is commonly used in the industry. The applied product to headspace volume ratio for the experiments in Tables 6.1 and 6.2 was approximately 1:9. The large headspace volume might cause only minor changes in headspace gas composition (oxygen partial pressure) to take place during storage. However, when packaging fresh meat products for retail sale, a large headspace volume is common. Furthermore, large amounts of oxygen have to permeate over the film or be used for meat/microbial metabolism before a noteworthy change in oxygen partial pressure takes place, and the meat colour becomes affected. A reduction in the applied oxygen level leaves the possibility of using more carbon dioxide or nitrogen in the package headspace.

6.5 Modelling in practice: cured ham

When packaging cured meat products it is important to keep the O_2 and light exposure at a minimum. Møller *et al.* (2002) investigated the colour stability of cured ham under different packaging and storage conditions according to Table

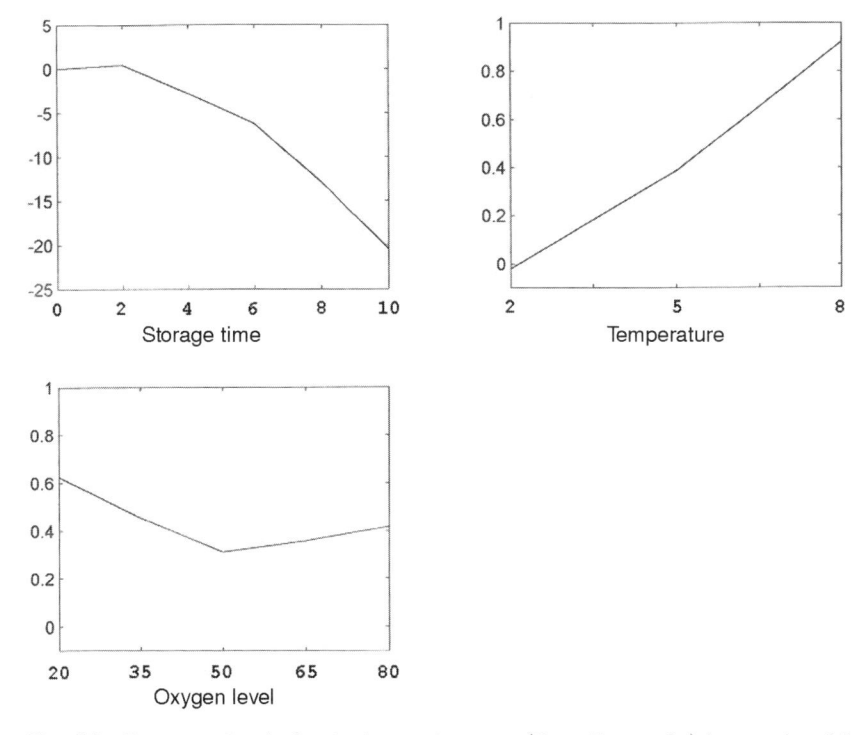

Fig. 6.3 Parameter levels for the interaction term (Day · Temp · O_2) in equation 6.7.

Table 6.3 Packaging and storage conditions used in the models developed in Møller *et al.*, (2002)

Modelling factor	Abbreviation	No. of levels	Setting of levels
Storage time (days)	Time	5	1, 3, 6, 9, 14
Residual O_2 level (%)	ResO2	3	0.1, 0.25, 0.5
Measured O_2 level (%)	MeasO2	–	Continuously
Oxygen Transmission Rate (ml/m²/24h/atm)	OTR	3	0.5, 10, 32
Volume ratio (product to headspace)	Vol	3	1:1, 1:3, 1:5
Light intensity (Lux)	Light	2	500, 1000
Nitrite content (ppm)	Nit	2	60, 150

6.3. Colour measurements were performed using a Minolta Colorimeter CR-300. The effect of light was evaluated as the light intensity from a fluorescent tube measured on the package surface. The resulting regression model (after removal of insignificant effects) considering only two-factor interactions is shown in equation 6.8.

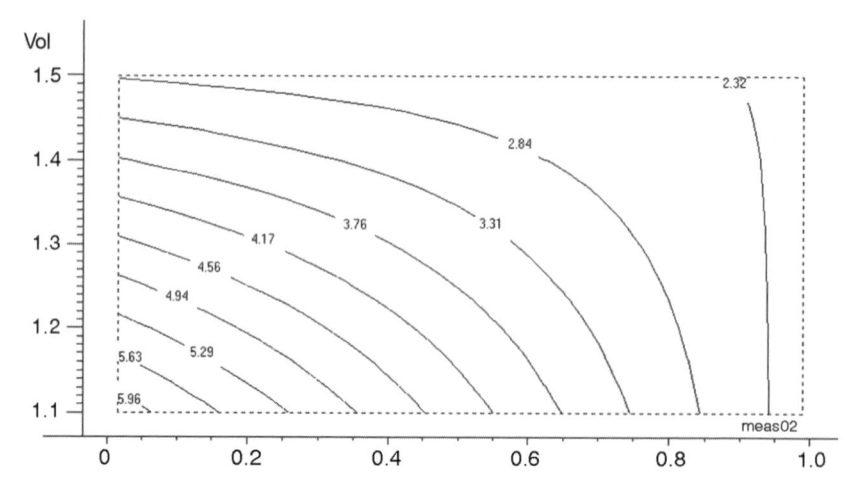

Fig. 6.4 Contour plot of the interaction effect between volume ratio (product:headspace) and measured O_2 level (%) after nine days storage (Adapted from Møller et al., 2002).

$$a\text{-value} = \beta_0 + \beta_1 \cdot \text{ResO2} + \beta_2 \cdot \text{Vol} + \beta_3 \cdot \text{Light} + \beta_4 \cdot \text{Nit} + \beta_5 \cdot \text{Time}$$
$$+ \beta_6 \cdot \text{MeasO2} + \beta_7 \cdot \text{ResO2} \cdot \text{Light} + \beta_8 \cdot \text{ResO2} \cdot \text{Time} + \beta_9 \cdot \text{ResO2} \cdot \text{MeasO2}$$
$$+ \beta_{10} \cdot \text{Vol} \cdot \text{MeasO2} + \beta_{11} \cdot \text{Light} \cdot \text{MeasO2} + \beta_{12} \cdot \text{Time} \cdot \text{MeasO2} \qquad 6.8$$

Where β is a regression coefficient.

As expected, the a-value decreases with increased time, increased residual O_2 level, increased OTR, increased light intensity and decreased nitrite content. However, the study also shows the importance of interactions between factors. Especially the interaction between O_2 level and product to headspace volume ratio is interesting. Normally, the focus is on the residual O_2 level (%) in the package and it is commonly overlooked that also the total amount of available oxygen molecules is important. The total amount of oxygen molecules available for colour deteriorative reactions is determined by the residual oxygen level after packaging, the meat to headspace volume ratio, and the amount of oxygen that permeates into the package headspace in combination. It is not sufficient to keep a low O_2 level in the package headspace. If the headspace volume is large there will still be plenty of oxygen molecules for colour deterioration.

Figure 6.4 shows a contour plot of the interaction between 'measured O_2 level' and 'volume ratio' (the remaining factors are fixed to the following settings: residual O_2 level $= 0.25\%$, illuminance $= 1000$ lux, nitrite $= 60$ ppm, storage time $= 9$ days). The a-value of the product for a given combination of 'measured O_2 level' and 'volume ratio' can be found from the plot by reading the a-value from the corresponding contour line, e.g., applying 0.10% 'measured O_2 level' and a 'volume ratio' of 1:1.3 results in an a-value of 5.6 after 9 days of storage. It appears that to maintain a high a-value, it is necessary to keep both the oxygen level and the headspace volume low (lower left corner of the plot), solely keeping the O_2 level low is not sufficient. The interaction between O_2

level and light intensity is also important. In order to maintain a good product colour it is necessary simultaneously to keep both the O_2 level and the light intensity low (Møller *et al.*, 2002).

6.6 Internal factors affecting colour stability

6.6.1 Fresh meat
Large variations in colour stability between meat of different origin can strongly influence the developed models. Different meat types show large variability due to different myoglobin content and different metabolic type (Renerre, 1990). The content of myoglobin is, e.g., largest in beef followed by lamb and pork, and the colour of pork is more stable than the other two species. Steaks of *Longissimus dorsi* muscles have high colour stability and steaks of *Semimembranosus* muscles have medium colour stability. Animals of different age, breed, feeding, etc., will also show differences in colour stability (Renerre, 1990; Jensen *et al.*, 1998)

It appears from Figs 6.5 and 6.6 that there is a huge variation in colour stability between meat from different sources. A range of intrinsic factors influence the oxidative balance in raw meat and thereby the colour stability of

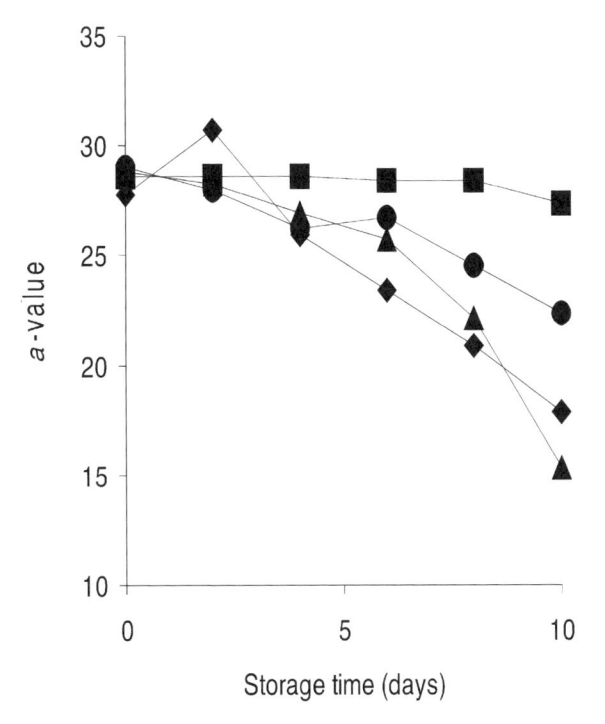

Fig. 6.5 Measured *a*-values for four different animals stored in 80% O_2 at 8°C (Adapted from Jakobsen and Bertelsen, 2000).

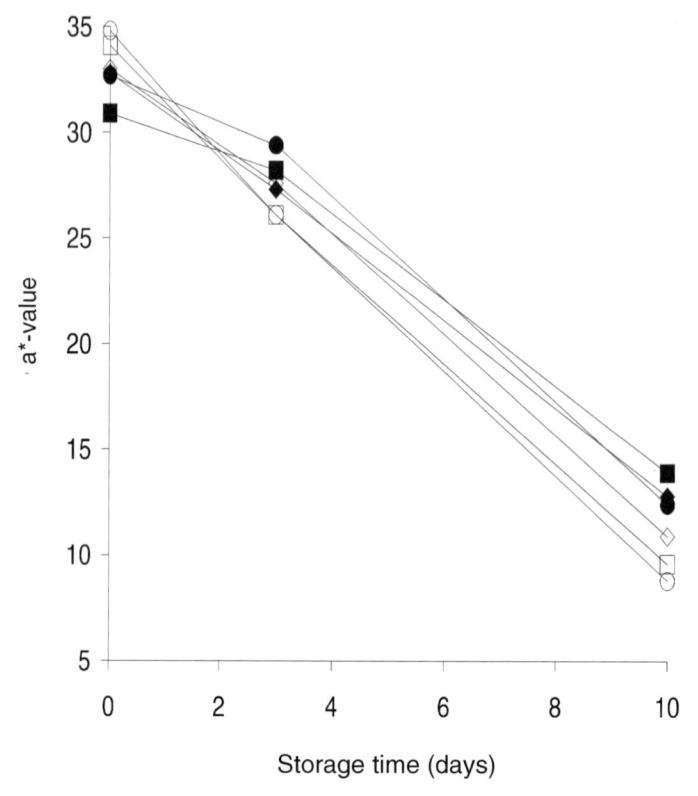

Fig. 6.6 Measured a-values for two muscle types from three different animals, stored in 80% O_2 at 8°C. *Longissismus dorsi* muscles (closed symbols) and *Semi-membranosus* muscles (open symbols).

the meat (Bertelsen *et al.*, 2000). Thus the oxidative stability of muscles is dependent on the composition, concentrations, and reactivity of (i) oxidation substrates (lipids, protein and pigments), (ii) oxidation catalysts (prooxidants such as transition metals and various enzymes) and (iii) antioxidants, e.g., vitamin E and various enzymes. For a review see Bertelsen *et al.* (2000).

Meat from different origins show different tendencies to undergo colour deterioration. It is therefore necessary to investigate meat from a large number of sources to be able to make general conclusions. Despite the large variations in colour stability of meat from the different animals and muscle types investigated in section 6.4 the pronounced effect of temperature and the constant interval of O_2 are common. Only the rate of colour deterioration differs.

6.6.2 Cured meat

A range of intrinsic factors affects the colour stability of nitrite cured meat products. The most important are the level of nitrite and the content of vitamin E

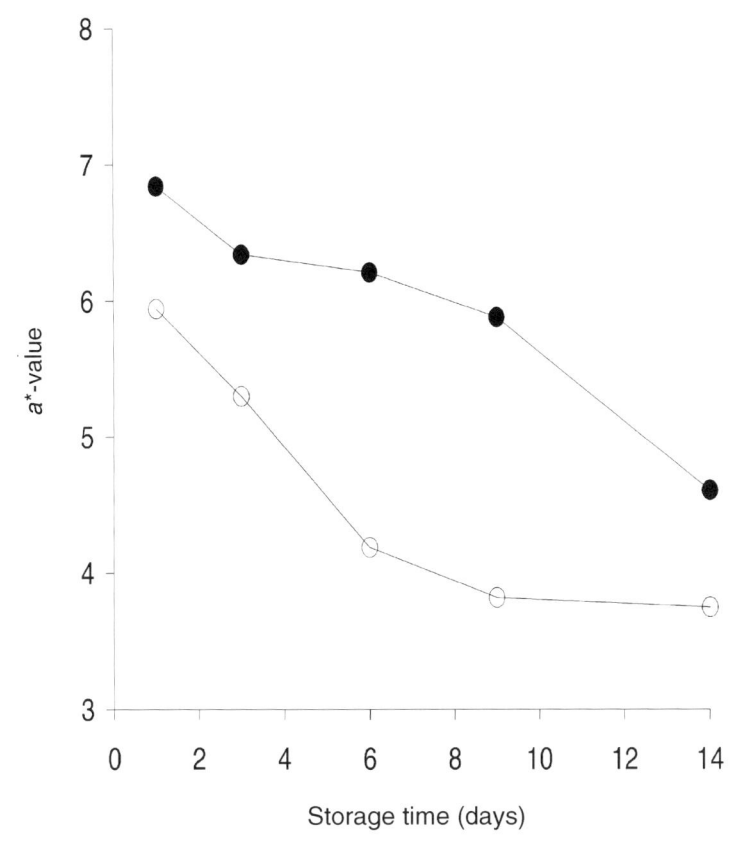

Fig. 6.7 Measured a-values for cured ham containing 150 ppm (filled circles) and 60 ppm (open circles) nitrite (each point is an average of 42 samples) (Data from Møller et al., 2002).

(Weber et al., 1999). Thus, optimum colour stability can be achieved only by using a multifactorial approach, where both intrinsic and extrinsic factors are considered (Bertelsen et al., 2000). From Fig. 6.7 the effect of the nitrite content on the rate of colour deterioration is evident. Increasing the nitrite content stabilises the colour. This result emphasises the necessity of investigating the specific product of interest in order to define critical levels of packaging and storage factors.

6.7 Validation of models

The examples in sections 6.4 and 6.5 clearly demonstrate the usefulness of modelling for identification of the most important factors/interaction of factors affecting colour deterioration. They also demonstrate how critical limits/ intervals of these factors can be identified. For fresh beef it is recognised that keeping a low storage temperature is the key parameter to obtain a long colour shelf-life. In addition a wide interval of oxygen partial pressure exists that result in optimal colour stability, leaving the possibility of optimising the gas composition with respect to other quality deteriorating reactions, e.g., lipid oxidation without compromising the colour stability. With respect to cured meat products it is important to realise that several factors influence the total amount of O_2 molecules available for oxidation.

Modelling of MAP systems shows great potential for optimising/tailoring storage and packaging parameters to maintain product quality, in this case a good meat colour stability (Jakobsen and Bertelsen, 2000; Lyijynen *et al.*, 1998; Pfeiffer *et al.*, 1999). As shown, modelling can be used to identify the most important factors affecting quality loss and to define critical levels of these factors. Multivariable experimental design is necessary to be able to investigate the large number of influencing factors on several levels as well as the interactions between factors. However, due to large biological differences between meat from different sources and to differences in handling and processing of the meat it is also important to recognise that internal factors have an effect on the developed models. The described models can be used to predict the general response of a meat product to changes in external factors, but not to predict the exact *a*-value for a certain piece of meat. That would require much more specific models (for each product type) and incorporation of knowledge on the internal factors into the models (Jakobsen and Bertelsen, 2000).

6.8 Future trends

The obvious tools for optimisation of product shelf life through controlling the packaging and storage conditions are computer simulations. Models of changes in headspace gas composition should be combined with models describing changes in the most important quality parameters. A computer program should be given inputs on:

- permeability of the different packaging films to be compared
- storage temperature
- relative humidity during storage
- gas composition measured after packaging
- the headspace and product volumes
- light conditions during storage.

By using computer simulations the time for reaching, e.g., an oxygen content critical for the colour stability of a given product can be predicted. Furthermore, demands for the permeability of the packaging material can be set, or the shelf life using a specific packaging film can be predicted. Such computer simulations were developed by Pfeiffer *et al.*, (1999) primarily for predicting quality changes, moisture gain and lipid oxidation in several dry products. The models described in the earlier sections are well suited for defining critical factors and levels for maintaining a good meat colour stability of fresh and cured meat products.

Computer simulations are an attractive supplement to storage experiments since it will not be necessary to test all combinations of the factors before the optimal packaging and storage conditions can be chosen considering both the product shelf life and minimisation of the packaging material.

6.9 References

BERTELSEN G, JAKOBSEN M, JUNCHER D, MØLLER J, KRÖGER-OHLSEN M, WEBER C and SKIBSTED L H (2000), Oxidation, shelf-life and stability of meat and meat products, *Proceedings 46th ICoMST*, Argentina.

BRO R (1997), PARAFAC. Tutorial and applications, *Chemometrics and Intelligent Laboratory Systems*, 38, 149–171.

BRO R and JAKOBSEN M (2002), Exploring complex interactions in designed data using GEMANOVA. Color changes in fresh beef during storage, *Journal of Chemometrics*, 16, 294–304.

DEVLIEGHERE F, DEBEVERE J and VAN IMPE J (1998), Concentration of carbon dioxide in the water-phase as a parameter to model the effect of a modified atmosphere on microorganisms. *International Journal of Food Microbiology*, 43, 105–113.

FAVA P and PIERGIOVANNI L (1992), Carbon dioxide solubility in foods packaged with modified atmosphere. II: Correlation with some chemical-physical characteristics and composition. *Industrie Alimentari*, 31, 424–430.

JAKOBSEN M and BERTELSEN G (2000), Colour stability and lipid oxidation of fresh beef. Development of a response surface model for predicting the effects of temperature, storage time, and modified atmosphere composition, *Meat Science*, 54, 49–57.

JAKOBSEN M and BERTELSEN G (2002), The use of CO_2 in packaging of fresh red meats and its effect on chemical quality changes in the meat: A review. *Journal of Muscle Foods*, 13, 143–168.

JENSEN C, FLENSTED-JENSEN M, SKIBSTED L H and BERTELSEN G (1998), Effects of rape seed oil, copper(B) sulphate and vitamin E on drip loss, colour and lipid oxidation of chilled pork chops packed in atmospheric air or in a high oxygen atmosphere, *Meat Science*, 50(2), 211–221.

LYIJYNEN T, HURME E, HEISKA K and AHVENAINEN R (1998), Towards precision food packaging by optimisation, *VTT research notes 1915*.

MCDONALD K and SUN D (1999), Predictive food microbiology for the meat industry: a review, *International Journal of Food Microbiology*, 52, 1–27.

MØLLER J K S, JAKOBSEN M, WEBER C J, MARTINUSSEN T, SKIBSTED L H and BERTELSEN G (2002), Optimization of colour stability of cured ham during packaging and retail display by a multifactorial design, *Meat Science*, In press.

PFEIFFER C, D'AUJOURD'HUI M, WALTER J, NUESSLI J and ESCHER F (1999), Optimizing food packaging and shelf life, *Food Technology*, 53, 6, 52–59.

RENERRE M (1990), Review: Factors involved in the discoloration of beef meat,

International Journal of Food Science and Technology, 25, 613–630.

ROBERTSON G L (1993), *Food packaging: Principles and practice*, Chapter 4, Marcel Dekker, Inc. New York.

WEBER J C, ALDANA L and BERTELSEN G (1999). Inhibition of oxidative processes by low level of residual oxygen in modified atmosphere packaged pre-cooked cured and non-cured meat products during chill storage. *Proceedings of the 11th IAPRI World Conference on Packaging*. Singapore. 114–120.

ZHAO Y, WELLS J H and MCMILLIN K W (1994), Applications of dynamic modified atmosphere packaging systems for fresh red meats: Review, *Journal of Muscle Foods*, 5, 299–328.

ZHAO Y, WELLS J H and MCMILLIN K W (1995), Dynamic changes of headspace gases in CO_2 and N_2 packaged fresh beef. *Journal of Food Science*, 60 (3), 571–557.

7

The fat content of meat and meat products

A. P. Moloney, Teagasc, Dunsany

7.1 Introduction

Fat is an essential component of meat for sensory perception of juiciness, flavour and texture. Fat in meat also supplies fatty acids that cannot be synthesised by humans. The perception of healthiness and sensory expectation are important quality criteria that influence the decision of a consumer to purchase a particular food product. Consumer perception of the influence of the content and composition of fat in meat, and in particular beef, for human health will be reviewed. Negative perceptions of beef as an excessively fat food have contributed to beef losing market share to competing meats and other protein sources throughout the developed world. Fresh meat production systems represent the combined and interacting effects of genotype, gender, age at slaughter and nutrition before slaughter, all of which can contribute to differences in the fat concentration of fresh meat. These influences will be briefly reviewed and it will be demonstrated that modern lean red meat can have an intramuscular fat concentration of 25–50 g/kg and can be considered a low-fat food.

The opportunities to exploit the diet of meat animals to produce flavoursome meat that has an increased concentration of conjugated linoleic acid (CLA), a compound that may protect against obesity, cancer and heart disease, a low fat concentration and a fatty acid profile more compatible with current human dietary recommendations will be illustrated. The chapter will end with a commentary on likely future trends in the fat content of meat and meat products including the possibility of meat being recognised as a functional food.

7.2 Fat and the consumer

The fat in meat supplies essential fatty acids and vitamins and plays an essential role in the sensory perception of juiciness, flavour and texture. Nevertheless, there is a perception among consumers and often the medical profession that red meat, in particular beef, is a food with an excessively high fat concentration. Further, meat fat is considered to cause a variety of human diseases, mainly because of the belief that it has a high proportion of saturated fatty acids (SFA) which raise blood cholesterol levels, a risk factor for cardiovascular disease (Department of Health, 1994). Historically, animal products were considered to be wholesome, versatile foods for humans and important for human health. From the 1960s however, attitudes towards fatty foods began to change and animal fats were linked to the onset of coronary heart disease and other diseases. In 1984, the Committee on Medical Aspects of Food Policy (COMA, 1984) published a report on diet in relation to cardiovascular disease and this report and its successors have become the basis of public policy in the UK. Among the evidence considered by the panel was the so-called direct evidence, i.e., the correlation found by Keys (1970) between mortality due to coronary heart disease and the proportion of dietary energy derived from saturated fat in seven countries, selected from 21 for which data were available. This conclusion has been subjected to increasing criticism (e.g. Blaxter and Webster, 1991) and Hegsted and Ausman (1988) demonstrated that if Keys' data for Japan and Italy are deleted then no statistically significant correlation remains between countries in the relationship between diet and coronary heart disease. Nevertheless, medical authorities world-wide recommend that energy intake from fat should not exceed 30–35%, that energy intake from SFA should not exceed 10% of total energy intake and that energy intake from monounsaturated fatty acids (MUFA) and polyunsaturated fatty acids (PUFA) should be approximately 16% and 7%, respectively, of energy intake. Furthermore, an increase in n-3 PUFA consumption such that the ratio of n-6:n-3 PUFA is <4:1 has also been recommended (Department of Health, 1994; Gibney, 1993, United States Department of Agriculture, 2000). It is likely that such recommendations contributed to the decline in red meat consumption and provided impetus to develop strategies to alter the total fat concentration and the fatty acid composition of meat fat to be more compatible with consumer requirements.

7.3 The fat content of meat

7.3.1 Fatness

In a recent briefing paper from the British Nutrition Foundation (1999), it was concluded that 'meat and meat products are an integral part of the UK diet and make a valuable contribution to nutritional intakes'. The fat content of meat varies with the choice of cut or meat product, the species of animal and the production system through which that animal has come (Sections 7.4 and 7.5).

Table 7.1 Total fat and fatty acid concentration of meat and meat products (g/100g) (adapted from Chan *et al.*, 1995, 1996)

	Fat	SFA*	MUFA*	PUFA*
Braising steak, lean, braised	9.7	4.1	4.1	0.6
Chicken breast, skinless, grilled	2.2	0.6	1.0	0.4
Lamb leg, lean, roasted, medium	9.4	3.8	3.9	0.6
Liver, pig, stewed	8.1	2.5	1.3	2.2
Minced beef, extra lean, stewed	8.7	3.8	3.8	0.3
Pork loin chops, lean, roasted	10.1	3.7	4.0	1.5
Turkey thigh, casseroled	7.5	2.5	2.7	1.8
Bacon, back, fat trimmed, grilled	12.3	4.6	5.2	1.6
Chicken korma	5.8	1.7	1.9	1.8
Chilli con carne, chilled/frozen, reheated	4.3	1.9	1.9	0.2
Ham, canned	4.5	1.6	2.0	0.4
Lamb kheema	14.5	3.8	5.3	4.2
Lamb kheema, reduced fat	9.7	3.4	3.6	1.8
Pork and beef sausages, grilled	20.3	7.5	9.1	2.2
Pork sausages, reduced fat, grilled	13.8	4.9	5.9	2.1
Salami	39.2	14.6	17.7	4.4
Steak and kidney pie, single crust	16.4	6.1	6.7	2.5
Turkey pie, single crust	10.3	4.5	3.7	1.5
Extra-lean meat				
Beef, extra-trimmed, lean	5.1			
Beef, mini-joint	3.4			
Beef, skirt steaks	6.9			
Lamb, extra-trimmed, lean	7.5			
Lamb, leg steaks	5.2			
Lamb, medallions	8.0			
Pork, extra-trimmed, lean	3.7			
Pork, escalopes	1.7			
Pork, medallions	3.9			

*SFA = saturated fatty acids, MUFA = monounsaturated fatty acids, PUFA = polyunsaturated fatty acids

Data on the fat content of a range of meat products are compiled and published in food composition tables by several agencies, world-wide, so selected examples only are shown in Table 7.1. Fat in meat can be present as intermuscular fat (between the muscles), intramuscular fat (or marbling, i.e., within the muscles) and subcutaneous fat (under the skin). Most of the fat is present as glycerol esters, but cholesterol, phospholipids and fatty acid esters are also present. Due largely to consumer preference for low-fat food products, the red meat industry began in the early 1980s to modify production systems to produce less fat in meat. The fat content of the carcass has decreased in Britain by over 30% for pork, making many pork cuts comparable with chicken, 15% for beef, and 10% for lamb, with further reductions anticipated for beef and lamb over the next 5–10 years (Higgs, 2000). These achievements are due to selective breeding and feeding practices designed to increase the carcass lean to

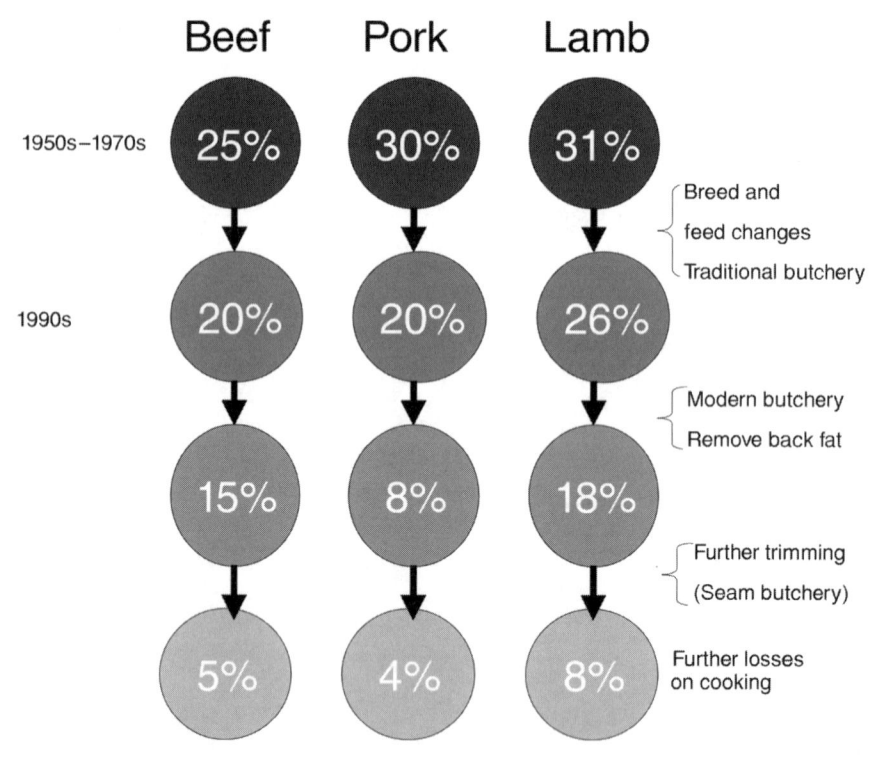

	Beef	Pork	Lamb	
1950s–1970s	25%	30%	31%	
	↓	↓	↓	Breed and feed changes / Traditional butchery
1990s	20%	20%	26%	
	↓	↓	↓	Modern butchery / Remove back fat
	15%	8%	18%	
	↓	↓	↓	Further trimming (Seam butchery)
	5%	4%	8%	Further losses on cooking

Fig. 7.1 Reduction in fat content of meat (After Higgs, 2000).

fat ratio; official carcass classification systems designed to favour leaner production; and modern butchery techniques (seaming out whole muscles, and trimming away all intermuscular fat). These changes are shown schematically in Fig. 7.1. Beef produced during our research had a marbling fat concentration in the order of 20–50 g/kg. This lean beef could therefore be considered a low-fat food, especially when compared to the fat concentration presented for meat in many tables of food composition (>70–100 g/kg).

The fat content of meat products can vary considerably, depending on the proportion of lean and fat from the original meat as well as the level of inclusion of other ingredients. Traditional meat products such as sausages, pastry-covered pies and salami are high in fat (up to 50%) but modern products include ready meals and prepared meats that can be low in fat (5%). The trend downwards in fat for red meat is reflected in the reduced fat content of a number of meat products, such as hams and sausages (Table 7.1). While reduced-fat meat products are now available, the potential for product development in this area has not been fully exploited.

While the meat industry continues to address consumer preference for lower-fat meat and meat products, the relationship between the fat content and sensory perception of meat must be considered. In some meat production systems (USA, beef), high intramuscular fat content (marbling) has been associated with superior tenderness, juiciness and overall satisfaction. Moreover, many of the flavour compounds of meat are contained in the fat component or are released due to chemical changes in the fat, alone and in interaction with the protein component, during ageing and cooking. The consensus of opinion now is that a decrease in intramuscular fat to 2% will not impair eating quality of meat (Wood, 1990). This has been observed with pork loin or lean chicken breast (Chizzolini et al., 1999). A reduction in intramuscular fat content to 2–5% with a relatively greater reduction in 'waste' fat depots such as subcutaneous and intermuscular would make a positive contribution to production efficiency and consumer health without negatively impacting on meat quality.

7.3.2 Fatty acids

The fatty acid composition of selected meat and meat products is also shown in Table 7.1. Most meats provide similar proportions of SFA and MUFA, making them an important source of the latter. While the ratio of PUFA to SFA is lower in ruminant tissue than non-ruminant tissue, SFA represents less than half of the total fatty acids of beef and of SFA, 30% are represented by stearic acid which has been shown to be neutral in its effect on plasma cholesterol in humans (Bonanome and Grundy, 1988). This indicates that the common reference to beef fat as very saturated is erroneous. Meat contributes to PUFA consumption, including docosahexaenoic acid and eicosapentaenoic acid of which there are few rich sources apart from oil-rich fish. Docosahexaenoic acid has an important role in the development of the central nervous system of the newborn while eicosapentaenoic acid is involved in blood clotting and the inflammatory response. Meat from ruminant animals in particular, but also monogastrics can be a source of CLA (Section 7.5.3). There is a growing body of evidence that a healthy diet which includes lean red meat can produce positive changes in lipid biochemistry. Blood cholesterol levels are increased by inclusion of fat, but not lean meat, in an otherwise low-fat diet. Equal amounts of lean beef, chicken and fish added to low fat, low saturated fat diets, similarly reduce plasma cholesterol and LDL-cholesterol levels in hypercholesterolaemic and normocholesterolaemic men and women.

7.4 Animal effects on the fat content and composition of meat

7.4.1 Fatness

An increase in fat deposition *per se* is generally accompanied by an increase in intramuscular fat concentration. The degree of fatness is determined by genotype, the weight of the carcass and how close the animal is to its ultimate mature size when slaughtered. In animal production systems which evolve to optimise economic efficiency, several of these factors may vary. The impact of these factors will be illustrated separately but likely interactions with the other

Table 7.2 Fat concentration (g/kg) of beef carcass and *longissimus dorsi* muscle (adapted from Keane, 1993)

	Carcass weight (kg)								
	280			340			400		
Sire Breed[a]	Sub. Fat[b]	IM. Fat[c]	LD[d]	Sub. Fat[b]	IM. Fat[c]	LD	Sub. Fat[b]	IM. Fat[c]	LD
Friesian	77	104	22	102	123	43	130	138	67
Hereford	91	114	26	121	134	50	155	150	77
MRI	76	102	22	98	120	46	123	135	73
Limousin	65	90	20	86	109	35	109	126	53
Blonde	53	74	16	72	90	25	92	105	37
Simmental	61	87	18	82	104	30	105	119	45
Belgian Blue	53	76	16	71	91	25	89	106	37
Charolais	55	80	16	74	96	28	95	110	43

[a] Mated to Friesian cows [b] Subcutaneous fat [c] Intermuscular fat [d] *Longissimus dorsi* muscle

factors and nutrition (Section 7.5) should also be considered. Across genotype, breeds that have light mature bodyweights mature earlier than those with a heavier mature bodyweight. Therefore at a constant time relative to birth, earlier maturing animals will be fatter than late maturing animals. This is illustrated by the data of Keane (1993) shown in Table 7.2 for different breeds of beef cattle. At 280 kg carcass weight, Friesians had 18% fat. The corresponding proportion for Herefords, an earlier maturing breed was 21%, and for the later maturing Limousin, Charolais and Belgian Blue breeds were 16%, 12%, and 13% respectively. As carcass weight increased, the proportions of fat increased and proportions of muscle and bone decreased. Compared with 280 kg, a 400 kg Friesian carcass had 27% fat. Corresponding proportions for Herefords and Charolais were 31% and 21%, respectively. Intramuscular lipid proportion increased with increasing carcass weight and did so more rapidly for earlier-maturing breeds. For example, over the carcass weight range 280–400 kg, lipid concentration increased by 51 g/kg for Herefords compared with an increase of only 21 g/kg for Belgian Blues. Similar lipid concentrations would be obtained from a Hereford carcass weighing 280 kg and a Charolais carcass weighing 340 kg. With respect to gender, heifers of the same breed grown together with steers achieved a similar carcass composition at a lighter carcass weight (267 vs. 326 kg) i.e., heifers are earlier maturing than steers (Keane, 1993). Similarly, castration of intact male animals renders the resulting castrates more early maturing with respect to body composition.

In general for any particular ratio, an increase in intake by a meat-producing animal will promote a higher growth rate and a fatter carcass (at a similar carcass weight) i.e., growth rate *per se* will increase fat deposition relative to protein deposition (Owens *et al.*, 1995). This seems to reflect some maximal rate of muscle growth which appears to be related to age as well as protein intake

(Bass *et al.*, 1990). However, there is some opportunity to decrease fatness by manipulating the growth path relatively close to slaughter. Thus Moloney *et al.* (2001) reported that compared to cattle finished on a grass silage and concentrate ration, feeding unsupplemented silage for 56 days followed by the same amount of concentrates offered *ad libitum* decreased internal fat weight and *longissimus dorsi* lipid concentration. Practical methods of decreasing fatness in farm animals have been reviewed (Bass *et al.*, 1990).

7.4.2 Fatty acids
Many comparisons of animal factors are confounded by differences in fatness. In general, increasing fatness results in greater unsaturation of lipid with the MUFA proportion increasing and SFA proportion decreasing (Duckett *et al.*, 1993). However, where corrections have been made for fatness, some differences in fatty acid composition due to genotype have been reported. Zambayashi *et al.* (1995) suggested that the Japanese Black breed of cattle has a genetic predisposition for producing lipids with higher MUFA concentrations than other breeds studied. The Wagyu beef breed is characterised by greater intramuscular than subcutaneous fat deposition and was found to have higher concentrations of MUFA and a higher MUFA:SFA ratio than other breeds in several studies (Xie *et al.*, 1996). Similarly for pigs, the Duroc breed, characterised by higher amounts of intramuscular fat relative to backfat, had higher intramuscular SFA and MUFA proportions and lower PUFA proportions than British Landrace pigs (Cameron and Enser, 1991). In both breeds, increasing intramuscular fat deposition caused a relatively greater increase in the MUFA proportion than the SFA proportion. Breed differences and effects of maturity or growth stage on the subcutaneous or intramuscular fatty acid composition of beef have been reviewed by de Smet *et al.* (2001). With regard to gender, fewer comparisons have been made but Malau-Aduli *et al.* (1998) reported phospholipid PUFA:SFA ratios of 0.27 and 0.54 for steer and heifers respectively, fed on pasture.

Specific breed differences in the n-6:n-3 PUFA ratio and in the concentration of longer chain n-3 PUFA that probably could not be attributed to differences in intramuscular fat concentration have also been reported. Choi *et al.* (2000) reported significantly higher proportions of $C_{18:3n-3}$ in neutral lipids and phospholipids and higher proportions of $C_{20:5n-3}$ and $C_{22:5n-3}$ in phospholipids of Welsh Black compared with Holstein Friesian cattle, resulting in a lower n-6:n-3 ratio in Welsh Black, whereas there were no differences in the concentrations of $C_{18:3n-3}$ and $C_{22:6n-3}$. The preferential deposition of n-3 PUFA was maintained on diets containing supplemental n-3 PUFA, indicating no breed by diet interaction. Itoh *et al.* (1999) found significant differences between Angus and Simmental cattle in the deposition of $C_{18:3n-3}$ and of the longer chain fatty acids, but breed by diet interactions were present for some of the fatty acids, making it difficult to interpret the breed effects. Despite the above, de Smet *et al.* (2001) concluded that much of the differences in fatty acid composition apparently due to genotype could be explained by variation in intramuscular fat concentration and that effects of genotype were in general much smaller than effects due to diet.

7.5 Dietary effects on the fat content and composition of meat

7.5.1 Fatness

When examining the effects of diet on the fat content of meat it is important to separate the direct effects of dietary ingredients from indirect effects of possible differences in energy intake on carcass weight and fatness. Carcass fatness in monogastrics and ruminants can be influenced by the energy and protein concentration in the diet. However, the extent to which the proportion of lean-to-fat is altered by dietary manipulations is limited without having a major impact on growth rate and feed efficiency. In pigs, restricting the energy intake by feeding a low-energy (low fat and/or high fibre) diet will reduce carcass fat deposition. Other nutrients must be supplied in sufficient amounts to support maximum lean tissue accretion or restriction in energy intake may result in protein being used for energy purposes. Feeding excess protein, i.e., excess essential amino acids, to pigs will result in a higher proportion of lean to fat in the carcass but the effect is primarily a result of energy restriction relative to protein. Changes in intramuscular fat concentration can also be accomplished by varying the energy and protein composition of the diet. Knowledge of energy and amino acid nutrition of ruminants is not as advanced as for monogastrics mainly due to pre-fermentation and transformation of dietary ingredients in the rumen of ruminants.

Nevertheless, there is a body of evidence that unwilted, extensively fermented grass silage can increase fatness relative to wilted silage/hay or non-silage-based diets and that starchy ingredients promote greater fatness than digestible fibre-based ingredients. In a grass silage-based ration, protein supplied in excess of requirement increased carcass fatness (Steen and Robson, 1995). Increasing propionate supply from the rumen by addition of sodium propionate to the diet decreased fat deposition (Moloney, 1998; 2002). Many studies have compared the effects of forage-based diets with concentrate (usually grain) -based diets. In a literature survey, Muir *et al.* (1998) found little difference in marbling between grain-fed and grass-fed beef at the same carcass weight. This conclusion is supported by French *et al.* (2000).

7.5.2 Fatty acids

Fatty acid deposition in monogastrics largely reflects dietary fatty acid composition (Wood and Enser, 1997, Rule *et al.*, 1995). This is illustrated by data from Verbeke *et al.* (1999) shown in Table 7.3. Intramuscular fat in pigs had high MUFA reflecting endogenous synthesis but incorporation of oilseeds in the diet can increase the PUFA:SFA and decrease the n-6:n-3 PUFA ratio. An

Table 7.3 Influence of fat sources on fatty acid composition of pig muscle (adapted from Verbeke *et al.*, 1999)

Fatty acids	Fat source				
	Tallow	Rapeseed	Soybeans	Linseed	Safflower
C18:1 (%)	44.06	46.55	38.75	38.17	48.8
C18:2 (%)	10.36	10.54	14.98	10.68	10.4
C18:3 (%)	0.52	1.11	1.04	4.41	1.40
PUFA:SFA	0.30	0.32	0.37	0.36	0.34
n-6:n-3	19.92	9.50	14.40	2.42	7.43

important difference between monogastrics and ruminants is that the long-chain n-3 PUFA, including eicosapentaenoic acid and docosahexaenoic acid, are not incorporated into triacylglycerols to any important extent in ruminants. They are incorporated mainly into membrane phospholipids and therefore, are found predominantly in muscle (Enser *et al.* 1996). This provides the opportunity to manipulate intramuscular fatty acid composition of ruminant meat without large increases in fatness *per se*.

In ruminants, dietary PUFA are hydrogenated to SFA but a proportion of dietary unsaturated fatty acids bypasses the rumen intact and is absorbed and deposited in body fat (Wood and Enser, 1997). Increasing the dietary supply of PUFA, particularly n-3 PUFA, is one strategy to increase PUFA concentrations in ruminant meat. In Table 7.4, inclusion of bruised whole linseed, a rich source of linolenic acid, resulted in 100% increase in concentration of linolenic acid in muscle while a linseed oil-fish oil treatment increased the marine n-3 PUFA concentrations (Scollan *et al.* 1997; 2000). The fatty acid composition of beef can be more efficiently modified by including in the diet, fatty acids that are protected from ruminal hydrogenation (Scott *et al.* 1971, Demeyer and Doreau, 1999). Scollan *et al.* (2001) showed that a protected lipid supplement markedly improved the PUFA:SFA ratio in muscle (Table 7.4).

Grass has higher PUFA and particularly higher n-3 PUFA, primarily as linolenic acid, than grain-based ruminant feeds. In general, grass-fed beef has higher concentrations of PUFA, particularly in the phospholipid fraction, than grain-fed beef (Griebenow *et al.*, 1997). An increase in the proportion of grass in the diet of finishing steers decreased the SFA concentration, increased the PUFA:SFA ratio, increased the n-3 PUFA concentration and decreased the n-6:n-3 PUFA ratio (French *et al.*, 2000). The n-3 PUFA detected in meat from the grass-fed cattle in this study was predominantly linolenic acid. The health benefits of n-3 PUFA from plant and marine (i.e. longer chain fatty acids) sources appear to differ. An expert workshop on this issue (de Deckere *et al.*, 1998) concluded that:

> there is incomplete but growing evidence that consumption of the plant n-3 PUFA, alpha-linolenic acid, reduces the risk of coronary heart disease. An intake of 2 g/d or 1% of energy of alpha-linolenic acid

Table 7.4 Influence of fat sources on the fatty acid composition (mg/100g tissue) of beef muscle (adapted from Scollan et al., 1997; 2000; 2001)

(i) Different sources of oil

Fatty acids	Control	Linseed	Fish oil	Linseed/fish oil	s.e.d.	Significance
C16:0	1029	1089	1305	1171	206.0	NS
C18:0	528	581	543	490	104.0	NS
C18:1	1209	1471	1260	1225	279.0	NS
C18:2	81	78	66	64	9.2	NS
C18:3	22	43	26	30	5.6	**
C20:4	23	21	14	17	1.5	***
C20:5	11	16	23	15	1.9	***
C22:6	2.2	2.4	4.6	4.9	0.52	***
Total fatty acids	3529	4222	4292	3973	741.0	NS
PUFA:SFA	0.07	0.07	0.05	0.05	0.011	NS
n-6:n-3	2.00	1.19	0.91	1.11	0.141	**

(ii) Oil protected from ruminal biohydrogenation

Fatty acids	Control	500g PLS†	1000g PLS†	s.e.d.	Significance
C16:0	986	843	598	117.8	*
C18:0	508	421	331	61.6	*
C18:1	1195	1144	759	177.0	*
C18:2	100	195	215	9.5	**
C18:3	23	46	46	4.3	**
C20:4	28	27	28	2.0	NS
C20:5	10	10	9	1.2	NS
C22:6	2	2	2	0.4	NS
Total fatty acids	3505	3260	2421	430.8	*
PUFA:SFA	0.06	0.19	0.28	0.029	**
n-6:n-3	4.6	4.4	4.7	0.48	NS

Notes
*=$p<0.05$, **=$p<0.01$, ***=$p<0.001$
†=protected lipid supplement.

appears prudent. The ratio of total n-3 over n-6 PUFA (linoleic acid) is not useful for characterising foods or diets because plant and marine n-3 PUFA show different effects, and because a decrease in n-6 PUFA intake does not produce the same effects as an increase in n-3 PUFA intake. Separate recommendations for alpha-linolenic acid, marine n-3 PUFA and linoleic acid are preferred.

Grass-fed beef can contribute to diets designed to achieve an increased consumption of n-3 PUFA.

7.5.3 Conjugated linoleic acid

Conjugated linoleic acid (CLA) refers to a mixture of positional and geometric isomers of linoleic acid (18:2 n-6). The *cis* 9, *trans* 11 form is believed to be the most common natural form of CLA with biological activity, but biological activity has been proposed for other isomers, especially the *trans* 10, *cis* 12 isomer. Conjugated linoleic acid has been shown to be an anticarcinogen, and to have antiatherogenic, immunomodulating, growth promoting, lean body mass-enhancing and antidiabetic properties (MacDonald, 2000; Whigham *et al.*, 2000). It is found in highest concentrations in fat from ruminant animals, where it is produced in the rumen as the first intermediate in the biohydrogenation of dietary linoleic acid. In the second step of the pathway, the conjugated diene is hydrogenated to *trans* 11 octadecenoic acid (trans-vaccinic acid) which is now thought to be a substrate for tissue synthesis of CLA via an enzymatic desaturation reaction. Because of the potential health benefits arising from CLA consumption, there is considerable research effort directed to increasing the CLA content of ruminant-derived food. Milk fat CLA concentrations are primarily influenced by linoleic acid supply to the rumen, by inclusion of grass in the diet and by the forage to concentrate ratio of the diet (Kelly *et al.*, 1998a, b; Jiang *et al.*, 1996).

For ruminant meat, an increase in the proportion of grass in the diet caused a linear increase in CLA concentration, while a grass silage/concentrate diet resulted in a lower CLA concentration than a grass-based diet with a similar forage to concentrate ratio (French *et al.*, 2000). Inclusion of sunflower oil in the supplementary concentrate to a silage-based diet also linearly increased muscle CLA (Noci *et al.*, 2002). Concentrations of CLA in Irish and Australian beef can be two to three times higher than those in United States beef. This presumably reflects the greater consumption of PUFA-rich pasture throughout the year by cattle in these countries. Protection of dietary CLA from ruminal biohydrogenation is being examined with equivocal results. Gassman *et al.* (2000) reported a 2.4 and 3.0-fold increase in intramuscular CLA concentration in rib and round muscle, respectively, in response to inclusion of 2.5% protected CLA in the diet of cattle. Dietary inclusion of CLA has also been shown to markedly increase the CLA concentration of pig muscle (from 0.09 to 0.55% total fatty acids in the study of Eggert *et al.*, 2001) and chicken muscle. In addition, there is evidence that the CLA concentration increases in foods that are cooked and/or otherwise processed.

7.6 Future trends

In the UK between 1989 and 1999, consumption of primary poultry meat and other meat products increased while that of carcass meat (beef, veal, mutton, lamb, and pork) declined (Robinson, 2001). Robinson (2001) considered this increase in consumption to be due mainly 'to increases in meat-based ready meats and takeaways eaten at home'. This clearly reflects consumer desire for convenience products and presents a major challenge to the non-poultry sector and to the prime red-meat sector, in particular. To regain market share, this sector of the meat industry will have to develop a more diverse range of products.

The decline in carcass meat consumption also reflects consumer preference for low-fat meat and meat products, guided by medical advice. Recent developments in decreasing the total fat concentration of meat and manipulating the fatty acid concentration have been illustrated in earlier sections of this chapter. There are clear opportunities to manipulate the animal component of fatness by integrating the various contributing factors, i.e., breed selection, use of non-castrated male animals, slaughtered young and fed appropriately, etc. It is likely that the nutrient requirements to optimise protein accretion while minimising adipose tissue accretion will be defined more precisely than at present. Current and future research will then focus on optimising both the supply of nutrients and the time when they are supplied (both diurally and during the lifetime of the animal) to allow the target animal to achieve its genetically determined body and muscle composition.

The data presented on factors affecting the fatty acid composition of the intramuscular fat of meat demonstrate that meat can be produced that has a fatty acid profile more compatible with current medical recommendations for human diet composition. In particular, red meats can now be produced that are low in fat, have a lower concentration of atherogenic SFA, higher MUFA and PUFA concentrations and lower n-6:n-3 PUFA ratio than was possible previously. Moreover, there is emerging evidence that palmitic acid and stearic acid do not contribute to coronary heart disease indicating that the perception that all SFA are 'unhealthy' is incorrect. Since meat intramuscular fat contains virtually no short chain SFA, only myristic acid contributes to elevated low density lipoprotein cholesterol, a risk factor for coronary heart disease and this typically represents just 3% of total fatty acids. Since n-6:n-3 PUFA ratio in meat is within the desirable range, future research will focus on enhancing the fatty acid profile of meat even further, in particular with the use of emerging rumen protection technology. Parallel research will be required to ensure adequate antioxidant protection in meat with an improved PUFA:SFA ratio.

The so-called 'lipid hypothesis' has guided medical advice for many years. This hypothesis is being increasingly criticised, particularly as on-going research on lipid metabolism in humans and its relationship to health and disease yields data inconsistent with this hypothesis. Moreover, the hypothesis that a low-fat, high-carbohydrate diet is best for preventing obesity, a disorder often considered to reflect fat consumption, is also being increasingly rejected. It is to be hoped that future medical guidelines will reflect the findings of recent dietary intervention-type research rather than be based predominantly on correlations arising from epidemiological, rather than retrospective-type studies. The discovery of CLA, together with the finding that ruminant fat is its primary natural source, is a positive advance for red meat, in particular. Clarification of the health-enhancing and disease-preventing properties of CLA in humans is the subject of extensive research world-wide. Reproduction in humans, of the observations made in laboratory animal models and tissue culture, will greatly add to the image of meat as a healthy food. Moreover, future meat could be considered a functional food, i.e., a food that has health benefits beyond basic nutrition. The American Dietetic Association has endorsed lean beef and lamb as functional foods (1999). Research on strategies to increase the CLA concentration in meat will continue and the observation that CLA is deposited in adipose tissue as well as the intramuscular phospholipid fraction will provide high CLA fat as a functional ingredient for healthy processed meats (see Jimenez–Colmenero *et al.*, 2001).

The positive contribution of meat to human diet and recognition that even at present, meat has a role in a healthy diet is not appreciated by consumers (Bruhn, 2000). Of consumers surveyed in the US (American Dietetic Association, 1997), fish was perceived as very healthful by 57% and poultry by 55%, but meat such as beef, pork, and lamb was seen as very healthful by only 13%; an additional 45% considered it somewhat healthful. Bruhn (2000) states that the potential for health-enhancing products is substantial. Enhanced nutritional components in animal products meet the preferences of consumers and health professionals. There is a need now and in the future for the meat industry to convey the positive nutritional contributions of meat products such as iron, zinc, 'healthy' fatty acids and CLA to both consumers and health professionals. The conclusion of Bruhn (2000) that 'communication is the key to correct consumer (and medical) myths and to increase awareness of new information or enhanced properties of healthful food' is most appropriate advice to all sectors of the meat industry for the future.

7.7 Sources of information and advice

Publications
GURR, M.I. (1999) *Lipids in Nutrition and Heath: A Reappraisal.* The Oily Press, Bridgwater.
ATKINS, R.C. *New Diet Revolution.* Vermilion, London.
ALLEN, P., DREELING, N., DESMOND, E., HUGHES, E., MULLEN, A.M. and TROY, D. (1999) *New Technologies in the Manufacture of Low Fat Meat Products. End of project report.* The National Food Centre, Dublin.
MEAT IN THE DIET. Briefing paper. The British Nutrition Foundation, London.
MCCANCE AND WIDDOWSON'S *The Composition of Foods.* The Royal Society of Chemistry and Ministry of Agriculture, Fisheries and Food (and supplements).
NEWMAN, C., HENCHION, M. and MATTHEWS, A. (2002) *Factors shaping expenditure on meat and prepared meals. End of project report.* The National Food Centre, Dublin.

Web sites
Enhancing the content of beneficial fatty acids in beef and improving meat quality for the consumer www.healthybeef.iger.bbsrc.ac.uk
Conjugated linoleic acid references http://www.wisc.edu/fri/clarefs.htm
British Nutrition Foundation http://www.nutrition.org.uk
Healthfinder – Gateway to reliable Consumer Health Information www.healthfinder.gov

U.S. Department of Agriculture, Nutrient data laboratory USDA. Nutrient database for standard reference, release 14 http://www.nal.usda.gov/fnic/foodcomp

7.8 References

AMERICAN DIETETIC ASSOCIATION (1997) *The ADA 1997 Nutrition Trends Survey.* Chicago, IL, USA.

AMERICAN DIETETIC ASSOCIATION (1999). Functional Foods – position of ADA. *Journal of the American Dietetic Association* 99, 1278–1285.

BASS, J.J., BUTLER-HOGG, B.W. and KIRTON, A.H. (1990) Practical methods of controlling fatness in farm animals pp. 145–200. In *Reducing fat in meat animals* (J.D. Wood and A.V. Fisher, editors). Elsevier Applied Science, London and New York.

BLAXTER, K.L. and WEBSTER, A.J.F. (1991) Animal production and food: real problems and paranoia. *Animal Production* 53, 261–269.

BONANOME, A. and GRUNDY, S.M. (1988) Effect of dietary stearic acid on plasma cholesterol and lipoprotein levels. *New England Journal of Medicine* 318, 1244–1248.

BRITISH NUTRITION FOUNDATION (1999) *Meat in the diet.* London: British Nutrition Foundation.

BRUHN, C. (2000) Consumer reactions, popular fancies, and scientific facts related to the healthfulness of meat products. *Proceedings of the American Society of Animal Science.* 1999. Available at http://www.asas.org/jas/symposia/proceedings/0910.pdf

CAMERON, N.D. and ENSER, M. (1991) Fatty acid composition of lipid in *longissimus dorsi* muscle of Duroc and British Landrace pigs and its relationship with lipid quality. *Meat Science* 29, 295–307.

CHAN, W., BROWN, J., LEE, S.M. and BUSS, D.H. (1995) Meat, Poultry and Game. Supplement to McCance and Widdowson's *The Composition of Foods.* The Royal Society of Chemistry and Ministry of Agriculture, Fisheries and Food.

CHAN, W., BROWN, J., CHURCH, S.M. and BUSS, D.H. (1996) Meat Products and Dishes. Supplement to McCance and Widdowson's *The Composition of Foods.* The Royal Society of Chemistry and Ministry of Agriculture, Fisheries and Food.

CHIZZOLINI, R., ZANARDI, E., DERIGONI, V. and GHIDINI, S. (1999) Calorific value and cholesterol content of normal and low-fat meat and meat products. *Trends in Food Science and Technology* 10, 119–128.

CHOI, N.J., ENSER, M., WOOD, J.D. and SCOLLEN, N.D. (2000) Effect of breed on the deposition in beef muscle and adipose tissue of dietary n-3 polyunsaturated fatty acids. *Animal Science* 71, 509–579.

COMMITTEE ON MEDICAL ASPECTS OF FOOD POLICY (1984) Diet and cardiovascular disease. *Department of Health and Social Security Report on health and*

social subjects No. 28. Her Majesty's Stationery Office, London.
DE DECKERE, E.A.M., KORVER, O., VERSCHUREN, P.M. and KATAN, M.B. (1998) Health aspects of fish and n-3 polyunsaturated fatty acids from plant and marine origin. *European Journal of Clinical Nutrition.* 52, 749–753.
DE SMET, S., RAES, K. and DEMEYER, D. (2001) Meat fatty acid composition as affected by genetics. *Proceedings Belgian Association for Meat Science and Technology,* 44–53.
DEMEYER, D. and DOREAU, M. (1999) Targets and procedures for altering ruminant meat and milk lipids. *Proceedings of the Nutrition Society* 58, 593–607.
DEPARTMENT OF HEALTH (1994) Nutritional Aspects of Cardiovascular Disease. Report on Health and Social Subjects no. 46, London: H.M. Stationery Office.
DUCKETT, S.K., WAGNER, D.G., YATES, L.D., DOLEZAL, M.G. and MAY, S.G. (1993) Effects of time on feed on beef nutrient composition. *Journal of Animal Science* 71, 2079–2088.
EGGERT, J.M., BELURY, M.A., KEMPA-STECZKO, A., MILLS, S.E. and SCHINCKEL, A.P. (2001) Effects of conjugated linoleic acid on the belly firmness and fatty acid composition of genetically lean pigs. *Journal of Animal Science* 79, 2866–2872.
ENSER, M., HALLETT, K., HEWETT, B., FURSEY, G.A.J. and WOOD, J.D. (1996) Fatty acid content and composition of English beef, lamb and pork at retail. *Meat Science* 44, 443–458.
FRENCH, P., STANTON, C., LAWLESS, F., O'RIORDAN, E.G., MONAHAN, F.J., CAFFREY, P.J. and MOLONEY, A.P. (2000) Fatty acid composition, including conjugated linoleic acid, of intramuscular fat from steers offered grazed grass, grass silage or concentrate-based diets. *Journal of Animal Science* 78, 2849–2855.
GASSMAN, K.J., BEITZ, D.C., PASSISH, F.C. and TRENKLE, A. (2000) Effects of feeding calcium salts of CLA to finishing steers. *Journal of Animal Science* 78 (Suppl 1), 275.
GIBNEY, M.J. (1993) Fat in animal products: facts and perceptions. In *Safety and Quality of Food from Animals. British Society of Animal Production Occasional Publication no. 17,* pp. 57–61 (J.D. Wood and T.L.J.. Lawrence, editors). Edinburgh: BSAP.
GRIEBENOW, R.L., MARTZ, F.A. and MORROW, R.E. (1997) Forage-based beef finishing systems: A review. *Journal of Production Agriculture* 10, 84–91.
HEGSTED, D.M. and AUSMAN, L.M. (1988) Diet, alcohol and coronary heart disease in men. *Journal of Nutrition* 118, 1184–1189.
HIGGS, J.D. (2000) Leaner meat: An overview of the compositional changes in red meat over the last 20 years and how these have been achieved. *Food Science and Technology Today* 14, 22–26.
ITOH, M., JOHNSON, C.B., COSGROVE, G.P., MUIR, P.D. and PURCHAS, R.W. (1999) Intramuscular fatty acid composition of neutral and polar lipids for heavy-weight Angus and Simmental steers finished on pasture or grain. *Journal*

of the Science of Food and Agriculture 79, 821–827.
JIANG, J., BJOERCK, L., FONDEN, R. and EMANUELSON, M. (1996) Occurrence of conjugated *cis*-9, *trans*-11-octadecadienoic acid in bovine milk: effects of feed and dietary regimen. *Journal of Dairy Science* 79, 438–445.
JIMENEZ-COLMENERO, F., CARBALLO, J. and COFRADES, S. (2001) Healthier meat and meat products: their role as functional foods. *Meat Science* 59, 5–13.
KEANE, M.G. (1993) Relative tissue growth patterns and carcass composition in beef cattle. *Irish Grassland and Animal Production Association Journal* 27, 64–77.
KELLY, M.L., BERRY, J.R., DWYER, D.A., GRIINARI, J.M., CHOUINARD, P.Y., VAN AMBURGH, M.E. and BAUMAN, D.E. (1998a) Dietary fatty acid sources affect conjugated linoleic acid concentrations in milk from lactating dairy cows. *Journal of Nutrition* 128, 881–885.
KELLY, M.L., KOLVER, E.F., BAUMAN, D.E., VAN AMBURGH, M.E. and MULLER, L.D. (1998b) Effect of intake of pasture on concentrations of conjugated linoleic acid in milk of lactating cows. *Journal of Dairy Science* 81, 1630–1636.
KEYS, A. (1970) Coronary heart disease in seven countries. *Circulation* 41:Suppl. 1.
MACDONALD, H.B. (2000) Conjugated linoleic acid and disease prevention: A review of current knowledge. *Journal of the American College of Nutrition* 19, 111s–118s.
MALAU-ADULI, A.E.O., SIEBERT, B.D., BETTEMA, C.D.K. and PITCHFORD, W.S. (1998) Breed comparison of the fatty acid composition of muscle phospholipids in Jersey and Limousin cattle. *Journal of Animal Science* 76, 766–773.
MINISTRY OF AGRICULTURE, FISHERIES AND FOOD (MAFF) (1991; 1998) *National Food Survey* 1990, 1997. London: HMSO.
MOLONEY, A.P. (1998) Growth and carcass composition in sheep offered isoenergetic rations which resulted in different concentrations of ruminal metabolites. *Livestock Production Science* 56, 157–164.
MOLONEY, A.P. (2002) Growth and carcass composition in sheep offered low or high metabolisable protein supplemented with sodium propionate. *Proceedings Agricultural Research Forum*, 20–21.
MOLONEY, A.P., KEANE, M.G., DUNNE, P.G., MOONEY, M.T. and TROY, D.J. (2001) Delayed concentrate feeding in a grass silage/concentrate beef finishing system: Effects on fat colour and meat quality. *Proceedings 47th ICOMST*, 188–189.
MUIR, P.D., DEAKER, J.M. and BROWN, M.D. (1998) Effect of forage- and grain-based systems on beef quality: a review. *New Zealand Journal of Agricultural Research* 41, 623–635.
NOCI, F., O'KIELY, P., MONAHAN, F.J., STANTON, C. and MOLONEY, A.P. (2002) The effect of sunflower oil in the supplementary concentrate on the conjugated linoleic acid (CLA) concentration in muscle from heifers fed a range of silages. *Proceedings Agricultural Research Forum* 18–19.
OWENS, F.N., GILL, D.R., SECRIST, D.S. and COLEMAN, S.W. (1995) Review of some aspects of growth and development of feedlot cattle. *Journal of Animal*

Science 73, 3152–3172.
ROBINSON, F. (2001) The nutritional contribution of meat to the British diet: recent trends and analyses. *Nutrition Bulletin* 26, 283–293.
RULE, D.C., SMITH, S.B. and ROMANS, J.R. (1995) Fatty acid composition of muscle and adipose tissue of meat animals. In: *The Biology of Fat in Meat Animals*, pp. 144–165 (S.B. Smith and D.R. Smith, editors) Illinois: American Society of Animal Science.
SCOLLAN, N.D., FISHER, W.J., DAVIES, D.W.R., FISHER, A.V., ENSER, M. and WOOD, J.D. (1997) Manipulating the fatty acid composition of muscle in beef cattle. *Proceedings of the British Society of Animal Science*, 20.
SCOLLAN, N.D., CHOI, N.J., KURT, E., FISHER, A.V., ENSER, M. and WOOD, J.D. (2000) Manipulating the fatty acid composition of muscle and adipose tissue in beef cattle. *British Journal of Nutrition* 85, 115–124.
SCOLLAN, N., GULATI, S., WOOD, J. and ENSER, M. (2001) The effects of including ruminally protected lipid in the diet of Charolais steers on animal performance, carcass quality and the fatty acid composition of longissimus dorsi muscle. *Proceedings 47th ICOMST*, p. 5.
SCOTT, T.W., COOK, L.J. and MILLS, S.C. (1971) Protection of dietary polyunsaturated fatty acids against microbial hydrogenation in ruminants. *Journal of the American Oil Chemists Society* 48, 358–364.
STEEN, R.V.J. and ROBSON, A. (1995) Effects of forage to concentrate ratio in the diet and protein intake on the performance and carcass composition of beef heifers. *Journal of Agricultural Science*, 125, 125–135.
UNITED STATES DEPARTMENT OF AGRICULTURE (USDA) (2000) Dietary Guidelines Advisory Committee, Dietary Guidelines for Americans.
VERBEKE, W., OECKEL, M.J. VAN, WARRANTS, N., VIAENE, J. and BOUCQUE, CH.V. (1999) Consumer perception, facts and possibilities to improve acceptability of health and sensory characteristics of pork. *Meat Science* 53, 77–99.
WHIGHAM, L.D., COOK, M.E. and ATKINSON, R.L. (2000) Conjugated linoleic acid: Implications for human health. *Pharmacological Research* 42, 503–510.
WOOD, J.D. (1990) Consequences for meat quality of reducing carcass fatness. In: *Reducing Fat in Meat Animals* pp. 344–397 (J.D. Wood and A.V. Fisher, editors). London: Elsevier Applied Science.
WOOD, J.D. and ENSER, M. (1997) Factors influencing fatty acids in meat and the role of antioxidants in improving meat quality. *British Journal of Nutrition* 78, S49–S60.
XIE, Y.R., BUSBOOM, J.R., GASKINS, C.T., JOHNSON, K.A., REEVES, J.J., WRIGHT, R.W. and CRONRATH, J.D. (1996) Effects of breed and sire on carcass characteristics and fatty acid profiles and crossbred Wagyu and Angus steers. *Meat Science* 43, 167–177.
ZAMBAYASHI, M., NISHIMURA, K., LUNT, D.K. and SMITH, S.B. (1995) Effect of breed type and sex on the fatty acid composition of subcutaneous and intramuscular lipids of finishing steers and heifers. *Journal of Animal Science* 73, 3325–3332.

Part II

Measuring quality

8
Quality indicators for raw meat

M. D. Aaslyng, Danish Meat Research Institute, Roskilde

8.1 Introduction

To be able to define the quality indicators for raw meat it is neccesary first to define what quality of meat is. The quality of raw meat can be defined as the suitability of meat for use in a specified product. If the meat is well suited for the product it is intended for, then the meat quality is defined as good. If the meat is less suitable for the product, then the meat quality is defined as poor.

The attributes of meat that determine the quality thus depend on the use for which the meat is intended. Quality can be defined as **technological quality**, describing meat for further processing like salting, curing, etc., or as **fresh meat eating quality** that describes meat for fresh meat consumption, and which includes all traits registered with our senses, both appearance, flavour and texture. The quality indicators for the two quality definitions to some degree overlap but some differences also exist. Other quality descriptions, however, exist like ethical quality and health quality but they will not be covered in this chapter.

8.2 Technological quality

In the processing of meat the yield is the main quality parameter as it determines the amount of available product for sale and is therefore of direct economic importance. The sensory quality of the processed product has an indirect economic importance as it might influence the amount of sold product, especially how often a consumer buys the same product again. Quality indicators in the raw meat that can predict the yield of the processed meat are especially pH and water-holding capacity, whereas the sensory quality of the processed meat can also be influenced by the colour, the meat/fat distribution and the fat quality in the raw meat.

8.2.1 pH and water-holding capacity

The yield of cured cooked products depends on the pH of the meat. The higher pH the higher yield (Müller, 1991). In an investigation of hams produced without phosphates the correlation between pH and total yield was around 0.4 (Table 8.1, unpublished data, Hviid, 2002 pers. comm.). Production of this type of ham includes a curing and a cooking step. An investigation has shown that the pH especially influences the yield by altering the cooking loss. The hams were from non-carrier and carrier pigs of the RN^--gene and were produced without phosphates. Pigs carrying the RN^--gene are recognised by having high glycogen content, high drip loss, and low pH. The curing yield was independent of genetic background and thereby of pH, whereas the carriers of the RN^--gene had a significantly higher cooking loss (Andersson et al., 1997). Not only the yield but also the colour can be influenced by pH. In cured bacon product the pH influenced the colour of the bacon. A pH below 5.3 resulted in bacon of an uneven color whereas pH above 5.7 gave a beefy, glazy appearance (Barton, 1971).

If the water-holding capacity is extreme like PSE (pale, soft exudative) or DFD (dark, firm, dry) meat it is also reflected in the yield PSE meat giving a lower yield and DFD meat giving a higher yield compared to normal meat (Barton Gade, 1984). Also dried hams with a 12-month season time show a higher weight loss if the raw ham was PSE (Maggi and Oddi, 1988). The relationship between drip loss and yield in production of hams from meat having a water-holding capacity in a more normal range is however relatively low as can be seen from Table 8.1 with a correlation below 0.3.

The water-holding capacity and pH is very dependent on the pre- and post-slaughter metabolism. A fast pH decline early post-mortem has often been shown to result in low water-holding capacity. In these works the pH has been determined 45 minute post-mortem and a low pH has been said to be due to a fast pH decline. Recent investigations where the pH was determined already 1 minute after debleeding have shown that it might not be the rate of the pH

Table 8.1 Correlation between meat quality and cooking loss/total yield in an investigation of hams produced as Jambon Superieur (without phosphates) of meat from LYD, LYDH and LYP pigs all free of the halothane gene and the RN^--gene ($n = 315$) (Hviid, 2002, unpubl. data).

	Cooking loss	Total yield
pH	−0.437	0.394
Drip loss	0.277	−0.286

decline but the time of acceleration of the pH decline which is crucial (Henckel *et al.*, 1999; Støier *et al.*, 2001). In animals with a high pre-slaughter stress the pH of the muscles was lower already at time of slaughter compared to less stressed animals (Henckel *et al.* 1999). When the concentration of creatinephosphate at the time of slaughter is high the anaerobic glycogen degradation and thereby the lactate production will occur later and the acceleration of the pH decline will therefore be delayed (Støier *et al.* 2001). Both investigations find, however, that the rate of pH decline was equally independent of pre-slaughter stress. pH_u is independent of the pH decline. Instead the limiting factor for the ultimate pH is the concentration of glycogen in the muscles (Henckel *et al.*, 1997). If the pigs are exhausted at slaughter the low glycogen content will result in a high ultimate pH and the meat will be DFD. In a more normal pH interval a linear correlation between pH_u and the glycogen content at slaughter has been seen at concentrations lower than about 53 mmol/kg. With more glycogen no correlation was seen (Henckel, Pers. Comm.).

pH is an easily determined raw meat quality attribute. It can be determined by direct measurement with an electrode in the meat, or a sample can be homogenised before determination. The water-holding capacity can be determined by three fundamentally different principles: (i) using external forces to drive out the water like the filter press method and by centrifugation, (ii) by letting the water drip out of the raw meat in a standardised way over a certain time period like the 'Honikel bag method' or the EZ-drip loss, or (iii) by heating the meat and measuring the cooking loss (Honikel, 1989; Honikel and Hamm, 1994; Rasmussen and Andersson, 1996; Christensen, 2002).

8.2.2 Colour
The colour of the raw meat is a combination of the content of myoglobin and the reflection from the protein denaturation. For the colour of the cured cooked meat only the content of myoglobin is important. During the curing process the myoglobin and oxymyoglobin are converted into nitrosomyoglobin in the presence of nitrite, forming a pink colour (Andersen *et al.*, 1988). During storage the colour might further change but it depends on both the package material and on the packaging and storage conditions (Andersen *et al.* 1988). The amount of myoglobin in the raw meat might therefore be an indicator of the raw meat quality with respect to further processing. The hem-group of both myoglobin and traces of hemoglobine can be quantified spectrophotometrically.

8.2.3 Meat/fat distribution and fat quality
The fat content and quality of the raw meat is also an indicator of the quality of processed meat. Cooked cured ham is in general regarded as a lean meat product in France. The amount of intramuscular fat (IMF) in French hams did not change other sensory attributes than colour and marbling when assessed by a trained sensory panel but acceptability by consumers decreased markedly with

increasing IMF (Fernandez et al., 2000). In dried, cured meat product a high amount of intramuscular fat is, however, demanded (Garcia et al., 1997). A study of dry cured pork ham comparing three terminal sires showed however that Duroc and Large White as terminal sire gave the same consumer acceptability independent of a larger IMF content in Duroc (Oliver et al., 1994). In bacon, especially, the lean/fat distribution is important to consumers. Most Irish consumers prefer a thin fat layer (44%) followed by medium fat (38%) and very thin fat (17%). On this background a back fat depth of 12–16 mm was recommended as optimal for most of this consumer group (Moss, 1993). As more focus has been on the health perspective of fat intake this might have changed towards a thinner fat layer since this investigation.

Fatty acid composition influences the quality of meat as well in respect to processed meat quality. A high content of polyunsaturated fatty acids like linoleic acid (C18:2) and n-3 fatty acids results in softening of the fat and a higher oxidative instability. As fat rancidity is one of the limiting factors in storage of products with fat-like sausages this is of importance as well (Woo and Maeng, 1983). If the animals are fed with vitamin E, the antioxidative status of the meat will however be increased and thereby limit the problem of rancidity while Cu in the feed will act as a prooxidant and thereby increase the problem with rancidity (Lauridsen et al., 1999).

IMF can be measured visually as fat marbling using different scales. This is a subjective method but with photos as references it is possible to standardise the assessment. By chemical analyses the content of IMF can be described objectively. Most chemical methods begin with an acid hydrolysis, to liberate the fat from proteins and other complexes, followed by an extraction step. This step determines which part of the lipids the later analysis will quantify. It is possible to get the neutral lipids (triglycerides) only or in addition a greater or smaller part of the phospholipids (polar lipids from the membrane). The phospholipid content ranges from about 0.5% in *longissimus dorsi* (LD) up to about 0.7% in *psoas major* and 0.9% in *masseter* in pig (Gandemer, 1999; Mourot and Hermier, 2001). In beef the content of phospholipids is about 0.7% in *longissimus dorsi* and 1.1% in *diaphragma* (Gandemer, 1999). The concentration of phospholipids in a muscle is rather invariable compared to the content of triglycerides, which can vary. After the extraction the amount of lipids can then be quantified gravimetrically. It is also possible to determine the amount of fat by methods like NMR. The fatty acid composition can be determined on an extract of the meat by GC.

8.3 Eating quality

When meat is used for fresh consumption the time from slaughter to counter can range from two days (in particular poultry and pork) to several weeks (in the case of beef). The appearance of the raw meat influences the consumer's willingness to buy the meat, and can therefore be regarded as an important quality factor for meat proposed for fresh meat consumption. However after cooking tenderness, juiciness, flavour, and appearance of the cooked meat together determine the eating quality. Quality indicators of raw meat for these parameters are the content of intramuscular fat and the composition of fatty acids, the collagen content and solubility, the sarcomere length, activity of proteolytic enzymes, pH and water-holding capacity, and the colour of the raw meat.

8.3.1 Intramuscular fat and fatty acid composition

The content of intramuscular fat or the degree of fat marbling has a great influence on the eating quality beginning when the consumers choose the meat in the supermarket. Many consumers will reject buying meat with a medium or high amount of visual fat marbling both in beef and pork even though they find it more palatable when eaten without knowing the amount of fat (Grunert, 1997; Bredahl et al., 1998; Brewer et al., 2001a), Bligaard, 2002, Pers. comm.).

There are conflicting results on the influence of IMF on tenderness. It is said to increase both the tenderness of the meat (DeVol et al., 1988; Cameron and Enser, 1991; Gwartney et al., 1996; Fernandez et al., 1999; Candek-Potokar et al., 1999; Laack et al., 2001; Brewer et al., 2001a; D'Souza and Mullan, 2002) and to have no effect or even a negative effect (Göransson et al., 1992; Kipfmüller et al., 2000). The reason for these results could be, that variations in IMF are always confounded with other variations, which also have significance for tenderness. The age of slaughter and the slaughter weight influence the content of IMF (Johnson et al., 1969; Candek-Potokar et al., 1999) but can also influence factors like the content and strength of connective tissue and in this way influence the tenderness. Feeding strategy not only influences the IMF content (Blanchard et al., 1999) but also the growth rate and thereby the proteolytic activity that is of significance for the tenderisation of the meat during ageing (Therkildsen et al., 2002). The genetic background also contributes to variations in IMF. In pork some breeds like the Chinese breeds and Berkshire have an extremely high fat content. In the more commercial breeds Duroc especially is known to have a higher content of IMF compared to the white breeds like Landrace and Large White. It has been shown however, that the correlation between IMF and sensory quality depends on breed (Fjelkner-Modig and Persson, 1986). The fatty acid composition can also influence the effect of IMF on tenderness. In pork the saturated and monounsaturated fatty acids are positively correlated to tenderness where polyunsaturated fatty acids are negatively correlated to tenderness (Cameron and Enser, 1991; Eikelenboom et al., 1996). The fatty acid composition is dependent on both breed (Garcia et al., 1986; Tejeda et al., 2001) and feed (Engel et al., 2001).

It has also been said that a high content of IMF would improve the robustness of the meat against a non-optimal cooking. This was shown in beef by Cummings et al., (1999) who found a decline in tenderness in meat with a low

IMF content when cooked to 80°C while meat with a high IMF content was still tender at this end-point temperature. The difference between the two groups at 70°C end-point temperature was only small. It was not possible to find a similar effect in another study (Rymill *et al.*, 1997) and the effect of IMF on the robustness of the meat might therefore interact with other matters, as the direct effect of IMF on tenderness is said to do.

Juiciness is the feeling of moisture in the mouth during chewing. It is a dynamic attribute changing during the chewing process. The content of IMF is positively correlated to juiciness (Savell and Cross, 1988; Gwartney *et al.* 1996; Flores *et al.*, 1999; Cummings *et al.*, 1999; Brewer *et al.* 2001a). Some investigations indicate especially that the sustained juiciness experienced during the last part of the chewing process, is increased by increasing amount of IMF (Savell and Cross, 1988; Aaslyng *et al.*, 2002). An increasing amount of IMF also implies a decrease in cooking loss (Aaslyng *et al.*, 2002). Juiciness is to some extent negatively correlated to cooking loss (Tornberg and Göransson, 1994; Toscas *et al.*, 1999; Aaslyng *et al.*, 2002) and the decreased cooking loss could explain part of the effect of IMF on juiciness.

The content of IMF also influences the flavour of meat (Candek-Potokar *et al.*, 1998; Fernandez *et al.* 1999). This might be due to production of a volatile component as the fatty acid composition is important to the flavour. In pork the content of polyunsaturated fatty acids is correlated with abnormal flavour while monounsatuated and saturated fatty acids are correlated with pork flavour and overall liking (Cameron and Enser, 1991; Cameron *et al.*, 2000). In beef it has been found that the meaty aroma was due to phospholipids and not to such a great extent to triglycerides (Mottram and Edwards, 1983). Flavour is a composition of volatile and nonvolatile components. It is not investigated how IMF influences the nonvolatile flavour components but part of the unspecified effect on flavour could be due to facilitating the contact between the flavour components and the taste buds.

8.3.2 Connective tissue

It is well known that muscles rich in connective tissue like *biceps femoris* are less tender than muscles containing less connective tissue like *psoas major* and the connective tissue has therefore often been in focus as a contributor to the toughness of the meat and therefore a raw meat quality indicator (Honikel, 1992). The main constitutent of the connective tissue is collagen. Collagen is a very strong protein polymere and it is said to make up about 2% of the total muscle protein in beef (Powell *et al.*, 2000). Collagen can be divided into a heat soluble and a heat insoluble fraction reflecting the degree of cross-linking of hydroxyprolin in the collagen (Powell *et al.*, 2000).

Ageing of the meat alters the connective tissue only marginally (Nordyke *et al.*, 2000) and part of the effect might actually be due to degradations of the proteoglycans in integrity with the connective tissue. The collagen molecules change during cooking and their influence on meat tenderness is much influenced by the end-point temperature and the heating rate (Powell et al., 2000).

The degree of cross-linking of collagen increases with the age of the animal. At the same time the shear force increases indicating a decrease in tenderness (Lebret et al., 1998; Fang et al., 1999). Also feeding can alter the degree of cross-linking. In beef a high energy level up to slaughter results in an increase in heat-soluble collagen and a decrease in shear force (Miller and Cross, 1987; Schnell et al., 1997). In a shear force determination the meat is heated in a controlled way, and depending on the end-point temperature the heat-soluble collagen will no longer add to the shear force. The effect of cross-linking on tenderness is, however, not that clear-cut any more, as it has been (Purslow, 1999). Looking across 12 beef muscles, a small correlation between the amount of collagen and tenderness at -0.36 was seen but within each muscle there was no significant correlation (McKeith et al., 1985) and others also find only a weak correlation between the amount of collagen or the degree of cross-linking and tenderness (Zgubic et al., 1998). The differences in tenderness according to age or feeding strategy might therefore be due to other variations as well.

A morphological analysis has demonstrated that the architecture of the intramuscular connective tissue in *semitendinosus* was dominated by thick collagen fibrils with parallel alignment, tightly bundled into fibres running in various directions. The fibres formed a tight network, which may predispose for tough meat. A characteristic feature for *psoas major* was thin collagen fibrils more randomly distributed forming a criss-cross pattern (Eggen et al., 2001). The aspects of organisation of the collagen fibrils might therefore explain some of the differences between muscles in tenderness rather than the degree of cross-linking. The use of collagen content or degree of cross-linking as a raw meat quality marker might therefore be reasonable to some degree across muscles, but within a muscle it might be more open to discussion.

8.3.3 Sarcomere length

The sarcomere length of the meat depends on the chilling and the metabolism post-mortem. If the temperature of a muscle is below approximately 10°C before the onset of rigor mortis cold shortening can occur. Also stretching of the muscle before rigor mortis influences the sarcomere length and longer sarcomere lengths have been found in *semitendinosus* and *biceps femoris* using pelvic suspension compared to achilles suspension (Møller et al., 1987). The sarcomere length has been shown in beef to correlate negatively to Warner-Bratzler shear force (Toscas et al. 1999). In a study across muscles in pork it has been shown that a sarcomere length above 2μm always implies tender meat whereas other factors influence the tenderness at lower sarcomere lengths. This could explain why some muscles like *semitendinosus* are not as tough as might have been expected from the collagen content (Wheeler et al., 2000). The sarcomere length can be analysed by laser diffraction. It is important to do several determinations on the same muscle as the sarcomere length not only on different slices from the same muscle, but also on several sites on the same slice (Honikel et al., 1986).

8.3.4 Enzymatic activity

It has long been known that ageing the meat increases the tenderness. The optimal length of ageing depends on the species – pork having a shorter recommended ageing time than beef. During the ageing period the protein structures are degraded beginning with a diffusion of the Z-lines. Two enzyme systems are said to be involved in this tenderisation – the calpains and the cathepsins. The exact role of the two enzyme systems during tenderisation is still a matter of discussion (O'Halloran et al., 1997).

The calpains are nonlysosomale enzymes. Between two and four different calpains are described depending on species (Dransfield, 1999). In beef and pork μ-calpain and m-calpain are expressed in the meat cell. The μ-calpain is activated by μM concentration of Ca^{2+} compared to m-calpain that is activated by mM concentrations of Ca^{2+}. μ-Calpain and not m-calpain is said to be the most important enzyme during the tenderisation process (O'Halloran et al., 1997; Dransfield, 1999; Geesink and Koohmaraie, 1999) even though many questions still remain (Dransfield, 1999). The activity of the calpains during ageing depends on the concentration of Ca^{2+}, on the pH, and on the concentration of their inhibitor calpastatin (O'Halloran et al., 1997). The proportion of μ-calpain to calpastatin has been shown to explain more than 50% of the change in myofibrillar fragmentation index during conditioning but only 30% of the change in shear force during the same period (McDonagh and Oddy, 1997).

The cathepsins are lysosomal enzymes with a low pH optimum – between 2.0 and 6.5 (Ertbjerg, 1996). As the pH of the meat decreases post mortem the membrane of the lysosomes becomes leaky and the enzymes are released. Even though most focus has been on μ-calpain other investigations have shown that the cathepsins also might contribute to tenderisation especially when the pH of the meat has dropped (Ertbjerg, 1996; O'Halloran et al., 1997).

The activity of the proteolytic enzymes at slaughter depends on the growth rate prior to slaughter. A high muscle protein synthesis pre-slaughter and thereby a high activity of the proteolytic enzymes in vivo could imply a faster rate of protein degradation post-mortem. This hypothesis has not yet been confirmed or rejected (Therkildsen, 1999). However, recent studies in pork indicate that this relationship exists and therefore the activity of μ-calpain prior to slaughter is a raw meat quality indicator for tenderness (Therkildsen et al., 2002).

8.3.5 pH and water-holding capacity

pH is an important raw meat quality indicator with respect to technological quality and influences the fresh meat eating quality as well. A quadratic connection between pH and tenderness and juiciness has been described in both pork (Dransfield et al., 1985) and beef (Cummings et al., 1999) but the main connection in both experiments occurred when pH was above 6.0. In normal meat – pH below 6.0 – there was no influence of pH on either tenderness or juiciness. This was confirmed in a recent experiment finding that even though meat with a pH below 5.4 had a higher drip loss than meat with pH between 5.4 and 5.8 there was no difference in juiciness (Aaslyng et al., 2002). Even in the normal pH range an effect on flavour can be expected. The beef flavour in a steak cooked to 60°C (rare) was significantly higher when pH was below 5.6 compared to a pH above 5.6. This is due to the fact that the Maillard reaction, which is responsible for many of the flavour components is very dependent on the pH (Tressl et al., 1989; Farmer and Mottram, 1990; Meynier and Mottram, 1995; Madruga and Mottram, 1995).

8.3.6 Colour

Colour is an important raw meat quality attribute as it influences the consumer in the choice of meat. A too pale or too dark colour often means that the consumer rejects the meat. The colour of fresh meat is a combination of the reflection due to protein denaturation as a result of the pH change and the concentration and oxidative status of myoglobin. A fast pH fall early post-mortem results in a pale colour whereas a high ultimate pH results in a dark, red colour. Myoglobin is purple but oxidation to oxymyoglobin gives a more red colour that for many consumers indicates freshness. During storage the oxymyoglobin can further oxidise to metmyoglobin, which causes a brown discolouring (Gutzke et al., 1997). The rate of oxidation depends on the species (Gutzke et al., 1997).

The colour of raw meat can be determined instrumentally or visually (Hunt, 1991). In an instrumental determination the values a*, b* and L* are measured. From these various other characteristics can be calculated like Hue angle, which is used for distinguishing colour families and chroma, which is the strength of a colour. a* represents the redness of the meat and is very dependent on the blooming time (Hunt, 1991). It is influenced by the pH of the meat because oxidation and reduction processes of myoglobin are pH-dependent. In comparison L* is independent of blooming time but still very dependent on pH (Brewer et al., 2001b). As about 65% of the variation in L* can be explained from variations in solubility of the sarcoplasmatic proteins (Joo et al., 1999) L* is a good indicator of degree of PSE/DFD (Brewer et al., 2001b).

A visual determination of meat colour must be standardised using only trained assessors. Various scales have been used depending on the species and on the exact purpose of the study (Hunt, 1991). In pork a much-used scale is the Japan colour scale whereas various scales can be used in beef. In general standards like photos are very helpful both in calibrating the assessors and in ensuring that the scale does not drift with time.

8.4 Determining eating quality

The eating quality of meat can be determined both sensorily and instrumentally. In a consumer test, where consumers are asked if they like or dislike the meat sample, the hedonic quality is described. For a more analytical assessment a trained sensory panel can be used. A descriptive analysis – a profiling – can be performed in different ways but in general, attributes are described quantitatively asking the question 'How intensive is this attribute in this sample?'. The attributes cover appearance, texture and flavour. By training the assessors it is possible to do this analysis objectively (Murray et al., 2001).

As meat is normally eaten warm the meat in a sensory profiling is served heated as well. This presents a challenge to the sensory laboratory as the meat samples presented to all assessors must reach the same end-point temperature at the same time and be served in a very standardised way. Furthermore, a biological variation can be found in texture not only between two slices of the same muscle but even within a single slice of the muscle. This emphasises the importance of very standardised procedures for slicing and serving of the meat.

In a sensory profile appearance, texture and flavour are determined simultaneously on the same slice of meat. This is not possible in an instrumental analysis. Texture can be determined by a shear force analysis. A meat sample is cooked in a standardised way and the necessary force required to cut the meat to a given extent is registered. Different types of knives as well as different ways of cutting the meat, e.g. until 80% compression, can be used. The correlation between a shear force analysis and tenderness assessed by a sensory panel depends both on the experimental conditions during the shear force analysis (Tornberg and Göransson, 1994; Tornberg, 1996) and on the cooking procedure used for the meat for the sensory panel (Aaslyng et al., 2001).

Flavour can be determined by GC-MS or GC-O analysis. In a GC-MS analysis the composition of volatile components in the meat headspace is identified and quantified. To translate this knowledge to an understanding of the flavour requires that some key component of the flavour be known. As meat flavour is a complex mix of many components this is rather difficult. In a GC-O analysis an assessor sniffs at the effluent of the GC and registers when an odour is apparent. In this way the most flavour-potent components can be picked out facilitating the interpretation of the GC-MS data (Grosch, 2001).

8.5 Sampling procedure

The analysis of a raw meat indicator in a sample is often used to predict the status of the whole muscle or even of the whole carcass. There is however a variation both within a muscle, between muscles and between animals and care must be taken in a prediction from one analysis. Figure 8.1 shows two slices of LD of beef from two different animals. The first slice is from the 11th thoracic vertebra. At this position the amount of fat marbling is very different between

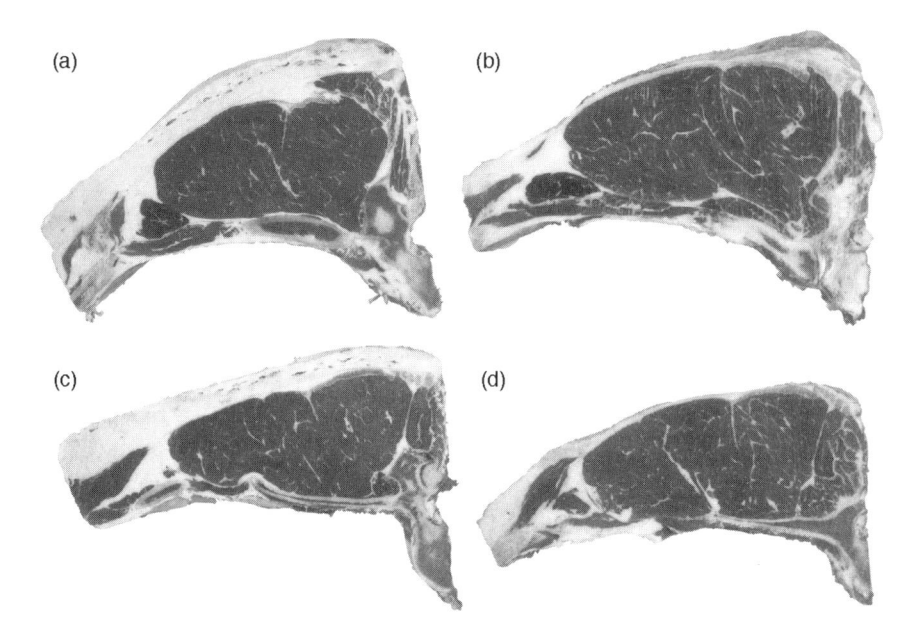

Fig. 8.1 Fat marbling in two slices from two different beefs. (a) animal 1, 11th thoracic vertebra, (b) animal 2, 11th thoracic vertebra, (c) animal 1, 13th thoracic vertebra, (d) animal 2, 13th thoracic vertebra (from N. T. Madsen, Danish Meat Research Institute).

the two animals. The second slice is from the 13th thoracic vertebra only 8 cm away from the first slice. At this position the difference between the two animals is much smaller. If the difference between the two animals in fat marbling was determined from only one sample the answer would very much depend on the sampling site.

The effect of sampling site when determining drip loss in pork has been investigated. On the left loin the drip loss was determined using the EZ-drip loss method where between one and three cylindrical cuts, 25mm in diameter and 25mm thick were used. The right loin (LD) was sliced as well and the drip loss was determined on the whole slices using the bag method to simulate the drip loss seen when a butcher is slicing the loin for sale. The results showed that a correlation between one EZ-drip loss determination and the whole drip of the other loin was 0.9. In this case one drip loss sample of LD was representative for the whole muscle (Christensen, 2002).

Looking across muscles it is even more difficult to predict the meat quality from one sample of one muscle. LD is often used as the sampling muscle because it is of economic importance and because it is a long muscle and easy to sample. The variation in tenderness of beef assessed by a sensory panel is however larger in LD than in four other muscles (Wheeler et al., 2000). Furthermore, the relationship between LD and other muscles in tenderness is not constant (Wheeler et al., 2000) – and for some muscles the correlation is very low (Matthews *et al.*, 1998). The LD muscle therefore cannot be used to predict the tenderness of other muscles. When deciding where a sample for an analysis is taken, it is therefore very important to consider what is the aim of the sampling and not to conclude more than actually possible from this sample.

8.6 Future trends

The future demands on raw meat quality reflect the use of the meat. Two key words exist: 'uniformity' and 'variety'. A future trend is the demand for a large amount of meat with a uniform raw meat quality for further processing. This makes it important that the raw meat quality can be predicted or determined early – if possible before the chilling begins. Another future trend is the demand for a greater variety in smaller amounts especially for the fresh meat consumption market. This makes it important that the raw meat quality can be controlled in order to design a specific quality.

8.7 References

AASLYNG, M. D., BEJERHOLM, C. and MØLLER, S. (2001). Correlation between shear force and sensory analysis in meat depending on cooking method. In *A Sense Odyssey, Conference Program Abstracts. The 4th Pangborn Sensory Science Symposium* Dijon, France, p. 240.

AASLYNG, M. D., BEJERHOLM, C., ERTBJERG, P., BERTRAM, H. C. and ANDERSEN, H. (2002). Cooking loss and juiciness of pork in relation to raw meat quality and cooking procedure. *Food Quality and Preferences*, Accepted.

ANDERSEN, H., BERTELSEN, G., BOEGH-SOERENSEN, L., SHEK, C. K. and SKIBSTED, L. (1988). Effect of light and package conditions on the colour stability of sliced ham. *Meat Science, 22,* 283–292.

ANDERSSON, M., OLSEN, E. V. and FROESTRUP, A.-B. (1997). Relationship between raw material quality and production yield of cooked hams manufactured without the use of phosphates. In *Proceedings 43rd. International Congress of Meat Science and Technology* Auckland, New Zealand, pp. 358–359.

BARTON, P. A. (1971). Abnormal pH_2 values in pigmeat and their effect on the quality of bacon. In *Proceedings 17th. European Meeting of Meat Research Workers* Bristol, UK, pp. 348–353.

BARTON GADE, P. (1984). Meat quality characteristics and their importance for processed pork products. In *The Consumer and Pork Quality Symposium* 19–20th September, Cape Town, Republic of South Africa.

BLANCHARD, P. J., ELLIS, M., WARKUP, C. C., HARDY, B., CHADWICK, J. P. and DEANS, G. A. (1999). The influence of rate of lean and fat tissue development on pork eating quality. *Animal Science, 68,* 477–485.

BREDAHL, L., GRUNERT, K. G. and FERTIN, C. (1998). Relating consumer perceptions of pork quality to physical product characteristics. *Food Quality and Preferences, 9*, 273–281.

BREWER, M. S., ZHU, L. G. and MCKEITH, F. K. (2001a). Marbeling effects on quality characteristics of pork loin chops: consumer purchase intent, visual and sensory characteristics. *Meat Science, 59*, 153–163.

BREWER, M. S., ZHU, L. G., BIDNER, B., MEISINGER, D. J. and MCKEITH, F. K. (2001b). Measuring pork color: effects of blooming time, muscle, pH and relationship to instrumental parameters. *Meat Science, 57*, 169–176.

CAMERON, N. D. AND ENSER, M. (1991). Fatty acid composition of lipid in longissimus dorsi muscle of Duroc and Bristish Landrace pigs and its relationship with eating quality. *Meat Science, 29*, 295–307.

CAMERON, N. D., ENSER, M., NUTE, G. R., WHITTINGTON, F. M., PENMAN, J. C., FISKEN, A. C., PERRY, A. M. and WOOD, J. D. (2000). Genotype with nutrition interaction on fatty acid composition of intramuscular fat and the relationship with flavour of pig meat. *Meat Science, 55*, 187–195.

CANDEK-POTOKAR, M., ZLENDER, B. and BONNEAU, M. (1999). Effects of breed and slaughter weight on longissimus muscle biochemical traits and sensory quality in pigs. *Ann. Zootech., 47*, 3–16.

CANDEK-POTOKAR, M., ZLENDER, B., LEFAUCHEUR, L. and BONNEAU, M. (1998). Effects of age and/or weight at slaughter on *longissimus dorsi* muscle: Biochemical traits and sensory quality in pigs. *Meat Science, 48*, 287–300.

CHRISTENSEN, L. B. (2002). Drip loss sampling in porcine longissimus dorsi. *Meat Science*, Accepted.

CUMMINGS, T. L., SCOTT, S. M. and DEVINE, C. E. (1999) *The effect of meat pH, Marbeling and cooking temperature on sensory attributes of beef striploin steaks*. Report 992 MIRINZ Food Technology and Research Ltd., Hamilton, New Zealand.

D'SOUZA, D. N. D. AND MULLAN, B. P. (2002). The effect of genotype, sex and management strategy on the eating quality of pork. *Meat Science, 60*, 95–101.

DEVOL, D. L., MCKEITH, F. K., BECHTEL, P. J., NOVAKOFSKI, J., SHANKS, R. D. and CARR, T. R. (1988). Variation in composition and palatability traits and relationships between muscle characteristics and palatability in a random sample of pork carcasses. *J. Animal Science, 66*, 385–395.

DRANSFIELD, E. (1999). Meat tenderness – the micro-calpain hypothesis. In *Proceedings 45rd. International Congress of Meat Science and Technology*, Yokohama, Japan (pp. 220–228). Tokyo, Japan.

DRANSFIELD, E., NUTE, G. R., MOTTRAM, D. S., ROWAN, T. G. and LAWRENCE, T. L. J. (1985). Pork Quality from Pigs fed on low glucosinate rapeseed meal: Influence of level in the diet, sex and ultimate pH. *J. Sci. Food Agric., 36*, 546–556.

EGGEN, K. H., PEDERSEN, M. E. and LEA, P. (2001). Structure and solubility of collagen and glycosaminoglycans in two bovine muscles with different textural properties. *J. Muscle Foods, 12*, 245–261.

EIKELENBOOM, G., HOVING-BOLINK, A. H. and VAN DER WAL, P. G. V. D. (1996). The

eating quality of pork. 2. Influence of the intramuscular fat. *Fleischwirtschaft*, 76, 517–518.
ENGEL, J. J., SMITH, J. W. I., UNRUH, J. A., GOODBAND, R. D., O'QUINN, P. R., TOKACH, M. D. and NELSSEN, J. L. (2001). Effects of choice white grease or poultry fat on growth performance, carcass leanness, and meat quality characteristics of growing-finishing pigs. *J. Anim. Sci.*, 79, 1491–1501.
ERTBJERG, P. (1996). The potential of lysosomal enzymes on post-mortem tenderisation. Ph.D. Thesis. The Royal Veterinary and Agricultural University, Copenhagen, Denmark.
FANG, S.-H., NISHIMURA, T. and TAKASHARAHI, K. (1999). Relationship between development of intramuscular connective tissue and toughness of pork during growth of pigs. *J. Animal Science*, 77, 120–130.
FARMER, L.-J. AND MOTTRAM, D.-S. (1990). Recent studies on the formation of meat-like aroma compounds. In *Flavour Science and Technology 6th Weurman Symposium*, Geneva, Switzerland, pp. 113–116.
FERNANDEZ, X., MONIN, G., TALMANT, A., MOUROT, J. and LEBRET, B. (1999). Influence of intramuscular fat content on the quality of pig meat – 1. Composition of the lipid fraction and sensory characteristics of *m. longissimus lumborum*. *Meat Science*, 53, 59–65.
FERNANDEZ, X., MOUROT, J., LEBRET, B., GILBERT, S. and MONIN, G. (2000). Influence of intramuscular fat content on lipid composition, sensory qualities and consumer acceptability of cured cooked ham. *J. Sci. Food Agric.*, 50, 705–710.
FJELKNER-MODIG, S. AND PERSSON, J. (1986). Carcass properties as related to sensory properties of pork. *J. Animal Science*, 63, 102–113.
FLORES, J., ARMERO, E., ARISTOY, M.-C. and TOLDRÁ, F. (1999). Sensory characteristics of cooked pork loin as affected by nucleotide content and post-mortem meat quality. *Meat Science*, 51, 53–59.
GANDEMER, G. (1999). Lipids and meat quality: lipolysis, oxidation, Maillard reaction and flavour. *Science Des Aliment*, 19, 439–458.
GARCIA, P. T., CASAL, J. J. and PARODI, J. J. (1986). Effect of breed-type on the relationships between intramuscular and total body fat in steers. *Meat Science*, 17, 283–291.
GARCIA, P. T., PENSEL, N. A., MARGARIA, C. A., PUEYRREDON, S. and DURSELEN, G. (1997). Effect of pig diet restriction on ham lipids. In *Proceedings 43rd. International Congress of Meat Science and Technology* Auckland, New Zealand, C-20 pp. 290–291.
GEESINK, G. H. AND KOOHMARAIE, M. (1999). Effect of Calpastatin on degradation of myofibrillar proteins by μ-Calpain under postmortem conditions. *J. Anim. Sci.*, 77, 2685–2692.
GÖRANSSON, A., SETH, G. V. and TORNBERG, E. (1992). The influence of intramuscular fat content on the eating quality of pork. In *Proceedings 38th. International Congress of Meat Science and Technology* Clermont-Ferrand, France, pp. 245–248.
GROSCH, W. (2001). Evaluation of the Key Odorants of foods by dilution

experiments, aroma models and omission. *Chemical senses, 26*, 533–545.
GRUNERT, K. G. (1997). What's in a steak? A cross-cultural study on the quality perception of beef. *Food Quality and Preferences, 8*, 157–174.
GUTZKE, D. A., TROUT, G. R. and D'ARCY, B. R. (1997). The relationship between the colour stability and the formation of the metmyoglobin in the surface of beef, pork and vension muscles. In *Proceedings 43rd. International Congress of Meat Science and Technology* Auckland, New Zealand, pp. 652–653.
GWARTNEY, B. L., CALKINS, C. R., RASBY, R. J., STOCK, R. A., VIESELMEYER, B. A. and GOSEY, J. A. (1996). Use of expected progeny differences for marbeling in beef: II. Carcass and palatability traits. *J. Anim. Sci., 74*, 1014–1022.
HENCKEL, P., KARLSSON, A. and PETERSEN, J. S. (1997). Exercise, fasting and epinephrine administration – effects on muscle glycogen degradation *in vivo* in pig *M. longissimus dorsi*. In *Proceedings 43rd. International Congress of Meat Science and Technology,* Auckland, New Zealand, pp. 294–295.
HENCKEL, P., KARLSSON, A., OKSBJERG, N. and PETERSEN, J. S. (1999). Control of post mortem pH decrease in pig muscles. Experimental design and testing of animal models. *Meat Science, 55*, 131–138.
HONIKEL, K. O. (1989). The meat aspects of water and food quality. In Hardman, T.M. *Water and food quality,* London, Elsevier Applied Science.
HONIKEL, K. O. (1992). The biochemical basis of meat conditioning. In Smulders, F. J. M.; Toldra, F.; Flores, J. and Prieto, M. *New technologies for meat and meat products,* ECCEAMST.
HONIKEL, K. O. AND HAMM, R. (1994). Measurement of water-holding capacity and juiciness. In Pearson, A. M. and Dutson, T. R. *Quality attributes and their measuremnet in meat, poultry and fish products.,* Glasgow, Blackie Academic and Professional, Chapman and Hall.
HONIKEL, K. O., KIM, C. J. and HAMM, R. (1986). Sarcomere shortening of prerigor muscles and its influence on drip loss. *Meat Science, 16*, 267–282.
HUNT, M. C. (1991). American meat science association committee on guidelines for meat color evaluation. *Proc. Recip. Meth. Conf., 44*, 1–14.
JOHNSON, H. R., GARRIGUS, R. R., HOWARD, R. D., FIRTH, N. L., HARRINGTON, R. B. and JUDGE, M. D. (1969). Dietary effects on beef composition. II. Quantity and distribution of fat. *J. Agric. Sci. Camb., 72*, 297–302.
JOO, S. T., KAUFFMAN, R. G., KIM, B. C. and PARK, G. B. (1999). The relationship of sarcoplasmic and myofibrillar protein solubility to colour and water-holding capacity in porcine longissimus muscle. *Meat Science, 52*, 291–297.
KIPFMÜLLER, H., BODIS, K., PESCHKE, W. and EICHINGER, H. M. (2000). Qualität von Schweinefleisch. *Ernährung-Umschau, 47*, 416–422.
LAACK, R. L. J. M. V., STEVENS, S. G. and STALDER, K. J. (2001). The influence of ultimate pH and intramuscular fat content on pork tenderness and tenderization. *J. Anim. Sci., 79*, 392–397.
LAURIDSEN, C., NIELSEN, J. H., HENCKEL, P. and SØRENSEN, M. T. (1999).

Antioxidative and oxidative status in muscles of pigs fed rapeseed oil, vitamin E and copper. *J. Anim. Sci.,* 77, 105–115.

LEBRET, B., LISTRAT, A. and CLOCHEFERT, N. (1998). Age-related changes in collagen characteristics of porcine loin and ham muscles. In *44th International Congress of Meat Science and Technology. Meat Consumption and Culture,* Barcelona, Spain, pp. 718–719.

MCDONAGH, M. B. AND ODDY, V. H. (1997). The relationship between variations in calpain system activity and postmortem rate of myofibre fragmentation is the same in both sheep and cattle. In *Proceedings 43rd. International Congress of Meat Science and Technology,* Auckland, New Zealand, pp. 580–581.

MCKEITH, F. K., DE VO, D. L., MILES, R. S., BECHTEL, P. J. and CARR, T. R. (1985). Chemical and sensory properties of thirteen major beef muscles. *J. Food Sci.,* 50, 869–872.

MADRUGA, M. S. AND MOTTRAM, D. S. (1995) The effect of pH on the formation of Maillard-derived aroma volatiles using a cooked meat system. *J. Sci. Agric.,* 68, 305–310.

MAGGI, E. AND ODDI, P. (1988). Prosciutti 'PSE': possibilita di stagionatura. Indagini preliminari. *Industrie Alimentari,* 27, 448–450.

MATTHEWS, K. R., HOMER, D. B. and WARKUP, C. C. (1998). Comparison of the eating quality of fourteen beef muscles. In *44th International Congress of Meat Science and Technology. Meat Consumption and Culture,* Barcelona, Spain, pp. 762–763.

MEYNIER, A. AND MOTTRAM, D.S. (1995) Effect of pH on the formation of volatile compounds in meat-related model system. *Food Chem.,* 52, 361–366.

MILLER, M. F. AND CROSS, H. R. (1987). Effect of feed energy intake on collagen characteristics and muscle quality of mature cows. *Meat Science,* 21, 287–294.

MÖLLER, A. J., KIRKEGAARD, E. and VESTERGAARD, T. (1987). Tenderness of pork muscles as influenced by chilling rate and altered carcass suspension. *Meat Science,* 21, 275–286.

MOSS, B. W. (1993). Factors affecting consumer preferences for bacon: A preliminary study. In *Flair sens meeting: Practical models for relationships between datasets,* Matforsk, As, Norway, pp. 80–82.

MOTTRAM, D. S. AND EDWARDS, J. E. (1983). The role of triglycerides and phospholipids in the aroma of cooked beef. *J. Sci. Food Agric.,* 34, 517–522.

MOUROT, J. AND HERMIER, D. (2001). Lipids in monogastric animal meat. *Reprod. Nutr. Dev.,* 41, 109–118.

MÜLLER, W.-D. (1991). Cooked cured products. Influence of manufacturing technology. *Fleischwirtschaft,* 71, 544–550.

MURRAY, J. M., DELAHUNTY, C. M. and BAXTER, I. A. (2001). Descriptive sensory analysis: past, present, future. *Food Research International,* 34, 461-471.

NORDYKE, J. G., KUBER, P. S., MARTIN, E. L., AVILA, D. M. D. and REEVES, J. J. (2000). A Research note. Shear force analysis as an indicator of changes in mechanical strength of isolated beef intramuscular connective tissue during aging. *J. Muscle Foods, 11*, 227–233.

O'HALLORAN, G. R., TROY, D. J., BUCKLEY, D. J. and REVILLE, W. J. (1997). The role of endogenous proteases in the tenderisation of fast glycolysing muscle. *Meat Science, 47*, 187–210.

OLIVER, M. A., GOU, P., GISPERT, M., DIESTRE, A., ARNAU, J., NOGUERA, J. L. and BLASCO, A. (1994). Comparison of five types of pig crosses. II Fresh meat quality and sensory characteristics of dry cured ham. *Livestock Production Science, 40*, 179–185.

POWELL, T. H., HUNT, M. C. and DIKEMAN, M. E. (2000). Enzymatic assay to determine collagen thermal denaturation and solubilization. *Meat Science, 54*, 307–311.

PURSLOW, P. (1999). The intramuscular connective tissue matrix and cell/matrix interactions in relation to meat toughness. In *45th International Congress of Meat Science and Technology Yokohama, Japan* Tokyo, Japan, pp. 210–219.

RASMUSSEN, A. J. AND ANDERSSON, M. (1996). New methods for determination of drip loss in pork muscles. In *Meat for the Consumer, 42nd. International Congress of Meat Science and Technology,* Lillehammer, Norway, pp. 286–287.

RYMILL, S. R., THOMPSON, J. M. and FERGUSON, D. M. (1997). The effect of intramuscular fat percentage on the sensory evaluation of beef cooked by different methods to two degrees of doneness. In *Proceedings 43rd. International Congress of Meat Science and Technology,* Auckland, New Zealand, pp. 212–213.

SAVELL, J. W. AND CROSS, H. R. (1988). The role of fat in the palatability of beef, pork and lamb. In *Designing foods: Animal product options in the marketplace,* Washington D.C., National Academy Press.

SCHNELL, T. D., BELK, K. E., TATUM, J. D., MILLER, R. K. and SMITH, G. C. (1997). Performance, carcass, and palatability traits for cull cows fed high-energy concentrate diets for 0, 14, 28, 42 or 56 days. *J. Anim. Sci., 75*, 1195–1202.

STØIER, S., AASLYNG, M. D., OLSEN, E. V. and HENCKEL, P. (2001). The effect of stress during lairage and stunning on muscle metabolism and drip loss in Danish pork. *Meat Science, 59*, 127–131.

TEJEDA, J. F., ANTEQUERA, T., MARTIN, L., VENTANAS, J. and GARCIA, C. (2001). Study of the branched hydrocarbon fraction of intramuscular lipids from Iberian fresh ham. *Meat Science, 58*, 175–179.

THERKILDSEN, M. (1999). Biological factors affecting beef tenderness with emphasis on growth rate and muscle protein degradation. Ph.D. Thesis. The Royal Veterinary and Agricultural University, Copenhagen, Denmark.

THERKILDSEN, M., RIIS, B., KARLSSON, A., KRISTENSEN, L., ERTBJERG, P., PURSLOW, P., AASLYNG, M. D. and OKSBJERG, N. (2002). Dietary induced changes in growth rate affect muscle proteolytic potential and meat texture. Effects of duration of the compensatory growth. *Animal Science*, in preparation.

TORNBERG, E. (1996). Biophysical aspects of meat tenderness. *Meat Science, 43*, S175–S191.

TORNBERG, E. AND GÖRANSSON, Å. (1994). The relationship between sensory and instrumental measurements of juiciness and tenderness in cutlets cooked to different temperatures. In *FLAIR SENS and QUEST meeting: Relating instumental, sensory and consumer data,* Ås, Norway, pp. 60–62.

TOSCAS P J., SHAW, F. D. and BEILKEN, S. L. (1999). Partial least squares (PLS) regression for the analysis of intrumental measurements and sensory meat quality data. *Meat Science, 52*, 173–178.

TRESSL, R., HELAK, B., MARTIN, N. and KERSTEN, E. (1989). Formation of amino acid specific Maillard products and their contribution to thermally generated aromas. In *Thermal generation of aromas,* Washington, D.C., Am. Chem. Soc.

WHEELER, T. L., SCHACKELFORD, S. D. and KOOHMARAIE, M. (2000). Variation in proteolysis, sarcomere length, collagen content and tenderness among pork muscles. *J. Anim. Sci., 78*, 958–965.

WOO, S.-J. AND MAENG, Y. S. (1983). Residual nitrite and rancidity of dry pork meat products. A rancidity and storability of homemade dry sausage and dry ham and public taste of dry ham. *Korean J. Food Sci. Technol., 15*, 6–11.

ZGUBIC, E., CEPIN, S. and ZGUR, S. (1998). Correlation between collagen traits and sensorial and physical traits in *M. Longissimus dorsi* of brown bulls. In *44th International Congress of Meat Science and Technology. Meat consumption and culture,* Barcelona, Spain, pp. 768–769.

8.8 Acknowledgement

I would like to thank Hanne Bang Bligaard, Marchen Hviid, and Susanne Støier for valuable discussions during the preparation of this chapter. Christel Dall and Inge Boje Brun are thanked as well for always getting me the literature I needed no matter when or from where.

9
Sensory analysis of meat

G. R. Nute, University of Bristol

9.1 Introduction

The science of sensory analysis is relatively young when compared with the traditional sciences such as physics and chemistry. An early systematic sensory test was the triangular test (*circa* 1940) which was used in Scandinavian countries. Parallel development was also in progress in the USA at about this time. The first book on sensory analysis was written by Tilgner[1] in Polish and this text was later translated into Czech, Hungarian and Russian. The second book on sensory analysis was written in Japanese (Masuyama and Miura),[2] whilst the third textbook, which most sensory analysts will recognise, was that published by Amerine, Pangborn and Roessler.[3] This book was based on the lectures given at The University of California at Davis as part of their Food Science course programme. A very practical book by Jellinek[4] is useful for teaching, and contains what are in effect menus of how to set up basic training courses as an introduction to sensory analysis. More recent books such as that by Meilgaard, Civille and Carr[5] build upon previous works and include applications of difference tests that have been developed in the intervening period.

Sensory analysis is very much an interdisciplinary subject and the aforementioned works cover in some detail the basis of human sensory perception related to thresholds of determination of basic tastes, the variance in individual sensory response and some examples of experimental design. The importance of good experimental design cannot be overemphasised in sensory experiments. Two useful texts are those published by Cochran and Cox,[6] a general statistics book on experimental design, and that by Gacula, Jr and Singh,[7] a statistics book aimed more specifically at sensory analysts. Many of the procedures used in sensory analysis are applicable to many different food items, including meat. This chapter will outline the methods used in the recruitment of sensory assessors, types of panel and training related specifically to meat and where possible published examples of where the different types of tests have been used in meat research.

9.2 The sensory panel

As far as the sensory analyst is concerned, sensory assessors are effectively the instruments used to determine the sensory attributes of a food item. The first step in sensory analysis is to assemble a panel. There is the option of using members of staff if there are sufficient numbers available. If this option is chosen, it will be necessary to impress upon line managers and individuals that it is an essential part of their duties to attend panel sessions. The disadvantage of this approach is that, whilst a considerable amount of time and effort is devoted to timetabling the sessions to ensure everyone is available, there can still be problems of non-attendance at panels. This raises the issue of missing data which is a major problem in meat tasting if it has taken up to two years to produce an animal. There is also the related problem of pre-conditioning the panel. This can occur in small companies and research groups where the assessors have a good idea of what the researcher is working on.

An alternative to the in-house panel is the recruitment of an external panel. These people are paid to attend sessions and often receive a bonus to encourage full attendance. The advantages and disadvantages of this type of panel have been addressed by Nally[8] who describes how they operate and their costs. Some of the sections on terms and conditions of employment given in her paper do not apply in the Euro zone. The advantage of this type of panel is that the sensory analyst can take more time to train the panel, and hold discussions for profiling without having to curtail sessions because assessors have other work commitments. The assessors are relaxed because their job is solely to attend the panel sessions. The disadvantage of this type of panel is the initial effort in advertising and setting up screening sessions for potential assessors and the further training that occurs with meat. When assessors leave there is often no quick way of recruiting assessors unless a reserve list of pre-screened assessors is kept and it remains available.

9.2.1 Screening criteria and training

Potential sensory assessors are required to undergo a series of screening tests. These procedures have been documented by various standards organisations, notably The British Standards Institution (BSI), The International Standardisation Organisation (ISO) and The American Society for Testing Materials (ASTM). All these organisations publish methods for selecting and training sensory assessors and give details of how to establish the basic taste acuity of assessors. As an example, assessors can be screened according to the methods outlined in the relevant BSI standard: BS 7667, Part 1.[9] This Standard outlines the methods of assessing an individual's ability to taste, assess and describe odours, assess and describe textures. These procedures will highlight any individuals that are unlikely to be efficient sensory assessors. When these tests have been completed it will be necessary to start training with meat. The methods outlined by Cross *et al.*[10] for basic training in meat are useful in training assessors. Their methods are based on utilising research results that show variations in sensory attributes. However, with the advent of BSE in the UK, it is no longer possible to use steaks from animals over 30 months of age as examples, and consequently it has been necessary to modify some of their suggested materials and methods.

9.2.2 Using samples in training

It is particularly important to prepare samples carefully when training assessors and assessing consistency of performance. A useful procedure for producing ranges of texture, juiciness and flavour in pork, for example, is to follow the methods outlined by Wood and Nute.[11] The procedures are based on varying the endpoint cooking temperature of pork steaks. Three endpoint temperatures were used: 65°C, 72.5°C and 80°C. The steaks (1.90cm) were cut in triplicate to create a series of uniform samples, an important aspect in reducing variation. Eight-point category scales were used for texture, juiciness and flavour where:

1 = extremely tough, extremely dry, extremely weak
2 = very tough, very dry, very weak
3 = moderately tough, moderately dry, moderately weak
4 = slightly tough, slightly dry, slightly weak
5 = slightly tender, slightly juicy, slightly strong
6 = moderately tender, moderately juicy, moderately strong
7 = very tender, very juicy, very strong
8 = extremely tender, extremely juicy, extremely strong

Analysis of variances revealed highly significant ($p < 0.001$) differences in texture related to endpoint temperature. There were means of 5.1, 4.5, 4.2 for 65°C, 72.5°C and 80°C respectively, with a standard error of the differences of means (s.e.d) of 0.153. Means for juiciness were 5.0, 4.3, 3.5 at 65°C, 72.5°C, 80°C respectively. Means for pork flavour intensity were 3.4, 3.5, 4.1 at 65°C, 72.5°C and 80°C respectively. This approach was tested using different sources of pigs and was found to be consistent in all cases.

The assessment of beef uses procedures based on modifying the conditioning period. To produce an example of extreme toughness, it is necessary to remove a section of hot m.longissimus dorsi, vacuum pack and immediately plunge into iced water, followed by blast freezing at −40°C. To produce tender beef, it is necessary to vacuum pack a section of m.longissimus dorsi and condition at +1°C for 20 days. Extremely tender beef can be produced by using m.psoas major, vacuum packing and conditioning for 10 days at +1°C. Dransfield,[12] has published a study in which instrumental toughness was measured in 18 beef muscles and heated in a waterbath to 60°C, 75°C and 90°C. These results can be used to produce further examples of toughness when training assessors.

9.2.3 General conditions for the assessment of samples
To conduct effective sensory panels, assessors need to be free of distractions and the tests should be carried out in a special room in which there is controlled lighting and good ventilation. Assessors should be seated at separate booths and should not be able to communicate with each other during the assessments. Samples of food should be uniform in size and of the same temperature at serving. They should be coded by a random three-digit number and presented in clean odour-free containers. If more than one sample is to be assessed, then care is needed to ensure that the assessors do not receive the samples in the same order, since this will introduce a bias. Assessors are instructed to rinse their mouths out with water between each sample to remove all traces of the previous sample.

9.3 Sensory tests

There are a range of different types of test:

- difference tests
- paired comparison tests
- triangular tests
- alternative forced choice/3-AFC tests
- duo-trio tests
- 'A'-'not A' test
- ranking tests
- two from five test.

9.3.1 Difference tests
Difference tests are tests where the assessor is presented with a choice situation, i.e., asked to select the odd sample or match the sample to a reference for example. A general consideration when using difference tests for meat is to understand the variances that occur naturally in meat and the complications that arise in the interpretation of the results. It is important not to bias the test. This means that, when presenting samples, it is necessary to ensure that the samples are at the same temperature, cut the same size, presented in the same way and, if colour differences are present, present the samples under red light to mask these effects.

9.3.2 Paired comparison tests (BS5929:part2:1982, ISO 5495)[13]

The paired comparison test is used to determine differences or preferences between two samples for a specified attribute, e.g., tougher or more tender. These differences may be directional or non-directional. A typical question in a directional test would be: 'Which sample is more tender?'. Here the method requires at least seven experts or 20 selected assessors. A typical question for a non-directional test would be: 'Which of these two meat samples do you prefer?'. Directional tests are one-sided (one-tailed tests) whereas non-directional tests are two-sided (two-tailed tests). It is necessary to balance the order of presentation of samples A and B as shown in Table 9.1. Since the BSI/ISO standard was published, further work on the operation of the test was published by Thieme and O'Mahony,[14] They suggested that if assessors were exposed to the range of samples likely to be encountered in the test, then the warmed-up paired comparison was the most sensitive test compared to the duo-trio and 'A' 'not A' test.

An investigation into the influence of a decontamination method using lactic acid on broilers was conducted by Van der Marel.[15] Broilers were submerged in 1% (v/v) lactic acid for 15s, pH 2.4, 15°C at three stages during processing. Carcasses were then stored for two days after which samples of thigh and drumstick were removed and then grilled for 30 minutes. Control and treated samples were presented as a pair to each assessor (12 assessors took part, 4 paired comparisons each) who were asked which sample they preferred. This application (pooling replicate results) of the paired comparison was a non-directional test and the expected number of choices in a particular direction would be 32/48. In this case there were 26 responses in the control direction and 22 in the treated direction. The assumption is that using lactic acid as a decontaminant would not be detected by a trained panel.

Table 9.1 Presentation of samples in directional and non-directional preference tests

Assessor	Directional test Presentation order		Non-directional Preference test Presentation order	
	Set 1	Set 2	Set 1	Set 2
1	AB	BA	AB	AA
2	BA	BA	BB	BA
3	BA	AB	BA	AB
4	AB	AB	AB	BA
5	BA	AB	AA	BB
6	AB	BA	BB	AA
7	AB	AB	AB	BB
8	BA	BA	AA	AB
9	BA	AB	BB	AB
10	BA	BA	AA	BA

9.3.3 Triangular tests (BS5929:part3:1984, ISO 4120-1983)[16]

In triangular tests, three coded samples are presented simultaneously, of which two are the same and one is different. Assessors are asked to select the odd sample. All six combinations are served (ABB,AAB,ABA,BAA,BBA,BAB). Some assessors will receive two samples of A and one of B, whilst others will receive two samples of B and one of A. If the number of assessors is not a multiple of six then it is necessary to present the six sets to each assessor on several occasions. The probability of selecting the correct odd sample by chance alone is 1/3. To analyse the test results, the number of correct replies at the agreed level of probability is compared with those in the reference table and checked to see if the number of correct replies exceeds that in the table of probability where $p = 1/3$. If the number is exceeded than the samples are significantly different.

In tests on meat it is usually not sufficient to assume that the difference between a pair of treatments will be consistent across all animals and treatments and usually the test is repeated using different pairs of animals. The question of whether it is permissible to combine the results of triangular tests has been discussed by Kunert and Meyners,[17] who stated that, if the experiment was properly randomised and controlled, then the assessments are independent and have a success probability of $p = 1/3$. Therefore, the sum of all correct judgements is binomial and the parameter $p = 1/3$ applies. This fulfils the criteria for the null hypothesis where A = B, i.e., no difference across n replications.

Dacremont and Sauvageot[18] have reported that, in some specific areas, it is good methodological practice to apply replications in triangular tests. This approach was applied by Dransfield et al.[19] in a study of bull versus steer meat. In an attempt to reduce variation between animals, pairs of twin bulls (dizygous) were obtained and one of the pair castrated at 80 days. Four different cooking procedures were used: roasting, casserole, mince and grilling using m.longissimus dorsi, m.supraspinatus, m.gastrocnemius and m.psoas major respectively. The results showed that bull meat could be significantly distinguished from steer meat. Later work by Dransfield et al.[20] on twin lambs from two different breeds, Dorset Down and Suffolk crosses, and comparing entires and castrates, showed that 52/106 assessments differentiated the sexes in the Dorset Down comparison and 49/110 in Suffolk crosses. Both results were significant ($p < 0.001$). In trials on lamb from the same experiment, but unpaired and using standard attribute tasting, no significant difference between ram and castrate lamb was observed. The inference is that, in the highly controlled twin situation, there is some difference that enabled assessors to identify ram from castrate, but this difference could not be established in the unrelated lamb system often encountered in sensory testing. It is important to remember that the triangular test is used for difference only, it is not permissible to ask assessors to identify the 'odd' sample on some criteria, e.g., flavour.

9.3.4 Alternative forced choice, 3 – AFC[21]

This test is similar in the way samples are presented to that of the triangular test with one important difference: the 'odd' sample is always the same throughout the test. The probability of selecting the 'odd' sample is still $p = 1/3$. This test has applications for estimating the sensory threshold of individuals and groups of assessors. This approach has been successfully utilised in studies on 'Boar taint' by Annor-Frempong et al.[22] incorporating the ascending method of limits method in conjunction with 3-AFC. Using a model system in which individual samples of androstenone and skatole were used in a neutral lipid base, assessors were asked to sniff the samples and indicate the 'odd' sample. The concentration on subsequent tests was gradually increased in the 'odd' sample until it was detected. This approach was successful in showing how individual responses to androstenone and skatole varied. It also enabled a group threshold response to be established, effectively the sensitivity threshold of the panel as a whole.

The range of sensitivity of individual assessors varied from 0.018 to 0.143 $\mu g\,g^{-1}$ for skatole with a group threshold of 0.026 $\mu g\,g^{-1}$, and 0.250 to 1.000 $\mu g\,g^{-1}$ for androstenone with a group estimate of 0.426 $\mu g\,g^{-1}$. In both series of trials, two different individuals had very low thresholds for androstenone and skatole and were excluded from further trials. Inclusion of these super-sensitive individuals would influence any panel test for using these compounds. Including non-sensitive assessors in a panel just to make up the numbers also presents a problem. These situations highlight the need to screen assessors before commencing a sensory trial, particularly when the response to an individual threshold for a compound is unknown.

9.3.5 Duo-trio test (BS5929:Part 8:1992; ISO 10399, 1991)[23]

The duo-trio test is an intermediate between the duo (paired) and the trio (triangular) test and is statistically less powerful than the triangle test. In this test the assessor receives one sample marked as a reference sample and two other coded samples and are asked which of the two samples matches the reference sample. The probability of selecting the correct sample by chance is 1/2. The presentation order for the duo-trio test is shown in Table 9.2.

Studies on the influence of chilling method, either water or brine chilled, on the eating quality of chicken breast and thigh meats were investigated by Jankey and Salman.[24] They used the duo-trio approach and gave between 20 and 25 assessors samples of deep fried chicken pieces. Assessors were able to detect the difference between chicken pieces from the two chilling treatments in both breast and thigh meats. Instrumental shear tests did not reveal this difference, although it was thought that textural differences would be the likely outcome from the chilling treatments.

9.3.6 'A'-'not A' test (BS5929:part5:1988)[25]

This test is used for evaluating samples having variations in appearance (when it is difficult to obtain strictly identical repeat samples). It can be also used as a

Table 9.2 Presentation order for the duo-trio test

Constant reference technique
Sample A is the reference sample

Assessor	Presentation order	
	Set 1	Set 2
1	R AB	R BA
2	R BA	R BA
3	R AB	R AB
4	R BA	R AB
5	R AB	R AB
6	R BA	R AB
7	R AB	R BA
8	R BA	R BA
9	R AB	R AB
10	R BA	R AB
11	R AB	R BA
12	R BA	R BA

Balanced reference technique*

Assessor	Presentation order	
	Set 1	Set 2
1	R^A AB	R^B AB
2	R^A BA	R^B BA
3	R^A AB	R^B AB
4	R^A BA	R^B BA
5	R^A AB	R^B AB
6	R^B BA	R^A BA
7	R^B AB	R^A AB
8	R^B BA	R^A BA
9	R^B AB	R^A AB
10	R^B BA	R^A BA
12	R^B BA	R^A BA

*Sometimes referred to as the 'alternating reference technique'.

perception test, to determine the sensitivity of an assessor to a stimulus. Assessors are asked to look, smell, touch or taste sample A and remember everything about the sample. The sample is then removed and replaced with a number of samples, (in some US textbooks, both 'A' and 'not A' samples are presented). The numbers of 'A' and 'not A' samples presented are unknown as far as the assessor is concerned. However, it is convenient to balance the number of 'A' and 'not A' samples so that, instead of calculating the chi-square, $p = 1/2$ tables can be used to investigate differences. However, it is important to add the number of correct results when 'A' is recognised to the number of the correct results for 'not A'. It is the total number of correct responses checked against the total number of responses that are used when using $p = 1/2$ tables.

If using the chi-squared test approach then the results from the test are recorded for each sample and used to construct a 2 × 2 contingency table. The number of correct and incorrect responses determine whether 'A' is recognised in a different way to 'not A'. Consider when a particular ingredient is used in a meat product and, for commercial reasons, it is necessary to change suppliers of this ingredient. The manufacturer wishes to know whether the replacement ingredient can be identified from the original. Twenty assessors take part in the test and are given five samples of 'A' and five samples of 'Not A'. The order of presentation is given below:

Assessor	Sample presentation order
1 to 5	A A B B A B A B B A
6 to 10	B A B A A B A A B B
11 to 15	A B A B B A B B A A
16 to 20	B B A A B A B A A B

Having completed the test, the results are assembled into a 2 × 2 contingency table:

Sample identified as	Sample presented		Total
	'A'	'Not A'	
'A'	60	35	95
'Not A'	40	65	105
Total	100	100	200

The chi-squared index is calculated using the expression:

$$X^2 = \sum_{i,j} \frac{(Eo - E_t)^2}{E_t}$$

where Eo is the observed number in box i,j (in which i is the number of the row and j is the number of the column); E_t is the theoretical number in the same box given by the ratio of the product of the number from the row and the number from the column to the total number. In this case the X^2 index is 12.53 and the critical value 3.84 at 5% probability. Therefore in this case there is a significant difference between the original ingredient and the replacement.

9.3.7 Ranking tests (BS5929:part 6 1989: ISO 8587:1988)[26]

Ranking tests are concerned with putting samples in order according to the strength of stimulus perceived. Ranking tests have the advantage that more than two samples can be compared at the same time, whereas only two samples can be compared in difference tests. However, the disadvantage is that inexact results can be obtained if the differences are very small or the samples themselves have wide variations between them. In this test, assessors are likely to suffer from sensory fatigue and this limits the usefulness of this test for meat.

Assessors evaluate a number of samples in random order and are asked to place them in rank order based upon a specified criterion. Assessors are instructed to avoid tied rankings where possible. Results are then collated and the rank sums for individual samples calculated. For an overall comparison of all the samples the Friedman value F is calculated using the following formula:

$$F = \frac{12}{JP(P+1)}(R^2 1 + R^2 2 + R^2 P) - 3J(P+1)$$

where J is the number of assessors, P is the number of samples, and R_1, R_2 are the rank sums given to P samples for J assessors. If F is equal or greater than the critical value corresponding to the number of assessors, the number of samples and the selected level of significance, it can be concluded that there is an overall difference between the samples.

If an overall difference has been established between samples, the rank sums of each sample can be used to identify the significant differences between sample pairs as follows. Let i and j be the samples with R_i and R_j their rank sums. Using the normal approximation, the two samples are different if:

$$|R_i - R_j| \geq 1.960 \frac{\sqrt{JP(P+1)}}{6} \quad \text{for 5\% level of probability}$$

Despite the problem of sensory fatigue there are still opportunities to use ranking tests in the meat area. Sheard et al.[27] investigated the factors that affect the 'white exudate' from cooked bacon. A range of 20 photographs were taken of the different levels of exudate from bacon and assessors ranked these photographs in order of increasing exudate. The results revealed differences between dry cured, Wiltshire untempered and rapid untempered, Wiltshire tempered and rapid tempered bacon. The dry cured bacon had the least exudate and the rapid tempered bacon the most. A second experiment examining just Wiltshire and rapid curing methods on loins from the same pig showed that the Wiltshire method had the least exudate and the rapid method the most.

9.3.8 Two from five test[28]

This is a general difference test which can be more efficient statistically than the triangular test since the probability of correctly guessing the two odd samples is 1/10 compared to the triangular test and paired comparison which are 1/3 and 1/2 respectively. Unfortunately the two from five test can be affected by sensory fatigue and this should be taken into consideration before this procedure is selected. The test consists of presenting assessors with five samples where three of the samples are A and two are B or three are A and two are B. There are 20 possible combinations of two from five samples as shown below:

AAABB	ABABA	BBBAA	BABAB
AABAB	BAABA	BBABA	ABBAB
ABAAB	ABBAA	BABBA	BAABB
BAAAB	BABAA	ABBBA	ABABB
AABBA	BBAAA	BBAAB	AABBB

If the number of assessors are less than twenty it is important to select at random from the above combinations, ensuring that there are an equal number of combinations that contain three As and three Bs. However, it has not been possible to find any applications of this test being used in meat science.

9.4 Category scales

Category scales are often used to rate a number of different stimuli, e.g., texture, flavour and juiciness. They have the advantage that they are easy to use and can cover a number of attributes, as well as up to six samples in a session. For the test to be successful it is necessary to have good experimental design. The results can then be analysed, for example, using analysis of variance techniques to establish the differences between and within the samples. The use of these types of category scales is useful for determining the gross differences in eating quality and is frequently used in animal production experiments that require sensory data. However, this approach may not yield sufficient information if there is a need to identify the individual characteristics of a food, in which case it is necessary to derive a profile of the food. There is often debate on how many scale points should be used in a category test and whether the category approach is the most efficient in terms of reproducibility and discrimination. Work by Lawless and Malone[29] examined four types of *rating* scales, nine-point category scales, line marking, magnitude estimation and a hybrid category and line scale. They showed that all methods were able to find significant differences between products. However, category scales had a small advantage over the other methods.

The term 'rating' has been introduced as against 'scoring'. These terms are often interchanged in sensory experiments on meat, but there are important differences between the two terms. Lawless and Malone[30] compared ratings scales for their sensitivity, replication and relative measurements. They concluded that that in spite of the small physical differences within the samples given to assessors, assessors used wide ranges within the scales. This conclusion supports the view that rating scales are relative. The term 'scoring' has often been used in sensory analysis of meat because much work on meat has involved carcass grading and judging. In these scenarios it has been possible to produce photographic standards depicting the various differences between the categories and then allocate a score. In meat tasting this is very difficult to achieve because of the difficulty of reproducing examples of meat at each scale point in a category scale.

There is a further aspect to consider on whether assessors are asked to rate or score a sample and this concerns the relationship between a physical measurement produced by an instrument and a psychological or sensory impression given by an assessor. An instrument that gave different values on different days for the same measurement would be quickly disregarded or repaired. Treating a sensory panel as an instrument of measurement has at its heart a belief that a particular point on a scale of physical measurement will always be associated with a physical stimulus in an assessor.[31] Psychological impressions depend on a number of factors related to the other samples present in the test and the stimulus that they generate in the assessor. They also depend on the intensity of the stimulus not reaching the point where response adaptation occurs i.e., the point at which an increasing intensity of the stimulus does not evoke an impression of increasing responses in the assessor.[32]

In meat science experiments, where the interest is in the major changes that occur, category scales are widely used and there are many examples of their application. In a study on the influence of breed, feed and conditioning time in pork, Wood et al.[33] used eight-point category scales to show that conditioning time affected the tenderness of pork, with ten-day conditioning producing more tender pork than one-day conditioning. Pork flavour was also increased after ten days and abnormal flavour decreased. Breed effects showed that Durocs had more pork flavour than Large Whites. In a study by Sheard et al.[34] on polyphosphate injection at 3% and 5% concentration in pork, eight-point category scales were used by a panel of ten assessors to rate tenderness, juiciness, pork flavour intensity and abnormal flavour intensity. The results showed that tenderness increased with increasing concentration of polyphosphate. Juiciness also increased with concentration to 3%, but increasing concentration beyond 3% did not increase juiciness perception. Pork flavour intensity decreased with increasing concentration of polyphosphate. The responses were similar in pork from both entires and females used in this trial.

9.5 Sensory profile methods and comparisons with instrumental measurements

There are various ways of carrying out descriptive analysis of foods. There are two basic methods of profiling: fixed-choice profiling and free-choice profiling. In the fixed-choice method, the assessors each develop a vocabulary to describe the food under test. One of the assessors acts as panel leader and their task is to discuss the words generated by the other assessors so that a consensus can be agreed. The agreed set of descriptors is then used to describe the food. The free-choice method involves the assessors being given the likely range of samples during a number of training sessions to derive a profile that is unique to them. The assessors then use this profile to describe the foods in the experiment. In both of these methods, it is important that the experiments are well designed statistically since, if they are not, it is unlikely that meaningful results will be obtained.

The relationship between sensory and instrumental results is an area of considerable interest. Relating the two is particularly important in such areas as new product development. Sensory analysis can help set the parameters for a new formulation, but instrumental measurements are needed to develop and ensure consistent product quality. Nute et al.[35] have explored the relationship between the sensory characteristics of ham and product composition. In this work 52 ham samples were obtained and subjected to sensory and instrumental analysis. The hams were subdivided into three groups, traditional, modern tumbled and massaged hams, and canned hams. A general descriptive profile covering appearance, texture and flavour was developed. The study used a method of statistical analysis known as Generalised Procrustes Analysis (GPA), originally described by Gower,[36] where the methods of assessors' usage of scales are taken into account in the analysis. There are basically three steps in GPA analysis:

1. Translation. This is a method of standardising the mean ratings for each assessor for each attribute. This effectively removes the variation between assessors locations along an adjectival scale.
2. Rotation/reflection. This takes account of assessors' usage of different adjectives to assess the same sensory stimulus. A good example is the confusion between the terms 'bitter' and 'astringent' as documented by Langron.[37]
3. Scaling. The overall variation among samples rated by each assessor is standardised and removes the effect of assessors rating over different ranges of the adjectival scales.

In this particular experiment all three steps were highly significant and demonstrated the effectiveness of GPA as determined by a procrustean analysis of variance. It should be mentioned that the calculation of the degrees of freedom is different from a normal ANOVA. With A assessors and D descriptive adjectives, the degrees of freedom are; D(A-1) for translation, A-1 for scaling and D(A-1)(D-1) for rotation/reflection.

The overall conclusion showed that the first principal axis accounted for 33% of the total variation and was a contrast between gelatinous appearance and firm texture. The second principal axis, accounting for 17% of the variation, was related to the appearance and flavour of the products whilst the third principal axis was related to colour, fatness and saltiness. Examining the relationship between mechanical and sensory properties showed that firmness increased with shear strength, and that gelatinous and plastic appearance was related to the amount of bound water in the hams.

In studies on the effects of fatty acid composition and its effect on eating quality, Fisher et al.[38] showed that the flavour of meat from lambs comprising Welsh Mountain, Soay, Suffolk off grass and Suffolk off concentrates could be characterised using a flavour profile. A profile containing 12 specific descriptors for flavour and five general descriptors adapted from category scales and converted to line scales was developed. Lamb meat from Soay was characterised by significantly higher values for 'livery 'and 'fishy' flavours. Other differences between the lamb breeds were related to the differences between those lambs on forage systems compared with those on concentrate systems.

Savage et al.[39] studied the changes that occur in adhesion between meat pieces using different concentrations of a crude myosin solution. Mechanical

188 Meat processing

Table 9.3 Line scales used in a texture profile of restructured beef steaks (each line 100 mm in length)

Descriptor	Anchor points	
	Left point	Right point
Tactile		
Firmness on cutting	Nil	Extreme
Crumbliness on cutting	Nil	Extreme
Fibrous particles on cutting	Nil	Extreme
Eating		
Rubberiness	Non-rubbery	Rubbery
Ease of fragmentation	Cohesive	Readily separating
Degree of comminution	Coarse	Fine
Tenderness	Extremely tough	Extremely tender
Moistness	Dry	Wet

Table 9.4 Effect of added myosin on the sensory attributes of meat products

	% added myosin					signf.	lsd
Sensory panel	0	1.75	3.5	5.25	7.0		
Tactile							
Firmness on cutting	50.5	57.5	61.0	64.5	68.7	***	4.2
Crumbliness on cutting	35.0	17.1	14.1	11.3	8.5	***	4.1
Fibrous particles on cutting	26.2	20.4	18.1	18.2	15.6	***	3.5
Eating							
Rubberiness	49.3	48.9	53.6	59.4	60.0	***	4.5
Ease of fragmentation	61.0	55.7	50.6	43.7	43.4	***	4.2
Degree of comminution	48.5	50.6	50.6	47.3	49.6	n.s.	4.0
Tenderness	60.4	61.6	60.7	56.4	56.4	**	3.5
Moistness	50.3	57.9	58.1	58.8	56.1	**	4.7

lsd = least significant difference; n.s. = not significant.
** = $p < 0.01$; *** = $p < 0.001$.

tests had established that it was possible to measure differences in total adhesive strength, but what was required was a description of how this was perceived by a trained sensory panel. A fixed profile using the descriptors shown in Table 9.3 was developed over a number of training sessions using the same assessors on each occasion. Using analysis of variance with myosin concentration as a factor, for each individual descriptor, differences that were related to myosin concentration were revealed as summarised in Table 9.4. Assessors were able to differentiate differences in adhesion caused by increasing myosin concentration and showed that samples with higher levels of myosin were firmer when cutting and less crumbly. When eating, higher levels of myosin resulted in samples that were more rubbery, less easy to fragment and less tender.

9.6 Comparisons between countries

The use of sensory panels in different countries often leads to the need to compare results across countries. Work has been done to investigate if results achieved with one panel in one country are the same as in another country. Two early comparisons were completed by Dransfield[40] et al. The first experiment compared meat from beef animals supplied by five research institutes. Each country supplied loin steaks from ten animals and eating quality was compared in Denmark, Ireland, the UK, France and Germany. The panels in each country found a wide variation in eating quality and many of the steaks were very tough. A common eight-point category scale was used for tenderness/toughness. Between the panels tenderness was highly correlated whilst flavour and juiciness were poorly related. The concern in this experiment centred on the fact that some of the samples were very tough and could dominant the other sensory attributes. The second experiment by Dransfield et al.[41] concentrated on only one laboratory supplying meat (UK) and producing two types of animal, Charollais cross steers and Galloway steers. Meat was exchanged between Belgium, Denmark, the UK, France, Germany, Ireland and Italy. Two product types were used, sirloin steaks and cubes of m.semimembranosus for a casseroling procedure. Again it was shown that texture and, in this experiment, juiciness were strongly related between countries, but again flavour could not be predicted accurately between countries. The authors suggested that different culinary practices could have affected this result since the endpoint temperatures that the steaks were cooked to varied between countries and it is known that this affects eating quality.

In a more recent study on lamb meat by Sanudo et al.,[43] samples of lamb were purchased in Spain and one loin from each lamb exported to the UK. Both panels used their own methods of sensory analysis, whereby UK used eight-point category scales and Spain used 100 mm line scales. The two panels used the same descriptors and both panels achieved the same results and interpretation of the results. In this trial a hedonic scale for preference was included and it is interesting that the panels agreed on terms such as lamb odour intensity, tenderness, juiciness, lamb flavour intensity. However, in hedonic (preference) ratings the Spanish panel preferred the Spanish lamb and the UK panel preferred British lamb.

9.7 Conclusions

Sensory analysis is an important tool in meat science and is becoming accepted as a necessary part of meat quality experiments. Advances in computer technology as a means of data capture and analysis facilitates more sophisticated statistical methods to be used by sensory analysts. As new instrumental technologies are developed, for example electronic noses, the interrelationships between these techniques and sensory methods need exploring. Work by Annor-Frempong[43] on 'boar taint' has already shown it is possible to classify 'boar taint' with an electronic nose in a way that is very similar to that obtained by a panel. A review by Schaller et al.[44] identifies nine product areas, including meat and fish, where this technology is providing useful information. Despite all these advances the basic procedures outlined earlier, where it is necessary to screen, train and monitor a panel, are still a vital requirement to produce effective sensory results.

9.8 References

1. TILGNER, D.J. *Analiza organoleptyczna zywnosci*, Warszawa: Wydawnictwo przemyslu Lekkiego I Spozywczego. 1957.
2. MASUYAMA, G. and MIURA S. *Handbook for Sensory Tests in Industry.* Tokyo, Japan: Juse Publishers, 1962.
3. AMERINE, M.A., PANGBORN, R.M. and ROESSLER, E.B. *Principles of Sensory Evaluation of Food.* New York, USA: Academic Press 1965.
4. JELLINEK, G. *Sensory Evaluation of Food, Theory and Practice.* Ellis Horwood, Chichester, UK. 1985.
5. MEILGAARD, M., CIVILLE, G.V. and CARR, B.T. *Sensory Evaluation Techniques.* Florida, USA, CRC Press 1991.
6. COCHRAN, W.G. and COX, G.M. *Experimental Designs*, second edition. Chichester, UK., John Wiley and Sons Inc., 1992.
7. GACULA, JR, M.C. and SINGH, J. *Statistical Methods in Food and Consumer Research*, London, Academic Press Inc., 1984.
8. NALLY, C.L. Implementation of Consumer Taste Panels. *Journal of Sensory Studies*, 1987, **2**, 77–83.
9. BSI ASSESSORS FOR SENSORY ANALYSIS. BS7667, part 1. *Guide to the selection, training and monitoring of selected assessors.*1993/ISO 8586–1:1993. London, BSI, 1993.
10. CROSS, H.R., MOEN, R. and STANFIELD, M. Training and testing of judges for the sensory analysis of meat quality. *Food Technology*, 1978 **32**, 48–52 and 54.
11. WOOD, J.D., NUTE, G.R., FURSEY, G.A.J. and CUTHBERTSON, A. The Effect of Cooking Conditions on the Eating Quality of Pork. *Meat Science*, 1995, **40**, 127–135.
12. DRANSFIELD, E. Intramuscular composition and texture of beef muscles. *Journal of Science and Food Agriculture*, 1977, **28**, 833–842.
13. BSI SENSORY ANALYSIS OF FOOD. PART 2. *Paired comparison test.* BS 5929,1982. London, BSI, 1982.
14. THIEME, U. and O'MAHONY, M. Modifications to sensory difference test protocols: The warmed up paired comparison, the single standard Duo-Trio and the A-Not A test modified for response bias. *Journal of Sensory Studies*, 1990, **5**, 159–176.
15. VAN DER MAREL, G.M., DE VRIES, A.W., VAN LOGTESTIJN, J.G. and MOSSEL, D.A. Effect of lactic acid treatment during processing on the sensory

quality and lactic acid content of fresh broiler chickens. *International Journal of Food Science and Technology*, 1989, **24**, 11–16.
16. BSI SENSORY ANALYSIS OF FOOD. PART 3. *Triangular test.* BS5929,1984 London, BSI, 1984.
17. KUNERT, J. and MEYNERS, M. On the triangle test with replications. *Food Quality and Preference*, 1999, **10**, 477–482.
18. DACREMONT, C. and SAUVAGEOT, F. Are replicate evaluations of triangular tests during a session good practice? *Food Quality and Preference*, 1997, **8**, 367–372.
19. DRANSFIELD, E., NUTE, G.R. and FRANCOMBE, M.A. Comparison of eating quality of bull and steer beef. *Animal Production*, 1984, **39**, 37–50.
20. DRANSFIELD, E., NUTE, G.R., HOGG, B.W. and WALTERS, B.R. 1990. Carcass and eating quality of ram, castrated ram and ewe lambs. *Animal Production*, 50, 291–299.
21. ASTM (AMERICAN SOCIETY FOR TESTING AND MATERIALS) *Standard practice for the determination of odour and taste thresholds by forced-choice ascending concentration series method of limits.* Philadelphia, PA, USA, ASTM, 1991.
22. ANNOR-FREMPONG, I.E., NUTE, G.R., WHITTINGTON, F.W. and WOOD, J.D. The problem of taint in pork:1. Detection thresholds and odour profiles of androstenone and skatole in a model system. *Meat Science*, 1997, **46**, (1) 45–55.
23. BSI SENSORY ANALYSIS OF FOOD PART 8. *Duo-trio test.* BS5929/ISO 10399. London, BSI, 1992.
24. JANKY, D.M. and SALMAN, H.K. Influence of chill packaging and brine chilling on physical and sensory characteristics of broiler meat. *Poultry Science*, 1986, **65**, 1934–1938.
25. BSI SENSORY ANALYSIS OF FOOD. Part 5. *'A' – 'not A' test.* BS 5929/ISO 8588-1987. London, BSI, 1988.
26. BSI SENSORY ANALYSIS OF FOOD. Part 6. *Ranking.* BS 5929, 1989. London, BSI, 1989.
27. SHEARD, P.R., TAYLOR, A.A., SAVAGE, A.W.J., ROBINSON, A.M., RICHARDSON, R.I. and NUTE, G.R. Factors affecting the composition and amount of 'white exudate' from cooked bacon. *Meat Science*, 2001, **59**, 423–435.
28. MEILGAARD, M.C., CIVILLE, G.V. and CARR, B.T. *Sensory Evaulation Techniques.* Boca Raton, USA, CRC Press, 1991.
29. LAWLESS, H.T. and MALONE, G.J. The discriminative efficiency of common scaling methods. *Journal of Sensory Studies*, 1986, **1**, 85–98.
30. LAWLESS, H.T. and MALONE, G.J. A comparison of rating scales: Sensitivity, replicates and relative measurement. *Journal of Sensory Studies*, 1986, **2**, 155–174.
31. RISKEY, D.R. Use and abuses of category scales in sensory measurement. *Journal of Sensory Studies*, 1986, **3/4**, 217–236.
32. O'MAHONY, M. Sensory adaptation. *Journal of Sensory Studies*, 1986, **3/4**, 237–258.

33. WOOD, J.D., BROWN, S.N., NUTE, G.R., WHITTINGTON, F.M., PERRY, A.M., JOHNSON, S.P. and ENSER, M. Effects of breed, feed level and conditioning time on the tenderness of pork. *Meat Science*, 1996, **1/2**, 105–112.
34. SHEARD, P.R., NUTE, G.R., RICHARDSON, R.I., PERRY, A.M. and TAYLOR, A.A. Injection of water and polyphosphate into pork to improve juiciness and tenderness after cooking. *Meat Science*, 1999, **51**, 371–376.
35. NUTE, G.R., JONES, R.C.D., DRANSFIELD, E. and WHELEHAN, O. Sensory characteristics of ham and their relationships with composition, viscoelasticity and strength. *International Journal of Food Science and Technology*, 1987, **22**, 461–476.
36. GOWER, J.C. Generalised Procrustes Analysis. *Psychometrika*, 1975, **40**, 33–51.
37. LANGRON, S.P. The Statistical Treatment of Sensory Analysis Data. Ph.D. Thesis, University of Bath, UK, 1981.
38. FISHER, A.V., ENSER, M., RICHARDSON, R.I., WOOD, J.D., NUTE, G.R., KURT, E., SINCLAIR, L.A. and WILKINSON, R.G. Fatty acid composition and eating quality of lamb types derived from four diverse breed X production systems. *Meat Science*, 2000, **55**, 141–147.
39. SAVAGE, A.W.J., DONNELLY, S.M., JOLLEY, P.D., PURSLOW, P.P. and NUTE, G.R. The influence of varying degrees of adhesion as determined by mechanical tests on sensory and consumer acceptance of a meat product. *Meat Science*, 1990, **28**, 141–158.
40. DRANSFIELD, E., RHODES, D.N., NUTE, G.R., ROBERTS, T.A., BOCCARD, R., TOURAILLE, C., BUCHTER, L., HOOD, D.E., JOSEPH, R.L., SCHON, I., CASTEELS, M., COSENTINO, E. and TINBERGAN, B.J. Eating quality of European beef assessed at five Research Institutes. *Meat Science*, 1982, **6**, 163–184.
41. DRANSFIELD, E., NUTE, G.R., ROBERTS, T.A., BOCCARD, R., TOURAILLE, C., BUCHTER, L., CASTEELS, M., COSENTINO, E., HOOD, D.E., JOSEPH, R.L. SCHON, I. and PAARDEKOOPER, E.J.C. Beef quality assessed at European Research Centres. *Meat Science*, 1984, **10**, 1–20.
42. SANUDO, C., NUTE, G.R., CAMPO, M.M., MARIA, G., BAKER, A., SIERRA, I., ENSER, M. and WOOD, J.D. Assessment of commercial lamb meat quality by British and Spanish taste panels. *Meat Science*, 1998, **48**, 1/2, 91–100.
43. ANNOR-FREMPONG, I.E., NUTE, G.R., WHITTINGHAM, F.W. and WOOD, J.D. The measurement of the responses to different odour intensities of 'boar taint' using a sensory panel and an electronic nose. *Meat Science*, 1998, **50**, 139–151.
44. SCHALLER, E., BOSSET, J.O. and ESCHER. F. Electronic Noses and Their Application to Food. *Lebensm.-wiss. u.Technol.*, 1998, **31**, 305–316.

10

On-line monitoring of meat quality

H. J. Swatland, University of Guelph

10.1 Introduction

On-line monitoring of quality for meat processing involves two main activities. Firstly, individual carcasses or primal cuts are selected whose meat reaches a required standard for premium products such as cured hams or in-flight meals. Probes, ultrasonics and video image analysis (VIA) are available. Obtaining access to interior muscles, and variation between and within carcasses are the main problems. Being able to predict meat quality from on-line measurements has great commercial potential, especially for tenderness and water-holding capacity (WHC), but we can allow no contamination, only imperceptible damage, and measurements must be very fast to keep pace with line speeds. Secondly, for the processing of comminuted meat, information from on-line sensors is used for process control in emulsion formation, curing or cooking. Operating conditions are relatively simple if the product can be accessed in homogenised batches or in a continuous stream. There is usually ample time for integrated measurements.

10.1.1 Apparatus

Despite the tremendous commercial advantages offered by on-line monitoring of meat quality, relatively few methods are available as off-the-shelf purchases. Apparatus that may be simple to build, develop and test in a laboratory with skilled operators and convenient operating conditions is seldom profitable to develop commercially for use in severe industrial conditions. Developers of scientific apparatus usually think in terms of hundreds or thousands of units to be manufactured, not ones and twos. The meat industry is reluctant to spend much unless apparatus is proven beyond doubt. Thus, small markets and lack of vision combine in a vicious circle to perpetuate vintage technology such as glass pH electrodes. However, a few meat processors have commissioned their own technology and are secretly reaping the rewards. Commercial secrets are seldom publicised in research papers, so this chapter can offer only an academic perspective on the subject.

Apparatus comes in several formats. VIA is remote. The camera is located some distance from the side profile of a carcass or a sectioned rib-eye. Ultrasonic transducers, however, are placed in contact with the sample, usually a carcass. Orthogonality must be maintained to form an image of sound waves reflected from muscle-fat boundaries. Spectrophotometric methods may be remote, as for a plate of comminuted product placed in a colourimeter or near-infrared (NIR) spectrometer. But, using optical fibres (Kapany, 1967), spectrophotometric methods may be adapted as probes to push into the carcass or form a window against a comminuted product (MacDougall and Jones, 1980). Optical fibres pushed into meat do not give the same reflectance spectrum as obtained with a conventional reflectometer (Fig. 10.1). When pushed into a carcass, the probe may be combined with a depth detector to give a spatial dimension to the data. For example, reflectance may be plotted versus depth in the carcass. This is called a transect. A thin probe minimises tissue distortion but supports only a small optical window. This reduces the volume of tissue through which reflectance is integrated, so that non-muscle tissues at the optical window may create anomalous spectra which do not represent muscle quality (Brøndum et al., 2000). Electromechanical probes have a long history for predicting meat toughness. A torque probe is currently available (Jeremiah and Phillips, 2000).

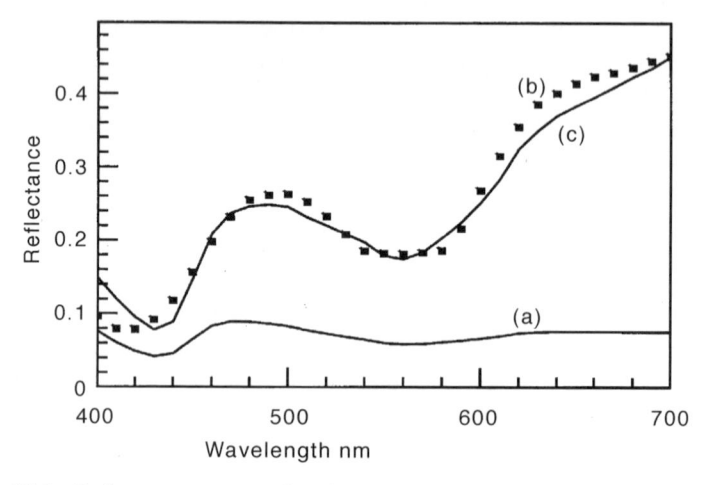

Fig. 10.1 Reflectance spectra of pork measured with optical fibres (a) and with a reflectance (b). Line c is line a raised to the third power of wavelength.

10.2 Measuring electrical impedance

Muscle has a relatively low electrical resistance while fat has a high resistance. Muscle contains continuous electrolytes with a relatively high conductivity, while fat is dominated by globules of insulating lipid. This enables the depth of subcutaneous fat to be found along a transect, overall lipid content to be detected electromagnetically, and exudative meat to be detected by high conductivity.

Electrical impedance measurements were first made in England in the 1920s and 1930s as DFD (dark, firm, dry) pork became a problem when shipping of pigs by rail to new centralised abattoirs replaced local slaughtering. Penetration of curing ingredients for traditional dry curing is dangerously slow but, when diffusion is further reduced by a paucity of extracellular fluid, it may allow deep spoilage. Banfield (1935) found the main factors to be: (i) curing salt concentration, (ii) volume of pickle pumped into the meat, (iii) intrinsic texture and moisture content of the meat, (iv) ambient temperature, (v) fat-free surface area for pickle penetration, and (vi) volume of meat relative to pickle. Penetration of curing salts was monitored on-line by testing the electrical resistance of the meat (Callow, 1936). This led to the discovery of pH-related variability in the resistance of fresh pork (Fig. 10.2), which now we may use to place pork on a scale from DFD to PSE (pale, soft, exudative).

10.2.1 Biophysical source

Skeletal muscle is composed of myofibres. These are extremely large cells, although reaching only about 0.1 mm in diameter they may be many centimetres in length. Like all other cells, myofibres are surrounded by a plasma membrane with strong dielectric properties and a capacitance of approximately 1 $\mu F\,cm^{-2}$. Thus, the cell membrane is a very effective insulator, and when meat probe

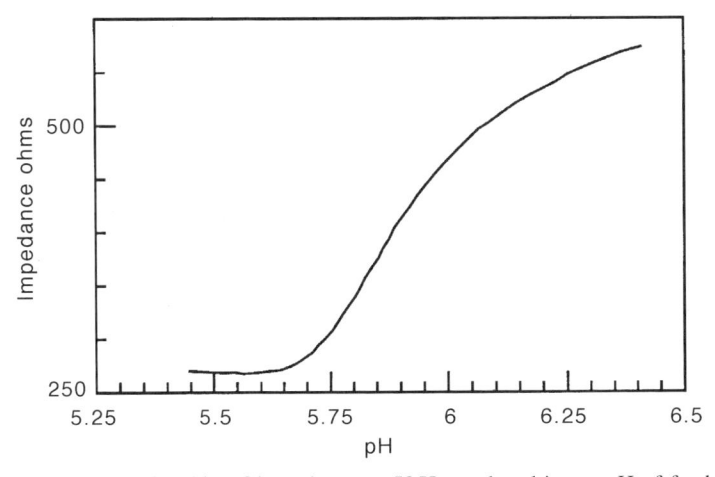

Fig. 10.2 The relationship of impedance at 50 Hz to the ultimate pH of fresh pork discovered by Callow in 1936.

electrodes are inserted into the muscle of a recently slaughtered carcass, strong overall capacitance is detectable.

10.2.2 Polarisation

For metal electrodes in contact with electrolytes within and between myofibres, an electrode discharges ions into the fluid while ions in the fluid tend to combine with an electrode. This may create a charge gradient across an electrical double layer so that the electrode-electrolyte interface behaves as a voltage source, and as a capacitor in parallel with a resistor. The magnitude of this polarisation depends on: (i) the metallic composition of the electrode; (ii) electrode area; (iii) electrode insulation by adipose cells; (iv) current density and electrode separation; (v) post-mortem changes in electrolyte composition; (vi) temperature; and (vii) the frequency of the test current, such that resistance and capacitance in parallel are usually inversely proportional to the logarithm of frequency. Polarisation in meat probes may be cancelled by using AC instead of DC, and by using separate source and detector electrodes. For example, with four electrodes, two supply the test current and two detect it.

10.2.3 Resistance, capacitance and anisotropy

Impedance is determined by resistance and capacitance (capacitive reactance), but these may be in series or parallel (Gielen *et al.*, 1986). Capacitance (C_S) and resistance (R_S) in series circuit are related to their equivalents in parallel as follows,

$$R_s = R_p/(1 + (2\pi f C_p R_p)^2) \tag{1}$$

$$C_s = C_p(1 + (1/(2\pi f C_p R_p)^2)) \tag{2}$$

As well as the intrinsic dielectric anisotropy of meat caused by myofibres being parallel (Epstein and Foster, 1983), other factors are involved in making on-line measurements. Plate electrodes may be used for excised or comminuted muscles, but carcass measurements require probes. A pair of probe electrodes may be inserted in three different planes relative to the longitudinal axes of myofibres. When electrodes and myofibres are coaxial, myofibres may be compressed concentrically around the electrodes to increase capacitance from membranes. But if electrodes open up channels of extracellular fluid along myofibres, this shorts the test current and reduces the current density on membranes. Thus, on-line use of electrical probes requires standardisation of electrode placement relative to myofibre orientation.

10.2.4 Electrode penetration

Depth of electrode penetration may determine the area of electrode contact and the current density between the electrodes. Contact area should be controlled by an insulated sleeve covering the base of the electrode so that exposure to the meat is constant. Surface wetness on a carcass may vary and it may be necessary to dry the carcass at points of measurement to avoid surface fluid shorting the electrodes. Even if the distance between electrodes is set by a manufacturer, electrodes may become bent. Moving the electrodes closer together decreases resistance, but may increase capacitance.

10.2.5 Freezing

Freezing obscures any initial differences in impedance, probably because ice crystals cut through cell membranes to create a continuous pool of electrolytes when the muscle is thawed. This provides a method for the detection of meat that has been frozen and thawed, relative to meat that has never been frozen (Salé, 1972).

10.2.6 PSE detection

Pioneer research was undertaken by Rowan and Bate Smith (1939), but their discoveries lay dormant until the late 1970s, when attempts were made to develop better methods than glass pH electrodes for sorting and detecting PSE pork carcasses. Several commercial systems became available in the 1980s to measure the electrical properties of pork, such as the MS-Tester and the Carnatest. As well as pork, impedance testing may be used to identify PSE turkey meat (Aberle et al., 1971). The MS-Tester measures the dielectric loss factor of pork using a frequency synthesiser combined with an analog to digital converter. With conductivity γ and dielectric constant ϵ_r, the dielectric loss factor d

$$d \approx \gamma/\epsilon_r \tag{3}$$

is found from the cotangent of the measured phase angle (Pfützner and Fialik, 1982). The Testron MS-tester uses two parallel scalpel blades 25 mm apart at 15 kHz (Kleibel et al., 1983).

DFD, normal and PSE categories are seldom discrete and usually overlap as a continuum. Separation is further complicated by the non-linear relationship of impedance with meat quality, as shown for the MS-Tester by Seidler et al. (1987). When capacitance in parallel (Cp) and resistance in parallel (Rp) are related to frequency (f), the quality factor (Q) is defined as,

$$Q = 2\pi f\, Cp\, Rp \tag{4}$$

The other term in common use is the dissipation factor,

$$D = 1/Q \tag{5}$$

The MS-tester may be capable of differentiating between normal and PSE pork as early as an hour or less post-mortem (Kleibel et al., 1983), but this is not always possible (Schmitten et al., 1984; Fortin and Raymond, 1988).

10.2.7 Adenosine triphosphate

Adenosine triphosphate (ATP) provides energy for muscle contraction and the maintenance of ion gradients across membranes such as those of the sarcoplasmic reticulum. The conversion of extensible living muscle to inextensible meat is determined by the length of time that anaerobic glycolysis can maintain ATP levels from stored glycogen. Impedance often appears to be correlated with pH, as in Fig. 11.2, but a more fundamental relationship may be between capacitance and ATP concentration. If lactate-induced damage to ion pumps in cell membranes is postulated as the primary cause of the decrease in impedance as the pH declines post-mortem, then DFD beef with a relatively high pH should have high impedance. This does not appear to be the case. In beef, Cp may be correlated with ATP, $r = 0.8$, whereas correlations of Cp with pH are weaker and sporadic in occurrence (Swatland and Dutson, 1984). Thus, the post-mortem decline of Cp is probably determined more by the leakage of ATP-driven ion pumps than by a direct effect of pH on cell membranes.

10.2.8 Electromagnetic scanning

When an animal or carcass is passed at a constant velocity through a magnetic field, skeletal muscles create a measurable perturbation in the magnetic field. For live animals, this principle was used in the EMME (model SA-1, electronic meat measuring equipment, EMME Corp., Phoenix, Arizona), then in the TOBEC (total body electrical conductivity) HA2 for measuring adiposity in humans. The manufacturer of the TOBEC (Agmed Inc., Springfield, Illinois) then established a subsidiary (Meat Quality Inc.) to produce equipment for on-line evaluation of the fat content of meat, the MQI Electromagnetic Scanner.

A perturbation in a magnetic field is difficult to standardise or express in scientific units. Thus, a phantom sample is used for standardisation. Carcass shapes are complex, so electromagnetic scanning is well suited for the analysis of boxed beef. The Emscan MQ27 system used in Australia uses a coil frequency of 2.5 MHz. The phase angle between the voltage and the current, and the amplitude of the current are measured at 50 Hz giving a lean meat computation output at 20 Hz.

As a pork carcass passes through the coil of the MQ25, a relative energy absorption curve is generated, and the curve is scaled from the height and position of the major peak. Predictions are possible to line speeds of 1,000 carcasses per hour. If temperature is taken into account, electromagnetic scanning results may be combined with carcass weight to give a strong predictor of total lean ($R^2 = 0.904$).

10.2.9 Bioelectrical impedance

A four-electrode bioelectrical impedance analyser (RJL Systems Detroit, Michigan) has been developed to predict carcass lean content. The electrodes are 21-gauge needles placed in an anterior to posterior sequence along the full length of the animal's back with a 10-cm separation between transmitter and detector electrodes at each end. For lamb, prediction equations ($R^2 = 0.97$) for total weight of retail cuts were developed using resistance, reactance, weight, distance between detector electrodes and temperature. Measured along the side of pork carcasses, the prediction of lean yield ($R^2 = 0.81$) is less accurate than TOBEC, but may be used for both live animals and carcasses (Swantek *et al.*, 1992). Marchello and Slanger (1992) found that the method was suitable for robotic sorting of pork shoulders according to lean content.

10.3 Measuring pH

The pH of living skeletal muscle is usually just above pH 7. It may decrease after slaughter to pH 5.4 to 5.7 in normal meat. If initial glycogen is limited, the pH stays high and the meat remains DFD (as it is in the live animal). If the pH decline is rapid (affecting muscle proteins while still warm) or extensive (giving a low ultimate pH), the meat becomes PSE. Thus, the pH of meat has a profound effect on colour, firmness and water-holding capacity, as well as subtle effects on taste, tenderness and rate of post-mortem conditioning. The pH of meat may be measured on-line with a gel-filled combination electrode, but the risk of broken glass may be unacceptable. Rugged portable equipment for routine on-line use in commercial plants is available (NWK Binär GmbH, Landsberg, Germany).

An unofficial terminology has developed among meat scientists. Subscripts are appended to indicate time of measurement. The two classical measuring times are 45 minutes and 24 hours post-mortem. These may be termed pH_1 and pH_2, respectively, or pH_{45} and pH_{24}. Other subscripts may appear, such as pH_{30} for a measurement at 30 minutes post-mortem. The context usually shows whether the subscript is in minutes or hours. This may not be precise, but it is convenient. Working in a commercial environment it is very difficult to ensure the accuracy of times post-mortem.

10.3.1 PSE in pork

A $pH_1 < 6.0$ is typically taken as the critical point below which commercially important PSE develops in pork (Bendall and Swatland, 1988). The pH_1-index is defined as the percent of pH_1 values below pH 6.0 at 45 minutes post-mortem. The following equations may be used to relate the mean pH_1 to the pH_1-index:

$$\text{for mean } pH_1 > 6.33: pH_1 - \text{index} = 37.8 \times (6.65 - pH_1) \quad (6)$$

$$\text{for mean } pH_1 < 6.33: pH_1 - \text{index} = 110.3 \times (6.39 - pH_1) \quad (7)$$

10.3.2 Conditions of measurement

Many factors affect pH measurements on-line, particularly the difficulty of establishing electrical contact with the reference cell of a glass electrode if the meat is dry. Meat may be homogenised in potassium chloride and iodoacetate solution to arrest glycolysis (Bendall, 1973). Corrections for temperature are required. The effect of temperature is to lower the pH by 0.1–0.15 units per 10°C rise in temperature and to raise it by the same amount per 10°C fall in temperature (Bendall and Wismer-Pedersen, 1962). This is caused by changes in the pK values of carnosine, anserine and histidine buffers in meat.

10.3.3 Light scattering

Meat with a high pH appears dark while meat with a low pH appears pale. This is caused by differences in light scattering. Incident light is transmitted deep into meat with a high pH and very little escapes to be seen by the observer. Incident light has limited penetration of meat with a low pH and most of it escapes from the meat to be seen as paleness by the observer. For measurements made by reflectance spectrophotometry therefore, meat with a high pH has deeper bands of selective absorbance (because the light path is longer than in meat with a low pH). Light scattering may be measured on-line for the direct prediction of the appearance of the meat, for either PSE or DFD (Gariépy et al., 1994), or as an indirect way of measuring other pH-related properties such as water-holding capacity (Andersen et al., 1999).

10.3.4 Biophysical source

Three factors may contribute to high light scattering in meat with a low pH. Firstly, Bendall and Wismer-Pedersen (1962) proposed that light scattering at a low pH is caused by denaturation of sarcoplasmic proteins, similar to heat denaturation of egg albumen (Bendall, 1962). Precipitated protein is detectable histologically in severe PSE pork (Bendall and Wismer-Pedersen, 1962). Secondly, another factor is that shrinkage of myofibrils at a low pH increases the refractive index difference between myofibrils and sarcoplasm so that scattering from the myofibrillar surface may be increased (Hamm, 1960; Offer et al., 1989). Myofibrils account for approximately 80% of the volume of pork, declining to 50% as the pH declines post-mortem. Thus, even small optical changes in the myofibrils have a major effect. A third factor is that increased myofibrillar refractive index caused by low pH may increase scattering by increasing the refractive deflection of light passing through myofibrils. Myofibrils are strongly birefringent, as indicated by the naming of A (anisotropic) and I (isotropic) bands, and this enables them to be investigated with polarised light. Measurements on individual myofibres show that the optical path difference (between rays following different refractive pathways allowed by birefringence) tends to increase when pH is decreased.

10.3.5 Laser scanning

Laser scanning of meat was introduced by Birth et al. (1978), based on the Kubelka-Munk analysis of light scattering. Illuminating the upper surface of a slice of muscle with a helium-neon laser gives:

$$\log M_T = A - Br \tag{8}$$

where M_T is the radiant excitance on the lower surface, A is the intercept and B is the slope of a regression (light intensity versus distance), and r is the path length through the meat. Birth *et al.* (1978) showed that

$$B = \log 2\,(S + K) \tag{9}$$

where S is a scatter coefficient (cm^{-1}), and K is the absorption coefficient (cm^{-1}). B may be called a spatial measurement of scattering, for the sake of convenience. Birth *et al.* (1978) showed that spatial measurements of scattering are related to meat quality.

10.4 Analysing meat properties using NIR spectrophotometry

NIR spectrophotometry has many certified applications in food analysis (Norris, 1984), especially for the fat content of mince or ground beef (Tøgersen *et al.*, 1999). It is ideal for plates and continuous streams of comminuted meat products (Isaksson *et al.*, 1996), but has also been used for carcasses (Chen, 1992; Chen and Massie, 1993). It may be used to predict functional properties of meat in processed products and to detect previously frozen meat (Downey and Beauchene, 1997).

10.4.1 Apparatus

Typical components are a tungsten-halogen source, a grating monochromator from 800 to 1100 nm, an optical system to direct the light through or onto a plate of comminuted meat and a photometer. Multiple scanning and rotation of the sample help to average heterogeneity, and smoothed absorbance measurements at about 100 wavelengths may be used to make predictions from a partial least squares chemometric calibration. Technical problems originate from drying of the sample surface, pH-related changes in light scattering, metmyoglobin formation, and sample temperature. These often limit the applicability of a prediction equation to situations exactly the same as the calibration population. Because NIR spectrophotometry is sensitive to fat content, and fat content has many effects on product quality, it may be difficult to extract information that relates directly to protein functionality.

10.5 Measuring meat colour and other properties

The oxygenation of purple myoglobin to red oxymyoglobin, and the oxidation of oxymyoglobin to brown metmyoglobin are vitally important for the appearance of meat products, as is the formation of heat-stable, pink nitrosylhaemochrome in curing (Fox, 1987). This latter reaction is so sensitive to nitrite that even nitrite from spices and tap water may prevent browning in cooked products.

Carbon monoxide may have a similar effect (even from exposure of transported animals to traffic fumes). Myoglobin is also important in connection with oxidative rancidity. Haemoglobin, although minimal in properly exsanguinated meat, may inadvertently be introduced from bone marrow in mechanically deboned meat.

10.5.1 Spectrophotometry

The Soret absorbance bands of deoxymyoglobin, oxymyoglobin, and metmyoglobin occur at 434, 416, and 410 nm, respectively (Bowen, 1949; Morton, 1975). The oxygenation of myoglobin causes a loss of the absorbance band at 555 nm and the appearance of two new absorbance bands at 542 and 578 nm with millimolar absorptivities of 13.2 and 13.3, respectively (Morton, 1975).

Working on-line with muscle surface reflectance is more difficult than the spectrophotometry of purified pigments in solution. Using reflectance spectrophotometry, Ray and Paff (1930) found that 542 and 578 nm absorbance bands were of equal intensity. But many published studies are of artificially stabilised surfaces. Subsurface oxidation of myoglobin may produce strange results on-line (because scattering is changed as well as the absorbance spectrum). From absorbance spectrophotometry of purified metmyoglobin, the secondary reflectance peak of metmyoglobin around 600 nm may encompass an additional small peak at 565 nm. If cut meat surfaces are exposed, product colour can be measured on-line by VIA (Lu et al., 2000).

Spectrophotometry via optical fibres may be used to assess the functional properties of comminuted meat (Swatland and Barbut, 1990). For mixtures of pork adipose tissue and beef muscle with a low connective tissue content, reflectance at 1000 nm may be correlated with lipid content ($r = 0.99$), at 930 nm with centrifugation fluid loss ($r = 0.77$), and at 1000 nm with cooking loss ($r = 0.99$). Correlations may be weakened when the meat is adjusted to a constant lipid content; for example, for pork muscle reflectance at 780 nm was correlated with pH ($r = -0.80$), at 690 nm with centrifugation fluid loss ($r = 0.74$), and at 710 nm with cooking loss ($r = 0.63$). Peak correlations with functional properties are highly wavelength dependent. In the first set of correlations given above, lipid content is the major variable and the strongest relationships are obtained with red or NIR. But if variation in lipid content is cancelled so that pH-dependent protein properties are dominant, then the peak correlations are with shorter wavelengths. Thus, in experimental studies aimed at finding the most suitable wavelengths for simple on-line sensors, the likely importance of lipid-related versus pH-related sources of variance should be considered.

With regard to cost-effectiveness, full-range spectrophotometry must necessarily be more expensive and more complex than a simple monochromatic sensor. But in choosing the less expensive and more robust option, it is essential to know the whole spectral range of correlations. Sometimes, correlations of reflectance with a functional property are strong at a certain wavelength, but weak at an adjacent wavelength. If both wavelengths are encompassed within the broad-band range of a simple sensor, then the results may be disappointing. But a narrow spectral range implies weak light intensity, thus either increasing the cost or the relative noise of the photometer.

10.6 Water-holding capacity

Water-holding capacity (WHC) may be defined as the ability of meat to retain its own water under external influences such as compression or centrifugation. In this case, water-binding capacity becomes the ability of the meat to bind extra water added to a product. Water absorption or gelling capacity may be defined as the ability of meat to absorb water spontaneously from an aqueous environment. The basic mechanism is the effect of pH on the myofilament lattice. As pH declines towards the isoelectric point of muscle proteins, a reduction in the negative electrostatic repulsion between myofilaments reduces the water space of the myofilament lattice. The release time of this fluid, however, is highly dependent on time-temperature interactions, muscle structure and handling.

The usual method for predicting WHC on-line is to exploit a correlation of WHC with pH and, indirectly, with light scattering (Jaud et al., 1992) or conductivity (Lee et al., 2000). The usual problem is weak correlations, often from curvilinear relationships. NIR reflectance measured with a probe at 30 min post-mortem can predict 24 hour fluid losses from pork, but the measurement takes six minutes (Forrest et al., 2000).

10.7 Sarcomere length

Sarcomere length has some important effects on meat quality. If a muscle sets in rigor mortis at a shortened length, there is a high degree of overlap between thick and thin myofilaments. The meat is tough and has a low WHC. A laser diffractometer may be used to measure sarcomere length,

$$\text{Sarcomere length} = \lambda/\sin\theta \tag{10}$$

where θ is the angle subtended by the first order diffraction band relative to the optical axis. Because the angles are relatively small, $\sin\theta \simeq \tan\theta$ (Rome, 1967). However, this method requires the dissection of individual or small bundles of myofibres, and is no more use on-line than is light microscopy. Bulk meat with high scattering from countless myofibres presents a real challenge.

With two plane polarisers (a fixed polariser and rotatable analyser), the maximum transmittance occurs when the analyser is rotated parallel to the polariser, and the minimum is when the analyser is perpendicular to the polariser. But if the polariser is at 0°, when the analyser is at 90°, then a birefringent muscle samples with myofibres at 45° between the polariser and analyser rotates light so that now it may pass through the analyser. The brightest

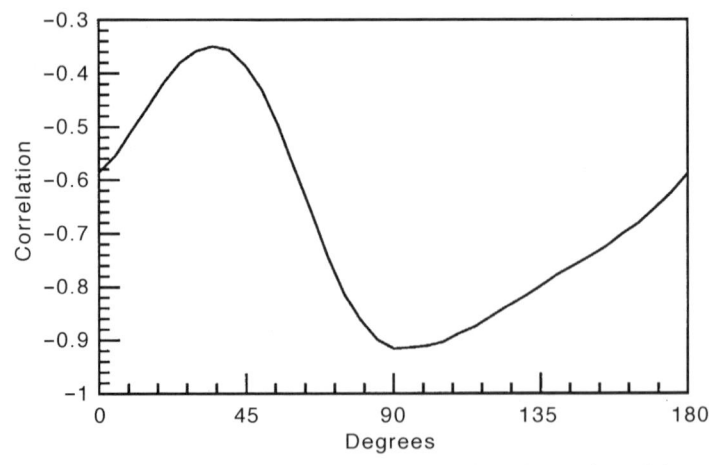

Fig. 10.3 With polarised NIR transmitted through pork, rotation of the analyser (degrees) affects the correlation of transmittance with sarcomere length.

bands are the A bands, where thick (myosin) and thin (actin) filaments overlap. But maximum transmittance now is at an analyser angle >90°, because of the optical path difference of the myofibre.

NIR is used to minimise scattering and enables measurements to be made on thin slices. The transmittance of NIR polarised light increases as sarcomere length increases from 1.2 to 1.5 μm, it peaks at sarcomere length 1.5 μm, then decreases as sarcomere length increases to 3.5 μm. Thus, if there are no cold-shortened sarcomeres, the transmission of polarised NIR may be correlated with sarcomere length (Fig. 10.3), and with functional properties in meat processing. For example, NIR birefringence is correlated with WHC and cooking losses in processed turkey meat ($r = 0.8$ and $r = -0.82$, respectively; Swatland and Barbut 1995), and a probe for on-line use has been developed (Swatland, 1996).

10.8 Connective tissue

Connective tissues such as elastin and Type I collagen are notorious sources of meat toughness. Elastin is always heat-stable, while Type I collagen may resist normal cooking if it is highly cross-linked, as in beef from an old cow. Gelatinised collagen is important in meat processing. It may be desirable in certain products, such as pork pies and cooked hams. However, in excess, when it has been added as a filler to a comminuted product, gelatin caps become a defect.

10.8.1 Carcass measurements

Both elastin and collagen, as well as heat-stable pyridinoline cross-links, are autofluorescent (Odetti *et al.*, 1994). When illuminated with UV light at 370 nm,

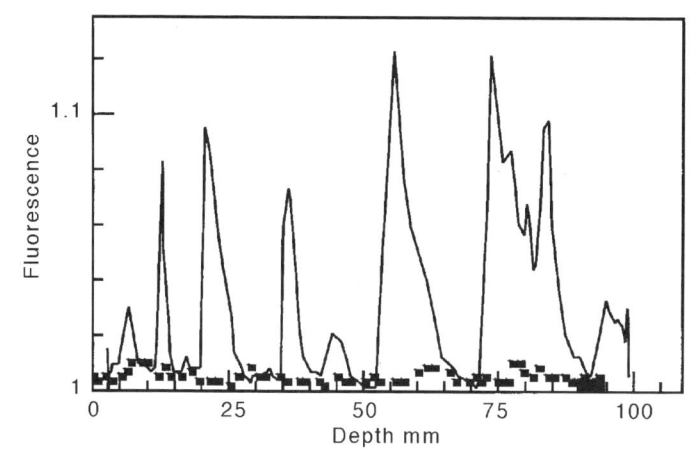

Fig. 10.4 UV fluorescence transects through beef semitendinosus with high sensory toughness (line) and *longissimus dorsi* with low sensory toughness (solid squares).

they fluoresce blue-white from about 400 to 550 nm. Measurements made with a UV fibre-optic probe may be correlated with sensory tenderness of beef (Fig. 10.4; Swatland *et al.*, 1994). The main problem is that connective tissue is only one source of beef toughness. The probe may be effective for sorting carcasses in a population where connective tissue toughness is dominant, but may fail if some other problem such as short sarcomere length is dominant.

10.8.2 In meat processing

Connective tissue levels in ground beef may be a problem if too many meat scraps with a high content of tendon are worked into a product. The result may be a gritty texture for hamburger, or excessive gelatin formation in a cooked product. Elastin derived from elastic ligaments has virtually the same fluorescence emission spectrum as Type I collagen from tendon and ligaments. This enables fluorescence emission ratios to be used to predict total connective tissue levels.

Under experimental conditions, collagen fluorescence in comminuted mixtures of chicken skin and muscle may be measured through a quartz-glass rod with a window onto the product (Swatland and Barbut, 1991). High proportions of skin decrease the gel strength of the cooked product ($r = -0.99$), causing high cooking losses ($r = 0.99$) and decreased WHC ($r = -0.92$). Fluorescence intensity may be strongly correlated with skin content ($r > 0.99$ from 460 to 510 nm) and, thus, may be strongly correlated with gel strength, cooking losses and fluid-holding capacity (Fig. 10.5). Correlations would be weaker in a practical application, but still adequate for feed-back control of product composition.

One of the problems in calibration is pseudofluorescence – reflectance of the upper edge of the excitation band-pass. This occurs because excitation and

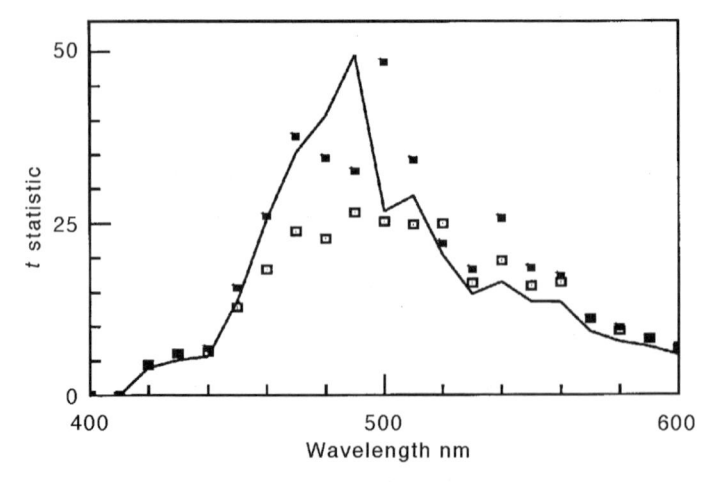

Fig. 10.5 Spectral distribution of the t-statistic for the correlation of fluorescence emission with skin content (line), gel strength (solid squares) and cooking losses (empty squares) in mixtures of chicken breast meat and skin.

emission maxima are fairly close, and the filters and dichroic mirrors used to separate excitation from emission are not perfect. Thus, the standard used to calibrate the apparatus for the measurement of relative fluorescence should have a similar reflectance to meat. Clean aluminium foil with a dull surface is a fairly close match to meat.

10.9 Marbling and fat content

Marbling is composed of clumps of adipose cells, mostly located between bundles of myofibres. It is notoriously difficult to measure objectively. Unless elaborate precautions are taken using polarised light and spectrophotometric identification, VIA may fail to separate marbling from connective tissue and glistening specular reflectance. Marbling distribution is anisotropic (because it follows muscle fasciculi) and subjective grades may have a non-linear relationship to extractable lipid. Marbling may be detected and distinguished from connective tissue using a multichannel probe (Swatland, 2000), but there is little or no application for meat processing. As increasing amounts of subcutaneous or intermuscular fat are comminuted together with lean muscle, reflectance increases at all visible wavelengths until reaching the spectrum of 100% adipose tissue (Franke and Solberg, 1971).

10.9.1 Yellow and soft fat

Factors in the animal's diet causing yellowing of fat (from carotene) and softening (low-melting point fatty acids) are readily detectable with a fibre-optic probe (Swatland, 1988; Irie and Swatland, 1992).

10.10 Meat flavour

On-line testing of meat aroma and, hence, flavour is possible (Linforth, 2000). The technology for the detection of volatiles continues to develop (Dickinson *et al.*, 1996; White *et al.*, 1996). Chemiluminescence may be used to assess the conditioning of beef (Yano *et al.*, 1996a) and for meat spoilage (Yano *et al.*, 1996b). Semiconductor gas sensors also are suitable for use on meat (Berdagué and Talou, 1993) and may be used to identify various types of meat (Neely *et al.*, 2001) and warmed-over flavour (Grigioni *et al.*, 2000).

10.11 Boar taint

In North America, male pigs for meat production are usually castrated, despite deleterious effects on rate of growth and feed efficiency. The primary reason is the risk of boar taint in the fat. This is an unpleasant odour that occurs when the fat is heated. However, boar taint is only detectable in pork from young males by a small percentage of consumers. Thus, the problem owes its notoriety to pork obtained from old boars that have been used for breeding, and it need not be a problem in intact young males slaughtered at a relatively light weight.

A major cause of boar taint is the concentration of sex steroids in the fat, such as 5α-androst-16-ene-3-one, commonly called androstenone. Androstenone smells strongly of animal urine. Other testicular steroids smell like musk. Other causes of boar taint include skatole and indole, with a faecal odour produced from tryptophan in the gut.

Carcasses may be tested subjectively on-line using a hot iron. Objective methods are available for skatole and androstenone. Both require a sample to be removed from the carcass for analysis, but the methods are sufficiently rapid to be considered commercially (Andersen *et al.*, 1993). On-line testing for boar taint may be possible using the Alabaster-UV semiconductor system (Berdagué and Talou, 1993). The sensor is mounted in a stainless-steel chamber with a UV source and gases are exposed to a cycle of ozone and then flushed by air.

10.12 Emulsions

10.12.1 Electrical method

The emulsifying capacity of meat may be evaluated by progressively adding oil during the formation of an experimental meat emulsion (Cunningham and Froning, 1972). Initially, the oil is trapped in a stable meat emulsion but further addition of oil causes the emulsion to break down. Emulsion break-down may be detected electrically (Webb *et al.*, 1970). Electrodes are located in an emulsion formed from salt-extracted meat protein solution using a mixing propeller (Fig. 10.6).

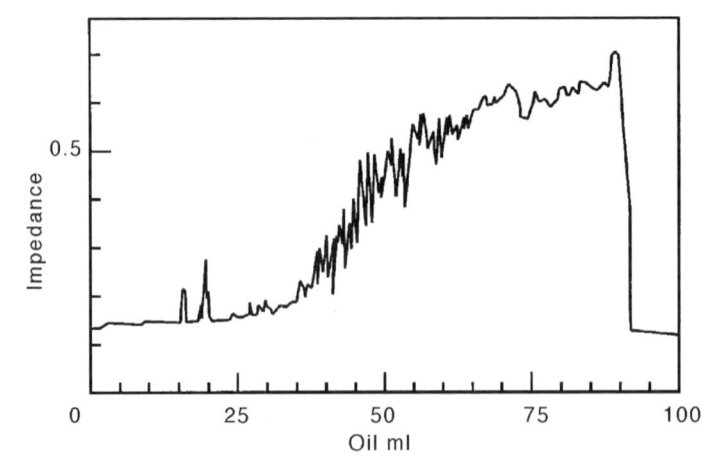

Fig. 10.6 Impedance changes during the formation and break of an emulsion as oil is added.

10.12.2 Optical method
During emulsion formation, the reduction in particle size caused by chopping has little optical effect. But the inclusion of air bubbles increases light scattering, thus increasing product paleness (Palombo et al., 1994). Scattering decreases if batters are stored or if there is a redistribution of air from small to larger bubbles. These changes may be monitored using fibre-optics, as in the gelation of whey proteins (Barbut, 1996).

10.13 Measuring changes during cooking

10.13.1 Temperature
Many off-the-shelf methods are available. Iron-constantan and chromel-alumel thermocouples are in common use. A thermocouple uses two dissimilar electrical conductors to generate a thermoelectric voltage proportional to the temperature difference between the two end junctions. Junctions may be exposed or covered, and/or grounded. Alternatively, a thermistor is a resistive circuit component, usually a two-terminal semiconductor, whose resistance decreases as temperature increases. Another method is to use a Callendar's thermometer (platinum RTD probe) with a thin film or wire of pure platinum. This has a high resistance and facilitates measuring the change in resistance with temperature. Microwave cooking may be monitored with a temperature-sensitive fluorescent cap on a quartz optical fibre. Infra-red radiometers calibrated as thermometers are available for remote sensing of temperature. The time constant is the length of time that a sensor takes to read the true temperature of the sample. Heat conduction along wires to the sensor must be minimal. Signal transmitters may be required for long wires.

10.13.2 Colour changes

Meat with an appreciable myoglobin concentration changes from red to greyish-brown when cooked (Pearson and Tauber, 1984). The brown pigments formed during cooking include denatured globin nicotinamide hemichromes (Tappel, 1957), denatured myoglobin (Bernofsky *et al.*, 1959), Maillard reaction products (Pearson *et al.*, 1962), metmyochromogen (Tarladgis, 1962) and haematin di-imadazole complexes (Ledward, 1974). Failure of meat to brown is a common commercial problem, particularly with the nicotinamide-denatured globin haemachromes of cooked poultry meat (Claus *et al.*, 1994).

Cooking increases reflectance at most wavelengths, but especially around 560 nm (Fig. 10.7). Reduction but not complete loss of myoglobin absorbance around 560 nm also occurs. Between 0 and 40°C there are small changes in reflectance around the Soret absorbance band for myoglobin but, for practical purposes, the colour may be regarded as constant. Above 40°C, however, reflectance starts to increase, peaking at 70°C, and then decreasing as the temperature is increased beyond this.

For each wavelength, the slope for the change in reflectance per degree of temperature ($\Delta R/\Delta t$) is influenced by the magnitude of the initial reflectance. The slope is high for reflectance peaks and low for valleys. This obscures what is really happening at different wavelengths. Corrections may be made by dividing the temperature slope ($\Delta R/\Delta t$) for each wavelength by the mean reflectance for each wavelength at all temperatures. This may be called the adjusted temperature coefficient for each wavelength and is useful in finding the temperatures at which major changes in colour occur. Changes around the Soret absorbance band are particularly complex (Fig. 10.8). To see how these changes affect subjective colour perception, chromaticity coordinates may be found from spectra using the weighted ordinate method. The results are not correct (for reasons evident in Fig. 10.1) but may be quite revealing. For example, Fig. 10.9,

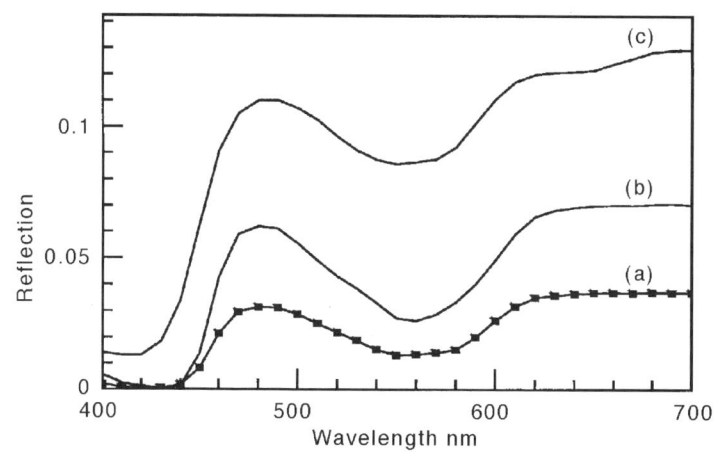

Fig. 10.7 Progressive changes in fibre-optic reflectance as lamb is cooked from 25° (a), to 50° (b), and to 80°C (c).

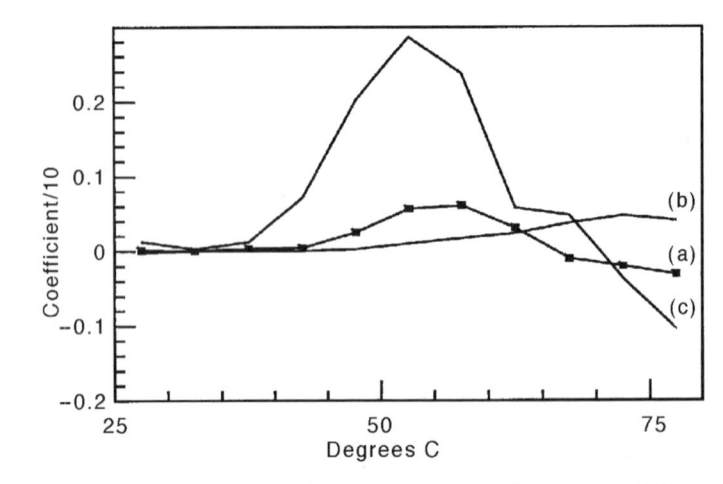

Fig. 10.8 Adjusted temperature coefficients 400 nm (a), 430 nm (b) and 480 nm (c) from the same experiment as Fig. 10.7.

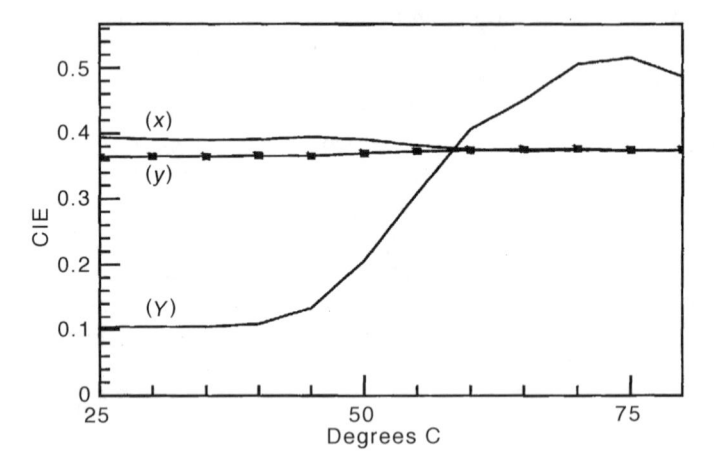

Fig. 10.9 CIE chromaticity coordinates x and y and paleness (luminosity, Y) calculated by the weighted ordinate method from spectra such as those shown in Fig. 10.7.

shows that cooking of lamb (as in Fig. 10.7), causes only minor changes in chromaticity coordinates x and y, but major changes in paleness (Y).

10.13.3 Gelatinisation
Initial heat-shrinkage of the endomysium during cooking causes the expression of fluid from the myofibre, while later gelatinisation of the perimysium creates the characteristic texture of cooked meat. Loss of collagen birefringence may be used to monitor the gelatinisation of collagen (Fig. 10.10).

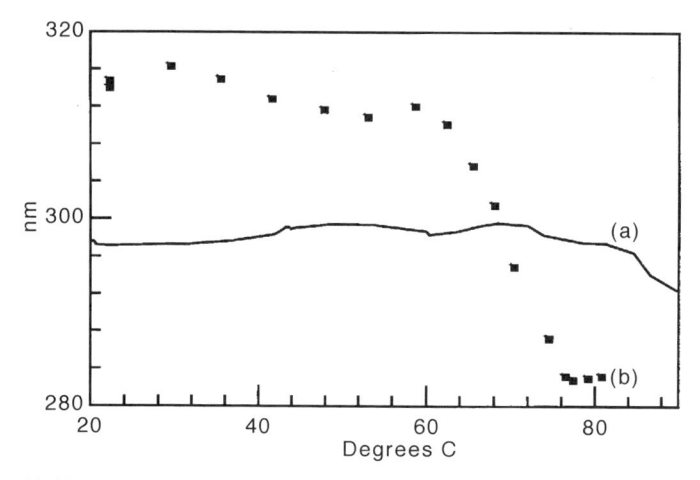

Fig. 10.10 Two extremes of gelatinisation – tough beef perimysium (a) versus perimysium from cooked turkey rolls with a crumbly texture (b). The optical path difference of birefringence (nm) is affected by the thickness of the sample (so differences in absolute value of a versus b are irrelevant), but turkey sample shows early gelatinisation starting at 60°C whereas the beef is unchanged until much higher temperatures.

10.14 Conclusion

There are clearly many ways to obtain useful information for meat processing from on-line sensors, both in selecting carcasses and in process control. The ready availability of software for multivariate analysis and neural networks makes it very tempting to adopt a strictly empirical approach – measuring a batch of known samples, then developing a prediction equation for on-line use. This may be unreliable if there is any change in the interaction between sample and apparatus, or in the nature of the samples evaluated. In the first category, simply moving the apparatus may necessitate a new prediction equation. In the second category, samples whose main variable is lipid content will not perform in the same way as samples dominated by pH-related protein functionality. The price of a trustworthy sensor is knowing how it works. Only then can the programmer anticipate the unexpected.

10.15 Sources of further information and advice

Further information on meat sensors may be obtained from Swatland (1995). For those working with fibre-optic sensors, many of the components and software concepts are the same as those used in microphotometry (Swatland, 1998). *Meat Science* is the obvious journal to watch for further developments. My own on-going studies usually appear in *Food Research International* which originates in Canada.

10.16 References

ABERLE E D, STADELMAN W J, ZACHARIAH G L and HAUGH C G (1971), 'Impedance of turkey muscle: relation to post mortem metabolites and tenderness', *Poultry Sci*, 50, 743–746.

ANDERSEN J R, BORGGAARD C, NIELSEN T and BARTON-GADE P A (1993), 'Early detection of meat quality characteristics; the Danish situation', *39th International Congress of Meat Science and Technology, Calgary, Alberta*, pp. 153–164.

ANDERSEN J R, BORGGAARD C, RASMUSSEN A J and HOUMØLLER L P (1999), 'Optical measurements of pH in meat', *Meat Sci*, 53, 135–141.

BANFIELD F H (1935), 'The electrical resistance of pork and bacon Part 1 Method of measurement', *J Soc Chem Ind Trans Comm*, 54, 411T.

BARBUT S (1996), 'Use of fibre optics to study the transition from clear to opaque whey protein gels', *Food Res Internat*, 29, 465–469.

BENDALL J R (1962), 'Some aspects of the denaturation of proteins of meat' in J M Leitch, *Proceedings of the First International Congress of Food Science and Technology*, Gordon and Breach Science Publishers, New York.

BENDALL J R (1973), 'Postmortem changes in muscle' in G H Bourne, *The Structure and Function of Muscle*, Vol II, Part 2, pp. 244–309, Academic Press, New York.

BENDALL J R and SWATLAND H J (1988), 'A review of the relationships between pH and physical aspects of pork quality', *Meat Sci*, 24, 85–126.

BENDALL J R and WISMER-PEDERSEN J (1962), 'Some properties of the fibrillar proteins of normal and watery pork muscle' *J Food Sci*, 27, 144–159.

BERDAGUÉ J-L and TALOU T (1993), 'Examples d'application aux produits carnés des senseurs de gaz à semi-conducteurs' *Sci Aliments*, 13, 141–148.

BERNOFSKY C, FOX J B and SCHWEIGERT B S (1959), 'Biochemistry of myoglobin VII The effect of cooking on myoglobin in beef muscle' *Food Res*, 24, 339–343.

BIRTH G S, DAVIS C E and TOWNSEND W E (1978), 'The scatter coefficient as a measure of pork quality', *J Anim Sci*, 46, 639–645.

BOWEN W J (1949), 'The absorption spectra and extinction coefficients of myoglobin', *J Biol Chem*, 179, 235–245.

BRØNDUM J, MUNCK L, HENCKEL P, KARLSSON A, TORNBERG E and ENGELSEN S B (2000), 'Prediction of water-holding capacity and composition of porcine meat by comparative spectroscopy', *Meat Sci*, 55, 177–185.

CALLOW E H (1936), 'The electrical resistance of muscular tissue and its relation to curing', *Special Report 75 Department of Scientific and Industrial Research Food Investigation Board London* pp. 75–81.

CHEN Y R (1992), 'Nondestructive technique for detecting diseased poultry carcasses', *Soc Photo-Optical Instrumentation Engineers*, 1796, 310–321.

CHEN Y R and D R MASSIE (1993), 'Visible/near-infrared reflectance and interactance spectroscopy for detection of abnormal poultry carcasses',

Trans Amer Soc Agric Engineers, 36, 863–869.

CLAUS J R, SHAW D E and MARCY J A (1994), 'Pink color development in turkey meat as affected by nicotinamide cooking temperature chilling rate and storage time', *J Food Sci*, 59, 1283–1285.

CUNNINGHAM F E and FRONING G W (1972), 'A review of factors affecting emulsifying characteristics of poultry meat', *Poultry Sci*, 51, 1714–1720.

DICKINSON T A, WHITE J, KAUER J S and WALT D R (1996), 'A chemical-detecting system based on a cross-reactive optical sensor array', *Nature*, 382, 697–700.

DOWNEY G and BEAUCHENE D (1997), 'Discrimination between fresh and frozen-then-thawed beef M longissimus dorsi by combined visible-near infrared reflectance spectroscopy', *Meat Sci*, 45, 353–356.

EPSTEIN B R and FOSTER K R (1983), 'Anisotropy in the dielectric properties of skeletal muscle', *Med Biol Eng Comput*, 21, 51–55.

FORREST J C, MORGAN M T, BORGGAARD C, RASMUSSEN A J, JESPERSEN B L and ANDERSEN J R (2000), 'Development of technology for the early post mortem prediction of water holding capacity and drip loss from fresh pork', *Meat Sci*, 55, 115–122.

FORTIN A and RAYMOND D P (1988), 'The use of the electrical characteristics of muscle for the objective detection of PSE and DFD in pork carcasses under commercial conditions', *Can Inst Food Sci Technol J*, 21, 260–265.

FOX J B (1987), 'The pigments of meat' in Price J F and Schweigert B S *The Science of Meat and Meat Products*, pp 193–216 Westport Connecticut: Food & Nutrition Press.

FRANKE W C and SOLBERG M (1971), 'Quantitative determination of metmyoglobin and total pigment in an intact meat sample using reflectance spectrophotometry', *J Food Sci*, 36, 515–519.

GARIÉPY C, JONES S D M, TONG A K W and RODRIGUE N (1994), 'Assessment of the Colormet fiber optic probe for the evaluation of dark cutting beef', *Food Res Internat*, 27, 1–6.

GIELEN F L H, CRUTS H E P, ALBERS B A, BOON K L, WALLINGA-DE-JONGE W and BOOM H B K (1986), 'Model of electrical conductivity of skeletal muscle based on tissue structure', *Med Biol Eng Comput*, 24, 34–40.

GRIGIONI G M, MARGARÍA C A, PENSEL N A, SÁNCHEZ G and VAUDAGNA S R (2000), 'Warmed-over flavour analysis in low temperature – long time processed meat by an electronic nose', *Meat Sci*, 56, 221–228.

HAMM R (1960), 'Biochemistry of meat hydration', *Adv Food Res*, 10, 355–436.

IRIE M and SWATLAND H J (1992), 'Assessment of porcine fat quality by fiber-optic spectrophotometer', *Asian-Austral J Anim Sci*, 5, 753–756.

ISAKSSON T, NILSEN B N, TØGERSEN G, HAMMOND P R and HILDRUM K I (1996), 'On-line, proximate analysis of ground beef directly at a meat grinder outlet', *Meat Sci*, 43, 245–253.

JAUD D, WEISSE K, GEHLEN K-H and FISCHER A (1992), 'pH and conductivity. Comparative measurements on pig carcases and their relationships to drip loss', *Fleischwirts*, 72, 1416–1418.

JEREMIAH L E and PHILLIPS D M (2000), 'Evaluation of a probe for predicting beef tenderness', *Meat Sci*, 55, 493–502.
KAPANY N S (1967), *Fiber Optics, Principles and Applications*, New York: Academic Press.
KLEIBEL A, PFÜTZNER H and KRAUSE E (1983), 'Measurement of dielectric loss factor. A routine method of recognizing PSE muscle', *Fleischwirts*, 63, 1183–1185.
LEDWARD D A (1974), 'On the nature of the haematin-protein bonding in cooked meat', *J Food Technol*, 9, 59–68.
LEE S, NORMAN J M, GUNASEKARAN S, VAN LAACK R L J M, KIM B C and KAUFFMAN R G (2000), 'Use of electrical conductivity to predict water-holding capacity in post-rigor pork', *Meat Sci*, 55, 385–389.
LINFORTH R S T (2000), 'Developments in instrumental techniques for food flavour evaluation', *J. Sci Food Agric*, 80, 2044–2048.
LU J, TAN J, SHATADAL P and GERRARD D E (2000), 'Evaluation of pork color by using computer vision', *Meat Sci*, 56, 57–60.
MACDOUGALL D B and JONES S J (1980), 'Use of a fibre optic probe for segregating pale soft exudative and dark firm dry carcasses', *J Sci Food Agric*, 31, 1371.
MARCHELLO M J and SLANGER W D (1992), 'Use of bioelectrical impedance to predict leanness of boston butts', *J Anim Sci*, 70, 3443–3450.
MORTON R A (1975), *Biochemical Spectroscopy*, Adam Hilger, Bristol.
NEELY K, TAYLOR C, PROSSER O and HAMLYN P F (2001), 'Assessment of cooked alpaca and llama meats from the statistical analysis of data collected using an electronic nose', *Meat Sci*, 58, 53–58.
NORRIS K H (1984), 'Reflectance spectroscopy' in Stewark K K and Whitaker J R *Modern Methods of Food Analysis,* AVI, Westport, Connecticut.
OFFER G, KNIGHT P, JEACOCKE R, ALMOND R, COUSINS T, ELSEY J, PARSONS N, SHARP A, STARR R and PURSLOW P (1989), 'The structural basis of the water-holding appearance and toughness of meat and meat products', *Food Microstruct*, 8, 151–170.
ODETTI P, PRONZATO M A, NOBERASCO G, COSSO L, TRAVERSO N, COTTALASSO D and MARINARI U M (1994), 'Relationships between glycation and oxidation related fluorescences in rat collagen during ageing. An *in vivo* and *in vitro* study', *Lab Invest*, 70, 61–67.
PALOMBO R, VAN ROON P S, PRINS A, KOOLMEES P A and KROL B (1994), 'Changes in lightness of porcine lean meat batters during processing', *Meat Sci*, 38, 453–476.
PEARSON A M and TAUBER F W (1984), *Processed Meats*, AVI, Westport Connecticut.
PEARSON A M, HARRINGTON G, WEST R G and SPOONER M E (1962), 'The browning produced by heating fresh pork. I. The relation of browning intensity to chemical constituents and pH', *J Food Sci*, 27, 177–181.
PFÜTZNER H and FIALIK E (1982), 'A new electrophysiological method for the rapid detection of exudative porcine muscle', *Zbl Vetmed A*, 29, 637–645.

RAY G B and PAFF G H (1930), 'A spectrophotometric study of muscle hemoglobin', *Am J Physiol*, 94, 521–528.
ROME E (1967), 'Light and x-ray diffraction studies of the filament lattice of glycerol-extracted rabbit psoas muscle', *J Mol Biol*, 27, 591–602.
ROWAN A N and BATE SMITH E C (1939), 'Changes in the electrical resistance of muscle associated with the onset of rigor mortis', *Ann Rep Food Invest Board, London*, 11–15.
SALÉ P (1972), 'Appareil de détection des viandes décongeelées par mesure de conductance électrique', *Bull Inst Internat Froid, Suppl*, 2, 265–275.
SCHMITTEN F, SCHEPERS K-H, JÜNGST H, REUL U and FESTERLING A (1984), 'Fleischqualität beim Schweine. Untersuchungen zu deren Erfassung', *Fleischwirts*, 64, 1238–1242.
SEIDLER D, BARTNICK E and NOWAK B (1987), 'PSE detection using a modified MS Tester compared with other measurements of meat quality on the slaughterline', in Tarrant P V, Eikelenboom G and Monin G, *Evaluation and Control of Meat Quality in Pigs*, pp 175–190, Martinus Nijhof, Dordrecht, Netherlands.
SWANTEK P M, CRENSHAW J D, MARCHELLO M J and LUKASKI H C (1992), 'Bioelectrical impedance: a nondestructive method to determine fat-free mass of live market swine and pork carcasses', *J Anim Sci*, 70, 169–177.
SWATLAND H J (1988), 'Carotene reflectance and the yellowness of bovine adipose tissue measured with a portable fibre-optic spectrophotometer', *J Sci Food Agric*, 46, 195–200.
SWATLAND H J (1995), *On-line Evaluation of Meat*, Technomic Press, Lancaster, Pennsylvania.
SWATLAND H J (1996), 'Effect of stretching pre-rigor muscle on the back-scattering of polarized near-infrared', *Food Res Internat*, 29, 445–449.
SWATLAND H J (1998), *Computer Operation for Microscope Photometry*, CRC Press, Boca Raton, Florida.
SWATLAND H J (2000), 'Connective and adipose tissue detection by simultaneous fluorescence and reflectance measurements with an on-line meat probe', *Food Res Internat*, 32, 749–757.
SWATLAND H J and BARBUT S (1990), 'Fibre-optic spectrophotometry for predicting lipid content, pH and processing loss of comminuted meat slurry. *Internat J Food Sci Technol*, 25, 519–526.
SWATLAND H J and BARBUT S (1991), 'Fluorimetry via a quartz-glass rod for predicting the skin content and processing characteristics of poultry meat slurry', *Internat J Food Sci Technol*, 26, 373–380.
SWATLAND H J and BARBUT S (1995), 'Optical prediction of processing characteristics of turkey meat using UV fluorescence and NIR birefringence', *Food Res Internat*, 28, 325–330.
SWATLAND H J and DUTSON T R (1984), 'Postmortem changes in some optical, electrical and biochemical properties of electrically stimulated beef carcasses', *Can J Anim Sci*, 64, 45–51.
SWATLAND H J, GULLETT E, HORE T and BUTTENHAM S (1994), 'UV fiber-optic

probe measurements of connective tissue in beef correlated with taste panel scores for chewiness', *Food Res Internat*, 28, 23–30.

TAPPEL A L (1957), 'Reflectance spectral studies of the hematin pigments of cooked beef', *Food Res*, 22, 404–407.

TARLADGIS B G (1962), 'Interpretation of the spectra of meat pigments. I. Cooked meats', *J Sci Food Agric*, 13, 481–484.

TØGERSEN G, ISAKSSON T, NILSEN B N, BAKKER E A and HILDRUM K I (1999), 'On-line NIR analysis of fat, water and protein in industrial scale ground meat batches', *Meat Sci*, 51, 97–102.

WEBB N B, IVEY F J, CRAIG H B, JONES V A and MONROE R J (1970), 'The measurement of emulsifying capacity by electrical resistance', *J Food Sci*, 35, 501–504.

WHITE J, KAUER J S, DICKINSON T A and WALT D R (1996), 'Rapid analyte recognition in a device based on optical sensors and the olfactory system', *Anal Chem*, 68, 2191–2202.

YANO Y, MIYAGUCHI N, WATANABE M, NAKAMURA T, YOUDOU T, MIYAI J, NUMATA M and ASANO Y (1996a), 'Monitoring of beef ageing using a two-line flow injection analysis biosensor consisting of putrescine and xanthine electrodes', *Food Res Internat*, 28, 611–617.

YANO Y, YOKOYAMA K and KARUBE I (1996b), 'Evaluation of meat spoilage using a chemiluminescence-flow injection analysis system based on immobilized putrescine oxidase and a photodiode', *Lebens Wissens Technol*, 29, 498–502.

11

Microbiological hazard identification in the meat industry

P. J. McClure, Unilever Research, Sharnbrook

11.1 Introduction

This chapter describes the main microbiological hazards associated with meat and meat products. Meat is associated with a variety of pathogenic microorganisms some of which we are relatively familiar with. Worryingly, the past two decades has seen the emergence of 'new' microbial agents capable of causing disease. These recent developments and the continued increase in the number of cases of foodborne illness have resulted in widespread concern for consumer safety. Many of the recently emerging foodborne pathogens are associated with meat from poultry, cattle and other animals, and they do not necessarily cause overt signs of illness in these animals. The appearance of these pathogens is, generally speaking, a global trend and is not restricted to particular geographic locations. The reasons for their emergence and spread are poorly understood and it is suspected that the shift to a global economy, international trade, and changes in the livestock industry may have contributed to these recent developments. No doubt, some of this is also due to improved surveillance, reporting and methods of detection.

The first principle of HACCP (conducting a hazard analysis) includes determination of the food safety hazards likely to occur and these may come from a variety of different sources. The list of hazards associated with meat and meat products includes protozoal parasites, helminths, arthropods, viruses, prions and bacteria, arguably the most important of these categories. Many bacteria are common inhabitants of animal intestines and their presence may be transient or long term. In addition to livestock being a source of infection, through internal carriage or hide contamination, pathogens may be introduced at any point in slaughter, processing, packaging, distribution and preparation of food. The bacteria and main protozoal parasites considered to be hazards in meat and meat products are discussed.

Understanding the origin of these different pathogens and their fate during processing is essential for control of the hazards and managing the risk posed by their presence. The analytical methods used to detect the presence of many of these pathogens have advanced significantly in recent years. These improvements in detection and characterisation methodologies now allow for the tracking of different pathogens through processing, enabling identification of the origin of these agents. These developments also allow links to be made between apparently unrelated (e.g. sporadic) cases. The specific methodologies used for enumeration and detection of particular pathogens are not within the scope of this chapter, but the general approaches and recent advances will be discussed. A number of future trends likely to impact on the hazards associated with meat and meat products are also discussed in this chapter. Genetic evolution will continue to contribute to the appearance of new pathotypes or pathovars of microorganisms and this will result in pathogens that possess new combinations of known and unknown virulence factors.

11.2 The main hazards

11.2.1 *Salmonellae*

The genus *Salmonella* is subdivided into over 2000 serotypes or serovars, based on unique antigenic structure. Further subdivision is possible through phage- and biotyping. Salmonellae are primarily intestinal parasites of humans and many animals, including rodents, wild birds and domestic animals. Recently, the nomenclature of salmonellae has been revised since modern taxonomic methods suggested that all serotypes of *Salmonella* probably belonged to one DNA-hybridisation group. *S. enterica* was originally subdivided into seven sub-groups, *S. enterica* subspp. *enterica, salamae, arizonae, diarizonae, houtenae, bongori* and *indica*. *S. enterica* subspp *bongori* has since been elevated to species level. Only serotypes of subsp. *enterica* are still named (e.g. *S. enterica* subsp. *enterica* serotype Typhimurium or *S.* Typhimurium or simply Typhimurium) indicating that the named serotype is a member of subsp. *enterica*.

Although many salmonellae are potentially pathogenic in animals, the response to infection by the same serotype in different animals may be different. Although a large number of serotypes have been identified, less than 10% have been isolated from man and other animals. Salmonellae are most often isolated from cattle and poultry. Serotypes are classified as either host-adapted or non-host-adapted, depending on their host range and the majority show no host specificity. Host adapted serotypes rarely cause disease in other hosts. *S.* Dublin is traditionally host adapted to cattle but in some case has shown a tendency to spread to swine and was originally isolated from a child. This serotype can cause severe disease (septicaemia, osteomyelitis, and meningitis) in some individuals.

Salmonellosis in animals
Generally speaking, young animals are more susceptible to salmonellosis than older ones. There are a number of factors that predispose animals to clinical salmonellosis and these include poor sanitation, overcrowding, parturition, transportation and concurrent infections with other pathogens (e.g. parasites, viruses, etc). Many animals, particularly swine and poultry are fed contaminated feed without developing any apparent clinical symptoms. Feed is usually contaminated through meat and bone meal, fish meal or soybean meal with organisms entering these materials during or after processing. Wild birds and rodents also provide a source of contamination from faeces contaminating feed or buildings, and other possible sources include contaminated poultry litter and water courses.

Salmonellosis in cattle usually begins as an enteric infection, commencing with colonisation of the intestine and invasion of the intestinal epithelium. This can be followed by septicaemia, abortion, meningitis, pneumonia or arthritis, after entry into the bloodstream. The two most important serotypes in cattle are *S*. Typhimurium and *S*. Dublin. Typhimurium is found worldwide and Dublin is found mainly in Europe, western US and South Africa. Antibiotic resistant strains of Dublin are now spreading to the north-eastern US.[1] Persistently shedding carrier animals are thought to be the primary reservoir of Dublin, with most infections occurring when animals are on pasture. Unlike Dublin, disease caused by Typhimurium is usually self-limiting, since persistent shedders are not the norm. Typhimurium is known for primarily enteric disease states whereas Dublin causes primarily septicaemia. The most important mode of transmission for Typhimurium is the faecal-oral route. Both serotypes cause serious disease with mortality rates sometimes as high as 50–75%. Other serotypes that have caused infection in cattle include Anatum, Montevideo, Newport and Saint-paul.

In sheep, serotypes associated with disease include Abortus ovis, Dublin, Montevideo and Typhimurium. Infection in flocks results from introduction of infected sheep and ingestion of the organism. Dublin and Typhimurium cause enteritis, septicaemia and abortion, similar to the conditions observed in cattle. For sheep and goats, Typhimurium infection can come from a variety of environmental and animal sources whereas cattle are the usual source of Dublin.

The serotypes most frequently associated with disease in pigs are Choleraesuis and Typhimurium and other serotypes that can cause disease in susceptible animals include Anatum, Derby, Heidelberg, Newport and Panama. Choleraesuis causes paratyphoid, Dublin causes enteritis and meningoencephalitis and Typhimurium and other serotypes cause enteritis and septicaemia. Heidelberg can also produce severe catarrhal enterocolitis.

In poultry, serotypes causing disease include Agona, Bareilly, Hadar, Oranienburg, Typhimurium, Gallinarum and Pullorum. The last two of these cause fowl typhoid and bacillary white diarrhoea. Pullorum is now rarely isolated in the United States and northern Europe, due to successful eradication programmes, but is of increasing importance in Latin America, the Middle East, the Pacific Rim, Africa and some parts of southern Europe. The incidence of Gallinarum has also been reduced due to changes in husbandry and through eradication of Pullorum, with which it shares common antigens. In the areas where Pullorum has been eradicated, Typhimurium is often found, causing paratyphoid. Typical conditions of paratyphoid in poultry include enteritis, diarrhoea and septicaemia.

During 1985/1986, *S.* Enteritidis PT4 emerged as a 'new' problem in poultry in Europe. In 1993, the first outbreak of PT4 occurred in the US, and the number of isolations from eggs and the farm environment of laying flocks suggests that eggs have had a major contribution to the dramatic increase in associated human illness. Enteritidis is an invasive serotype and has achieved prominence because of its association with poultry eggs. Although eggs have been recognised as a source of infection for Typhimurium, the incidence of food poisoning cases from this source has always thought to have been low.

S. Typhimurium DT104 has recently emerged in cattle populations in particular parts of the world and causes severe diarrhoea, with an associated mortality rate of 50–60%. Long-term carriage (up to 18 months following an outbreak) has been observed in many species including cattle.[2]

Salmonellosis in man
Non-typhoidal *Salmonella spp.* are one of the most commonly reported foodborne pathogens in industrialised countries. Symptoms of human salmonellosis include nausea, vomiting, diarrhoea, abdominal cramps and fever, with illness lasting for 3–12 days.[3] Associated clinical conditions also include reactive arthritis, Reiters syndrome, septic arthritis and septicaemia.

Certain serotypes are being increasingly reported as the cause of salmonellosis. In 1989, Typhimurium, Enteritidis, Heidelberg, Hadar and Agona accounted for 57.9% of all serotypes isolated from human infections and accounted for 46.5% of isolations obtained from poultry, in the US. One serotype that has increased significantly in recent years is *S.* Enteritidis. Before 1990, Typhimurium was the most common cause of reported salmonellosis in a number of geographic regions. In 1990, this serotype was overtaken by Enteritidis and is now a major cause of human food poisoning in many countries.[4] In recent years in the UK and western Europe, the predominant phage type responsible for egg-borne salmonellosis is PT4 whereas in the US, although there is no predominant phage type associated with egg-borne infection, PT8 and PT13a are the most commonly isolated phage types.[5] The emergence of other phage types, such as PT6 in the UK, continues to occur, as does the emergence of other types such as *S.* Typhimurium DT104, which is now appearing in Europe, north America and elsewhere.[6] The main reservoir of this pathogen, which often exhibits resistance to multiple antibiotics, is thought to be cattle, but there are reports of increasing incidence in poultry, sheep, pigs and goats. This is in contrast to *S.* Enteritidis, which is mainly associated with eggs and poultry. The invasiveness of DT104 in humans does not appear to be any different to other salmonellae, but an increase in occurrence of severe illness has been reported, with higher proportions of those infected requiring hospitalisation.

11.2.2 *Escherichia coli*
Like salmonellae, the primary habitat of *E. coli* is the intestinal tract of man and other warm-blooded animals. Many *E. coli* are commensal organisms and cause no harm but there are some types that are pathogenic to man and other animals and these are not regarded as part of the normal flora of the human intestine. *E. coli* is also commonly found in external environments (e.g. soil and water) that have been affected by human and animal activity. *E. coli* is divided into more than 170 serogroups based on the somatic (O) antigens, and over 50 flagellar (H) and 100 capsular (K) antigens allow further subdivision into serotypes. Serogrouping and serotyping are used with biotyping, phage typing and enterotoxin production to distinguish strains able to cause infectious disease in man and animals.

There are many types of disease caused by *E. coli* and these depend on the virulence factors present. The known virulence factors include adhesins and colonisation mechanisms, haemolysin, ability to invade epithelial cells and production of a number of toxins including heat labile enterotoxins, heat stable enterotoxins, cytotoxic necrotising factors and vero cytotoxins (or Shiga toxins, Stx1 and 2). The adhesion and colonisation factors include fimbrae, haemaglutinnins and specific adhesins such as the F4 (K88) antigen. The encoding genes of these and other virulence factors may be carried on transmissible plasmids or on the chromosome. There are currently six recognised virulence groups comprising enteropathogenic *E. coli* (EPEC), enterotoxigenic *E. coli* (ETEC), enteroinvasive *E. coli* (EIEC), verotoxigenic *E. coli* (VTEC or Shiga-like toxin producing *E. coli* or SLTEC, which include enterohaemorrhagic *E. coli* or EHEC), enteroaggregative *E. coli* (EAggEC) and diffusely adherent *E. coli* (DAEC).

Disease in animals
E. coli infections occur frequently in many farm animals, including poultry.[7,8] In younger animals, there are principally two types of disease which are systematic colibacillosis, caused by a range of O-serogroups and enteric colibacillosis, caused by a few host-specific enteropathogenic strains. In older animals, a third group of diseases, caused by a number of O-serogroups, causes mastitis in cows and sows. Other, sporadic infections, such as urinary tract infections, can also occur.

Enteric colibacillosis involves oral infection, followed by site specific adhesion to intestinal mucosa, allowing colonisation and release of toxins, which causes damage to intestinal cells or other organs. Colibacillary diarrhoea is an acute disease and occurs most frequently in calves, lambs and piglets, soon after birth and is mainly caused by ETEC. The OK groups in calves and lambs tend to be the same, whilst the OK groups associated with pigs are rarely isolated from other species. In cattle, common serotypes include O8, O9 and O101. In pigs, the most common serotype is O149 and other commonly isolated serotypes include O8, O138, O147 and O157. There appears to have been little change in the serotypes that cause colibacillary diarrhoea or their virulence factors, in recent years. Other forms of enteric colibacillosis are colibacillary toxaemia in pigs, associated with a few serotypes that cause shock in weaner sydrome, haemorrhagic colitis and oedema disease. These forms of disease are thought to relate to production of enterotoxin, endotoxin or a neurotoxin.

Systemic colibacillosis is caused by invasive strains and involves their survival and multiplication in extra-intestinal sites. This occurs frequently in calves, lamb and poultry, but not in pigs, and develops by passage of *E. coli* from the alimentary or respiratory mucosa to the bloodstream. From there, a localised infection, such as meningitis or arthritis in calves and lambs or air sacculitis and pericarditis in poultry, or a generalised infection (colisepticaemia) can develop. O78 and O2 are commonly isolated from poultry, and these serotypes are rarely observed in human isolates. Strains causing colibacillosis in calves belong to relatively few serotypes such as O15:K, O35:K, O137:K79, and O78:K80. The last of these is the most frequently isolated and is also associated with similar conditions in lambs and poultry.

Bovine mastitis is still an important disease and can range from mild forms, which cause clots, milk discoloration and udder swelling, to severe illness that can result in death of the affected animal. Mastitis is caused by a large number of serotypes that are not easily distinguished from strains in normal faeces. Endotoxin and necrotising cytotoxin are thought to play significant roles in this disease.

It is generally thought that VTEC do not cause overt disease in animals, but there is increasing evidence of illness caused by some VTEC in neonatal calves and older animals.[9,10] Bovine VTEC strains share many of the virulence markers with VTEC strains causing infection in man, but in Germany, the intimin-positive strains (those strains causing attaching and effacing lesions, encoded by the *eae* gene) are thought to be restricted to the stx_1 genotype, only capable of producing Stx1. In Brazil, however, a recent study has shown that 60% of the VTEC strains isolated from cattle possess both stx_1 and stx_2. Serogroups O5 and O118 are mainly associated with disease in calves, serogroups O26, O103 and O111 cause disease in calves and humans, but O157 is generally considered to be carried by healthy animals and only associated with disease in humans.

Disease in man
Worldwide, the importance of diarrhoeal and other diseases caused by *E. coli* is immense, particularly in children in developing countries. In developed countries, the incidence of foodborne illness associated with *E. coli* is also significant and appears to be increasing. More worryingly, recent years have seen the emergence of particularly virulent *E. coli*, such as *E. coli* O157:H7 (the predominant VTEC serotype), that are able to cause serious illness in man, with low infectious doses, e.g. fewer than 100 cells. The severity of

disease caused by VTEC can vary from asymptomatic carriage to haemorrhagic colitis (HC), to life-threatening conditions such as haemolytic uraemic syndrome (HUS) in children and thrombotic thrombocytopaenic purpura (TTP) in adults. HUS is the most common cause of acute renal failure in children. For the other pathotypes of *E. coli*, such as EPEC, ETEC and EIEC, clinical studies suggest that more than 10^5 EPEC are necessary to produce diarrhoea, 10^8 ETEC are necessary for infection and diarrhoea and 10^8 EIEC are required to produce diarrhoeal symptoms in healthy adults. EPEC cause a bloody diarrhoea in infants (commonly referred to as infantile diarrhoea), which in some cases may be prolonged; ETEC cause self-limiting diarrhoea, vomiting and fever, and travellers' diarrhoea; EIEC cause shigella-like dysentery; EAggEC cause persistent diarrhoea in children, particularly in developing countries; and DAEC cause childhood diarrhoea.

Even though there are many *E. coli* responsible for disease in animals, most of the *E. coli* pathogenic in man are not the same as those causing illness in animals. Indeed, the principal reservoir for many human pathogenic *E. coli* is believed to be man. However, this is not true of VTEC, including *E. coli* O157:H7, where the main reservoirs are thought to be cattle and other ruminants.[11] Dairy cattle, particularly young animals within herds, have been identified as a reservoir of *E. coli* O157:H7 and other VTEC, and this serotype has also been isolated from other ruminants such as sheep and goats. Hence raw foods of bovine or ovine origin are likely to be vehicles of *E. coli* O157:H7 and other VTEC through faecal contamination during slaughter or milking procedures. In one survey, four per cent of cattle were contaminated prior to slaughter, and after processing, 30% of the carcasses were contaminated.[12] The most frequently implicated vehicle of infection for *E. coli* O157:H7 is undercooked ground beef. Surveys of raw meats for sale have revealed *E. coli* O157:H7 in 2–4% of ground beef, 1.5% of pork and poultry, and 2% of lamb.[13,14] Other studies suggest contamination rates for VTEC in some raw foods of between 16 and 40%. The incidence of *E. coli* O157:H7-related illness is worldwide.

Human infections with VTEC O157:H7 are under nationwide surveillance in a number of countries, but detection of other non-O157 VTEC types is more difficult and performed only by specialist laboratories. Humans are likely to be more exposed to non-O157 VTEC because these strains are more prevalent in animals and as contaminants in foods. The growing number of non-O157 serogroups associated with human disease now include O26, O103, O111, O118 and O145.[15] It is thought that both horizontal gene transfer and intragenic combination are important for evolution of VTEC. Particular regions of the globe show patterns of emergence that appear to be unique, e.g. 20–25% of *E. coli* O157 isolates in Germany are sorbitol +ve. The most common non-O157 serotypes in Germany are O26:H$^-$, O103:H$^-$, O111:H$^-$ and O145. In Italy, the HUS cases caused by O26 strains now outnumber those caused by O157:H7. Studies in animals demonstrate that some of these serogroups, such as O118, are also prevalent in farm animals, and are a likely reservoir.

Human pathogenic strains of VTEC vary in their ability to cause illness, and this depends on virulence attributes and other unknown factors.[16] Pathogenic VTEC O26, O103 and O111 belong to their own lineages and possess unique profiles of virulence determinants that are different from the virulence profile of *E. coli* O157:H7, which is said to contain a more complete repertoire of virulence traits. This may explain why *E. coli* O157:H7 is the predominant VTEC serotype.

One of the problems that is becoming increasingly recognised is that the terminology used to describe diarrhoeagenic *E. coli* is complex and by no means definitive. Since it was first recognised that *E. coli* could cause diarrhoea, an array of virulence factors have been discovered and a number of categories of diarrhoeagenic *E. coli* have been proposed, generally based on the presence of non-overlapping virulence factors. However, EPEC strains and EHEC strains are often regrouped under the name of attaching/effacing *E. coli* (AEEC) on the basis of the ability to produce common attaching and effacing lesions in their hosts.

There are already a number of documented studies describing isolates that do not fit neatly into any of the recognised categories of diarrhoeagenic *E. coli*. This should not be surprising considering that the virulence factors are encoded on 'pathogenicity islands', bacteriophage, transposons and transmissible plasmids. Some of these elements have also been found in other members of the *Enterobacteriaceae*. Therefore, we should anticipate that there will be other combinations of known and currently unknown virulence factors appearing in the group of organisms we currently call *E. coli*, and other members of the *Enterobacteriaceae*.

11.2.3 *Campylobacter jejuni*

C. jejuni was not recognised as a cause of human illness until the late 1970s but is now regarded as the leading cause of bacterial foodborne infection in developed countries.[17] It is one of 20 species and sub-species within the genus *Campylobacter* and family *Campylobacteriaceae*, which also includes four species in the genus *Arcobacter*. Despite the huge number of *C. jejuni* cases currently being reported, the organism does not generally trigger the same degree of concern as *E. coli* or salmonellae, since it rarely causes death and is rarely associated with newsworthy outbreaks of food poisoning. It is among the most common causes of sporadic bacterial foodborne illness. *C. jejuni* is associated with warm-blooded animals, but unlike salmonellae and *E. coli* does not survive well outside the host. *C. jejuni* is susceptible to environmental conditions and does not survive well in food and is, therefore, fortunately relatively easy to control. Food associated illness usually results from eating foods that are re-contaminated after cooking or eating foods of animal origin that are raw or inadequately cooked. The organism is part of the normal intestinal flora of a wide variety of wild and domestic animals, and has a high level of association with poultry.[18] The virulence of the organism, as suggested by the relatively low infectious dose of a few hundred cells and its widespread prevalence in animals, are important features which explain why this relatively sensitive organism is a leading cause of gastroenteritis in man.

Campylobacteriosis in animals
C. jejuni is a commensal organism of the intestinal tract of a wide variety of animals.[19] In cattle, young animals are more often colonised than older animals, and feedlot cattle are more likely to be carriers than grazing animals. Colonisation of dairy herds has been associated with drinking unchlorinated water. Day-old chicks can be colonised with as few as 35 organisms, and most chickens in commercial operations are colonised by four weeks. Reservoirs in the poultry environment include insects, unchlorinated drinking water and farm workers, but probably not feeds, since these are thought to be too dry for survival of campylobacters. It has been proposed that *C. jejuni* is a cause of winter dysentery in calves and older cattle, and experimentally infected calves have shown some clinical signs of disease such as diarrhoea and sporadic dysentery. Nevertheless, the aetiology of naturally occurring disease in animals remains unconfirmed. *C. jejuni* is, however, a known cause of bovine mastitis, and the organisms associated with this condition have been shown to cause gastroenteritis in persons consuming unpasteurised milk from affected animals. Other campylobacters, such as *C. fetus*, are known to cause abortions in sheep and cattle and some strains of *C. sputorum* are known to cause porcine intestinal adenomatosis and regional ileitis in pigs, but these appear to be host-specific diseases.

Campylobacteriosis in man
C. jejuni and *C. coli* are the most common campylobacters associated with diarrhoeal disease in man and are clinically indistinguishable. Also, most laboratories do not attempt to distinguish between the two organisms. It is thought that *C. coli* constitute 5–10% of cases reported as caused by *C. jejuni* in the US. Campylobacteriosis in man is usually characterised by an acute, self-limiting enterocolitis, lasting up to a week. A small proportion (5–10%) of affected individuals suffer relapses. Symptoms of disease often include fever, abdominal pain and diarrhoea, which may be inflammatory, with slimy/bloody stools, or non-inflammatory, with watery stools and absence of blood. Reactive arthritis and bacteraemia are rare complications and infection is also associated with Guillain-Barré syndrome, an autoimmune peripheral neuropathy causing limb weakness. This condition is thought to be associated with particular serotypes (e.g. O:19, O:4 and O:1) capable of producing structures that mimic ganglioside motor neurons. There are a number of pathogenicity determinants that have been suggested for *C. jejuni*, including motility, adherence, invasion and toxin production, but little is known about the mechanism causing disease in man.

There is considerable evidence that poultry is the main vehicle for transmitting *Campylobacter* enteritis in man. Poultry typically has populations of 10^4–10^8 *C. jejuni* per gram of intestinal content and more than 75% of chickens and turkey often carry the organism in their intestinal tract. It is estimated that 30% of retail poultry is contaminated with *C. jejuni* at levels of 10^2–10^4 per gram. Also, serotypes associated with poultry are also frequently associated with illness in humans.

Prior to 1991, *Arcobacter butzleri* and *A. cryaerophilus* were known as aerotolerant *Campylobacter*. These organisms have been associated with abortions and enteritis in animals and enteritis in man. Although both species are known to cause disease in man, most human isolates come from the species *A. butzleri*. There is very little known about the epidemiology, pathogenesis and real clinical significance of Arcobacters, but it is thought that consumption of contaminated food may play a role in transmission of this group of organisms to man. Although Arcobacters have never been associated with outbreaks of foodborne illness, they have been isolated from domestic animals, poultry, ground pork and water.

11.2.4 *Yersinia enterocolitica*
Surveillance data suggest that *Yersinia enterocolitica* is an increasing cause of gastroenteritis in man in Europe and the US.[20] The main cause of yersiniosis during the 1970s and 1980s was thought to be milk and the main serotype associated with disease was O:8. Since then, O:3 has become the predominant serotype in developed countries. The main reservoir of this serotype and other important serotypes, such as O:9, is pigs and consumption of pork is an important risk factor for infection. *Y. enterocolitica* is a component of the intestinal flora of red meat animals, particularly pigs. Poultry is known to carry significant levels of yersinias. Although meat and meat products from goats and sheep have never been implicated in outbreaks of foodborne yersiniosis, small ruminants can harbour the pathogen.

Y. enterocolitica causes gastroenteritis in man and can also cause persistent arthritis. Infection does not, however, always result in diarrhoea. Yersiniosis is usually characterised by abdominal pain, accompanied by fever, with or without diarrhoea. Because of its ability to multiply at refrigeration temperatures, *Y. enterocolitica* is of special interest to particular areas of the food industry. There is relatively little known about the mechanisms of pathogenicity but the genes for invasion of mammalian cells lie on the chromosome and all the other known pathogenicity determinants are found on a plasmid. The other member of this genus that can cause gastroenteritis is *Y. pseudotuberculosis* and large outbreaks of gastroenteritis caused by this organism have been reported in Japan. Disease associated with *Y. pseudotuberculosis* resembles typhoid and is often fatal. *Y. enterocolitica* is not known to cause disease in animals. *Y. pseudotuberculosis* is rarely associated with infections in cattle and sheep, with those in cattle manifesting as pneumonia or abortion.

11.2.5 Staphylococcus aureus

Meat or meat products are not thought to be a major source of *S. aureus* infection in man even though *S. aureus* is an important pathogen in animals. The principle source of transmission between animals and man is unpasteurised milk and cheese made from unpasteurised milk. Outbreaks of staphylococcal food poisoning in man are frequently associated with improper food handling and temperature abuse of foods of animal origin, but it is generally believed that the main source of contamination is food handlers. Nevertheless, strains of *S. aureus* can become endemic in food processing plants and meat can be contaminated from animal or human sources. *S. aureus* has been isolated from cattle carcasses and is also found in raw beef. There is a high correlation between coagulase production and production of enterotoxins, of which there are at least seven heat-stable types associated with food poisoning. In animals, *S. aureus* causes a number of different diseases. The most relevant disease for transmission of the organism to man is bovine and ovine mastitis.

11.2.6 Listeria monocytogenes

Listeriosis is an atypical foodborne disease that has attracted a great deal of attention since the early 1980s mainly because of the severity, high mortality rate and non-enteric nature of the disease. Listeriosis is caused by *L. monocytogenes*, which is found in many environments and is frequently carried in the intestinal tract of many animals, including man. *L. monocytogenes* is often found in healthy animals and humans, with a carrier rate of 10–50% in cattle, poultry and swine. The organism has been isolated from a variety of foods, at levels of 13% in raw meat, 3–4% raw milk and 3–4% of dairy products.[21] Some of the major outbreaks in man have been attributed to meat products such as pork tongue and meat paté. Foods associated with outbreaks have largely been refrigerated, processed and are ready-to-eat. The disease in man is commonly associated with meningitis, septicaemia and abortion. Recent outbreaks, however, have been associated with a milder form of disease characterised by gastroenteritis and flu-like symptoms. In these recent outbreaks, serogroup 1/2 has been implicated whereas many of the human strains isolated previously belong to serovar 4b and to one major ribovar. Serogroup 1/2 accounts for most of the food and environmental isolates and together, serotypes 1/2a, 1/2b and 4b account for up to 96% of the isolates in man. Host factors are likely to play an important role in the susceptibility to listeriosis, together with presence of virulence factors in the organism. Many individuals frequently ingest *L. monocytogenes* without any apparent ill effects. Although listeriosis is a severe disease, the number of cases, compared to some of the other foodborne diseases, is relatively low.

Since *L. monocytogenes* is widely distributed in soil, vegetation and faeces, most animals are exposed to it during their lifetime. *L. monocytogenes* is also commonly found in large numbers in poor-quality silage, and ruminants fed this material are more likely to develop listeriosis. As in humans, predisposing

228 Meat processing

factors are important for disease in animals. The clinical conditions associated with animal listeriosis are similar to the human disease, and include septicaemia, abortion, enteritis and meningoencepahalitis. Interestingly, the isolates associated with processed meats more often originate from the processing environment than from the animal itself.

11.2.7 *Clostridium perfringens*

Strains of *C. perfringens* are classified into 5 types, A–E, according to the extracellular toxins that are formed. Type A is responsible for almost all cases of foodborne disease in humans. Type C very rarely causes foodborne disease and results in necrotic enteritis, but is only a concern in individuals who are nutritionally impaired or whose intestinal proteolytic enzyme activity is reduced. Type A *C. perfringens* is usually present in the soil at concentrations of 10^3–10^4/g. The other types are obligate parasites of domestic animals and do not persist in the soil. Type A strains occur widely in raw and processed foods, but at numbers too low to cause infection. The organism is found in the alimentary tract of nearly all species of warm-blooded animals.

C. perfringens is primarily associated with outbreaks of food poisoning involving handling problems and meat, meat products and poultry are frequently implicated in outbreaks. Illness usually results from ingestion of heavily contaminated food and typical symptoms are diarrhoea and severe abdominal pain. Occasional reports of illness within 2 h of ingestion indicate ingestion of preformed toxin. Sporulation of ingested bacteria is also associated with production of enterotoxin, which is destroyed by heating (e.g. 60 °C for 10 min).

In animals, type A causes yellow lamb disease in sheep and a similar illness (toxin produced in the small intestine) in goats. Type B is known to cause lamb dysentery, and haemorrhagic dysentery in sheep, goats and calves. Type C causes enterotoxaemia in a variety of animal species and type D causes the same disease, but apparently only in sheep. Type E is believed to cause haemorrhagic necrotic enteritis in calves.

11.2.8 *Clostridium botulinum*

Strains of *C. botulinum* are classified into several types (A–G) depending on the antigenic properties of the toxin produced. Types A, B, E and F are responsible for most cases of human botulism, whereas types C and D cause illness in animals. The outbreaks of foodborne botulism associated with meats, such as home-cured hams, tend to occur mainly in Germany, France, Poland and Italy. The incidence of foodborne botulism is extremely low, but the severity of disease and its heat resistance mean that it is the target microorganism of many preservation processes used for foods. Spores of *C. botulinum* are present in the soil and environment, but to a lesser extent than *C. perfringens*. Spores may be present in meat, but this is usually at levels between 0.1–10 spores/kg. The disease in man is an intoxication and causes general weakness of limbs and respiratory muscles, and often nausea and vomiting. Like humans, botulism in animals almost always arises from ingestion of food contaminated with preformed toxin. There is evidence that animals carry spores and this may lead to internal contamination and contamination of meat processing environments. In Europe, occurrence of spores is generally infrequent but when it occurs, levels can reach 7/kg of sample whereas incidence in meats in the US is much lower, probably reflecting the incidence of meat-associated botulism in the two areas.

11.2.9 Other bacteria

Other bacteria associated with meat animals include brucellae (e.g. *Brucella melitensis*) and *Bacillus anthracis*, which can cause disease in man but are regarded as a relatively low risk from meat and meat products. *B. cereus* is a ubiquitous organism and has been found in raw beef and milk, and the organism is directly linked to dairy cows, being incriminated in abortions and mastitis. Therefore, contamination of carcasses of dairy cows is possible but is not thought to constitute a significant risk in foods of animal origin. Foodborne illness caused by *B. cereus* generally results from improper handling of foods. Other organisms, such as *Corynebacterium pseudotuberculosis*, *Mycobacterium paratuberculosis* and *Pasteurella spp.* are responsible for diseases in animals and have been linked to disease in humans but transmissibility from animals to man has yet to be proven.

11.2.10 Parasites

Giardia duodenalis and G. lamblia
G. duodenalis is one of the most common protozoal infections in man, causing diarrhoeal disease in infants and young children, in both industrialised and developing countries. The parasite is also found in many domestic animals including cattle, sheep and goats, particularly young animals. There is evidence of zoonotic transmission, but the major sources of contamination are thought to be water or food contaminated with water that has been in contact with faecal material.[22]

Cryptosporidium parvum
C. parvum has a wide spectrum of animal hosts, including cattle, goats, other farm animals and man. It is an intracellular parasitic protozoan responsible for self-limiting diarrhoeal illness in its hosts.[23] Symptoms include watery diarrhoea, nausea, anorexia, abdominal cramps, fever and weight loss. The life-cycle is completed within one host and large numbers of oocysts are then transmitted, in faeces, to the environment, where they may survive for long periods of time. In man, if individuals are young or immunocompromised, more serious gastroenteritis can occur and this can be fatal. In diarrhoetic young goats and sheep, there is a high prevalence of *C. parvum*, suggesting a strong association between infection and disease. In surveys looking for presence of *C.*

parvum in animals, oocysts were found in calves at levels of up to 22% of animals tested. The reservoirs and routes of transmission suggest that meat and meat products may be a source of infection in humans. Sausage and tripe have been shown to contain *C. parvum* oocysts.[23] Contaminated water is known to be the cause of large outbreaks of disease. Poor diagnosis of disease in man and the small numbers of oocysts (100s) necessary to cause infection mean that many cases of cryptospordiosis may go undetected.

Other parasites
Toxoplasma gondii is a protozoal parasite well known for causing abortions in sheep. The organism is also known to cause acute primary infection in man and is a particular risk to pregnant women. Consumption of raw or undercooked mutton is thought to be responsible for transmission to man.

Trichinella spiralis is reponsible for trichinellosis, which, in man, begins as an acute gastrointestinal condition and is followed by fever and myalgias. Chronic illness may result since 10–20% of cases develop neurologic or cardiac symptoms. Illness in man results from consumption of raw or undercooked pork, wild boar or horse meat, with most cases occurring in Europe. *Taenia saginata* also causes outbreaks of disease in Europe through consumption of infected beef.

Cyclospora spp. cause very similar disease to *C. parvum* and are also similar in other respects such as biology and pathogenesis, but there is only one species, *Cyclospora cayetanensis*, known to cause illness in man. This species is not known to have any other animal host. Other members of the genus cause disease in other animals.

Echinococcus granulosus is another parasite that can cause infection in man. The larval stage is found in sheep, goats, cattle, pigs and man. The final host for the parasite is the dog. Contamination of meat is not thought to occur directly; the main route of infection is through contamination of eggs from dogs.

11.2.11 Other agents

Transmissible spongiform encephalopathies (TSEs)
Scrapie has been prevalent in sheep and goats in particular parts of the world for many years. This disease is regarded as the prototype of TSEs, found in humans and other animals. These TSEs cause progressive degenerative disorders of the nervous system and result in death. There is no doubt that these are infectious diseases but the nature of the infectious agents remains elusive. Theories about the causative agent vary and there is continuous debate about the presence of nucleic acids and the importance of a protease resistant protein (prion theory), derived from a normal host protein. In the early 1980s, an epidemic of bovine spongiform encephalopathy (BSE) began in the UK, and the recycling of infected cattle material is thought to have continued driving this epidemic.

The recent emergence of a new variant of Creutzfeldt-Jakob disease (vCJD) in humans in the UK has led to the belief that this new disorder is related to the transmissible agent causing BSE. The working hypothesis is that transmission has occurred through contaminated material entering the food chain. This, in turn, has focused attention back on scrapie as a potential source of infection, despite the fact that large quantities of contaminated material must have been consumed without any apparent ill effects. It is not known how many vCJD cases are likely to emerge as a consequence of the BSE epidemic and there is still much to learn about all aspects of this group of diseases.

Viruses
Viruses are not generally considered to be transmitted to man via meat and meat products, although caliciviruses infect humans and other animals. Within the family *Caliciviridae*, there are four distinct genera comprising vesiviruses and lagoviruses, which contain a broad range of animal viruses and Norwalk-like viruses (NLV or small round structured viruses) and Sapporo-like viruses, which until recently have only been associated with man. NLVs are the main cause of gastrointestinal illness in restaurants and institutions. Recent data suggest that NLV infections often occur in calves and sometimes in pigs.[24] The significance of this recent finding is unknown at the present time.

11.3 Analytical methods

This section of the chapter provides a brief overview of the types of methods available for detection of foodborne pathogens. Detailed description of methods for each of the pathogens discussed above are not included. Conventional methods for the detection and characterisation of bacteria associated with foods rely on specific media. These methods tend to be relatively cheap, sensitive and can provide both quantitative and qualitative information. However, they can be lengthy procedures, are labour intensive, rely on multiplication of the target organism and do not use genetic information, which can be used to discriminate between closely related organisms. Nevertheless, there have been advances in recent years that facilitate some of these conventional procedures such as the introduction of chromogenic or fluorogenic media, removing the need to do further sub-culturing and biochemical steps. Modifications to particular media have also been made to improve performance and cut down some of the other steps involved in conventional culture methods. Other improvements include availability of automated colony counting, using image analysis, and availability of automated biochemical identification systems. These advances provide results directly comparable to conventional tests but make testing much more convenient.

Reliance on particular methods for the detection of pathogens can lead to problems where atypical types or responses are evident. For example, *E. coli* O157:H7 isolates are routinely distinguished from other *E. coli* because of their inability to ferment sorbitol. This means that sorbitol +ve *E. coli* O157:H7 strains, such as those found in Germany, would go undetected during routine testing. Selective media, because of their inclusion of inhibitory agents, may also underestimate target organisms if they are injured. In such cases, inclusion of a recovery stage is critical to the detection procedure.

Alternative approaches for the detection of specific microorganisms have also been developed in recent years and these include flow cytometry, impedimetry, immunological techniques and nucleic acid based assays. Flow cytometry is an optically-based approach that can detect low numbers of cells (e.g. 10^2–10^3 bacteria) rapidly (within minutes), but food matrices can interfere with the technique and distinction between live and 'dead' cells can be problematic. It has been used for the enumeration of viruses in water and is also used to enumerate *Cryptosporidium* oocysts. Impedimetry is based on changes in the electrical conductivity of liquid media caused by growth of the target organism. Although this method is not 'rapid', it is convenient for high throughput since it is fully automated and can deal with multiple samples simultaneously. Specificity is dependent on the media used to grow the target organisms.

Immunological methods are based on the specific binding of an antibody to an antigen. The advent of monoclonal antibodies now provides a consistent and reliable source of characterised antibodies. Immunoassays are divided into homogeneous and heterogeneous assays. There is no need for markers with homogeneous assays, since the antibody-antigen complex is directly measurable and the test time is short. Examples of this type of assay are agglutination reactions, immunodiffusion and turbidimetry, and tests are available for most pathogens. Heterogeneous assays are more complex procedures and use immobilised antibodies on a variety of supports and reporting systems. These precedures can be carried out without the need for special equipment. Detection limits are between 10^3–10^5 cell/ml for most pathogens. Direct detection in foods is not possible and enrichment is required. Immunoassays can also detect bacterial toxins. Automated immunoassays are also now commercially available.

Developments in genetically-based techniques in recent years provide a step-change in analytical capability for detection and characterisation of pathogens. These techniques are based on the hybridisation of target DNA or RNA with a specific DNA probe. The specificity of this probe is dependent on its nucleotide sequence. When hybridisation has occurred, detection can be via a number of methods, similar to those used in immunoassays. Commercial assays are now available for a number of pathogens. The detection limit for bacteria is 10^3 cells, so enrichment is sometimes required. Alternatively, an amplification step may be used. Examples of this are polymerase chain reaction, involving denaturation of the target DNA and annealing of primers to the single strand, followed by extension of the primers using a thermostable polymerase, or RNA amplification through the concerted action of enzymes (NASBA®). Use of amplification methods requires clean samples, and availability of commercial kits now enables routine laboratories to carry out procedures which until recently were regarded as complex and only carried out in specialist laboratories. Because these methods are based on genetic elements, results only indicate the potential to produce toxin or express virulence. There are also problems with false positives (e.g. 'dead' cells) and negatives (polymerase inhibitors or accessibility to the target organism).

Molecular typing is also possible now, allowing identification to sub-species level, aiding epidemiological and taxonomic studies. These techniques are often referred to as fingerprinting methods. They include restriction fragment length polymorphism (RFLP), random amplified polymorphic DNA (RAPD), pulsed field gel electrophoresis (PFGE) and AFLP® which combines PCR and RFLP. Whilst conventional methods still have an important role to play, molecular methods are likely to become more commonly used. The next breakthrough in diagnostic methodology is likely to come from 'DNA chip' technology, which combines semiconductor manufacturing with molecular techniques. This technology will allow rapid and cheap analysis of multiple sequences, using large arrays of nucleotides, making it possible to detect and type different organisms in the same food sample. There are, however, significant hurdles to be overcome, with viruses and parasites posing their own particular problems.

11.4 Future trends

Foodborne pathogens that have emerged in recent years share a number of characteristics. Nearly all of these have an animal reservoir from which they spread to man, i.e. they are foodborne zoonoses, but unlike established zoonoses, they do not often cause illness in the animal host. Another worrying trend is that these pathogens are able to spread globally in a short period of time. Many of the emerging pathogens are becoming increasingly resistant to antibiotics and this has been attributed, partly, to the use of antibiotics in animals. The practice of using antibiotics in animal production is coming under increasing pressure and there have been recent legislative changes that address this issue in particular parts of the world. Unfortunately, it is likely that some of these practices will continue in those areas that are not properly regulated or policed.

New food vehicles have been identified in recent years. These new vehicles include foods that were once thought to be 'safe' such as eggs, apple juice, fresh fruit, fresh vegetables and fermented meats. With consumer preferences for fresher, less heavily processed foods likely to continue, it is possible that new food vehicles for foodborne disease will continue to emerge. Alternative processes, if incorrectly assessed, may also provide an additional source of infection. Continued consolidation within the food industry is likely to lead to increasingly large markets and wider distribution from centralised manufacturing operations. With increasing demand from increasing populations, we are likely to see more re-use and recycling of water and waste, and this may have an impact on the microbiological hazards we have to face.

Fortunately, improved epidemiological capability, provided through better detection methods and better cooperation/coordination between different surveillance networks, is likely to allow quicker detection of geographically widespread outbreaks of foodborne disease. Molecular methods are transforming taxonomy and our understanding of the genomes of particular pathogens and groups of pathogens, such as the *Enterobacteriaceae*. This has already let us gain some insight into evolutionary processes and should allow us to better anticipate the potential of microorganisms to incorporate new genetic material and develop new virulence characteristics. Better understanding of pathogenesis of foodborne disease and colonisation of animals may also allow development of new intervention strategies.

With anticipated increases in the average life expectancy, through improved medical treatment of chronic disease and other advances, there is likely to be an increase in the proportion of persons with age-related susceptibility to foodborne disease. Also, there is likely to be a continuing increase in the number of immuno-suppressed individuals, due to infection with HIV and other chronic illnesses.

At the present time we are seeing a decrease in the number of cases of some common foodborne pathogens, such as salmonellae, in developed countries like the US, UK and other parts of Europe. This is encouraging and suggests that some disease prevention strategies may be beginning to take effect. Despite this, the incidence of foodborne illnesses and deaths caused by unsafe food are increasing. The genetic plasticity of the microorganisms poses a serious threat for the future, and will undoubtedly lead to the emergence of novel infectious diseases. At the genetic and molecular level, the virulence traits of pathogens clearly show us that pathogenicity does not arise by slow adaptive evolution but rather by step changes.

11.5 Sources of further information and advice

General articles describing members of the *Enterobacteriaceae*, such as salmonellae, *E. coli* and *Y. enterocolitica* are available.[3,20] *Enterobacteriaceae* and *E. coli* infections in animals have been reviewed in a number of articles.[7,8] Specific articles describing foodborne listeriosis, campylobacteriosis and the emergence of *E. coli* O157:H7 are also available.[11,18,21] Foodborne parasites are reviewed in several articles.[22,23,25] A review of analytical methods used in microbiology has been published recently.[26]

11.6 References

1. McDONOUGH P L, FOGELMAN D, HIN S J, BRUNNER M A and LEIN D H, 'Salmonella enterica serotype Dublin infection: an emerging infectious disease for the Northeastern United States', *J Clin Micro*, 1999 **37** 2418–27.
2. BARLEY J P, '*S. typhimurium* DT104 in cattle in the UK', *Vet Rec*, **140** 75.
3. D'AOUST J, '*Salmonella* species' in *Food Microbiology – fundamentals and frontiers*, ASM Press, Washington, 129–58, 1997.
4. RODRIGUE D C, TAUXE R V and ROWE B, 'International increase in

Salmonella enteritidis: a new pandemic?' *Epid Inf*, 1990 **105** 21–7.
5. MISHU B, KOEHLER J, LEE L A, RODRIGUE D, BRENNER F H and BLAKE P *et al.*, 'Outbreaks of *Salmonella enteritidis* infections in the United States, 1985–1991', *J Inf Dis,* 1994 **169** 547–52.
6. CENTERS FOR DISEASE CONTROL AND PREVENTION, 'Multi-drug resistant *Salmonella* serotype Typhimurium – United States', *Mor Mort Wkly Rep*, 1997 **47** 308–10.
7. LINTON A H and HINTON M H, 'Enterobacteriaceae associated with animals in health and disease', *J App Bact Sym Supp*, 1988 71S-85S.
8. WRAY C and WOODWARD M J, '*Escherichia coli* infections in farm animals' in Escherichia coli: *mechanisms of virulence*, ed SUSSMAN M 1997, 49–84.
9. DEAN-NYSTROM E A, BOSWORTH B T and MOON H W, 'Pathogenesis of O157:H7 *Escherichia coli* in neonatal calves' in *Mechanisms in the Pathogenesis of Enteric Diseases*, Plenum Press, New York, 47–51, 1997.
10. PEARSON G R, BAZELEY K J, JONES J R, GUMMING R F, GREEN M J, COOKSON A and WOODWARD M J, 'Attaching and effacing lesions in the large intestine of an eight-month-old heifer associated with *Escherichia coli* O26 infection in a group of animals with dysentery', *Vet Rec*, 1999 **25** 370–2.
11. ARMSTRONG G L, HOLLINGSWORTH J and MORRIS J G, 'Emerging foodborne pathogens: *E. coli* O157:H7 as a model of entry of a new pathogen into the food supply of the developed world', *Epid Rev*, 1996 **18** 29–51.
12. CHAPMAN P A, WRIGHT D J, NORMAN P, FOX J and CRICK E, 'Cattle as a possible source of verocytotoxin-producing *Escherichia coli* O157:H7 infections in man', *Epid Inf*, 1993 **111** 439–47.
13. DOYLE M P and SCHOENI J L, 'Isolation of *Escerichia coli* O157:H7 from retail fresh meats and poultry', *Appl Env Micro*, 1987 **53** 2394–6.
14. SEKLA L, MILLEY D, STACKIW W, SISLER J, DREW J and SARGENT D, 'Verotoxin-producing *Escherichia coli* in ground beef – Manitoba', *Can Dis Weekly Rep*, 1990 **16** 103–5.
15. WORLD HEALTH ORGANISATION, 'Report on a WHO working group meeting on shiga-like toxin producing *Escherichia coli* (SLTEC) with emphasis on zoonotic aspects', Bergammo, Italy, Rep no WHO/CDS/VPH/94.136, 1994.
16. NATARO J P and KAPER J B, 'Diarrheagenic *Escherichia coli*', *Clin Rev Micro*, 1996 **34** 2812–14.
17. TAUXE R V, 'Emerging foodborne diseases: an evolving public health challenge', *Emer Infect Dis*, 1997 **3** 425–34.
18. KETLEY J M, 'Pathogenesis of enteric infection by *Campylobacter*', *Micro*, 1997 **143** 5–21.
19. STERN N J and KAZMI S U, '*Campylobacter jejuni*' in *Foodborne Bacterial Pathogens*, Marcel Dekker, New York, 71–110, 1989.
20. OSTROFF S '*Yersinia* as an emerging infection: Epidemiologic aspects of yersiniosis', *Contrib Micro Immunol*, 1995 **13** 5–10.
21. FARBER J M and PETERKIN P I, '*Listeria monocytogenes*, a foodborne pathogen', *Micro Rev*, 1991 **55** 476–511.

22. TREES A J ,'Zoonotic protozoa', *J Med Micro*, 1997 **46** 20–4.
23. HOSKIN J C and WRIGHT R E, '*Cryptosporidium*: an emerging concern for the food industry', *J Food Prot*, 1991 **54** 53–7.
24. VAN DER POEL W H M, VINJÉ J, VAN DER HEIDE R, HERRERA M-I VIVO A and KOOPMANS M P G, 'Norwalk-like calicivirus genes in farm animals', *Emerg Inf Dis*, 2000 **6** (1).
25. GOODGAME R W, 'Understanding intestinal spore-forming protozoa: cryptosporidia, microsporidia, isospora and cyclospora', *Ann Intern Med*, 1996 **124** 429–41.
26. DE BOER E and BEUMER R R, 'Methodology for detection and typing of foodborne microorganisms', *Int J Food Micro*, 1999 **50** 119–30.

Part III

New techniques for improving quality

12

Modelling beef cattle production to improve quality

K. G. Rickert, University of Queensland, Gatton

12.1 Introduction

This chapter refers to computer models that simulate beef cattle production. Such models consist of mathematical equations and instructions which mimic the roles, interactions and influences of the various inputs to beef cattle production. The chapter recognises that modelling is a term which refers to both building and using models, and that beef cattle production includes the complex interactions between the physical environment, financial environment, management, feed supply, and animal reproduction and growth. The chapter considers the challenge faced by model builders in dealing with such complexity, overviews possible applications, and gives an example of a simple beef production model.

Pasture and animal scientists started to model beef cattle production after computers first became available for research in the 1960s. A rapid expansion in the range, scope and role of models followed in response to the even more rapid expansions in the power and accessibility of computers. Insight into the progress and philosophy of modelling pasture and animal production are obtained from recent reviews.[1,2] Models have been a valuable aid to research, extension, and management at the farm, industry or government levels because of the following three attributes.

1. If each equation in a model is regarded as a hypothesis pertaining to a specific process or component, then a model can be regarded as a collection of hypotheses, derived from past research that can be further modified and developed through new research. In this way, a model becomes a repository for past research and a precursor for future research.[3,4] Model construction is now a common activity that gives research direction and focus.

2. Models provide a quantitative description of the many interacting components which may have conflicting responses in a beef production system.[4] This is a powerful and unique attribute that greatly exceeds the analytical capacity of the human mind. For example with beef cattle, as stocking rate increases (the number of animals per unit area of land), the liveweight and value per animal decrease, variable costs increase and production per hectare at first increases and then decreases.[5] A manager must balance the trade-offs between profit, risk, pasture degradation and premium prices.[6] Similar trade-offs between productivity, stability and sustainability are common in farming systems[7] and a model allows users to experience 'virtual' reality in managing grazing systems.
3. Models can give a quantitative extrapolation in space and time of information derived from past research and experiences. For example, by processing historical records of daily weather data, estimates of variability in output can be expressed as probability distributions.[8] Similarly, by processing the historical weather for different land units in a region, and thereby estimating spatial and temporal variations in forage production, estimates of safe stocking rates can be compared against trends in actual regional stocking rates to indicate periods of overgrazing.[9] Further, if the spatial model uses current weather data as input, the output is a near real-time display of pasture and/or animal production[10] that can influence government or industry policies. All of these applications rely on a model's ability to extrapolate information in temporal and spatial dimensions, and this attribute is fundamental to the role of models in information transfer.[11]

Today a wide range of models on different aspects of plant and animal production are being used as aids to research, farm management, and to determine government or industry policies.[1]

12.2 Elements of beef cattle production

Beef cattle production deals with the conversion of climatic and edaphic inputs into plant products, which are consumed by various classes of animals in a beef cattle herd to give meat for human consumption. This beef production system consists of four interacting biophysical and bioeconomic subsystems, which are manipulated through the management subsystem in response to the climate subsystem (Fig. 12.1). The structure and significance of the various subsystems are described in more detail below.

The climate subsystem is largely outside the management subsystem but it directly affects the four subsystems influenced by a manager. For example, rainfall supplies soil water for plant growth, may cause soil erosion, and influences the rate of waste decomposition in soil. Further, prevailing temperature, humidity and radiation influence plant growth, and the incidence of plant and animal pests and diseases. Climatic inputs also display seasonal and

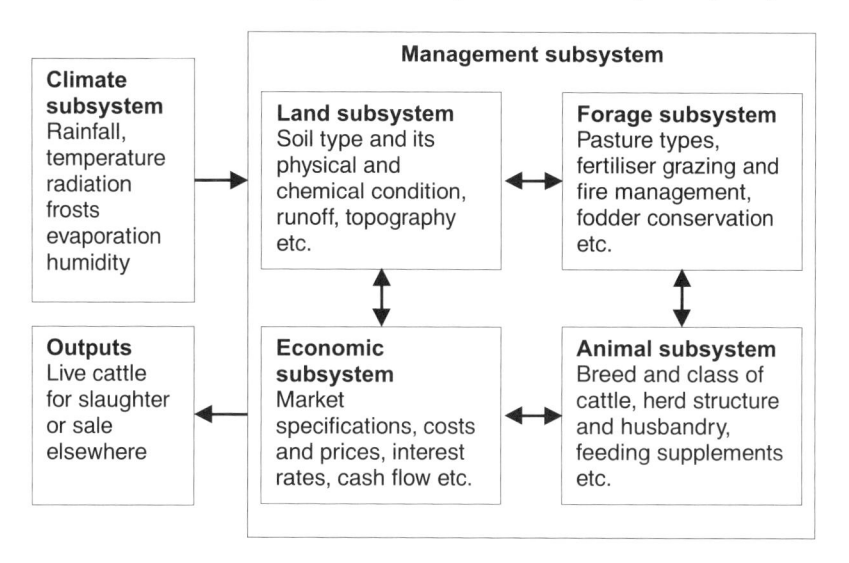

```
┌─────────────────┐  ┌─────────────────────────────────────────────────────┐
│ Climate         │  │              Management subsystem                   │
│ subsystem       │  │ ┌─────────────────┐      ┌─────────────────────┐    │
│ Rainfall,       │  │ │ Land subsystem  │      │ Forage subsystem    │    │
│ temperature     │─▶│ │ Soil type and its│◀────▶│ Pasture types,      │    │
│ radiation       │  │ │ physical and    │      │ fertiliser grazing and│   │
│ frosts          │  │ │ chemical condition,│   │ fire management,    │    │
│ evaporation     │  │ │ runoff, topography│    │ fodder conservation │    │
│ humidity        │  │ │ etc.            │      │ etc.                │    │
└─────────────────┘  │ └─────────────────┘      └─────────────────────┘    │
                     │         ▲▼                         ▲▼               │
┌─────────────────┐  │ ┌─────────────────┐      ┌─────────────────────┐    │
│ Outputs         │  │ │ Economic        │      │ Animal subsystem    │    │
│ Live cattle     │  │ │ subsystem       │      │ Breed and class of  │    │
│ for slaughter   │◀─│ │ Market          │◀────▶│ cattle, herd structure│   │
│ or sale         │  │ │ specifications, costs│ │ and husbandry,      │    │
│ elsewhere       │  │ │ and prices, interest │ │ feeding supplements │    │
│                 │  │ │ rates, cash flow etc.│ │ etc.                │    │
└─────────────────┘  │ └─────────────────┘      └─────────────────────┘    │
                     └─────────────────────────────────────────────────────┘
```

Fig. 12.1 Interrelationships between biophysical and bioeconomic subsystems (rectangles) with the management subsystem of the farmer. The biophysical and bioeconomic subsystems contain processes that determine their status. The interface between two subsystems (arrows) represents a conversion of materials into a new form. The manager is constantly responding to the climate subsystem, which impacts to varying degrees on the soil, pasture, animal and economic subsystems.

year-by-year variations and a manager must devise strategies to cope with these variations. Indeed, matching the farming system to the level and variability of climate inputs is a big challenge for a farm manager.[12] Seasonal variations in climate give rise to seasonal variations in quality and type of forage which may trigger fodder conservation (e.g. hay) to offset periods of forage deficiency. Wide year-by-year variations in climate inputs, often expressed as droughts or floods which lead to major perturbations in forage supply and market prices, need to be handled through skillful and resourceful management.[13] However, long-term weather forecasts now give managers prior warning of likely climatic extremes. For example, in northern Australia seasonal forecasts indicate the probability of rainfall in the forthcoming three to six months exceeding the historical median value, thereby permitting managers to make an early response to a likely distribution of rainfall.[14] Also extremely hot or cold temperatures can cause deaths in plants and animals, and computer models such as GRAZ-PLAN,[15] coupled to weekly weather forecasts, give early warning of likely mortalities in susceptible classes of animals. In both cases, recent improvements in the reliability and skill of weather forecasting are helping farmers to cope with wide variations in climate.

The land subsystem supplies water and nutrients for plant growth. Since it includes many of the ecological processes that sustain the whole system, both the manager and interest groups in the wider community are keen to keep the land subsystem in good condition. Land degradation through soil erosion, desertification, salinisation, acidification and nutrient decline is a major concern in many of the world's grazing lands and has led to the notion of landscape management. With this approach, managers in a region with a common attribute, such as a river catchment, are encouraged to adopt strategies that enhance sustainable development rather than exploitation of the land subsystem. Landscape management also recognises that grazing lands produce food as well as ecosystem services, such as water and biodiversity that are needed to sustain the cities where most people live. Preferred management strategies for a landscape may arise through different management options being assessed by government agencies or local communities, and computer models are often useful tools in this process.[16]

Plants within the forage subsystem supply digestible nutrients when grazed by cattle. Forage accumulates through plant growth and forage not eaten, together with faeces and urine from cattle, return to the soil subsystem through the detritus food chain. The quality of forage on offer varies with the growing conditions and type of plant species in the system. New growth is the most digestible and there is a steady decline in quality as plant parts age, die and senesce. Since temperate grasses have a higher digestibility than tropical grasses, grazing systems in temperate zones tend to display higher animal performance than tropical zones, Leguminous species tend to have higher digestibility than gramineous species.[17] If a grazing system is based on sown pastures the manager may select to grow a mixed-pasture which usually consists of a few species that are well suited to a particular situation. This contrasts with native rangelands where the system consists of many different species, often including trees. Here a manager aims to keep the pasture in good condition by maintaining adequate plant cover to reduce soil erosion and a predominance of desirable rather than undesirable plant species.[18] In both sown pasture production systems and native rangelands, forage condition and animal performance can be manipulated by management options such as the choice of stocking rate, type and amount of fertiliser application, periods of grazing and conservation, level of supplementary feeding, and fire in the case of rangelands.[19, 20]

The cattle subsystem produces animals for sale through the processes of reproduction and growth within a herd consisting of different animal classes. The number of different animal classes on a farm largely depends on the quality of the pasture subsystem and on the objectives of a manager. In essence, breeding cows produce calves and after weaning these move into different classes as they grow and age (Table 12.1). Usually young female cattle (heifers) are selected to replace aged or culled cows and are mated for the first time when they reach maturity and a specific weight that depends on the breed and prevailing nutrition. Under good nutrition, heifers may be mated first at 15–18 months of age, but with the poorer nutrition in extensive rangelands, mating usually takes place at 24–30 months. Heifers that are not required for replacing cows might be sold for slaughter or for breeding purposes elsewhere. Male cattle are commonly castrated before weaning although a small number of high-

Table 12.1 Classes of cattle commonly found in beef cattle herds in extensive grazing systems. Adult equivalent, being the ratio of the energy requirement of a class to the energy requirement of an adult animal, is a coefficient for equating animal numbers in each class to a common base. Intensive grazing systems with a higher level of nutrition will have fewer classes since cattle are sold at a younger age

Animal class	Adult equivalent	Age years	Comments
Cows and calves	1.3	2–12	Managers aim to have breeding cows calve annually. Calves are usually weaned at about 6 months of age.
Yearling heifers	0.55	0.5–1.5	Heifers are females that have not had one calf. When mature at 1.5 to 2.5 years, depending on breed and growing conditions, some are mated to replace culled cows. Surplus heifers may be sold for slaughter or as breeding stock.
2-year-old heifers	0.75	1.5–2.5	
Yearling steers	0.55	0.5–1.5	Steers, or castrated males, are sold for finishing elsewhere, or for slaughter. Age and weight at sale depends on the level of nutrition they experience, the specifications of available markets, and on the price advantage of different markets. Within limits set by prevailing climatic and economic conditions, a manager can target a specific market by manipulating feed supplies in the pasture subsytem.
2-year-old steers	0.8	1.5–2.5	
3-year-old steers	1.0	2.5–3.5	
4-year-old steers	1.1	3.5–4.5	
Culled cows	1.0	3–12	Cows no longer suitable for breeding due to age or infertility. Usually conditioned and sold for slaughter.
Bulls	1.1	3–7	Male animals for mating with cows. One bull is required for every 20 to 25 cows.

performing males may be retained to replace aged bulls. Depending on the prevailing nutrition and markets, male cattle may be retained for one to three years after weaning, to be sold for slaughter or for finishing elsewhere on another farm or in a feedlot. Thus, which market to target, and how the cattle should be fed to meet the market, are key strategic decisions for a manager. Deciding when to sell specific groups of cattle is a key tactical decision for a manager.

The different classes of cattle in a beef herd have different nutritional requirements because they differ in weight and age. The term adult equivalent (AE) relates the energy requirement of different classes to a common base, the energy requirement for maintenance of an adult animal, such as a non-lactating cow. The AEs of Table 12.1 can be determined from feeding tables but a first approximation for growing cattle is given by:

$$AE = LW^{0.75}/105.7 \qquad (12.1)$$

where LW and $LW^{0.75}$ are the liveweight and metabolic weight of animals in a specific class and 105.7 is the metabolic weight of a non-lactating bovine with a liveweight of 500 kg/head.[21]

The market subsystem refers to the different markets for beef cattle available to a manager along with the prices and profit margins associated with each market. Specifications for markets vary with location. In an extreme case there is no specification, and all cattle are sold as beef with no separation of cuts at retail outlets. At the other extreme, individual animals are prepared for a specific market and traced through the supply chain, with carcasses being graded for quality and various cuts of meat separated and sold at prices that reflect consumer preferences and the grade. Farmers in countries that export beef, such as USA, Australia, Canada and New Zealand, commonly have a range of market options that are specified in terms of age, gender, weight and fat thickness of a carcass. However, the classification scheme is not standardised internationally, although there is an international trend to reduce the allowable limits for residues of pesticide and growth promotants in export beef. Penalties for farmers in not meeting specifications for chemical residues are usually severe, including condemnation of all meat in the case of excess chemical residues.

12.3 Challenges for modellers

The above description of beef production is deceptively simple. In practice a model builder is faced with the challenge of expressing the complex interactions between components of the system (Fig. 12.1). Specific challenges include

- how to match the primary purpose of the model to the most appropriate structure
- how to handle natural variability in the biophysical components and the interface between the subsystems, and
- how to validate the completed model.

Answers to these questions are interrelated and reflect back to the history and philosophy of model building.

12.3.1 Matching purpose and structure

Models of beef production systems are commonly built as aids to research, farm management or policy evaluation and their structure may be mechanistic, empirical or a combination of both.[1] Empirical models estimate outputs by empirical equations developed from experimental observation of output in relation to one or more influencing variables, while mechanistic models reflect a theoretical understanding of the factors that control outputs. The relative merits of mechanistic and empirical structures have been hotly debated and the choice of structure is a critical and often difficult decision for a model builder.[2, 4, 22, 23] Mechanistic models, because of their stronger theoretical base, tend to be more versatile and are more likely to explain responses than empirical models, but they may not be more accurate and often contain parameters that are difficult to determine in practical situations. Conversely, the robustness of an empirical model depends on the range of experimental data used in its derivation, and spurious results might occur if it is applied outside this range. Thus model builders should specify the derivation and application of an empirical model, and users should adhere to these specifications. As a variation on the above distinction, some models combine both empirical and mechanistic elements, such as an empirical model being used to process and interpret the results previously stored from many simulation experiments with a mechanistic model.

Research models are built by researchers to analyse the complex interactions in beef production systems. They can be regarded as a repository for past research since they collate and integrate information from past research. They are also a precursor for future research since gaps in knowledge and understanding are highlighted. Because research models focus on processes and their interactions, they are often mechanistic in structure and have a limited distribution. However, GRAZE is an exception to this statement, being a comprehensive mechanistic model of forage and animal growth that is widely distributed and well documented.[24] Sometimes a research model evolves into a management or policy model, thereby reducing development costs.

Models for farm management are usually designed to evaluate management options pertaining to one or more components of the system. They aid management by evaluating different scenarios thereby allowing preferred strategies to be identified, but importantly, a manager is free to accept or reject the output. Developing this type of model requires considerable time and effort, since to be accepted by potential users, the package needs to operate in a convenient and reliable manner, have a high degree of validity or skill, and have a commercial arrangement for distribution and after-sales service.[1] FEEDMAN[25] is an example of many commercial decision support systems that focus on farm management. However, history suggests that experienced farmers do not readily use such software for common routine decisions unless its use is clearly beneficial and it is promoted by a trusted product champion.[26–28] On the other hand, professional farm advisors who are paid to recommend preferred management options are likely to use the software to justify a recommendation. Because a farm advisor may have many clients, decision support software that is regularly used by a few farm advisors may still have a big impact on farm management. Both mechanistic and empirical sub-models are widely used in management software.

Policy models serve government or industry leaders by estimating outcomes to possible scenarios and initiatives in policy. Both mechanistic and empirical sub-models are used in policy models dealing with pasture and animal production. Policy models range from those that provide a one-off analysis of a specific problem to those that provide a regular ongoing service. An example of a one-off analysis that influenced policy was the rejection of a plan, based on results from field research over ten years, to construct farm dams and use the stored water to irrigate crops to improve the forage supply in north western Queensland.

Simulation studies based on long-term records of climate showed that the plan was not viable because rainfall was too variable.[29] Apparently the field study that supported the plan coincided with a run of high-rainfall years. An example of a regular ongoing service is the monthly maps of relative pasture yield, adjusted for prevailing stocking rates, which are derived from a pasture production model operating on a 5×5 km grid for the State of Queensland.[30] The maps provide an objective assessment of drought status for government and industry. Constructing and maintaining a policy model of this scale requires an integrated team of scientists, programmers and support staff. As with management models, a policy model's credibility depends on its scientific base and validity.

12.3.2 Coping with linkages between components

With regard to Fig. 12.1, the status of each subsystem is expressed by several different terms, which reflect the purpose of the overall model and the structure of the sub-models that simulate each subsystem. Since the subsystems are interdependent, they need to be linked in an appropriate manner, an issue in model building that is often called the interface problem. As an illustration, simple expressions of the status of each subsystem might be:

1. climate subsystem – inputs of solar radiation and/or temperature on plant growth and rainfall on soil water supply;
2. land subsystem – amount of soil water (mm) available for plant growth in response to daily rainfall runoff, drainage and evapotranspiration;
3. pasture subsystem – yield (kg/ha) of leaf and stem, potentially for each plant species in the pasture, in response to daily plant growth less consumption and senescence;
4. animal subsystem – liveweight (kg/head) of each animal class, in response to an initial liveweight and accumulated daily liveweight gain; and
5. economic subsystem – farm profit ($ or $/ha) in response to value of animals sold less variable costs.

Interface between climate, land and pasture subsystems
Mechanistic models often estimate plant growth as the product of intercepted solar radiation and radiation use efficiency. Intercepted radiation depends on leaf area of the forage, and radiation use efficiency links the soil and climate subsystems, being dependent on prevailing climate, soil nutrient status and soil water supply.[31] In practice, radiation interception and radiation use efficiency are difficult to simulate in pastures in rangelands that are a mixture of C3 and C4 species growing as spaced plants under trees in a semi-arid environment, and are grazed selectively by cattle. Under these complex circumstances an empirical model based on field observations can be a useful tool. For example, pasture growth (PG kg/ha) can be estimated as:

$$PG = WUE * WU \qquad (12.2)$$

where *WUE* is water use efficiency, another term that links the two subsystems for a specified site (kg/mm), and *WU* is water use over a specified time step (e.g. mm/day).

Equation (12.2) avoids the difficulties associated with radiation interception by recognising the strong direct relationship between water use via transpiration and forage growth via photosynthesis, two gaseous transfer processes that are controlled by leaf stomata. It can be applied at different temporal and spatial scales.[32] On a daily time step, *WUE* becomes transpiration efficiency and *WU* is daily transpiration estimated by a sub-model of soil water balance, but on monthly or seasonal time step, *WUE* becomes rainfall use efficiency and effective rainfall (actual rainfall less runoff) is an approximation of *WU*. Although *WUE* varies with fertility status of the soil, seasonal conditions and the number of trees present, it is a parameter that can be determined simply for a site from measurements of plant growth in relation to *WU*. The FEEDMAN decision support system estimates monthly plant growth through this approach and the default values of *WUE* for many different soil-forage combinations were either obtained from field experiments or by integrating output from a daily plant growth model. In either case, the default values can be customised to reflect local conditions.

Interface between pasture and animal subsystems
This interface must account for nutritional demands of different classes of animals, all of which have the ability to move and select a preferred diet from a pasture that exhibits wide spatial and temporal variation in yield and quality.

In mechanistic terms, animal production is dependent on intake of digestible nutrients, and once the amount and quality of diet is known, models for estimating different forms of production (e.g. liveweight change, milk production, wool growth) in different animal classes already exist.[33] Thus the interface problem becomes how to estimate, either directly or indirectly, two interdependent terms, the amount (intake) and quality (digestibility) of diet. Actual intake is usually less than a potential intake, which depends on the breed and liveweight of animals, due to constraints arising from the amount and quality of forage on offer. Forage digestibility declines with age, is greater in leaf than stem, and varies across species. Mechanistic models commonly simulate diet selection by partitioning the forage on offer into digestibility or age categories with animals then selecting progressively from high to low categories until their appetite is satisfied.[34] Whilst this approach tends to mimic diet selection in temperate pastures reasonably well, the descriptive functions are essentially empirical relationships derived from field experiments. The approach has been less successful in rangelands with a more heterogeneous botanical composition and sward structure.[35] However, a more realistic algorithm for diet selection in heterogeneous forages places plant species into broad preference categories (e.g. preferred, desirable, undesirable, toxic, emergency and non-consumed) and then computes the proportion of each preference class in the diet.[36] The algorithm assumes that an animal has experience with the vegetation, and has learned to avoid toxic species and non-consumed species. The 'emergency' category accounts for species that are only eaten after the preferred, desirable and undesirable species are depleted.

The above 'mechanistic' models are essentially based on 'empirical' expressions derived from diet selection studies with parameters that are rather abstract and site specific. To avoid these difficulties, the FEEDMAN package used the notion of potential liveweight gain to characterise the seasonal variation quality of different forages. Potential liveweight gain is the monthly liveweight gain of a standard animal (a 200 kg cross-bred steer, *Bos taurus* by *Bos indicus*) grazing the forage at a low stocking rate in a good season. It is a bioassay for forage quality that can be measured, but more importantly, it is meaningful to farmers and can be adjusted to reflect local experience and knowledge. With potential liveweight gain for a standard animal given, the energy concentration of the forage can be estimated and applied to different animal classes, after taking account of the impact of high stocking rate on reducing intake and dry conditions reducing forage quality.[25] Because this approach uses a bioassay to characterise forage quality, and a mechanistic model to estimate animal performance, it can be readily adapted to herds of different species, breeds and classes of livestock.

Interface between animal and economic subsystems
Operating profit of a beef cattle enterprise on a farm is given by:

$$Gross\_profit = Number\_sold \times (Animal\_value - Variable\_costs)(\$) \quad (12.3)$$

where *Number_sold* is the number of animals sold, *Animal_value* is the average value of sale animals, *Variable_costs* are average variable or operating costs per animal associated with different management options. Comparison of the gross profit for different management options indicates the relative profitability of the options.

Estimation of *Variable_costs* is a simple arithmetic exercise, but since there is wide variation in local costs, a model must allow a user to modify and recall this information, and a user must update the information as required. On the other hand, estimation of *Animal_value* is a two-step process where animals are first allocated to a market category (if more than one exists), each with a corresponding sale price that usually exhibits spatial and temporal variation. Thus, tables of market prices for use in the calculation of *Animal_value* need to be updated regularly. The determination of market categories is location specific since there is wide national and international variation in the title and specifications for each category. In countries with well developed beef markets, categories may be specified by age, sex and breed of cattle, by weight expressed as liveweight or carcass weight, and by an indication of the degree of 'finish' expressed as a condition score in live cattle or fat thickness for carcasses. However, markets are not necessarily mutually exclusive in that while a premium market may have narrow specifications, cattle suited to a premium market may also be suited to a lower-priced market with wider specifications. Mechanistic models attempt to estimate animal growth and development, and the associated fat deposition.[37,38] Condition score has been derived empirically from the history and status of animal performance,[39] but neither approach has been applied to a full range of market specifications. FEEDMAN uses a simple approach to estimate *Animal_value* in that the characteristics of each herd are compared against entries in a table of markets, specified in terms of monthly sale price, and breed, age, class and liveweight of cattle. The highest price match is then selected and used to calculate *Animal_value*.

12.3.3 Coping with natural variability
On-farm complexity
Creating a 'user friendly' presentation of software that mimics pasture and animal production on a farm is a challenge because a multi-dimensional scenario must be described through a keyboard and monitor. The multi-dimensional scenario might consist of descriptions of fields in the farm, pastures in the fields, number and class of animals in herds, grazing management of herds, and period, type, and amount of supplementary feeding (Fig. 12.1). In addition, potential users commonly prefer the software to have keystrokes and a screen layout similar to other familiar software. Also, outputs must be clear, easily understood, and suitable for further analysis or storage. One approach used by model builders to meet these requirements is to consult with a panel of potential users on a regular basis and progressively modify the software in response to suggestions from the panel.[11] Such 'interactive prototyping' is a time-consuming task that can lead to major changes in the layout of screens for entering data and displaying results, but experience has shown that model builders, who know a package intimately, are not experts in 'user friendly' presentations. In practice, there are tradeoffs between the capacity of a decision support package to handle wide variations in farm production systems and the need for the package to be 'user friendly'. Extensive help notes, default values for input parameters, and training exercises and examples all assist a novice user in mastering a package. In addition to complexity due to on-farm variations mentioned above, climate and prices are off-farm inputs that display wide spatial and temporal variations.

Climate
In the case of climate a user may wish to evaluate management options over a range of seasonal conditions contained in historical records of climate. One approach is to use all historical data as an input and then express key outputs, such as farm profit, as a probability distribution. Another approach is to use a probability distribution of historical annual rainfall to establish categories of 'seasons' that reflect natural variations, such as:

250   Meat processing

very dry,   rainfall likely to be less than this category in 10% of years;
dry,   rainfall likely to be less than this category in 30% of years;
median,   rainfall likely to be less than this category in 50% of years;
wet,   rainfall likely to be less than this category in 70% of years; and
very wet,   rainfall likely to be less than this category in 90% of years.

The former approach demands access to a large database of historical records of climate, particularly if a model is to apply to a wide range of locations, each with a different climate history. The second approach, to select from the same comprehensive database a relatively small number of typical climate categories for each location, thereby eliminates the need for regular access to a large database of historical records. Both approaches are an attempt to assess management options simulated by the model in terms of the risk or likelihood of certain outcomes. This is a key attribute of models of beef production in variable climates, which is not obtained by using average or median climate data. Indeed, if only median climate data is used, animal production at high stocking rates is overestimated because year-by-year variations and interactions are ignored.[8]

In addition to analysing historical records of climate, model users are frequently interested in evaluating management options in relation to the current status of cattle and forage on a farm and future climate scenarios that are based on long-term weather forecasts.[13] Currently long-term weather forecasts indicate the probability of rainfall in the next three or six months being above or below median rainfall, and the skill of the forecasts is improving.[40] To cater for this requirement, models must allow users to enter potential future rainfall.

12.3.4   Verification and validation

Model verification ensures that the computer programs on which a model is based are free of 'bugs' and perform properly within specific limits. Usually a model builder uses special input data and parameters to test components of a model and their interactions under a wide range of operating conditions. The program needs to be corrected if values of the various variables and processes exceed an acceptable range. Problems may arise from a flaw in the algorithm describing a process, particularly as upper or lower limits are approached, or from a typing error in the program code. A sensitivity analysis is another component of verification that indicates the relative importance of accuracy in model inputs. Here a simulation experiment is designed to test the relative sensitivity of inputs and parameters that influence a system. Obviously accuracy is more important with sensitive than with insensitive inputs. The relative sensitivity of different inputs is indicated by comparing the change in output caused by a specific change in the different inputs (e.g. percent change in output after a 5, 10 or 20% change in an input parameter). Whilst verification is primarily the responsibility of model builders, simple exercises on these lines give model users a good appreciation of the operation and limitations of a model.

Model validation refers to how well a model mimics the system it is meant to represent. Validation is commonly demonstrated by first instructing a model to mimic a wide range of scenarios that have been actually observed, and then by comparing predictions from a model against the observations. The validation data should be independent of the data used in developing a model. Linear regressions of observations against predictions are commonly used to make the comparisons. The closer the slope and coefficient of determination for a regression are to unity, and the intercept to zero, the better the validity of a model. However, there are theoretical and practical problems with validation based on regression analysis,[41] and the confidence of the model builders should be recognised as a model undergoes development and modification.[42,43] Of course, serious users also develop confidence in a model through less formal validations as they compare predictions against their own observations and experiences. In practice, validation is an ongoing activity that warrants considerable effort by the model builders and independent experts, particularly when the model attempts to mimic large variation in production systems and is used as an aid to politically or financially sensitive decisions.[44] In essence a model is 'valid' when it sufficiently mimics the real world to fulfil its objectives, and when decisions based on the model are superior to those made without the model.[45]

12.4 Simple model of herd structure

It is obvious from Fig. 12.1 and Table 12.1 that for a given farm, the number and class of cattle in the animal subsystem depends on the amount and quality of growth in the forage subsystem. These interactions are captured in the following simple empirical model of herd structure in relation to broad management options. It also illustrates how a model that incorporates a few basic parameters can be a powerful analytical tool.

The notion of farm carrying capacity (CC) is a good starting point. This is the long-term safe stocking rate for a farm, one that does not cause ecological deterioration of the production system. It is a vital concept for managed grazing systems that incorporate the biological, commercial and social elements pertaining to good land care. It is commonly used to quantify a farm for sale or leasing in Australia and the USA, and because different classes of cattle have different nutritional requirements, it is commonly expressed as adult equivalents (see equation (12.1)).

In rangelands where forage growth is dependent on rainfall, carrying capacity is largely dependent on the amount of forage growth and on the proportion of growth that can be eaten (utilisation, U) without causing degradation of the pasture. Thus, based on the report by Johnston et al.[32]

$$CC = R * WUE * A * U/I \quad (AE) \tag{12.4}$$

where CC is farm carrying capacity, R is effective rainfall (mm/year, in subtropical climates this is annual rainfall less runoff), WUE is water use efficiency (e.g. 5 kg/ha/mm), A is area of the farm (ha), U is safe utilisation (e.g. 0.25) and I is annual intake for an adult animal (e.g. 4000 kg/year). Whilst WUE varies with the inherent fertility of the soil, fertiliser applications and presence of trees, it is simple to measure. On the other hand U is not simply measured but studies have shown it ranges from about 0.1 in arid infertile environments to about 0.5 in moist fertile environments. Although equation (12.4) demonstrates the derivation of CC from first principles, in practice farm CC is usually determined from local knowledge and experience.[32] The next task is to determine herd structure or the distribution of carrying capacity across the various animal classes.

When all cattle on a farm originate from the breeding cows (i.e. no off-farm purchases) the system is characterised by three performance indicators, which underpin a simple but versatile mathematical model of herd structure.

(1) Weaning rates refer to the number of calves weaned per hundred cows mated. This key indicator depends on the nutritional health status of cows and on the number and fertility of bulls. It commonly ranges from 95% in high-performing herds to less than 50% in herds of poor performance, a value that will not sustain the herd in the long term.
(2) Survival rates refer to the proportion of each class of cattle that survive a year. Mortality from poor health, accident or predators is common, particularly in extensively-managed beef production systems. The animal classes most prone to mortality are breeding cows and calves soon after weaning. Clearly high survival rates are desirable and susceptible classes of cattle commonly receive special feeding to avoid mortality from poor nutrition.
(3) Culling rates refer to the proportion of breeding cows culled annually for age, infertility, or other imperfections. Hence, if the effective breeding life of a beef cow is about ten years, culling helps to maintain high weaning rates. The rate of culling, plus the mortality of breeding cows defines the number of replacement heifers required to maintain a constant number of breeding cows.

The following model, which is suitable for a spreadsheet, provides a 'steady state' estimate of number in the various classes of cattle in a herd (herd structure, Table 12.1), in response to a few key assumptions and parameters, and local knowledge of performance criteria. The model depends on four assumptions.[46] First, all animal classes on a farm with breeding and growing cattle can be specified by a manager, and are related numerically to the number of cows mated, provided extra animals are not purchased. Second, the overall carrying capacity of a farm, in terms of number of adult equivalents, is either known or can be estimated by equation (12.4). Third, for simplicity, cows and calves are regarded as a single animal class until the calves are weaned. Fourth, the number of cows mated (CM) is fixed for each situation because if one dies or is culled from the breeding herd it is replaced with a heifer. Thus the 'n' classes of cattle on a farm can be represented as

$$CC = A1 \times CM + A2 \times CM + A3 \times CM + \ldots An \times CM \qquad (12.5)$$

and after collection of terms and simplification

$$CM = CC / \sum Ai \qquad (12.6)$$

where Ai is a coefficient that relates the number of animals in the 'i'th class of cattle to CM, the numbers of cows mated. Ai is the product of four factors:

$$Ai = PFi \times CFi \times SRi \times BRi \qquad (12.7)$$

where PFi is a flag to indicate if the ith class of animal is present (1, present; 0, absent); CFi is a factor to convert the ith class of animal to adult equivalents (Table 12.1); SRi is the proportion of the original number surviving in the ith class; and BRi is the ratio of the number of animals in the ith class to the number of breeders when survival in the class is 100%.

WR is weaning rate, expressed as a percentage of the number of calves weaned to number of cows mated. If half the weaners are assumed to be female, it follows that $BRi = WR/2$ for each class of steers in the herd, and for heifer cattle BRi is similar to steers until replacement heifers enter the breeding herd.

Replacement heifers enter the breeding herd when two or three years of age by adjusting PFi accordingly. First dead cows are replaced ($DEATHS$ = percentage of CM dying each year), then culled cows are replaced according to a specified culling policy ($CULL$ = preferred percentage of CM replaced each year). If there are too few heifers for the culling policy, all available heifers are used as replacements and the shortfall is noted by the lack of surplus heifers for subsequent sale and a reduced ratio for culling. If there are too few heifers to replace the dead cows the herd cannot be sustained. Thus for culled cows:

$$BR_{cull\ cows} = MAX(0, MIN(CULL, WR/2 - DEATHS)/100) \qquad (12.8)$$

and for any surplus females

$$BR_{surplus\ females} = MAX(0, (WR/2 - CULL - DEATHS)/100) \qquad (12.9)$$

Once the number of cows mated have been calculated using equation (12.6), the number of cattle in the remaining animal classes is given by

$$Ni = CM * PFi * SFi * Bri \qquad (12.10)$$

where $i > 1$ since for cows, being class 1, $Ni = CM$.

Table 12.2 illustrates the application of this model to four scenarios pertaining to breeding and growing beef cattle on extensive rangelands. Case 1 represents a herd where disease and/or poor nutrition severely restricts performance of the breeding herd and this limitation is removed in Case 2. Case 3 is similar to Case 2 except for a 50% increase in farm carrying capacity, which might occur through farm development options such as buying more land, controlling woody weeds or sowing improved pasture. Case 4 illustrates the effects on herd structure of a further improvement in performance of breeding cows along with a reduction in age of selling steers and mating heifers, as might

Table 12.2 Herd structures generated by the simple model given above in response to changes in key parameters that might occur as health, nutrition and management improves in a 'closed' herd consisting of breeding and growing cattle on extensive rangeland

Key parameters	Case 1	Case 2	Case 3	Case 4
Farm carrying capacity (*CC*) adult equivalents	1000	1000	1500	1500
Weaning ratio (*WR*) (% of cows mated)	50	80	80	90
Cow mortality rate (*DEATHS*) (%)	15	5	5	3
Ideal culling ratio for cows (*CULL*) (%)	20	20	20	20
Age of steers at sale: years	4	4	4	3
Age of surplus heifers at sale: years	3	3	3	2
Simulated results				
Total number of cattle in herd	1088	1121	1682	1700
Number of breeding cows	421	303	455	532
Proportion of herd as breeding cows (%)	39	27	27	31
Number of culled cows	42	61	91	106
Proportion of breeding cows culled (%)	10	20	20	20
Number of surplus heifers sold	0	44	66	112
Number of steers sold	99	114	171	227
Total number of cattle sold	141	219	328	446
Proportion of sale cattle in herd (%)	13	20	20	26

occur from a further improvement in herd nutrition and management. Whilst Table 12.2 is a static representation that ignores the transitional states that would occur when changing from Case 1 to Case 4, it shows the broad implications of management options on herd structure and number of cattle for sale. It also illustrates that simple 'spreadsheet' models can be a useful first step in selecting broad management options that warrant a more detailed evaluation.

12.5 Future developments

Modelling pasture and animal production has come a long way in three decades. Its future as an aid to research is assured since it provides direction and context to research programs.

While farmers have been slow to adopt decision support packages that aid routine decisions, professional advisors who need to give good advice to many clients are more receptive to new tools that assist in evaluating management options within complex systems across a wide range of environments. Future developers of farm management models will probably regard farm advisors or service agencies rather than farmers as the primary customers. Also, the models will be more user-friendly through the use of improved graphics and visualisation techniques, and the provision of support and upgrades via the World Wide Web.

The scope and range of policy models are expanding rapidly because they provide policy makers with an objective assessment of complex problems. This trend will continue, but policy models are likely to expand from the traditional biophysical base to include socioeconomic components and estimates of the impact of policies on the 'triple bottom line' – ecological sustainability, profitability and social acceptability.[47–49] Indeed, a future challenge will be how to better integrate the technologies pertaining to hard and soft systems, such as pasture and animal production models being part of participatory action research, and thereby involving stakeholders in defining and evaluating policies.[16,50]

A global network of information for model development and proven software modules is expanding through the World Wide Web. Model developers will have increasing access to libraries of algorithms, and computer operating environments which will encourage more rapid development of new models and a rich set of shared applications and experiences. However, since models are repositories for information and results from past research, there remains a global need for scientists and government agencies to organise creditable databases of information, which are critical to the future development of decision support systems and integrated policy models.[51]

12.6 References

1. RICKERT K G, STUTH J W, MCKEON G M, 'Modelling pasture and animal production'. In *Field and Laboratory Methods for Grassland and Animal Research*, L T Mannetje and R M Jones (Eds), pp. 29–65, Wallingford, CAB Publishing, 2000.
2. HERRERO M, DENT J B, FAWCETT R H, 'The plant/animal interface in models of grazing systems'. In *Agricultural Systems Modeling and Simulation*, R M Peart and R B Curry (Eds), pp. 495–542, Marcel Dekker Inc, New York, 1998.
3. EBERSOHN J P, 'A commentary on systems studies in agriculture'. *Agric Syst*, 1976 **1**(3) 173–84.
4. BLACK J L, DAVIES G T, FLEMING J F, 'Role of computer simulation in the application of knowledge to animal industries'. *Aust J Agric Res*, 1993 **44**(3) 541–55.
5. JONES R J, SANDLAND R L, 'The relation between animal gain and stocking rate. Derivation of the relation from the results of grazing trials'. *J Agric Sci UK*, 1974 **83**(2) 335–42.
6. RICKERT K G, 'Stocking rate and sustainable grazing systems'. In *Grassland Science in Perspective,* A Elgersma, P C Stuik and L J G van de Maesen (Eds), Wageningen Agricultural University Papers 96.4, pp. 29–63. Wageningen Agricultural University, Wageningen, 1996.
7. CONWAY G R, 'Sustainability in agricultural development: trade-offs between productivity, stability, and equitability'. *J Farm Syst Res Ext*, 1994 **4**(2) 1–14.
8. RICKERT K G, MCKEON G M, 'Models for native pasture management and development in south east Queensland'. In *Ecology and Management of*

the World's Savannas, J.C. Tothill and J.C. Mott (Eds), pp. 299–302, 1985.

9. MCKEON GM, DAY KA, HOWDEN SM, MOTT JJ, ORR DM, SCATTINI WJ, WESTON EJ, 'Northern Australian savannas: management for pastoral production'. *J Biogeogr*, 1990 **17**(4–5) 355–72.
10. HALL WH, Near-real Time Finacial Assessment of the Queensland Wool Industry on a Regional Basis. PhD Thesis, University of Queensland, 1997.
11. STUTH JW, HAMILTON WT, CONNER JC, SHEEHY DP, BAKER MJ, 'Decision support systems in the transfer of grassland technology'. In *Grasslands for our World*, MJ Baker (Ed.), pp. 234–42, SIR Publishing, Wellington, 1993.
12. LANDSBERG RG, ASH AJ, SHEPHERD RK, MCKEON GM, 'Learning from history to survive in the future: management evolution on Trafalgar Station, North-East Queensland'. *Rangel J*, 1998 **20**(1) 104–18.
13. JOHNSTON PW, MCKEON GM, BUXTON R, COBON DH, DAY KA, HALL WB, QUIRK MF, SCANLAN JC, 'Managing climate variability in Queenslands's grazing lands – new approaches'. In *Applications of Seasonal Climate Forecasting in Agricultural and Natural Ecosystems – the Australian Experience*, G Hammer, N Nicholls and C Mitchell (Eds), Kluwer Academic Press, Amsterdam, 2000.
14. MCKEON GM, ASH AJ, HALL WB, STAFFORD SMITH DM, 'Simulation of grazing strategies for beef production in north-east Queensland'. In *Applications of Seasonal Climate Forecasting in Agricultural and Natural Ecosystems – the Australian Experience*, G Hammer, N Nicholls and C Mitchell (Eds), Kluwer Academic Press, Amsterdam, 2000.
15. DONNELLY JR, MOORE AD, FREER M, 'GRAZPLAN: decision support systems for Australian grazing enterprises–I. Overview of the GRAZPLAN project, and a description of the MetAccess and LambAlive DSS.' *Agric Syst*, 1997 **54**(1) 57–76.
16. ABEL N, 'Resilient rangeland regions'. In *People and Rangelands: Building the Future*, D Eldridge and D Freudenberger, (Eds), *Proceedings of the VI International Rangeland Congress, Townsville, Queensland, Australia, 19–23 July, 1999* **1**, 21–30.
17. NORTON BW, 'Differences between species in forage quality. Temperate and tropical legumes and grasses, digestibility'. In *Nutritional Limits to Animal Production from Pastures*, JB Hacker (Ed.), Farnham Royal, Commonwealth Agricultural Bureaux, 1982.
18. BROWN JR, 'State and transition models for rangelands. 2. Ecology as a basis for rangeland management: performance criteria for testing models'. *Trop Grassl*, 1994 **28**(4) 206–13.
19. HODGSON JG, *Grazing Management. Science into Practice*, Harlow, Longman, 1990.
20. HEITSCHMIDT RK, WALKER JW, 'Grazing management: technology for sustaining rangeland ecosystems?'. *Rangel J*, 1996 **18**(2) 194–215.

21. MINSON D J, WHITEMAN P C, 'A standard livestock unit (SLU) for defining stocking rate in grazing studies'. *Proceedings of the XVI International Grassland Congress, 4–11 October 1989, Nice, France, 1989, 1117–18*, 1989.
22. SELIGMAN N G, BAKER M J, 'Modelling as a tool for grassland science progress.' In *Grasslands for our World*, M J Baker, (Ed.), pp. 228–33, SIR Publishing, Wellington, 1993.
23. MONTEITH J L, 'The quest for balance in crop modeling'. *Agron J,* 1996 **88**(5) 695–7.
24. LOEWER O J, 'GRAZE: a beef-forage model of selective grazing'. In *Agricultural Systems Modeling and Simulation*, R M Peart and R B Curry (Eds), pp. 301–417, New York, Marcel Dekker Inc., 1998.
25. SINCLAIR S E, RICKERT K G, PRITCHARD J R, FEEDMAN – A feed-to-dollars beef and deer management package, QZ00004, Brisbane, Queensland Department of Primary Industries, 2000.
26. COX P G, 'Some issues in the design of agricultural decision support systems'. *Agric Syst,* 1996 **52**(2) 355–81.
27. BUXTON R, STAFFORD-SMITH M D, 'Managing drought in Australia's rangelands: four weddings and a funeral'. *Rangel J,* 1996 **18**(2) 292–308.
28. RICKERT K G, 'Experiences with FEEDMAN, a decision support package for beef cattle producers in south eastern Queensland'. *Acta Hort,* 1998 **476** 227–34.
29. CLEWETT J F, *Shallow Storage Irrigation for Sorghum Production in Northwest Queensland*. Bulletin QB85002, Brisbane, Queensland Department of Primary Industries, 1985.
30. HASSETT R C, WOOD H L, CARTER J O, DANAHER T J, 'Statewide groundtruthing of a spatial model'. In *People and Rangelands: Building the Future*, D Eldridge and D Freudenberger (Eds), *Proceedings of the VI International Rangeland Congress, Townsville, Queensland, Australia, 19–23 July, 1999,* **2**, 763–4.
31. HAMMER G L, WRIGHT G C, 'A theoretical analysis of nitrogen and radiation effects on radiation use efficiency in peanut'. *Aust J Agric Res,* 1994 **45**(3) 575–89.
32. JOHNSTON P W, MCKEON G M, DAY K A, 'Objective "safe" grazing capacities for south-west Queensland, Australia: development of a model for individual properties'. *Rangel J,* 1996 **18**(2) 244–58.
33. FREER M, MOORE A D, DONNELLY J R, 'GRAZPLAN: decision support systems for Australian grazing enterprises – II. The animal biology model for feed intake, production and reproduction and the GrazFeed DSS'. *Agric Syst,* 1997 **54**(1) 77–126.
34. DOVE H, 'Constraints to the modelling of diet selection and intake in the grazing ruminant'. *Aust J Agric Res,* 1996 **47**(2) 257–75.
35. HALL W B, RICKERT K G, MCKEON G M, CARTER J O, 'Simulation studies of nitrogen concentration in the diet of sheep grazing Mitchell and mulga grasslands in western Queensland'. *Aust J Agric Res,* 2000 **51** 163–72.

36. QUIRK M F, STUTH J W, WEST N E, 'Verification of POPMIX preference algorithms for estimating diet composition of livestock'. In *Rangelands in a Sustainable Biosphere*, Denver, Society for Range Management, 1996.
37. SCA, *Feeding Standards for Australian Livestock: Ruminants. Standing Committee on Agriculture*, East Melbourne, CSIRO Publications, 1990.
38. WILLIAMS C B, JENKINS G M, 'Predicting empty body composition and composition of empty body weight changes in mature cattle'. *Agric Syst*, 1997 **53**(1) 1–25.
39. RICKERT K G, MCKEON G M, 'A computer model of the integration of forage options for beef production'. *Proc Aust Soc An Prod*, 1984 **15** 15–19.
40. DAY K A, AHERNS D G, PEACOCK A, RICKERT K G, MCKEON G M, 'Climate tools for northern grassy landscapes.' *Proceedings of Northern Grassy Landscapes Conference Katherine, 29–31 August 2000*, Darwin, Northern Territory University, 2000.
41. MITCHELL P L, 'Misuse of regression for empirical validation of models'. *Agric Syst*, 1997 **54**(3) 313–26.
42. HARRISON S R, 'Regression of a model of real-system output: an invalid test of model validity'. *Agric Syst*, 1990 **34**(3) 183–90.
43. HARRISON S R, 'Validation of agricultural expert systems'. *Agric Syst*, 1991 **35**(3) 265–85.
44. SCANLAN J C, MCKEON G M, DAY K A, MOTT J J, HINTON A W, 'Estimating safe carrying capacities of extensive cattle-grazing properties within tropical, semi-arid woodlands of north-eastern Australia'. *Rangel J*, 1994 **16**(1) 64–76.
45. DENT J B, EDWARD JONES G, MCGREGOR M J, 'Simulation of ecological, social and economic factors in agricultural systems'. *Agric Syst*, 1995 **49**(4) 337–51.
46. RICKERT K G, ESPIE N J, STOCKUP: a program to assess the structure of beef cattle herds, Brisbane, Department of Primary Industries, 1990.
47. PANDEY S, HARDAKER J B, 'The role of modelling in the quest for sustainable farming systems'. *Agric Syst*, 1995 **47**(4) 439–50.
48. BELLAMY J A, LOWES D, ASH A J, MCIVOR J G, MACLEOD N D, 'A decision support approach to sustainable grazing management for spatially heterogeneous rangeland paddocks'. *Rangel J*, 1996 **18**(2) 370–91.
49. ABEL N, ROSS H, WALKER P, 'Mental models in rangeland research, communication and management'. *Rangel J*, 1998 **20**(1) 77–91.
50. PARK J, SEATON R A F, 'Integrative research and sustainable agriculture'. *Agric Syst*, 1996 **50**(1) 81–100.
51. BESWICK A, JEFFREY S J, MOODIE K B, 'Climate databases: fundamental to the effective management of Australia's rangelands'. In *People and Rangelands: Building the Future*, D Eldridge and D Freudenberger (Eds), *Proceedings of the VI International Rangeland Congress, Townsville, Queensland, Australia, 19–23 July, 1999*, **2**, 850–1.

13

New developments in decontaminating raw meat

C. James, Food Refrigeration and Process Engineering Research Centre (FRPERC), University of Bristol

13.1 Introduction

Throughout the European Union (EU) consumers are requiring the food industry to provide them with an increasing range of safe, nutritious and healthy chilled foods of high sensory quality and an increased shelf-life. To meet the demand for healthier food of high sensory quality, the use of additives and preservatives is being reduced or eliminated and minimal processing techniques introduced.

To increase food safety whilst maintaining or increasing storage life, a considerable amount of time, effort and money has been spent in adopting HACCP techniques, including the use of mathematical modelling of microbial growth, better packaging methods and improved temperature control within the chill chain. Nevertheless there is little, if any, sign within official statistics of significant reductions in the incidence of food-borne illnesses within EU countries. Over the period 1982–94 cases have risen by 330, 50, and 200% in Spain, Norway and the United Kingdom respectively. In one year, 1988–89, cases rose by 28% in Belgium, 75% in France and between 20 and 26% in Germany. Poultry, red meat and meat products together make up the largest single source of food poisoning in the EU. For example, 47% of the outbreaks in Belgium, 38% in Sweden and 45% in the UK were attributed to poultry and red meat. Wastage due to microbiological spoilage and poor appearance due to desiccation is also high from meat and meat products.

There is no terminal step (such as cooking) to eliminate pathogenic organisms from many raw products such as red and white meat until it reaches the consumer. The consumer is relied upon to adequately cook the meat sufficient to kill any bacteria injurious to health prior to ingestion. Several of the pathogens present on meat are psychrotrophic and can grow at refrigeration temperatures. Centralised processing and preparation of these products is growing, increasing the distance and time between initial preparation and the consumer, thus increasing the risk of growth of pathogens. Ideally, some form of terminal step should be introduced, failing that any step that reduces the microbial load would be advantageous to public health and of economic significance to the industry. That is provided such a step did not change the intrinsic nature of the food, i.e., the 'raw' produce or meat must remain 'raw'.

It would be expected that improvements to slaughtering procedures should result in a significant reduction in bacterial contamination of carcasses. It is therefore very disappointing that the last published scientific survey (Hinton *et al.*, 1998) reported that 'It can ... be safely concluded that there is little evidence of any major change in the bacteriological quality of British beef during the last 10–15 years'. There would seem to be no basis for believing that the situation is different in the UK from that of other EU member states. The main effect of introducing new procedures, however, appears to be a change in the distribution of contamination rather than a substantial reduction in total number (James *et al.*, 1999). In the few cases where a generally lower microbial contamination has been reported, the reductions were quite small. For example, comparisons of two dressing methods in New Zealand reported changes ranging from a 1.41 decrease to a 0.5 $\log_{10}$ Colony Forming Units cm^{-2} increase in bacterial counts at different positions (Bell *et al.*, 1993). The application of 'strictly hygienic' procedures such as surgical gloves and disinfected knives under near laboratory conditions have been found to have a significant effect on levels of contamination. However, attempts at transferring laboratory technology into commercial operations have met with limited success.

Many studies have shown that at the time of slaughter the muscle tissue of a healthy animal is essentially sterile (Gill, 1979). The surface of the meat is contaminated with pathogenic and spoilage organisms during slaughter and subsequent handling. Exposed surfaces of the hide, fleece and skin of cattle, sheep and pigs, and the feathers of poultry are covered with dust, dirt and faecal matter. Further contamination can occur from exposure to intestinal contents, which like faeces contain salmonellas and campylobacters – the two most common causes of food-borne disease. If microbes on the surface of meat could be eliminating or substantially reduced immediately after slaughter the risk of cross-contamination during processing would be substantially reduced. An efficient method of surface decontamination therefore offers substantial advantages in terms of food safety, spoilage and economics.

13.2 Current decontamination techniques and their limitations

Many decontamination techniques have been suggested and studied over the years. However, many of these have only been attempted on a laboratory scale. Methods of decontaminating meat can be divided into those that rely on the activity of physical treatments and others that use chemicals to either remove or destroy the microorganisms.

13.2.1 The problems of decontaminating raw meat
Animal carcasses are not ideal shapes to decontaminate. Most decontamination treatments rely on physical contact and uniform coverage of the meat surface. This is difficult, as the surfaces of many produce and whole animal carcasses are very irregular. For example, the outer surface of a carcass has many crevices and folds. These areas are very difficult to treat and provide protection to attached bacteria. They slow down the penetration of aqueous and gas treatments and cause shadowing problems for radiation treatments such as ultraviolet (UV) light. As well as protecting bacteria, these areas often clog up with physical contamination, such as dirt and hair, and do not drain well. Pools of water or chemical solutions lying in these areas can have a detrimental affects on the visual quality of the meat and cause difficulties in controlling the contact time of treatments.

There is much evidence that the time at which products are treated greatly affects the efficacy of decontamination processes. The longer bacteria reside on product surfaces, the more difficult removal becomes, because of the ability of bacteria to attach to tissue. Bacteria differ in their ability to attach to different surfaces and the time they require to become fully attached. The formation of bio-films may increase the resistance of bacteria to disinfectants such as chlorine. Surfactants such as 'Tween 80' have been used to increase surface wetting, in theory allowing the disinfectant to 'get at' the bacteria. 'Tween 80' is not used for food production because it causes unacceptable organoleptic changes. Two surfactants, 'Orenco Peel 40' and 'Tergitol', are used for fruits and vegetables in the USA (Zhang and Farber, 1996).

13.2.2 The difference between decontamination methods and treatments
There is rarely any distinction made in the literature between decontamination 'methods' (i.e., the method of applying a treatment) and decontamination 'treatments'. This often clouds the practical issues of decontamination. There is often too much emphasis placed on the treatment rather than the method of application. Decontamination is not a matter of simply dipping or spraying the product with chemicals or water, or giving it a quick flash of light. For example, many factors affect the efficiency of aqueous spray systems. In automated spray cabinets the position and number of the sprays, the shape of the spray, and spray pressures, all have a significant effect on the treatment irrespective of the nature of the substance being pumped through the sprays. Many studies have shown that the method of decontamination is often more important than the treatment.

Most abattoirs have relied in the past on manual sprays to wash red meat carcasses; thus, automated spray cabinets have been a natural development. Some studies, however, have shown that a deluge method of application where the carcass is passed under a waterfall offers a more effective method of coverage (Davey and Smith, 1989). The use of water sprays is currently the most common method of cleaning carcasses. Many studies have been carried out to optimise spraying systems and investigate their efficiency. Extensive related studies have been carried out in the UK by Bailey (1971), in Ireland by Kelly *et al.*, (1981), in the USA by Anderson and co-workers on the CAPER system (Anderson *et al.*, 1984), and in Australia by Smith and Davey (1990). In a number of these studies the effect of adding organic acids and chlorine to the water systems was evaluated. Together these studies provide essential information on the parameters that affect spray washing. Physical parameters include spray pressure and flow rate, and nozzle type, configuration and the angle of spray. As well as these physical parameters variables such as tissue type, inoculation menstruum, inoculation amount, or temperature of treatment all affect the result of decontamination procedures.

Heat treatments, with or without chemicals, are very reliant on the method of application. To prevent cooking the product, such treatments have to provide a uniform heating of all surfaces for a short period. This is not particularly difficult to achieve on a laboratory scale, spraying or dipping small samples using hot water for example. Similarly, laboratory studies using steam have shown that if very high temperatures are applied for very short times, followed by cooling the surface rapidly, high bacterial reductions can be achieved on meat without affecting the surface appearance. However, successfully applying such techniques to carcasses in an abattoir, for instance, presents many engineering challenges.

13.3 Washing

Washing meat or produce with water can effectively remove physical contaminates such as soil, hairs and other debris, however its affect on bacterial numbers is marginal. The temperature at the surface and the method of applying the water are the two most important factors in bacterial removal. While it is generally accepted that washing is an effective method of removing visible contamination from meat carcasses, there is persistent criticism that it may redistribute bacteria over the carcass. At present, most washes utilise cold water and the evidence is that cold water has little effect on microbial numbers. Trials on sheep have shown that washing led to bacterial contamination of the dorsal area, which was uncontaminated before washing (Ellerbroek *et al.*, 1999). Contamination on the ventral area was not reduced, an area most likely to be contaminated during dressing operations. Residual water remaining on the carcass was believed to enhance bacterial multiplication during storage. Cold water washing of beef carcasses has been shown by one study to be ineffective and tending to bring about a 'posterior to anterior redistribution' (Bell, 1997). However, another study (Charlebois *et al.*, 1991) found that there was little difference in the distribution of faecal coliforms before and after trimming and washing on beef carcasses in three abattoirs.

Washing with water alone usually obtains a reduction in microbial numbers of 1- to 2-log-units (10 to 10^2) on meat. Increasing the temperature of the water increases the reduction. However, a spray jet rapidly loses heat by evaporation. Studies have shown that the maximum impact temperature on the carcass of a spray placed 30 cm away and supplied with water at 90ºC is approximately 63ºC (Bailey, 1971). Abattoirs have always been worried about the effect of hot water on the appearance of carcasses. However, studies (Smith, 1992) have shown that treatments of 80ºC for 10 s not only significantly reduce bacterial levels but do so without any permanent damage to the surface tissue.

Automated washing systems for meat carcasses have long been seen to be the way forward. The most comprehensive publicly documented studies to date have been on the CAPER (Carcass Acquired Pathogen Elimination Reduction) system developed in the USA and the Australian 'Deluge' system. The Australian system depends on the action of hot water solely to decontaminate. While the CAPER system has been designed as a two-stage process involving a water stage to remove physical contamination and organic acids to sanitise the carcass. Commercial spray cabinets similar to the CAPER system are available in the USA, while trials have been carried out on a commercial version of the deluge system in Australia.

13.4 The use of chemicals

Many studies have been carried out to test groups of chemicals for antimicrobial activity against specific pathogenic and food spoilage organisms. A wide range of chemicals are known which will destroy or severely limit the growth of pathogenic and spoilage bacteria. However, the number of chemicals that are likely to be approved for use on meat is severely limited (Table 13.1), not least because of legal restrictions. While chlorine has been an accepted part of washing fruits and vegetables for many years, chemical washing of red meat has not generally been accepted. The poultry industry has utilised chlorine to keep chiller water clean, which has had a knock-on affect on microbial counts, but its use in the EU is being stopped following health concerns. More recently trisodium phosphate (TSP) has been used for poultry. There is also growing interest in the use of ozone and naturally occurring antimicrobials. The effectiveness of most chemical treatments depends on concentration, application temperature and exposure time.

13.4.1 Application of chemicals

When considering all chemical treatments the method of application must also be considered. In many cases, these are 'drop-in' additions to the washing process rather than an integral part of the washing system. Most chemicals are applied in the form of aqueous solutions therefore as with water treatments the method of application will have a significant influence on how effective a treatment will be.

Table 13.1 Chemicals investigated, with varying success, to decontaminate meat

Organic acids	Lactic, acetic, fumaric, citric, ascorbic, formic, propionic, benzoic, sorbic
Chlorine	Gaseous chlorine, sodium hypochlorite, calcium hypochlorite
Chlorine dioxide	
Sorbates	Potassium sorbate, sorbic acid
Polyphosphates	(Trisodium orthophosphate (TSP), sodium hexametaphosphate, sodium tripolyphosphate, tetrasodium pyrophosphate)
Ozone	
Hydrogen peroxide	
Potassium chloride	
Lysozyme	
Disinfectants	(Glutaraldehyde (1,5-pentanedial), Poly hexamethylenebiguanide hydrochloride (PHMB), Iodophor, Cetylpyridinium chloride (1-hexadecylpyridinium chloride) (CPC), Carntrol(active ingredient copper sulfate pentahydrate), Timsen(40% N-alkyldimethylbenzylammonium chloride in 60% stabilised urea))

Most of the chemicals described have been investigated in laboratory studies by dipping small samples of meat into solutions of the chemicals. Immersion is a very effective method of ensuring full coverage of a product. However, there are a number of practical problems with immersion. Aside from the logistical problem of immersing a side or whole carcass, maintaining chemical concentration is difficult. As well as being lost through spillage and absorption by the meat, the activity of the solution will change as the chemical reacts with the microorganisms and other organic material. Acid solutions lose activity as the anions are easily bound by peptides and proteins released by the meat. Chlorine also reacts with organic material. Ozone and hydrogen peroxide in solution rapidly decompose. While immersion may be practical for cuts of meat, (sub)primals and poultry carcasses it is unlikely to be adopted for treating sides and carcasses.

Spraying is the most common way of applying chemicals to carcasses. Most studies have used manual spray devices. The effectiveness of a manual system, whether it is using just water or a chemical spray, depends very much on the skill of the operator and will vary from operator to operator. This means that any results are difficult to quantify. Even the effectiveness of automated cabinet systems depends upon the influence of various physical parameters. These parameters have been covered in the earlier section on spraying with water. Most of the cabinet studies have used equipment based on the CAPER system or one made by US CHAD Co., though a number of groups have used purpose-built systems.

13.4.2 Chlorine

The various forms of chlorine are probably the most widely used sanitisers in the food industry. They include gaseous chlorine (Cl_2), sodium hypochlorite (NaOCl), calcium hypochlorite ($Ca(OCl)_2$), and chlorine dioxide (ClO_2). Apart from ClO_2, which has a different mode of action, these compounds form hypochlorous acid (HOCl) in aqueous solution, and it is this form that is active against microorganisms. The anti-microbial action of all chlorine compounds is due to their oxidising affect. However, while chlorine is widely used in the EU by the food industry to wash vegetables, particularly salad vegetables, it is not permitted to be used on meat. Despite this, scientific trials have been carried out on its applicability. Many studies have shown that applying chlorine at concentrations of 200 ppm and above to meat carcasses can produce a 2 log $(10)^2$ reduction in bacterial numbers. These reductions can be further increased by raising the temperature of the chlorine. Most pathogens can be readily controlled, though not eliminated, by chlorine but some would require concentrations higher than 200 ppm. It is very unlikely that chlorine concentrations above this level would be allowed legally for meat.

Numerous concerns are increasingly being expressed about the use of gaseous chlorine and hypochlorite solutions. Among these are the fact that they react with phenolic compounds and the resultant chlorophenols can cause tainting at very low concentrations, as well as possible human health risks associated with chlorinated lipids and proteins. Chlorine dioxide has been proposed as a safe alternative since it does not react in this way. There are also many practical problems in terms of control of chlorine levels, protection of delivery systems from corrosion, etc.

13.4.3 Organic acids

There are many commercially available organic acids. The effects of different concentrations, temperatures and mixtures of many of these have been studied on meat micro-flora. Acetic and lactic acid have been the most widely studied of the organic acids, while propionic, citric and fumaric acids have also been investigated. Organic acids are naturally present in many foods, and are relatively cheap as they are the principal products of many natural fermentation reactions.

The effect of organic acids depends on three factors (Ingram et al., 1956); (i) the effect of pH, (ii) the extent of dissociation of the acid, and (iii) a specific effect related to the acid molecule. In general the antimicrobial action of organic acids is due to pH drop. The lower the pH the greater the effect. Lowering the pH, however, through the addition of an inorganic acid is ineffective (Reynolds and Carpenter, 1974). Dissociation of the acid is also a factor. Undissociated weak acids are 10 to 600 times as effective in inhibiting and killing microorganisms as dissociated forms (Eklund, 1983). Organic acids are mainly undissociated when dissolved in water. They therefore have a stronger antimicrobial action than inorganic acids that are totally dissociated in water. Buffering the acids (through the addition of a soluble salt of the base acid) will increase their effectiveness, as more undissociated molecules will be present. Even under the same conditions of pH and acid dissociation there are differences in the antimicrobial action of various organic acids (de Koos, 1992), this is due to the nature of the anion (Smulders, 1995).

Washes and sprays containing organic acids have been successfully used in decontaminating beef, lamb, pork and poultry carcasses. Researchers agree that organic acids can reduce the numbers of pathogenic and spoilage organisms on meat by typically 1 to 3.5 log microorganisms per g producing an extension of shelf life of 7 to 17 days respectively. In investigations where the temperature of the acid is varied, greater reductions in bacterial numbers are achieved at higher temperatures. However, in many cases the meat has been immersed in the acid mixture and it is difficult to separate the effect of the temperature from that of the acid.

Studies have generally used concentrations of between 2 to 4% with some as high as 24% and it is not clear what should be the maximum concentration. In some studies concentrations of 2% acetic acid were reported to produce discolouration on pork loins (Cacciarelli *et al.*, 1983). In others, at 3% no adverse effects were found on lean samples but slight off flavours and grey discolouration was reported on fats (Anderson *et al.*, 1979a). Overall, treatment with 2% lactic acid solutions applied at a meat surface temperature of 37°C have been described as optimal (Anderson and Marshall, 1989). Some researchers advocate a mixture, others single acids.

It is disappointing that the reductions produced in commercial trials are often significantly lower than those found in laboratory studies. In laboratory trials the samples have often been inoculated with high levels of bacteria and in these situations the acids may be more effective. Also, producing an even surface coverage of acid is far easier on a small sample in the laboratory than over a whole carcass in the abattoir.

13.4.4 Polyphosphates

Trisodium phosphate (TSP) was developed in the US for the control of salmonella on poultry. TSP (Na_3PO_4) possibly works by removing a thin layer of fat from the carcass surface and in doing so removing the microorganisms attached to the surface (Giese, 1992), it then causes rupture of the bacterial cell membrane. Ruptured cells are not protected and succumb to the ionic strength and high pH of the medium.

There are conflicting reports on the sensitivities of Gram-positive and Gram-negative bacteria to polyphosphates. It has been reported that Gram-positive bacteria are generally more sensitive to polyphosphates than are Gram-negative bacteria (Lee *et al.*, 1994), but TSP has been reported to be more active against Gram-negative bacteria, such as *Salmonella* spp., *Campylobacter* spp. and *Pseudomonas* spp. (Corry and Mead, 1996). There are conflicting reports on its effectiveness on reducing microorganisms on red meat tissues (Dickson *et al.*, 1994; Gorman *et al.*, 1995).

13.4.5 Ozone

Ozone (O_3) is a water-soluble naturally occurring gas that is a powerful oxidising agent. It is also very unstable, on exposure to air and water it rapidly decomposes to form oxygen, hence generation is usually at the point of use. In general, bacteria are more susceptible than yeasts or moulds; Gram-positive bacteria are more sensitive than Gram-negative; bacterial spores are more resistant than vegetative cells. Temperature, relative humidity, pH, stage of microbial growth and organic matter present have all been shown to affect ozone antimicrobial action.

Gaseous ozone was used commercially in the 1940s to extend the refrigerated storage life of meat (Ewell, 1943). The use of gaseous ozone in meat-conditioning coolers has long been accepted by the FDA in the USA (Graham, 1997). However, meat pigments and fats are sensitive to oxidation by high concentrations of ozone (≥ 10 ppm). Studies using ozonated water have reported conflicting results, some reporting advantages over other chemical treatments, others showing no advantages over washing with water alone (Reagan *et al.*, 1996).

13.5 New methods: steam

Steam at 100°C has a substantially higher heat capacity than the same amount of water at that temperature. If steam is allowed to condense onto the surface of meat it will rapidly raise the surface temperature of the meat. One very attractive feature of condensing steam is its ability to penetrate cavities and condense on any cold surface. The basis for why steam treatment need not cook raw meat while killing bacteria and the penetrative ability of gases has been dealt with by Morgan *et al.* (1996a). Heat kills bacteria mainly by inactivating the most sensitive vital enzymes. Typically the heat of activation of these enzymes is 8.38 to 50.28 kJ(g.mol)$^{-1}$. The heat of activation for irreversible muscle cooking is 209.5 to 419 kJ(g.mol)$^{-1}$, substantially higher. Only micrograms of enzyme need to be inactivated compared to the grams of muscle denatured during cooking. 'For a square centimetre of surface contaminated with 100 bacteria, 15 million times as much heat is needed to cook the surface to a depth equal to the length of a bacterium compared to the heat needed to kill all the bacteria' (Morgan *et al.*, 1996a). Since bacteria are present only on the surface of the meat even assuming that heating rates are the same theoretically the bacteria should die earlier than the meat would cook. In fact the meat will take longer since it requires conductive heat transfer through the muscle. Exposure times for chicken meat in air-free thermally saturated steam at various temperatures are shown in Figure 13.1, an equivalent time in 100°C water would be about 1000 ms.

Water vapour molecules are much smaller than bacteria, for example 2 by 10^{-4} μm in diameter compared with 0.7 by 4 μm for *Salmonella* cells (Morgan *et al.*, 1996a). Therefore steam is capable of reaching any bacteria in cavities.

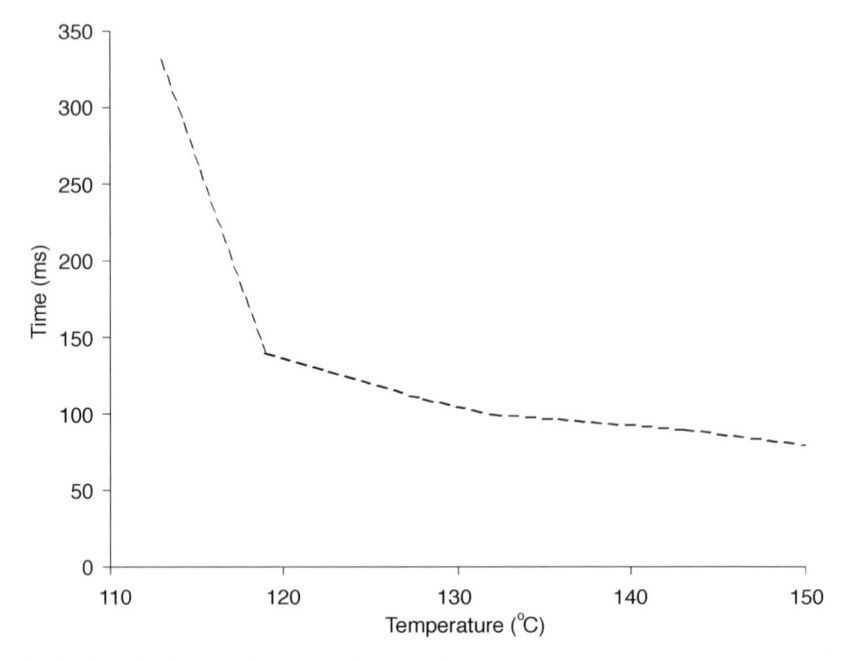

Fig. 13.1 Time for cooking to begin on broiler meat pieces exposed to steam at various temperatures (adapted from Morgan et al., 1996a).

Although the velocity of steam is reduced by cavities of diameter less than the mean free path of the gas density this does not restrict steam reaching bacteria. In 140°C saturated steam, the mean free path of the steam molecule is 0.4 μm, half the diameter of the smallest cavity capable of containing a *Salmonella* cell.

To prevent cooking the steam must condense on the surface rapidly, and re-evaporate equally rapidly. Gases move by either flow or diffusion. Flow is rapid, motivated by a pressure gradient. Diffusion is much slower and motivated by a concentration gradient of the gas through other gases. During steam treatment air, and any other non-condensable gas present, is concentrated by the inrush of condensing steam forming a layer around the product surface. This prevents steam flow, slowing condensation as the steam diffuses through the layer. Non-condensable gases can come from three sources; gases around the meat when enclosed in the chamber; gases entering with the treatment steam; and gases which have been desorbed by heat from the meat or other surfaces. The temperature at which water boils is a function of pressure. At atmospheric pressure, steam will initially be created at 100°C. At lower pressures the generation temperature will be lower, at higher pressures it will be higher.

Two laboratory studies on the direct application of steam through a hose to a meat carcass report conflicting results. In one study, direct treatment of pork carcasses showed a reduction of total bacterial counts of 6 log micro-organisms per cm^2 (Biemuller et al., 1973). However, the steam marred the appearance of the carcasses. In contrast a study on beef carcasses showed direct application to be ineffective and to reduce storage life (Anderson *et al.*, 1979b).

The effects of various steam treatments on the appearance, shelf-life and microbiological quality of chicken portions have been investigated at the University of Bristol (James *et al.*, 2000a). Application of steam at atmospheric pressure (100°C for 10 s) on naturally contaminated chicken breast portions resulted in a 1.65 $\log_{10} \text{cfu cm}^{-2}$ reduction in the numbers of total viable bacteria. However, in comparison with untreated controls, this treatment did not extend the shelf-life. Steam treatment for up to 10 s on chicken portions inoculated with a nalidixic acid resistant strain of *Escherichia coli* serotype O 80 resulted in a maximum reduction of 1.90 $\log_{10} \text{cfu cm}^{-2}$. Overall, results indicated that significant reductions in microbiological numbers could be achieved on chicken meat using steam. However, the reductions achieved were less than would be expected from the time temperature cycles achieved.

Additional work at the University of Bristol has compared steam condensation (100°C for 8 s), hot water immersion alone (90°C for 8 s), and chlorinated hot water (250 ppm, 90°C for 8 s) for treating lamb carcasses (James *et al.*, 2000b). All three treatments produced carcasses with lower aerobic plate counts than untreated controls (average count of 3.2 $\log_{10} \text{cfu cm}^{-2}$). There was no significant difference between the steam and hot water treatments with both treatments reducing counts by approximately 1 $\log_{10} \text{cfu cm}^{-2}$. Overall the chlorinated hot water treatment reduced counts by 1.6 $\log_{10} \text{cfu cm}^{-2}$. Although chlorine proved the most effective, the authors felt that current attitudes towards the use of chemicals relegated its use in comparison with the other two treatments.

Steam can be produced under vacuum at temperatures substantially below 100°C without substantially reducing its heat capacity. Sub-atmospheric steam has been shown to be an effective way of decontaminating poultry drumsticks and carcasses, surface temperatures of 75°C for four minutes achieving reductions of the order of 5.5 and 3 log, respectively (Klose *et al.*, 1971).

EU and UK government funded studies, involving the University of Bristol, have been carried out on the use of sub-atmospheric steam pasteurisation systems for treating a range of food products (Evans, 1999). During trials each food type was inoculated with a pathogen and a spoilage organism (*Salmonella enteritidis* and *Pseudomonas fluorescens* on poultry, *Escherichia coli* O157:H7 and *Pseudomonas fragi* on beef and *Yersinia enterocolitica* and *Pseudomonas fragi* on pork). The samples were treated in a decontamination apparatus (developed as part of the work,) at temperatures between 55 and 85°C for times between ten seconds and ten minutes and the reduction in microbial contaminants determined. As an additional treatment, organic acids were added before heat treatment and their effect quantified. The effects of three acids applied at 10 or 55°C were investigated (acetic (0.15, 0.23 or 0.3M), lactic (0.1, 0.15 or 0.2M), buffered lactate and a mixture of acetic and lactic acids). Water applied at 55°C was used as a control.

In the absence of acid or water sprays, steam at 75 or 85°C for 40 seconds was required in order to reduce levels of *S. enteritidis* on chicken by 3–4 log cycles.

Treatments of 75°C for ten seconds reduced *Y. enterocolitica* on pork skin and *E. coli* O157:H7 on beef by 2–3 log cycles. The addition of organic acids increased microbial reductions in all of the three meats investigated. The application of acids at high temperature (55°C) and stronger molarities was found to be most effective. Improved effects of the acids were also found when the time of contact between the acid and the food prior to steam treatment was increased. Contact times of between four and six minutes were required to achieve reductions in the pathogens studied by 4 $\log_{10}$ cfu cm^{-2} on pork skin and between 5–6 $\log_{10}$ cfu cm^{-2} on beef or chicken skin. On all meats acetic, lactic or a mixture of these acids were most effective, although when water was applied as control, reductions above those achieved by steam alone were recorded. This indicated that some of the action of the acids was to wash microbes from surfaces.

When steam alone was used to decontaminate beef samples the shelf-life of samples stored at 0 or 10°C in either vacuum packs or air was extended only if the storage temperature was maintained at 0°C. When acids were applied in addition to steam treatment the storage life of pork skin and chicken was extended. On the pork skin four acids were compared (0.3M acetic, 0.2M lactic, an acetic/lactic mixture and buffered lactic). Of these acids, buffered lactic was capable of retaining microbial counts below the pre-decontamination level for 14 days if stored at 0°C and five days if stored at 10°C. Extensions in the storage life of decontaminated chicken samples when treated with organic acids (lactic) and steam were also found. On the inoculated and treated samples of poultry the levels of TVCs and pseudomonads remained below 6 log for 2–3 times longer than the control samples (for three days at 10°C and 12 days at 0°C). Total numbers of microbes on the treated samples required longer periods of time to reach a set level. This was primarily because they had lower initial levels of contamination. The rate of growth of salmonella surviving the heat treatment over the 20-day storage period at 0°C was negligible as would be expected at this temperature. At 10°C growth was slow, but the overall increase did not exceed one log cycle during the five-day storage.

This study has shown that re-contamination, after decontamination, of pork with *Y. enterocolitica* did not present a higher risk of increased growth during aerobic storage (at 0 or 10°C) on decontaminated than on untreated pork when the numbers of background flora were low compared to the inoculated bacteria. The risk was slightly greater when the samples were vacuum packaged. On chicken, *Salm. enteritidis* grew well in aerobic conditions at 10°C both on the decontaminated and the untreated meat.

As *E. coli* O157:H7 has a very low infectious dose and causes very severe disease the consequences of growth could be especially serious. In this study a storage temperature of 10°C was used as a 'worst case'. When a high inoculum (3.6 $\log_{10}$ cfu cm^{-2}) was used, the multiplication was much more rapid on the decontaminated beef when vacuum packaged. This was also the case for the three experiments with low inoculum, but there was a greater variability in the results. These observations need to be investigated further, particularly as *E. coli* O157:H7 is known to be resistant to acids. However, it should be borne in mind that post-process recontamination is likely to include mainly non-pathogenic (competitive) microorganisms, at least some of which are likely to compete successfully with *E. coli* O157:H7. In addition, raw meat should be stored at temperatures significantly below 7°C, which is the minimum temperature for growth of this pathogen.

In all cases some degree of cooking was apparent on the meats investigated. On the chicken and beef the outer layer of muscle was slightly cooked, although the skin of the chicken was barely affected. With pork skin the steam treatment slightly influenced the surface colour but did not affect consumer acceptability of the samples. When acids were added to the pork skin consumers were able to detect an acid odour immediately after treatment, although the acid odour decreased during storage.

Morgan *et al.* (1996a, b) have developed a device that enables very high temperature surface treatment of meat without cooking. This system utilises very rapid cycling (for milliseconds) of heating and cooling using steam under pressure and vacuum cooling. Meat samples are placed in a rotating chamber that as it rotates is exposed to three other chambers, a vacuum, steam, and final vacuum. This allows temperatures of up to 145°C. Tests, using inocula of *L. innocua* on chicken meat, have shown that substantial reductions can be achieved. Subsequent work showed that treatment at 145°C for 25 ms produced a 4 log reduction on raw chicken meat. Treatment at 121°C for 48 ms produced a 2.5 and 1.9 log reduction on beef and pork samples, respectively.

The most successful steam process yet, in terms of industrial application, has been that developed in the USA by Frigoscandia, the Steam Pasteurisation System (SPS). Studies on this commercially available system for treating red meat carcasses have been conducted and published by Kansas State University (Nutsch *et al.*, 1995; 1996, 1997, 1998; Phebus *et al.*, 1996a, b, c, 1997a, b, 1999;). Significant reductions of the order of 3.5 log-units for specific bacteria have been reported. The full commercial system (SPS 400 Steam Pasteurisation system) consists of a three-stage cabinet. Washed carcasses pass through an air-drying stage to remove residual water from the carcass before an enclosed-steam treatment stage followed by spray cooling. The full unit is very large, 'the size of a subway car' (Smith, 1996), and a single-steam unit (SPS 30 Steam Pasteurisation system) for small abattoirs has also been marketed. Approval of the use of steam pasteurisation as an antimicrobial step in the beef slaughter process was granted by the USDA in 1995.

Commercial evaluations of the SPS have been carried out in the UK under a MAFF LINK scheme, using a SPS SC100 cabinet (Eveleigh, 2000). Initial trials showed that both natural TVC and Enterobacteriaceae counts on carcasses produced in the abattoir involved in the study were too low to show any significant differences in process treatments before and after the SPS. Thus bacteria were surface inoculated prior to treatment. The TVC results showed that 85°C had little effect, even for 12 s, compared to 90°C, while there was little difference at higher temperatures. This was taken to indicate that there were a number of thermotolerant organisms present. While results showed that a 90°C treatment for 8 s would satisfy requirements for an effective treatment with the least effect on carcass colour, the participants chose to carry out further trials using 90°C for 10 s, even though there was a noticeable colour change.

13.6 Other new methods

A whole range of more novel techniques, such as microwaves (Paterson *et al.*, 1995) ultra-violet light (Stermer *et al.*, 1987) or visible light (Mertens and Knorr 1992), have been suggested for treating meats, and in some cases demonstrated to be viable alternatives. Most of these methods depend on heat to destroy the bacteria present though a number of non-thermal treatments have been proposed (Mertins and Knorr, 1992).

Many of the alternative physical decontamination treatments rely on the effect of radiant energy on surface bacteria (Table 13.2). These methods include ultraviolet radiation (UV), visible light, and lasers. Others rely on the effect of electromagnetic fields and include microwave, electrical stimulation (ES), high voltage pulsed electric field (PEF) and oscillating magnetic field pulses (OMF). In addition high pressures, air ions and ultrasound have been investigated.

UV has been used to extend the storage life of chilled meat. Many reports show that exposure to UV can reduce surface contamination of meat by 2 to 3 $\log_{10}CFUcm^{-2}$ and it would appear to have no deleterious effects on the appearance of the meat. Under high intensity UV, exposure times would be <10 s. UV appears suitable for on-line decontamination of meat either as carcasses, primals or retail cuts. To achieve an even exposure of all points on the surface of the meat appears to be the main technical problem. The use of robotics and automation to orientate the meat with the source or move the source(s) over the surface to prevent 'shadowing' needs to be examined. Very brief high intensity pulses of visible light produce a >1 log reduction in bacterial numbers. Few data, however, are currently available on the process. Attaining very high surfaces temperatures for a very short period using lasers might also have much to offer in the future.

The number of papers on the use of microwave energy for meat decontamination look promising. Papers report that a 40 second exposure can

Table 13.2 Mode of anti-microbial action of different novel decontamination treatments

Method	Mode of action on microbial cells
Microwave	Thermal effect
Ultraviolet light	UV effect
Pulsed light	UV (or thermal) effect
Ultrasound	Rupture of cell membrane
Ultra high pressure	Denaturing of protein
Pulsed electric fields	Rupture of cell membrane

reduce bacterial counts on chicken pieces by 2 log (Cunningham, 1978, 1980). However, recent work at the University of Bristol (Göksoy et al., 1999, 2000) refute these claims. This work using domestic microwave ovens showed microwave heating to be too uneven and unreproducable to surface heat meat cuts without cooking. Although there have been numerous publications claiming a possible non-thermal antimicrobial affect of microwave exposure most reviewers of the literature have concluded that any destruction of microorganisms in a microwave is purely thermal.

Electrical stimulation appears to produce a small reduction in bacterial levels in laboratory studies on model foods. Reductions of up to 6 $\log_{10}CFUcm^{-2}$ have been reported after the application of high voltage PEF (Zhang et al., 1994). The technique, however, is in its early stages and will require considerable development before it can be applied to small pieces of meat. Similarly the application of OMF appears to be an effective means of destroying bacteria, especially for treating liquid foods, but is not likely to have applications in the decontamination of meat in the near future (Mertens and Knorr, 1992).

Ultrasound is effective only in a liquid medium and therefore has limited application for red meat carcasses though it may prove useful for treating cuts of meat. Laboratory trials have shown that very high pressure processing is an effective method of extending the chilled storage life of highly contaminated minced meat (Carlez et al., 1994). It also significantly reduces the risk of survival of pathogenic microorganisms. The cost of high-pressure equipment that could process substantial quantities of meat, however, appears to limit its commercial uptake.

13.7 Future trends

Meat carcasses typically contain between 10^1 and 10^4 microorganisms per g (James et al., 1999). To achieve any significant improvement in the microbiological condition of such products we require a 4 log-unit reduction in total bacterial numbers. To date no adequate method of achieving this has been found without affecting the sensorial quality of meat. No treatment, as yet, can be relied upon to eliminate all pathogens. Typical reductions for non-chemical and chemical decontamination treatments are shown in Table 13.3 and Table 13.4, respectively.

With the increase in commercial interest and use of decontamination treatments (particularly in the USA) more and more studies are being carried out in operational abattoirs unlike much of the earlier work that was often on a bench scale. In the US commercial abattoirs are utilising a wide range of decontamination treatments, often sequentially.

While steam is being applied commercially in America, and undergoing trials in the UK, for beef carcasses, experimental work is still ongoing on its use for other meats. Much work at FRPERC has concentrated on poultry while others have even applied it to delicate meats such as fish. Despite the success and

Table 13.3 Typical microbial reductions achieved by non-chemical meat decontamination treatments

Treatment	Log reduction (APCs)
Water – cold	1–2
Water – hot	1–3
Steam	2–4 (6)
Ultraviolet	0–2
Visible light	1–3
Microwave	1–2
Ultrasound	0–1.5

Table 13.4 Typical microbial reductions achieved by chemical meat decontamination treatments

Treatment	Log reduction (APCs)
Organic acids	1–3.5
Chlorine	1–2
Chlorine dioxide	1–2
Trisodium phosphate	1–3
Ozone	0.5–3
Hydrogen peroxide	2–3

commercial realisation of steam pasteurisation systems there are still holes in the understanding of these systems. To realise the full potential of steam surface pasteurisation it is necessary to understand the relationship between heating/cooling cycles and appearance/quality changes for foods of interest. The conditions that will maximise bacterial reduction without significant quality changes need to be identified along with the engineering understanding to produce those required conditions consistently over the surface of food when presented at industrial throughputs. While there are effective commercial systems available there is much evidence that these systems still require further development to achieve full efficiency.

Many studies still concentrate on the efficacy of different chemicals, often as combined chemical solutions. In the EU legal restrictions on the use of chemicals for treating raw meat remain. Thus it is still unlikely that chemicals will find widespread application in the meat industry. However, chemical washing of fruits and vegetables is widespread and a growing research topic. Internationally, the least controversial methods of treating meat involve washing or some form of heat treatment.

The adoption of surface decontamination treatments by the meat industry in the EU remains restricted by legislation on what should or should not be permitted. Meanwhile the adoption of an *E. coli* 'zero tolerance' requirement in the USA, has effectively forced the American meat industry to use anti-microbial systems. Thus the introduction of efficient anti-microbial systems in other countries may be required just to maintain current export markets. In response to American policy, Australian and New Zealand plants are also investigating and developing anti-microbial systems.

It is debatable whether consumers will be willing to pay extra for safer food. They logically believe that the food is already safe. Processes that eliminate pathogens should also produce a substantial reduction in the number of spoilage organisms and hence an extension of storage life. This will help the economics of food production, allowing longer production runs, delivery to more distant markets, and reduced waste. However, the introduction of surface pasteurisation systems does not directly improve profitability by cost savings or increased throughput. Consequently, despite the obvious advantages to the industry as a whole and the consumer, it will be introduced only if it is cheap, reliable and has low running costs. Atmospheric steam surface pasteurisation has the potential to meet these requirements.

The majority of previous studies into surface decontamination techniques have been conducted by laboratory-based microbiologists interested mainly in the effects of such treatments on specific bacteria. This research cannot be successfully scaled up to industrial usefulness because of the engineering problems in recreating the effective conditions as used in the laboratory in an industrial environment. Involving food engineers *and* skilled microbiologists from the outset has significantly greater chance of successful scale up. Engineers building the laboratory equipment will be fully aware of the conditions used and by direct interaction with microbiologists, have complete knowledge of the bactericidal effects. Because of the close collaboration and awareness of the other disciplines' limits, the construction of effective large-scale equipment is possible.

Particular engineering challenges exist in the development of handling systems for non-laboratory decontamination treatment of meats. For treatments to be effective, all surfaces need to be exposed to the steam environment. This requires non-contact or minimal-contact handling systems. Most food products are delicate and require gentle handling to avoid bruising and damage. Handling systems must also integrate into the industrial line and the throughput rate must be equivalent to current production rates. The handling and transport systems will differ for each product type and will be influenced by the treatment times required.

The main aim should be to concentrate on pathogens and harmful microorganisms on real food surfaces. Typical spoilage organisms are less hardy than pathogens. Reductions in spoilage microorganisms should be seen as a beneficial 'side-effect' of pathogen destruction conditions. Since food poisoning results from ingestion of an infectious dose of pathogens, the absolute levels of microorganisms remaining on products after treatment should be used as the main measure of success. However, inoculation microbiology may be used to evaluate process parameter relationships for specific products within each scale of processing system.

Where work has been done with real food, it has often looked at reductions in inoculated bacterial levels as the measure of effectiveness. Whilst inoculation microbiology allows the relative effectiveness of different processes and process variations to be evaluated, it is not a true representation of 'real-world' microorganisms. Inoculated microorganisms are usually at much higher levels, differently attached, and in differing growth stages from microorganisms typically found on a food product during an industrial production process. Whilst inoculation microbiology is of use in developing equipment, more trials should be carried out using naturally occurring contamination. Whilst this will involve greater experimental effort, the results will be directly applicable to the industrial problem because natural contamination on food will be considered at the technology development stage.

In conclusion, any decontamination system for meat adopted in the EU will depend on a perceived need by government and food retailers, and will probably require changes to current legislation. The EU Scientific Committee on Veterinary measures relating to Public Health have published views regarding decontamination for poultry carcasses (EU, 1998). It is probable that any decontamination system for red meat will need to address the recommendations of this report. The Committee recommended that:

1. Antimicrobial treatment should be used only as part of an overall strategy for pathogen control throughout the whole production chain.
2. Before any decontamination compound or decontamination technique is authorised for use it should be fully assessed.
3. The person/company proposing such a decontamination compound or decontamination technique must demonstrate that all aspects are covered.
4. The person/organisation using a decontamination compound or decontamination technique must demonstrate that effective control of parameters critical for efficacy and safe use are in place and that good practice and appropriate HACCP plans are implemented.
5. Based on the conclusions of their report, a framework is established for the assessment of decontamination compounds or decontamination techniques proposed.

13.8 Sources of further information and advice

An extensive review of decontamination has been published by James and James (1997) at the University of Bristol and regular updates made in 1999 and 2000. Good general reviews of decontamination and contamination issues have been published in recent years by Bolder (1997), Corry and Mead (1996), Dorsa (1997), Sofos *et al.*, (1999), Bjerklie (2000). An overview of contamination and decontamination issues involving poultry meat has been discussed by Smulders (1999). Useful books include those edited by Gould (1995), Smulders (1987), and Ellerbroek (1999).

Reviews of specific subjects include Jeyamkondan *et al.*, (1999) on pulsed electric field processing, Kim *et al.*, (1999) on ozone applications and Mertens and Knorr (1992) and Palmieri *et al.*, (1999) on non-thermal preservation methods (such as pulsed electric fields, pulsed light and oscillating magnetic fields).

13.9 References

ANDERSON, M E and MARSHALL, R T (1989), 'Interaction of concentration and temperature of acetic acid solution on reduction of various species of microorganisms on beef surfaces', *Journal of Food Protection*, 52(5), 312–315.

ANDERSON, M E, MARSHALL, R T, STRINGER, W C and NAUMANN, H D (1979a), 'Evaluation of a beef carcass cleaning and sanitising unit', Presented at the 1979 Summer Meeting of the ASAE and CSAE, Paper No. 79-6014.

ANDERSON, M E, MARSHALL, R T, STRINGER, W C and NAUMANN, H D (1979b), 'Microbial growth on plate beef during extended storage after washing and sanitising', *Journal of Food Protection*, 42(5), 389–392.

ANDERSON, M E, NAUMANN, H D and COOK, N K (1984), 'Design specifications of a red meat carcass washing and sanitising unit', Presented at the 1984 Winter Meeting of the American Society of Agricultural Engineers, Paper No. 84-6546.

BAILEY, C (1971), 'Spray washing of lamb carcasses', *Proceedings of the 17th European. Meeting of Meat Research Workers*, Bristol, Paper B16, 175–181.

BELL, R G (1997), 'Distribution and sources of microbial contamination on beef carcasses', *Journal of Applied Microbiology*, 82, 292–300.

BELL, R G, HARRISON, J C L and ROGERS, A R (1993), 'Preliminary investigation of the distribution of microbiological contamination on lamb and beef carcasses', *MIRINZ Technical Report 927*.

BIEMULLER, G W, CARPENTER, J A and REYNOLDS, A E (1973), 'Reduction of bacteria on pork carcasses', *Journal of Food Science*, 38, 261–263.

BJERKLIE, S (2000), 'Intervention overview', *Meat Processing: North American Edition*, June, 94, 96–97.

BOLDER, N M (1997), 'Decontamination of meat and poultry carcasses', *Trends in Food Science & Technology*, 8(7), 221–227.

CACCIARELLI, M A, STRINGER, W C, ANDERSON, M E and NAUMANN, H D (1983), 'Effects of washing and sanitising on bacterial flora of vacuum-packaged pork loins', *Journal of Food Protection*, 46(3), 231–234.

CARLEZ, A, ROSEC, J, RICHARD, N and CHEFTEL, J (1994), 'Bacterial growth during chilled storage of pressure treated minced meat', *Lebensmittel-Wissenschaft und Technologie*, 27, 48–54.

CHARLEBOIS, R, TRUDEL, R and MESSIER, S (1991), 'Surface contamination of beef carcasses by faecal coliforms', *Journal of Food Protection*, 54(12), 950–956.

CORRY, J E L and MEAD, G C (1996), *Microbial Control in the Meat Industry: 3.*

Decontamination of Meat, Concerted Action CT94-1456, University of Bristol Press.
CUNNINGHAM, F E (1978), 'The effect of brief microwave treatment on numbers of bacteria in fresh chicken patties', *Journal of Food Protection*, 57, 296–297.
CUNNINGHAM, F E (1980), 'Influence of microwave radiation on psychrotrophic bacteria', *Journal of Food Protection*, 43(8), 651–655.
DAVEY, K R and SMITH, M G (1989), 'A laboratory evaluation of a novel hot water cabinet for the decontamination of sides of beef', *International Journal of Food Science and Technology*, 24, 305–316.
DE KOOS, J T (1992), 'Lactic acid and lactates. Preservation of food products with natural ingredients', *Food Marketing and Technology*, March, 1–5.
DICKSON, J S, CUTTER, C G N and SIRAGUSA, G R (1994), 'Antimicrobial effects of trisodium phosphate against bacteria attached to beef tissue', *Journal of Food Protection*, 57(11), 952–955.
DORSA, W J (1997), 'New and established carcass decontamination procedures commonly used in the beef-processing industry', *Journal of Food Protection*, 60(9), 1146–1151.
EKLUND, T (1983), 'The antimicrobial effect of dissociated and undissociated sorbic acid', *Journal of Applied Bacteriology*, 54, 383–389.
ELLERBROEK, L (ed.) (1999) *COST Action 97 – Pathogenic micro-organisms in poultry and eggs. 8. New Technology for Safe and Shelf-stable Products*, Luxembourg: Office for Official Publications of the European Communities.
ELLERBROEK, L I, WEGENER, J F and ARNDT, G (1993), 'Does spray washing of lamb carcasses alter bacterial surface contamination?', *Journal of Food Protection*, 56:5, 432–436.
EUROPEAN UNION (1998), *Benefits and Limitations of Antimicrobial Treatments for Poultry Carcasses*, Report of the Scientific Committee on Veterinary Measures relating to Public Health, 30 October 1998.
EVANS, J A (1999), 'Novel decontamination treatments for meat', in Ellerbroek, L, *COST Action 97 – Pathogenic micro-organisms in poultry and eggs. 8. New Technology for Safe and Shelf-stable Products*, Luxembourg: Office for Official Publications of the European Communities, 14–21.
EVELEIGH, K (2000), 'Carcass decontamination', *Meat and Poultry 2000 – Seminar 1*, Campden and Chorleywood Food Research Association, UK, 17 July 2000.
EWELL, A W (1943), 'Allies of refrigeration in food preservation', *Refrigerating Engineering*, 43, 159–162.
GIESE, J (1992), 'Experimental process reduces *Salmonella* on poultry' *Food Technology*, 46(4), 112.
GILL, C O (1979), 'A review: Intrinsic bacteria in meat', *Journal of Applied Bacteriology*, 47, 367–378.
GÖKSOY, E O, JAMES, C and JAMES, S J (1999), 'Non-uniformity of surface temperatures after microwave heating of poultry meat', *Journal of*

Microwave Power and Electromagnetic Energy, 34(3), 149–160.

GÖKSOY, E O, JAMES, C and CORRY, J E L (2000), 'The effect of short-time microwave exposures on inoculated pathogens on chicken and the shelf-life of uninoculated chicken meat', *Journal of Food Engineering*, 45, 153–160.

GORMAN, B M, SOFOS, J N, MORGAN, J B, SCHMIDT, G R and SMITH, G C (1995), 'Evaluation of hand-trimming, various sanitising agents, and hot water spray-washing as decontamination interventions for beef brisket adipose tissue', *Journal of Food Protection*, 58(8), 899–907.

GOULD, G W (1995), *New Methods of Food Preservation*, Blackie Academic and Professional, Chapman & Hall, London.

GRAHAM, D M (1997), 'Use of ozone for food processing', *Food Technology*, 51(6), 72–75.

HINTON, M H, HUDSON, W R and MEAD, G C (1998), 'The bacteriological quality of British beef 1. Carcasses sampled prior to chilling', *Meat Science*, 50(2), 265–271.

INGRAM, M, OTTAWA, F J H and COPPICE, J B M (1956), 'The preservative action of acid substances in food', *Chemistry and Industry*, 42, 1154–1163.

JAMES, C and JAMES, S J (1997) *Meat Decontamination – the State of the Art*. EU Concerted Action Programme: CT94 1881, University of Bristol.

JAMES, C, NICOLSON, M and JAMES, S J (1999), *Review of microbial contamination and control measures in abattoirs*, FRPERC, University of Bristol.

JAMES, C, GÖKSOY, E O, CORRY, J E L and JAMES, S J (2000a), 'Surface pasteurisation of poultry meat using steam at atmospheric pressure', *Journal of Food Engineering*, 45, 111–117.

JAMES, C, THORNTON, J A, KETTERINGHAM, L and JAMES, S J (2000b), 'Effect of steam condensation, hot water or chlorinated hot water immersion on bacterial numbers and quality of lamb carcasses', *Journal of Food Engineering*, 43, 219–225.

JEYAMKONDAN, S, JAYAS, D S and HOLLEY, R A (1999), 'Pulsed electric field processing of foods: A review', *Journal of Food Protection*, 62(9), 1088–1096.

KELLY, C A, DEMPSTER, J F and MCLOUGHLIN, A J (1981), 'The effect of temperature, pressure and chlorine concentration of spray washing water on numbers of bacteria on lamb carcasses', *Journal of Applied Bacteriology*, 51, 415–424.

KIM, J-G, YOUSEF, A E and DAVE, S (1999), 'Application of ozone for enhancing the microbiological safety and quality of foods: A review', *Journal of Food Protection*, 62(9), 1071–1087.

KLOSE, A A, KAUFMAN, V F, BAYNE, H G and POOL, M F (1971), 'Pasteurisation of poultry meat by steam under reduced pressure', *Poultry Science*, 50, 1156–1160.

LEE, R M, HARTMAN, P A, OLSON, D G and WILLIAMS, F D (1994), 'Bactericidal and bacteriolytic effects of selected food-grade phosphates, using *Staphylococcus aureus* as a model system', *Journal of Food Protection*,

57(6), 276–283.

MERTENS, B and KNORR, D (1992) 'Developments of non-thermal processes for food preservation', *Food Technology*, 46(5), 124–133.

MORGAN, A I, GOLDBERG, N, RADEWONUK, E R and SCULLEN, O J (1996a), 'Surface pasteurization of raw poultry meat by steam', *Lebensmittel -Wissenschaft und Technologie*, 29, 447–451.

MORGAN, A I, RADEWONUK, E R and SCULLEN, O J (1996b), 'Ultra high temperature, ultra short time surface pasteurisation of meat', *Journal of Food Science*, 61(6), 1216–1218.

NUTSCH, A L, PHEBUS, R K, SCHAFER, D, PRASAI, R K, UNRUH, J, WOLF, J and KASTNER, C L (1995), 'Use of steam for reduction of *Escherichia coli* O157:H7, *Salmonella typhimurium* and *Listeria monocytogenes* populations on raw meat surfaces', *Food Safety Consortium, Annual Meeting, Agenda, Presentations and Progress Reports*, 25–26 October, Kansas City, Missouri, 102–108.

NUTSCH, A L, PHEBUS, R K, SCHAFER, D, WOLF, J, PRASAI, R K, UNRUH, J and KASTNER, C L (1996), 'Steam pasteurization of beef carcasses', *Cattlemen's Day 1996, Report of Progress 756, Agricultural Experiment Station, Kansas State University*, 1–3.

NUTSCH, A L, PHEBUS, R K, RIEMANN, M J, SCHAFER, D E, BOYER, J E, WILSON, R C, LEISING, J D and KASTNER, C L (1997), 'Evaluation of a steam pasteurisation process in a commercial beef processing facility', *Journal of Food Protection*, 60(5), 485–492.

NUTSCH, A L, PHEBUS, R K, RIEMANN, M J, KOTROLA, J S, WILSON, R C, BOYER, J E and BROWN, T L (1998), 'Steam pasteurisation of commercially slaughtered beef carcasses: Evaluation of bacterial populations at five anatomical locations', *Journal of Food Protection*, 61(5), 571–577.

PALMIERI, L, CACACE, D and DALL'AGLIO, G (1999), 'Non-thermal methods of food preservation based on electromagnetic energy', *Rivista Italiana EPPOS*, 27, 5–11.

PATERSON, J L, CRANSTON, P M and LOH, W H (1995), 'Extending the storage life of chilling beef: microwave processing', *Journal of Microwave Power and Electromagnetic Energy*, 30(2), 97–101.

PHEBUS, R K, NUTSCH, A L and SCHAFER, D E (1996a), 'Laboratory and commercial evaluation of a steam pasteurisation process for reduction of bacterial populations on beef carcass surfaces', *Proceedings of the 49th Annual Reciprocal Meat Conference*, 49, 121–124.

PHEBUS, R K, NUTSCH, A L, SCHAFER, D E and KASTNER, C L (1996b), 'Effectiveness of a steam pasteurisation process for reducing bacterial populations on beef carcasses in a commercial slaughter facility', *Food Safety Consortium, Annual Meeting, Agenda, Presentations and Progress Reports*, October 21–22, Kansas City, Missouri, 198–202.

PHEBUS, R K, NUTSCH, A L, SCHAFER, D E and KASTNER, C L (1996c), 'Effectiveness of steam pasteurisation at reducing naturally occurring bacterial populations at five anatomical locations on commercially slaughtered beef carcasses',

Food Safety Consortium, Annual Meeting, Agenda, Presentations and Progress Reports, October 21–22, Kansas City, Missouri, 203–206.

PHEBUS, R K, NUTSCH, A L, SCHAFER, D E and KASTNER, C L (1997a), 'Steam pasteurization to reduce bacterial populations on commercially slaughtered beef carcasses', *Cattlemen's Day 1997, Report of Progress 783, Agricultural Experiment Station, Kansas State University*, 4–5.

PHEBUS, R K, NUTSCH, A L, SCHAFER, D E, WILSON, R C, RIEMANN, M J, LEISING, J D, KASTNER, C L, WOLF, J R and PRASAI, R K (1997b), 'Comparison of steam pasteurisation and other methods for reduction of pathogens on freshly slaughtered beef surfaces', *Journal of Food Protection*, 60(5), 476–484.

PHEBUS, R. K., TRUAX, A., SPORING, S., RUEGER, S. A., SCHAFER, M., BOHRA, L. K., HARRIS, L. and RETZLAFF, D. D. (1999) Antibacterial effectiveness of a second generation steam pasteurization system for beef carcass decontamination. *Cattlemen's Day 1999, Report of Progress 831, Agricultural Experiment Station, Kansas State University*. 4–6.

REAGAN, J O, ACUFF, G R, BUEGE, D R, BUYCK, M J, DICKSON, J S, KASTNER, C L, MARSDEN, J L, MORGAN, J B, NICKELSON II, R, SMITH, G C and SOFOS, J N (1996), 'Trimming and washing of beef carcasses as a method of improving the microbiological quality of meat', *Journal of Food Protection*, 59(7), 751–756.

REYNOLDS, A E and CARPENTER, J A (1974), 'Bactericidal properties of acetic and propionic acids on pork carcasses', *Journal of Animal Science*, 38(3), 515–519.

SMITH, G (1996), 'Steam is the theme in the war on pathogens', *Meat Processing: North American Edition*, 35(2), 32–34.

SMITH, M G (1992), 'Destruction of bacteria on fresh meat by hot water', *Epidemiology and Infection*, 109, 491–496.

SMITH, M G and DAVEY, K R (1990), 'Destruction of *Escherichia coli* on sides of beef by a hot water decontamination process', *Food Australia*, 42(4), 195–198.

SMULDERS, F J M (ed.) (1987), *'Elimination of Pathogenic Organisms from Meat and Poultry'*, Elsevier: Amsterdam-New York-Oxford.

SMULDERS, F J M (1995), 'Preservation by microbial decontamination; the surface treatment of meats by organic acids', in Gould, G W, *New Methods of Food Preservation*, Elsevier: Amsterdam-New York-Oxford, Chapter 12, 253–282.

SMULDERS, F J M (1999), 'Contamination and decontamination of foods of animal origin, with special reference to poultry meat', in Ellerbroek, L, *COST Action 97 – Pathogenic micro-organisms in poultry and eggs. 8. New Technology for Safe and Shelf-stable Products*, Luxembourg: Office for Official Publications of the European Communities, 4–13.

SOFOS, J N, BELK, K E and SMITH, G C (1999), 'Processes to reduce contamination with pathogenic microorganisms in meat', *45th International Conference of Meat Science and Technology (ICoMST '99), Yokohama, Japan*, 45(2), 596–605.

STERMER, R A, LASATER-SMITH, M and BRASINGTON, C F (1987), 'Ultraviolet radiation – an effective bactericide for fresh meat', *Journal of Food Protection*, 50(2), 108–111.

ZHANG, Q, CHANG, F -J, BARBOSA-CÁNOVAS, G V and SWANSON, B G (1994), 'Inactivation of microorganisms in a semisolid model food using high voltage pulsed electric fields', *Lebensmittel -Wissenschaft und Technologie*, 27, 538–543.

ZHANG, S and FARBER, J M (1996), 'The effects of various disinfectants against *Listeria monocytogenes* on fresh-cut vegetables', *Food Microbiology*, 13, 311–321.

14

Automated meat processing

K. B. Madsen and J. U. Nielsen, Danish Meat Research Institute, Roskilde

14.1 Introduction

The meat industry has trailed far behind other industries as far as process automation is concerned. The reason is that until recently the biological variation in the raw material made it very difficult to automate. However, the development of faster computer techniques, more sophisticated sensors and more advanced vision techniques has facilitated progress of automation in the meat industry during the last decade. In many industrialised countries there is an increasing shortage of labour. Compared to most other industries the working environment in the meat industry is characterised by noise, draft, humidity, cold and repetitive work under pressure. Dealing with the slaughtering of animals, separation and movement of the various components of the carcass, and disposal of waste is also both physically demanding and, for some, unpalatable. The meat industry is not, therefore, very attractive for prospective employees, especially young people. To overcome potential labour shortages and get better returns from its existing workforce, the meat industry has come under increasing pressure to automate the most labour-intensive parts of meat production. These pressures have been increased in Europe by relatively static overall consumption of meat over the last decade and the increasing pressure on margins in meat processing.

Automation, however, always requires a high level of investment. A reasonable payback time is realistic only in industries with high production volumes and high labour costs. There is the additional risk that automation creates more repetitive and uninteresting work, exacerbating potential labour shortages. Profitable automation of processes in the meat industry requires a certain degree of uniformity in the raw material. This is the main reason why automation has come much further in the pig, poultry and lamb industries than in

284 Meat processing

the beef sector. Automation has the potential to create end products of more uniform quality. If machinery is properly cleaned and sterilised, it also becomes possible to produce safer products. However, if these high hygienic standards are not maintained, automated production may severely compromise meat safety.

In beef production the first use of robotic equipment was in splitting a complete carcass into carcass sides. A number of manufacturers produce such equipment which replaces this particularly arduous manual process. However, the performance of such equipment is still not satisfactory in terms of the accuracy of splitting the carcass down the centre of the spinal column, and the hygiene problems associated with the deposition of bone dust and other detritus on the edible surfaces of the carcass. These problems have been highlighted recently by the BSE crisis in Europe. During the 1980s a very comprehensive and ambitious Australian development project called 'Fututech' was started with the aim of automating the complete beef slaughtering process. For a number of reasons this project failed and was abandoned (Purnell, 1998). Other attempts at comprehensive automation of the beef slaughtering process have been made in the US by the Texas Beef Group and in the UK by the University of Bristol. None of them has, however, yet resulted in commercial equipment for the industry (Purnell, 1998). Semi-automatic equipment for grading of beef carcasses has been developed by companies in Denmark, Germany, Australia and Canada. They are all based on vision image analysis (VIA). Some of them are used in daily production even though none of them has yet been approved by the EU.

Automation of sheep and lamb slaughtering has mainly been carried out in Australia and New Zealand. The Meat Industry Research Institute of New Zealand (MIRINZ) has in particular been very active in this area. Several manual operations have been mechanised, improving both process efficiency and hygiene, though lamb production is still not fully automated (Templer *et al.*, 1998). In contrast, automated slaughtering and preparation of poultry is carried out at very high line speeds, though it requires birds to be of uniform size and shape. Recently research using sensor-driven robotic techniques to improve the cutting process has been carried out in UK and US, although these processes are not yet commercialised (Purnell, 1998).

This chapter will, however, deal mainly with the automation of pig slaughtering processes. Denmark has for many years been a leader in the automation of primary and secondary processes in pig slaughtering. This has been partly due to high labour costs, but is also due to the co-operative structure of the Danish meat industry which has enabled joint financing of the high development and investment costs of process automation.

14.2 Current developments in robotics in the meat industry

Standard robots have been in use for many years in many industries. However, the meat sector has been reluctant to introduce standard industrial robots for a number of reasons including:

- the harshness of the environment in the meat industry
- the speed, reliability and cost of the robots
- the complexity of the processes involving handling biological materials.

The use of dedicated automatic equipment has therefore been the dominant feature for the meat sector. However, standard modern multi-axis robots, used in combination with advanced vision and high-speed digital video techniques, have provided the foundation for developing dedicated automatic machinery for the meat industry. These techniques have speeded up the development process and helped to find solutions which were simply not possible a few years ago.

During the last few years industrial robots with names such as 'Clean Room Robot', 'Envirobot' and 'Shiny Robot' have been introduced to the meat industry. Industrial Research Ltd., New Zealand, has marketing the Envirobot for handling organic products in harsh environments such as the food industry. It has been developed especially for slaughterhouses in Australia and New Zealand, where it is used for the critical so-called Y-cuts in connection with dehiding and cutting of lamb carcasses and for sawing through the breastbone on cattle carcasses. The Envirobot is made of stainless steel and resists harsh cleaning materials and the corrosive chilled environment in slaughterhouses (Purnell, 1998).

The German company KUKA Roboter GmbH has patented a cabinet for their robots, which is capable of resisting the harsh slaughterhouse environment. At present it is installed at the Gilde HedOp slaughterhouse in Norway and at the FACCSA slaughterhouse in Spain. The robot, which performs certain cuts and jet-ink printing on pig carcasses, is guided by a vision camera. The KUKA robot seems to provide a cheaper solution to the environmental problems presented by slaughterhouse conditions than the Envirobot (Khodabandehloo, 1999). The German company imt-Peter Nagler GmbH has patented a robot packaging system 'Ulixes Sortierer' for high-speed packaging of hot dogs, bacon and similar products. This system should have good market penetration as it replaces what was a time-consuming and arduous manual operation with significant potential hygiene risks (Wolf, 1999). Traditional industrial robots are increasingly used for packaging and palletisation in the food industry because products at that stage have a very well defined shape and are easy to handle with traditional robot programming solutions.

14.3 Automation in pig slaughtering

Table 14.1 provides an overview of the handling of slaughter pigs from transportation of the live animals through the slaughter process to the packaging of the final products. The pre-slaughter handling of the animals is important both for animal welfare and for ensuring good meat quality. The transportation of animals to the slaughterhouse has to be performed as gently as possible. Often the pigs are delivered from the farmer in the same groups in which they are reared. At the most advanced slaughterhouses the pigs are held in the lairage and

Table 14.1 Schematic overview of the pig slaughter process

Process section	Main subprocesses
Live animal handling	Transportation to slaughterhouse
	Unloading of pigs
	Veterinary ante-mortem inspection
	Rest period in lairage
	Transfer to stunning
	CO_2 stunning/electrical stunning
Sticking and bleeding	Shackling
	Sticking and bleeding
UNCLEAN — Surface treatment	Scalding
	Deshackling
	Mechanical dehairing
	Gambrelling
	Singeing
	Rind scraping/polishing
CLEAN — Evisceration and trimming	Carcass opening
	Removal of secondary organs
	Removal of stomach and intestines
	Removal of pluck
	Preparation for splitting
	Carcass splitting
Inspection and quality measurements	Veterinary inspection of carcasses and cuts
	Carcass weighing
	Carcass classification
Downstream processes	Carcass chilling
	Sorting into quality groups
	Storage and temperature Equilibration
Secondary processes	Carcass cutting
	Boning and trimming of cuts
	Packaging

driven to stunning in these groups, thus preventing fighting and stress. CO_2-stunning is becoming more and more widespread due to advantages with respect to animal welfare and meat quality. Equipment for automatic CO_2-stunning of groups of 5–6 pigs, developed by the Danish Meat Research Institute (DMRI), is being installed in many European and American slaughterhouses at the moment (Christensen *et al.*, 1997).

Shackling and sticking of the stunned animals will probably never be fully automated, as the animals are still alive at the time of these operations. Apart from these two operations and the gambrelling operation, all other operations on the unclean slaughterline (scalding, singeing, scraping and polishing) have been carried out automatically for several years. Dehiding, which is rarely done outside the Far East, is only partly automated.

Carcass splitting either by chopping or by sawing has been automated for many years. A few years ago the Dutch company STORK MPS launched the F-line modular series of dedicated robots, which are now operational in a number of countries. They have been installed in slaughterhouses with line speeds of 600 pigs per hour (or more) in a single line layout or 1200 pigs per hour in a double line (Van Ochten, 1999). In the F-line concept, the following processes have been automated:

- cutting of the pelvic symphysis
- opening of the abdomen and the thorax
- removal of the fat end
- neck cutting
- removal of the leaf lard.

The Danish company SFK-Danfotech has, in co-operation with the Danish Meat Research Institute (DMRI), developed a series of dedicated robots for automation of slaughterline processes. The robot series is manufactured as separate modules, each robot having its separate hydraulic station and PLC control. All robots have capacities for between 360 and 400 carcasses per hour and may be installed separately or as a line dependent on the requirements of the slaughterhouse. The series of robots consists of:

- a measuring unit to determine the length of the pig carcass
- throat cutter
- fat end dropper and ham divider
- carcass opener
- evisceration equipment
- equipment for back finning
- a carcass splitting machine.

All the robots are operational in a number of slaughterhouses either individually or as more or less complete lines (Anon., 2001). Automatic weighing of dressed carcasses in connection with carcass classification is performed either semi-automatically or fully automatically (ultrasound, vision or optical probes) on most modern slaughterlines. The only two fully automatic classification systems are the Danish Carcass Classification Centre developed by the DMRI and the AUTO-FOM system developed by the Danish company SFK-Technology (Madsen *et al.*, 1992; Brandscheid *et al.*, 1997).

14.4 Case study: the evisceration process

The evisceration process, which involves the cutting and removal of the pig's internal organs, is one of the most demanding operations on the slaughterline. Even when slaughterhouses try to ease individual operator strain through job rotation and a practical design of the work place, intestine and pluck removal are still the most difficult processes to man on the slaughterline. Until recently these operations have been carried out manually as no automatic equipment has existed.

Manual evisceration at Danish slaughterlines with capacities of 360–400 pigs per hour is normally carried out by three operatives. The first operative cuts free the intestinal tract from the abdominal cavity and then separates the pluck set and the intestinal tract inside the carcass by cutting the oesophagus close to the stomach. Finally he lifts the intestines weighing about 10 kg into a conveyor tray. The second operative loosens the diaphragm and leaf fat. The third operative cuts the pluck set loose from the thoracic cavity and neck and lifts the pluck set weighing about 5 kg onto a conveyor hook. The evisceration process, particularly the cutting of the oesophagus close to the stomach inside the carcass, risks causing contamination of the carcass with pathogenic bacteria.

Both from a working environment, a cost and a hygiene point of view, the evisceration process is an obvious candidate for automation. This has been done successfully by the DMRI in co-operation with the Danish company SFK-Danfotech. The automatic evisceration equipment (Fig. 14.1) is capable of handling 360 carcasses per hour including the necessary cleaning and disinfection. The pluck set and the intestinal tract are removed together by the robot, allowing separation to be done manually outside the carcass, thus improving hygiene compared with existing manual methods. The equipment also eliminates the heavy work of lifting the intestinal tract and the pluck set.

Basically, the automatic evisceration system consists of a measuring station and a handling station. The measuring station measures the position of the elbow of the pig carcass, which is anatomically close to where cutting operations occur, allowing the handling station to identify where to cut into the carcass. The measuring station is positioned on the slaughter line prior to carcass opening. The handling station contains seven tools (as shown in Fig. 14.1):

1. A thorax opener.
2. Units for fixing the carcass.
3. An integrated tenderloin cutter, resistance bracket and lung loosener.
4. Conveyor systems for transporting the pig carcass into and out of the handling station.
5. A special unit cutting the intestinal attachment to the spine and cutting through the diaphragm at the spine.
6. An intestine shovel lifting up the intestines.
7. Brackets for detachment of the diaphragm and leaf fat. A motorised knife integrated in each bracket cuts through the diaphragm.

A complete cycle for the handling of a carcass consists of the following steps:

1. The carcass arrives at the handling station.
2. The carcass is pushed into the machine by the conveyor system (4), data from the measuring station are transferred to the handling station and the tools take up their starting position.

Automated meat processing 289

Fig. 14.1 Automatic evisceration system.

3. The carcass is held in the fixing unit (2).
4. The intestine shovel (6) lifts up the intestines hanging out from the opened carcass thereby exposing the insertion points for the leaf fat brackets.
5. The leaf fat brackets (7) are pushed into the carcass and opened.
6. The thorax is opened by the thorax opener (1).
7. The intestine shovel (6) is withdrawn. Intestines hanging out will fall down through the opening between the leaf fat brackets.
8. The knives built into the leaf fat brackets cut free the diaphragm.
9. The back cutter (5) is moved along the spine and penetrates the diaphragm. The back cutter is then moved upwards along the spine and cuts through the diaphragm and the connective tissue between spine and the intestinal tract. (Operations 9 and 10 are carried out simultaneously with operations 4 to 8).

290   Meat processing

10. The tool containing the tenderloin knife (3) moves downwards. Having passed the hind legs the knife is turned in. The knife cuts the tenderloins along the spine. The knife is moved out. The tool is now placed on top of the diaphragm with a predetermined force thereby acting as resistance during the following leaf fat loosening.
11. The leaf fat brackets (7) are moved upwards inside the pig. The brackets pass between the leaf fat and abdominal wall thereby detaching the leaf fat completely.
12. The tool (3) continues downwards into the thoracic cavity of the carcass thereby loosening possible adhesions between the lungs and the thoracic wall. At the same time the thorax is opened further by the thorax opener (1) to facilitate the detachment process.
13. The tool (3) is taken out horizontally from the thoracic cavity of the carcass, pulling the organs out.
14. The thorax opener (1) and the fixing units (2) are moved back to their starting positions.
15. The carcass is pulled out of the handling unit.
16. The tools are washed first in cold and then in hot water before the next carcass arrives.

As mentioned previously, the machine is capable of handling 360 carcasses per hour. This implies a total operation cycle of 10 seconds for each carcass. At present the cutting and division of the pluck set into the individual organs is also in the process of being automated by the DMRI. This is a difficult task as the pluck set is not very well defined. Automatic equipment for this purpose is expected to be available by 2004.

14.5 Automation of secondary processes

The main secondary processes are cutting and boning of carcasses. These processes include:

- carcass cutting
- cutting of the middle
- boning of the fore-end
- hind-leg boning
- boning and trimming of bellies
- boning and trimming of the loin.

14.5.1 Carcass cutting

After chilling and temperature equalisation for about 20 hours the carcass has a temperature of approximately 2°C. For further processing the carcass is most often cut into the primary cuts: hind leg, middle and fore-end. Fully automatic equipment performing this operation was developed by the DMRI and is marketed by the Danish company ATTEC. The equipment consists of a conveyor and three stations:

1. a laying-down station
2. a measuring station
3. a sawing station with a hind leg saw and a fore-end saw.

At the laying-down station the two half carcasses are positioned horizontally on a conveyor. Simultaneously the gambrel is taken off and the hind feet are sawn off. The conveyor brings the carcass to the measuring station where a hook takes hold of the pubic bone of each half carcass and brings it to a fixed position relative to the hind leg saw. During this movement the position of the elbow of the two half carcasses is recorded. The conveyor afterwards takes the carcass to the sawing station. During transport to the sawing station the fore-end saw is positioned according to the recorded position of the elbow of the carcass. At the sawing station the half carcasses are divided into fore-ends, middles and hind legs.

14.5.2 Cutting of the middle

In many current slaughter lines the rib tops are cut off manually with a circular saw and the loin and the belly are subsequently divided manually by a band saw. These processes, however, can soon be automated fully. The Canadian company LeBlanc is marketing equipment for splitting the loin and the belly automatically. Quebec Industrial Research Centre (CRIQ) recently presented a prototype for cutting off the rib tops automatically (Robocut^{TM}). During 2002 ATTEC will market equipment for cutting off the rib tops and dividing the loin and the belly automatically. This equipment (Fig. 14.2) has been developed by the DMRI. The equipment consists of:

- a conveyor
- a measuring station
- a sawing station.

The conveyor is equipped with a grasping device, which pulls the rib tops out to their full length. This straightens out the backbone, simplifying the subsequent sawing off of the rib tops. After fixing in the conveyor, the middle passes through measuring stations, which measures the length and any inaccuracy in the earlier carcass splitting. Based on these measurements, the saws are adjusted to cut off the belly from the loin and subsequently cut off the loin from the rib tops.

14.5.3 Boning of the fore-end

Boning of fore-ends has most often been done manually. However, the Dutch company STORK MPS is marketing equipment for press boning of fore-ends. The equipment has a tool consisting of two halves with indentations

Fig. 14.2 Automated secondary processes: cutting of the middle.

corresponding to the bones in the fore-end. The fore-end is placed between the two halves. These are pressed together and the meat is squeezed off the bones. The disadvantage of using this method is that the texture of the meat is changed, and the meat can contain splinters of bone. As a result, the use of this equipment has not become widespread in the industry.

Against this background the DMRI has developed equipment which can cut off the riblet and neck-bone automatically from fore-ends (Fig. 14.3). The Townsend company will market the equipment during 2002. The equipment operates as follows:

- a conveyor, equipped with a grasping device similar to that described in the previous section, catches the neck-bone of the fore-end and straightens it out to simplify the later removal of the neck-bone
- the fore-end then passes a number of measuring stations and knives
- the first knife loosens the meat behind the neck bone
- the next knife loosens the deeper portion of the neck-bone
- the riblet is cut free by a sword-like knife mounted in a 3-axis robot
- the fore-end now passes a saw, which cuts off the riblet and saws through the first neck joint
- finally, the meat is pulled off the neck bone before the last knife cuts the neck bone free.

The remaining processes of removing the shank bones, humerus and shoulder blade will be carried out by equipment which is currently being developed at the DMRI and is expected to be marketed by the middle of 2004.

Fig. 14.3 Automated secondary processes: boning of the fore-end.

14.5.4 Hind leg boning
Boning of hind legs has to be done very precisely as it is the most expensive cut. For this reason it is still most often done manually. However, in 1998 the Japanese company MYCOM at the IFFA exhibition in Frankfurt demonstrated equipment called Hamdas, which bones hind legs automatically. However, this equipment has a low yield and it will probably take some years before this process is completely or partly automated.

14.5.5 Boning and trimming of bellies
Boning of bellies is also most often carried out manually. It may also take years before this process is completely or partly automated. However, CRIQ has developed equipment which can cut the ribs off the belly by using a vision-guided robot. Trimming of bellies is also still normally carried out manually. At the beginning of 1990 the Danish company Lumetech marketed equipment which trimmed bellies by using water jets and vision guided robots, but the system never achieved widespread use in industry because of its high price.

14.5.6 Boning and trimming of the loin
Boning of loins is another predominantly manual process, particularly as loin is an expensive cut requiring a high degree of accuracy and skill. It will probably take a number of years before this process is completely or partly automated. Fat

trimming of loins has, however, been automated for a number of years. A number of companies are marketing so-called loin pullers where the loin is pulled through a fixed shaped knife. The Danish company CARNITECH has improved this concept by controlling the shape of the knife with a computer. On the basis of information from measuring the thickness of the fat, a computer adjusts the position and shape of the knife cutting off the fat, making the fat trimming of the loin more precise.

14.6 Future trends

Within the next few years slaughtering of pigs as well as secondary processing will be partly automated. Fig. 14.4 shows the layout of an automated slaughter line of the future. There are, however, a number of issues which still need to be resolved before automation can progress to this point.

14.6.1 Automation and hygiene

In automated, high-speed production, sterilisation of tools and equipment will be of great importance in preventing cross-contamination. This is an area which requires major attention in the future. In many cases the present disinfection of machinery between the processing of individual products is not sufficient. More attention has to be paid at the design stage to the ease of cleaning the tools that get into contact with the products. Often a dedicated disinfection system has to be used for each individual type of equipment. Application of several steps for washing and disinfection of automatic equipment has proved to be very effective. In addition, varying water pressures and temperatures helps to optimise cleaning. A possibility might also be application of lactic acid for disinfection of especially exposed tools. This disinfection method is, however, not yet permitted everywhere.

14.6.2 Automation and management

With the increased automation of previously manual operations at slaughterhouses, it is also necessary to focus on the new challenges faced by line and supervisory staff. Increased automation will influence and change the current role of operatives, their level of responsibility and the qualifications they need. In the future:

- Supervisors will have fewer and more demanding operatives who want more influence on their working situation. The supervisor will therefore have to spend more time on management and establishment of good co-operation and training of the production team.
- The quality of products will to a large extent depend on the automatic production equipment. Much more attention will therefore have to be paid to the surveillance and control of the automatic machinery in order to secure

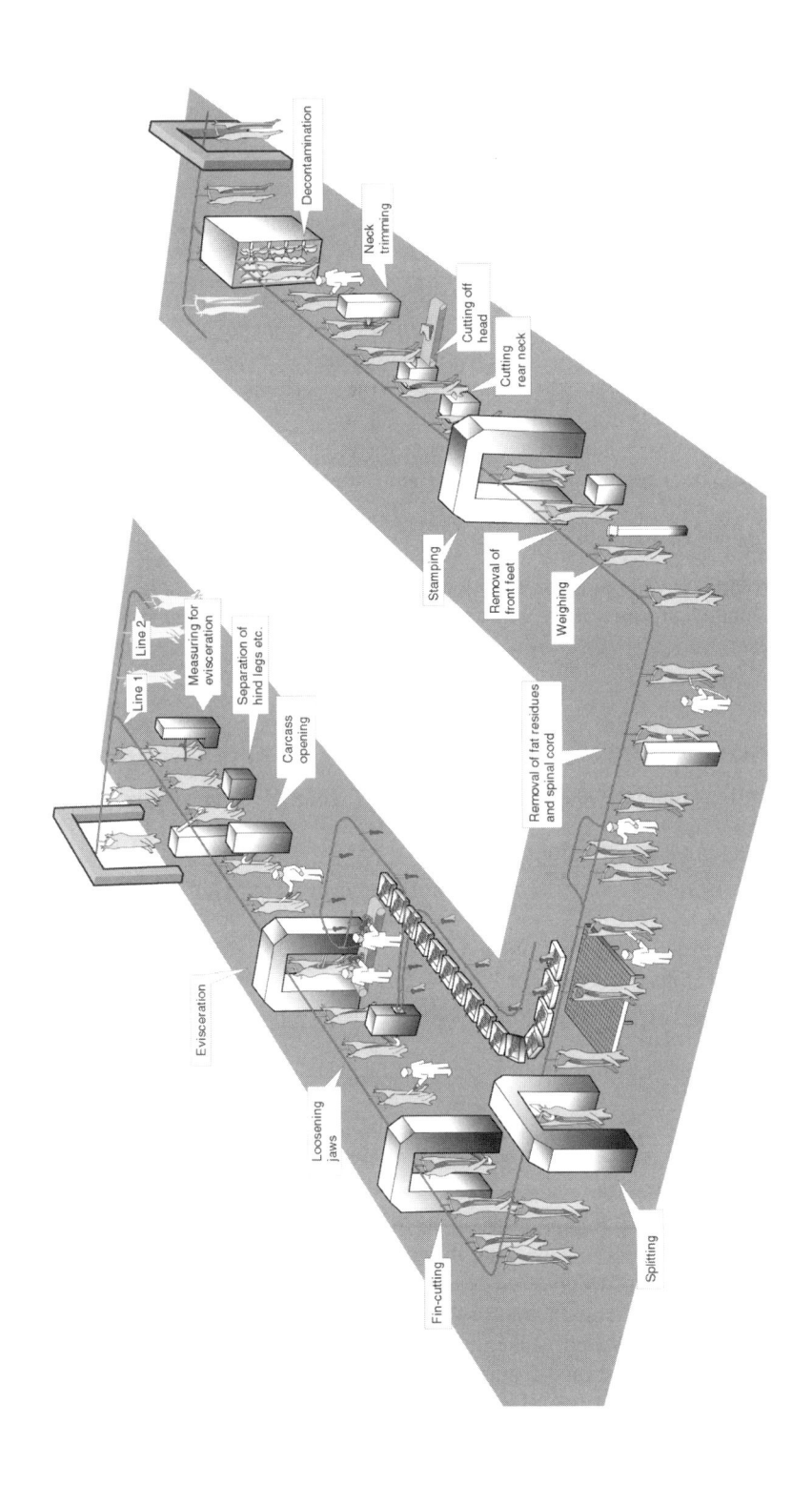

Fig. 14.4 Layout of an automated slaughter line of the future.

high yield and short downtimes. The operatives will not only have to monitor the automatic equipment and make minor adjustments, but must also check the quality of the products. This requires wider knowledge both for operatives and for production management.

Automation will also change the present organisation and communication structure:

- It may require production teams of butchers, maintenance staff and machine operatives working separately for the production manager. Payment of the employees will probably be on a team basis. This might ease integration and co-operation between the traditional trade groups, ensuring higher productivity and better quality of end products.
- The supervisor will have to act as coach for the production team.
- There will be a requirement for fuller and more rapid communication with management as well as a continuous dialogue with the production and maintenance departments.

On the whole, the change from mainly manual craft-based production to almost fully automated production will give the production management many new challenges.

14.7 References and further reading

ANON. (2001), Automatic slaughterline for pigs. *Fleischwirtschaft International*, March.
BRANDSCHEID, W. et al. (1997), Bestimmung der Handelsklassen und des Handelswertes von Schweinchälfter mit dem Gerät Autofour. *Fleischwirtschaft* 77(7).
CHRISTENSEN, L. et al. (1997), New Danish developments in pig handling at abbatoirs. *Fleischwirtschaft* 77: 604–607.
KHODABANDEHLOO, K. (1999), Advancing robots in the meat industry. *Meat Automation* No. 2.
MADSEN, K. B. (1994), Automation in the Pig Slaughterline. ECCEAMST course, September.
MADSEN, K. B. et al. (1992), Fremgangsmåde oganlag til behandling eller undusøgelse af slagtokjoppe, Patent 6295/86.
NIELSEN, J. U. (1999), Automatic evisceration of pigs. *Meat Automation* No. 2.
PURNELL, G. (1998) Robotic Equipment in the Meat Industry. *Meat Science*, Vol. 49.
TEMPLER, R. et al. (1998) New Automation Techniques for Sheep and Beef Processing, *Meat Automation* No. 2.
VAN OCHTEN, S. (1999), Automation for the slaughter line. *Meat Automation* No. 2.
WOLF, A. (1999), High-speed packaging using robots. *Meat Automation* No. 2.
ZINK, J. (1995), Application of automation and robotics to pig slaughtering. Proceedings of 41st ICOMST.

15

New developments in the chilling and freezing of meat

S.J. James, Food Refrigeration and Process Engineering Research Centre (FRPERC), University of Bristol

15.1 Introduction

The American Food and Drug Administration (FDA) reckons – 'conservatively', according to one official – that there may be anywhere between 24 and 81 million cases and 10,000 'needless deaths' from food poisoning in the US every year. In England and Wales reported cases of food poisoning have increased nearly fivefold over the last decade. In addition to the human cost the economic loss in terms of working days and medical treatment in the US is between $5 billion and $6 billion. A study of one outbreak of *Salmonella enteritidis* in the UK calculated that it cost the country between £224 and £321 million. Even a superficial examination of the problem reveals that temperature control is the prime factor that determines the safe distribution life of many foods including meat. Temperature control also determines the ultimate eating quality and economic yield of the product.

Temperature is one of the major factors affecting microbiological growth. Microbiological growth is described in terms of the lag phase and the generation time. When a microorganism is introduced to a particular environment there is a time (the lag phase) in which no increase in numbers is apparent followed by a period when growth occurs. The generation time is a measure of rate of growth in the latter stage. Microorganisms, have an optimum growth temperature at which a particular strain grows most rapidly, i.e. the lag phase and generation time are both at a minimum. They also have a maximum growth temperature above which growth no longer occurs. Above this temperature, one or more of the enzymes essential for growth are inactivated and the cell is considered to be heat-injured. However, in general, unless the temperature is raised to a point substantially above the maximum growth temperature then the injury is not lethal and growth will recommence as the temperature is reduced. Attaining temperatures substantially above the maximum growth temperature is therefore critical during cooking and re-heating operations.

Of most concern during storage, distribution and retail display of meat and meat products is a third temperature, the minimum growth temperature for a microorganism. As the temperature of an organism is reduced below that for optimum growth then the lag phase and generation time both increase. The minimum growth temperature can be considered to be the highest temperature at which either of the growth criteria, i.e., lag phase and generation time, becomes infinitely long. The minimum growth temperature is not only a function of the particular organism but also the type of food or growth media that is used for the incubation. Although some pathogens can grow at 0°C, or even slightly lower, from a practical point of view the risks to food safety are considerably reduced if meat is maintained below 5°C.

Meat may also become microbiologically unacceptable as a result of the growth of spoilage microorganisms. Their growth can produce unacceptable changes in the sensory quality of many foods and their rate of growth is also very temperature dependent. The development of off odours is usually the first sign of putrefaction in meat and occurs when bacterial levels reach approximately 10^7 per cm^2 of surface area. When bacterial levels have increased a further tenfold, slime begins to appear on the surface and meat received in this condition is usually condemned out of hand. At 0°C beef with average initial contamination levels can be kept for at least 15 days before any off odours can be detected. Every 5°C rise in the storage temperature above 0°C will approximately halve the storage time that can be achieved. So from a spoilage point of view, and the underlying economic consequences of extended storage/distribution life, temperature is again a very important factor in food production.

Meat exhibits other particular quality advantages as a result of rapid cooling. In meat the pH starts to fall immediately after slaughter and protein denaturation begins. The result of this denaturation is a pink proteinaceous fluid, commonly called 'drip', often seen in pre-packaged joints. The rate of denaturation is directly related to temperature and it therefore follows that the faster the chilling rate the less the drip. Investigations using pork and beef muscles have shown that rapid rates of chilling can halve the amount of drip loss.

A final, but important, quality and economic advantage of temperature control is a reduction in weight loss, which results in a higher yield of saleable material. Meat has a high water content and the rate of evaporation depends on the vapour pressure at the surface. Vapour pressure increases with temperature and thus any reduction in the surface temperature will reduce the rate of evaporation. The use of very rapid chilling systems for pork carcasses has been shown to reduce weight by at least 1% when compared with conventional systems.

In a world where consumers require 'safe food' of high quality without additives and preservatives then temperature is the critical control factor. Most of the previous discussion has centred around chilled foods, but to provide longer safe shelf-lives than can be produced by chilling, temperature control in the form of freezing is again the answer.

15.2 The impact of chilling and freezing on texture

Whilst a number of characteristics affect the overall quality and acceptability of both fresh and frozen meats, tenderness is the major characteristic of eating quality because it determines the ease with which meat can be chewed and swallowed. The tenderness of meat is affected by both chilling/freezing and storage. Under the proper conditions, tenderness is well maintained throughout the chilled/frozen storage life, but improper chilling/freezing, can produce severe toughening and meat of poor eating quality. Refrigeration has two critical roles in meat tenderness. One is in the prevention of muscle shortening in the period immediately following slaughter. The second is in the conditioning of the meat so that the desired degree of tenderness is obtained.

Chilling has serious effects on the texture of meat if it is carried out rapidly when the meat is still in the pre-*rigor* condition, that is, before the meat pH has fallen below about 6.2 (Bendall, 1972). In this state the muscles contain sufficient amounts of the contractile fuel, adenosine triphosphate (ATP), for forcible shortening to set in as the temperature falls below 11°C, the most severe effect occurring at about 3°C. 'Cold-shortening' first became apparent in New Zealand, when tough lamb began to be produced routinely by the improved refrigeration techniques that were introduced after the Second World War (Locker, 1985). As 'rules of thumb', cooling to temperatures not below 10°C in ten hours for beef and lamb (Offer *et al.*, 1988) and in five hours for pork (Honikel, 1986) can avoid cold-shortening.

Thaw shortening, which occurs when a rapidly frozen muscle is thawed, resembles cold-shortening in that it sets in while the level of contractile fuel (ATP) is still high. However, it differs because the amount of work done and force developed are much higher. With 'thaw-shortening' the temperature is raised through the danger zone from 0 to 10°C, whereas in cold-shortening it is reduced through this zone. The rate of contraction depends entirely on the rate of thawing. Rapid thawing of a freely suspended, unloaded muscle strip causes very dramatic shortening often to less than 40% of the 'frozen' length. The terms 'conditioning', 'ageing', 'ripening', 'maturing' and 'the resolution of rigor' have all been applied to the practice of storing meat for periods beyond the normal time taken for cooling and setting, to improve its tenderness after cooking. Conditioning imposes a severe limitation on processing conditions because it is a slow process.

The bulk of investigations to determine the time required for tenderising changes to take place in beef have been carried out in North America. Deatherage and Harsham (1947) investigated the changes in beef at 0.5 to 2°C and found that the tenderness of cooked sirloin (*Longissimus dorsi*) increased up to 17 days storage with some additional improvement up to 31 days. They concluded that unless beef is to be aged beyond four weeks, it need only be aged 2.5 weeks. Doty and Pierce (1961) also showed that conditioning for two weeks at 0.5 to 2°C improved texture and caused very substantial reductions in the shear strength of cooked meat, but much less change occurred during the next two weeks. Increasing the delay period before freezing, enhances tenderness because meat ages at chill temperatures but not at normal freezer temperatures. Meat which has been conditioned prior to freezing is more tender than that frozen within one or two days and the difference is maintained throughout frozen storage for nine months.

Freezing rate affects the rate of tenderising after thawing but not the ultimate tenderness. Freezing at $-10°C$ more than doubles the rate; freezing in liquid nitrogen almost trebles the rate. Freezing is known to cause structural damage by ice crystal formation. It seems likely that ice crystals, particularly small intracellular ice crystals formed by very fast freezing rates, enhance the rate of conditioning probably by release of enzymes (Dransfield, 1986). Repeated freeze-thaw cycles using relatively low freezing rates does not seem to cause any enhanced tenderising (Locker and Daines, 1973).

15.3 The impact of chilling and freezing on colour

The appearance of meat at its point of sale is the most important quality attribute governing its purchase. The ratio of fat to lean and the amount of marbled fat are important appearance factors and another is the colour of the meat. The changes in colour of the muscle and blood pigments (myoglobin and haemoglobin, respectively) determine the attractiveness of fresh red meat, which in turn influences the consumers' acceptance of meat products (Pearson, 1994). Consumers prefer bright-red fresh meats, brown or grey-coloured cooked meats and pink cured meats (Cornforth, 1994). Red colour is more stable at lower temperatures because the rate of oxidation of the pigment decreases. At low temperatures, the solubility of oxygen is greater and oxygen consuming reactions are slowed down. There is a greater penetration of oxygen into the meat and the meat is redder than at high temperatures.

Changes in colour have been reported resulting from chilling treatment. Taylor et al., (1995) found that electrical stimulation of pork produced higher lightness (L), i.e., paler, values than those measured in non-stimulated sides. Spray chilling of pork has some effect on its colour during the initial chilling period (Feldhusen et al., 1995a). After four hours of chilling the musculature of sprayed ham becomes lighter and red and yellow values decrease. However, after 20 hours there was no significant difference in the colour values. The surface of the skin becomes lighter after spray chilling.

Newly cut conditioned meat is known to show a brighter surface after a short exposure to air than unconditioned meat (Doty and Pierce, 1961; Tuma et al., 1962; 1963). MacDougall (1972) studied the effects of conditioning on colour and on subsequent storage in packages of high oxygen permeability typical of those used for display and in vacuum packages of low oxygen permeability. Meat, when cut and exposed to air, changed from dull purple red to a bright cherry red, which is measured as an increase in 'lightness', a 'hue' change towards red and an increase in 'saturation'. The magnitude of the change on blooming for conditioning meat as compared with unaged was the same size for 'lightness' but was twofold greater for 'hue' and threefold greater for 'saturation'. Conditioned meat, when freshly cut, was lighter but more purple than the unconditioned. After one hour exposure to air, conditioned meat had a redder 'hue' which was considerably more saturated and intense than the unconditioned samples. These changes in lightness, hue and saturation produced by conditioning result in a brighter, more attractive appearance. The overall colour improvement was of a similar magnitude to that which occurred on blooming.

The colour of frozen meat varies with the rate of freezing. Taylor (1930, 1931) reported that, as the speed of freezing diminished, the appearance of the product changed and at very low rates there is a marked development of translucence. Later experiments have demonstrated a direct relationship between freezing rate and muscle lightness; the faster the rate the lighter the product. Guenther and Henrickson (1962) found that 2.5 cm thick steaks frozen at $-9°C$ were dark. Those frozen at -34 to $-40°C$ had the most desirable colour and that those frozen at -73 to $-87°C$ tended to be pale. Jakobsson and Bengtsson (1969, 1973) obtained similar results; very rapid freezing in liquid nitrogen spray at a freezing rate of about 13 cm hr^{-1} produced meat which was unnaturally pale. Air blast freezing at 2 cm hr^{-1} gave the best frozen appearance while very slow freezing at 0.04 cm hr^{-1} resulted in a darker colour and the formation of ice on the product surface. Zaritzky et al. (1983) reported that the surface of liver frozen at high rates was lighter in colour. These differences in frozen meat lightness result from the dependence of ice crystal growth on the freezing rate. Small crystals formed by fast freezing scatter more light than large crystals formed by slow freezing and hence fast frozen meat is opaque and pale and slow frozen meat is translucent and dark. 'Freezer burn' is the main appearance problem that traditionally affected the appearance of meat in frozen storage. Desiccation from the surface tissues produces a dry, spongy layer that is unattractive and does not recover after thawing. This is commonly called 'freezer-burn'. It occurs in unwrapped or poorly wrapped meat. The problem is accentuated in areas exposed to low humidity air at high velocities, and by poor temperature control.

In thawed meat the rate of pigment oxidation is increased (Cutting, 1970) and therefore the colour will be less stable than in fresh. On prolonged frozen storage, a dark brown layer of metmyoglobin may form 1–2 mm beneath the surface so that on thawing the surface colour will rapidly deteriorate. Meat, which has lost its attractiveness during frozen storage because of oxidation of oxymyoglobin on the surface, will remain brown after thawing. Unwrapped meat thawed in high humidity air, water or in steam under vacuum appears very white and milky after thawing. However, if then stored in a chill room for 10 to 24 hours it will be almost indistinguishable from fresh meat. Unwrapped meat thawed in air at high temperatures and low humidities will take on a dark, dry, tired appearance. It will not recover its appearance during chilled storage and will often require extensive trimming before sale.

15.4 The impact of chilling and freezing on drip loss and evaporative weight loss

15.4.1 Drip loss

The quality of fresh meat exposed for retail sale is initially judged on its appearance. The presence of exudate or 'drip', which accumulates in the container of pre-packaged meat or in trays or dishes of unwrapped meat, substantially reduces its sales appeal (Malton and James, 1983). Drip can be referred to by a number of different names including 'purge loss', 'press loss' and 'thaw loss' depending on the method of measurement and when it is measured.

In general, beef tends to lose proportionately more drip than pork or lamb. Since most of the exudate comes from the cut ends of muscle fibres, small pieces of meat drip more than large intact carcasses. The protein concentration of drip is about 140 mg/ml, about 70% of that of meat itself. The proteins in drip are the intracellular, soluble proteins of the muscle cells. The red colour is due to the protein myoglobin, the main pigment of meat.

The problem of drip loss is not, however, confined to retail packs. The meat industry uses large boneless primal cuts, which are packed in plastic bags, for distribution throughout the trade. These may be stored under refrigeration for many weeks before use and during this time a considerable volume of drip may accumulate in the bag. Not only does this exudate look unattractive but it also represents an appreciable weight loss to the user when the meat is subsequently removed from its container.

Excessive drip could have a small effect on the eating quality of meat. Perceived juiciness is one of the important sensory attributes of meat. Dryness is associated with a decrease in the other palatability attributes, especially with lack of flavour and increased toughness (Pearson, 1994). However, moisture losses during cooking are typically an order of magnitude higher than most drip losses during refrigeration. Consequently, small differences in drip loss will have little effect on eating quality. The potential for drip loss is inherent in fresh meat and is influenced by many factors. These may include breed, diet and physiological history, all of which affect the condition of the animal before it is slaughtered. After slaughter, factors such as the rate of chilling, storage temperatures, freezing and thawing can all influence the drip produced.

Rapid cooling of meat immediately after slaughter will reduce drip loss after subsequent cutting operations. The potential for drip loss is established in the first period of cooling, the temperature range conducive to drip is down to about

30°C or perhaps a little lower. There are a number of publications showing that rapid cooling can reduce drip production. Taylor (1972) compared two cooling treatments for pig carcasses. In 38 out of 40 paired legs, the drip loss was less after the quicker cooling. The difference varied between breed and ranged from approximately 1.6–2 fold.

Gigiel et al. (1985) removed cylindrical samples of muscle from freshly slaughtered beef. The curved surface and one end of the cylinder were surrounded by insulation and the free end placed in contact with solid CO_2. Since heat was extracted from only one end this produced a wide range of cooling rates through the length of the cylinder. After cooling and equalisation, the cylinder was cut into discs and the drip potential of each disc measured using a centrifuge technique described by Taylor (1982). Close to the surface in contact with the CO_2 the rate of cooling was highest but freezing occurred and the drip was high. Minimum drip potential was measured in the next region where high cooling rates were achieved without freezing. Drip then increased as cooling time to 7°C increased. In meat drip loss increases with length and temperature of chilled storage. Work by Lee et al. (1985) clearly showed the effect of both. Drip loss from pork cubes increased substantially during 21 days of storage at 0, 3 and 7°C. The rate of increase was greater at the higher temperatures. In storage at 0 and 3°C no increase in drip loss with time was measured after 21 days. At 7°C drip was still increasing between 21 and 28 days.

A number of scientific investigations, which can be compared to commercial practice, have defined the effect of freezing rate on drip production. Petrovic et al. (1993) stated that the optimal conditions for freezing portioned meat are those that achieve freezing rates between 2 and 5 cmh^{-1} to $-7°C$. Grujic et al. (1993) suggest even tighter limits 3.33 to 3.95 cm h^{-1}. These results are scientifically very interesting, however, in industrial practice most meat is air frozen in the form of large individual pieces or cartons of smaller portions. In commercial situations, freezing rates of 0.5 cm h^{-1} in the deeper sections would be considered 'fast' and there would be considerable variation in freezing time within the meat. The samples frozen by Sacks et al. (1993) were much smaller (77.6 g in weight) than most commercial products. Even with such small samples there was no significant difference in drip after 48 hours between cryogenic freezing at $-90°C$ and a walk-in freezer operating at $-21°C$.

15.4.2 Evaporative weight loss

From the moment an animal is slaughtered the meat produced begins to lose weight by evaporation. Under typical commercial distribution conditions, it has been estimated that lamb and beef lose from 5.5 to 7% by evaporation between slaughter and retail sale (Malton, 1984). Weight losses from pork are probably of the same magnitude. In addition to the direct loss in saleable meat there are also secondary losses. Excessive evaporation during initial chilling and chilled storage produces a dark unattractive surface on the meat. Either this has to be removed by trimming, or the meat is downgraded and sold at a reduced price.

Freezing does not stop weight loss. After meat is frozen, sublimation of ice from the surface occurs. If the degree of sublimation is excessive, the surface of the meat becomes dry and spongy, a phenomenon called 'freezer burn'. In the United States weight loss resulting from a combination of direct evaporative loss and freezer burn in pork bellies stored for one month before curing, was estimated to be 500 000 kg (Ashby and James, 1974). Since then developments in the use of moisture impervious packaging materials have significantly reduced sublimation in frozen meat.

15.5 The cold chain

The cold chain links temperature changing operations, such as chilling, freezing, thawing, handling and cooking, and temperature maintenance operations, such as chilled and frozen storage, transportation and display. In general, as meat moves along the cold chain it becomes increasingly difficult to control and maintain its temperature. Temperatures of bulk packs of meat and meat products in large storerooms are far less sensitive to small heat inputs than single consumer packs in transport or open display cases.

15.5.1 Primary chilling/freezing

After slaughter meat carcasses are at a temperature which is close to the optimum growth temperature for many microorganisms and chilling is required before the meat can be processed or distributed. The first stage in the cold chain will therefore be a cooling operation to reduce the temperature of the meat to a value that limits microbiological and quality changes. If the meat is to be distributed in a chilled form then this value will be above the initial freezing point and in the range -1 to $+15°C$. In many countries the legal limit is $+7°C$. If a frozen food is required then the value will typically be between -12 and $-30°C$.

15.5.2 Secondary processing

Any meat that is processed after its initial primary chilling or freezing operation is likely to gain heat and consequently rise in temperature. This rise can range from a few degrees in a packing operation to 100s in cooking. To maintain product quality it is often important to remove this added heat. Industrial cooking processes cannot be guaranteed to eliminate all pathogenic organisms and if cooling rates are slow microbial spores that survive the cooking process will germinate and grow. Systems that produce a rapid reduction in the temperature of the meat will retard microbial growth and consequently extend shelf life. This is especially important when chilling cooked products that will eventually be consumed cold or in a warm reheated state.

15.5.3 Storage and transport

After the temperature of the meat has been reduced to a desired value it is likely to remain at that temperature for a period which may range from a few hours, for chilled products, to a number of years in the case of frozen foods. During that period it may remain in a single store or be transported around the world. Theoretically, there should be few problems in the storage and transport of either chilled or frozen foods. If the meat is at the correct temperature when it is loaded into the storage room or transport vehicle then all the refrigeration is required to do is to insulate the meat from sources of heat. Most of the time the main source will be a slow movement of heat from the outside ambient through the walls, which can be easily catered for. Coping with door openings and people moving around inside the refrigerated chamber is more difficult. However, if the standard of management is good, and the meat bulk stacked, then temperature rise in the meat will be small and restricted to exposed surfaces.

In practice there are many problems in meat storage and transport because the meat and meat products are not at the correct temperature when they are loaded. If the temperature of the meat is too high when it is wrapped and bulk packaged then it becomes very difficult to reduce that temperature within the storage or transport compartment. Since these systems are not designed to extract heat then the average room temperature can rise together with that of any other foods already stored. Failure to remove the required heat before loading can be due to a number of causes including, (i) insufficient time allowed, (ii) insufficient refrigeration capacity to cater for high initial product load, (iii) overloading, (iv) variability in size of products or (v) incorrect environmental conditions.

15.5.4 Retail display

Chilled meat products spend periods ranging from a few minutes to a week in retail display and in extreme cases with frozen products a few months. During that time there is a conflict between the need to protect the food from extraneous heat sources and those to display it to its best advantage. Retail display cabinets can have integral or remote refrigeration units, air movement can be gravity or forced air and the displays can be single-tier, multi-tier or well.

15.5.5 Domestic transport and storage

Even if food producers and retailers maintain acceptable product temperatures during the distribution chain they lose control when the product leaves the retail store. After the meat is removed from a display cabinet it spends a period outside a refrigerated environment whilst it is carried around the store and then transported home. Temperatures of foods, especially thin sliced products, can rise considerably during these journeys.

Consumers have considerable faith in the temperature maintenance properties of domestic refrigerators. However, measurements taken at five positions at five-minute intervals over an average of six days in refrigerators in 250 homes have revealed considerable variation in performance. The highest recorded mean temperature was 11.37°C and the lowest −0.89°C, producing a range in mean temperatures of 12.3°C. The overall mean air temperature for all the refrigerators in the survey was 6.04°C. Consequently, over half operated at mean temperatures that would support the growth of salmonella.

15.6 Temperature monitoring

It is often stated by those in the meat and refrigeration industries that 'anyone can measure a temperature'. Many millions of measurements are made of both meat and environmental temperatures in the meat industry. However, in many cases the measurements made are an unreliable guide to the effectiveness of the refrigeration process. Even when the correct temperatures have been obtained the data are often poorly analysed and rarely acted upon. The industry needs to measure temperatures accurately, reliably, meaningfully, simply and cheaply. It needs to be able to analyse the data and respond when required. It needs the correct instrumentation and the expertise to collect and interpret the temperature data.

The first consideration is the range of temperatures to be measured. For the meat industry, a range from −40 to +150°C would cope with the temperatures found in freezers, chillers, storage rooms, retail display cabinets, and in water used for cleaning or scalding tanks in the abattoir. If they produce cooked meat products then the upper temperature may rise to 250 to 300°C. As well as the measuring range the range of ambient temperatures over which the instrument will work needs to be considered. The electronics of many temperature measurement instruments are designed to work to the specified accuracy only within certain ambient temperature ranges, usually 0 to 40°C. If temperatures in a cold store are to be measured the instrument itself may need to be kept warm until it is used.

Any temperature measuring system should be tested over the operating range at regular intervals to ensure accuracy and should also have a current calibration certificate from its manufacturer or official standards laboratory. The system can be checked by means of a calibration instrument, or against a reference thermometer that is known to be accurate. Melting ice (which if made from distilled water should read 0°C or −0.06°C if made from tap water with 0.1% salt) may be used to check sensor accuracy. The ice should be broken up into small pieces and placed in a wide-necked vacuum flask with a depth of more than 50 mm. The system should be agitated frequently and the temperature read after a few minutes when stable. If differences of more than 0.5°C are found the instrument should either be very carefully adjusted, or sent for calibration.

Accurately determining the temperature of chilled meat throughout the cold chain is difficult. Training and experience are required to locate positions of maximum and minimum temperature in abattoirs, stores, vehicles and display cabinets. The problem is further exaggerated by changes in position with time caused by loading patterns and the cycling of the refrigeration plants. Obtaining a relationship between environmental temperatures (that can be measured relatively easily) and internal meat temperatures is not a simple process. Relating temperatures obtained in a non-destructive manner with internal meat temperatures again poses problems. Determining the temperature of cuts of meat with regular shapes is quite simple but doing so for irregular cuts of meat is more difficult.

All the temperature measurement problems associated with chilled foods will equally apply to quick-frozen foods. In addition, there are a number of other problems. Many instruments have sensors that will accurately measure temperatures of $-20°C$ and below, but the instruments themselves become inaccurate or fail to operate at low temperatures. If frozen foods are removed from their low temperature environment to one suitable for the instrument the surface temperature rises very rapidly. However, the main problem is that of actually inserting a temperature sensor into frozen meat.

The surface temperature of a food or pack can be measured by placing a temperature sensor (such as those discussed above) in contact with the surface. In practice there are very large temperature gradients on both sides of the surface and the presence of the sensor can influence the temperature being measured. Extending the surface of the sensor to measure the average temperature over a larger surface area is one method used to minimise these problems. This method is recommended for such applications as between-pack measurement.

Since it is impossible to measure the temperature of an exposed surface accurately, the next best thing is to take a measurement of the temperature between two food items. As long as good thermal contact is achieved between the temperature sensor and the packs, a between-pack method should provide an accurate measurement of the pack temperature. If the thermal conductivity of the packaging material is high and the food makes a good thermal contact with the pack then the temperature measured will be close to that of the product. With a product such as skin-wrapped chilled sausages, the above requirements are satisfied. A temperature sensor, especially a flat-headed probe, can be sandwiched between two packs. An accurate measurement is obtained due to the combination of a flexible food and a thin wrapping. With chilled food in cartons or bubble packs the accuracy is much lower.

The contact problems are much greater with a frozen product. Since the surface of a frozen product is not flexible only point contact can be achieved between the surface of the product and that of the pack or probe. Using a flat probe with extended contact surfaces does not necessarily improve the accuracy of temperature measurement. In extreme cases, for example with frozen sausages, the contact surfaces may extend out into the air stream and measure air not product temperature. With packs of small items such as diced meat the accuracy will be much better. Care must also be taken to pre-cool the probe before temperatures are measured. This is especially important with low heat capacity packaging materials.

Non-contact temperature measurement devices measure the amount of energy in an area of the infra-red spectrum that is radiated from the surface being measured. Basic instruments measure the average temperature of the area in a small field of view. More complicated systems of thermal imaging provide a temperature picture of all the objects over a much wider area. A certain amount of knowledge is needed in order to interpret the values that such instruments give (Evans *et al.*, 1994; James and Evans, 1994). The first point to bear in mind when using infra-red thermometry is that the temperature measured is the surface temperature. If the meat has been in surroundings that have not changed in temperature for a long period of time, then it is likely that the surface temperature will be very close to that of the meat beneath the surface. However, if the temperature of the surroundings is changing or has changed over the past 24 hours, then it is likely that the surface temperature will not be the same as the temperature deep within the meat. It is also necessary for the operator to know how much of the surface is 'seen' by the infra-red instrument, as it will measure the 'average' temperature over the whole of this area. The target area can vary significantly from instrument to instrument and with the distance between the instrument and the surface.

There is a further complication in the use of infra-red thermometers; reflected radiation. The instruments will see the radiation emitted from a surface and also an amount of radiation from the surroundings that is reflected by that surface. The reflected radiation will therefore constitute an error. For warm objects at a temperature greater than their surroundings, the amount of reflected radiation will be small in relation to that from the surface and consequently the error will be small. With frozen meat, the temperature of the meat is no warmer than, and often colder than, the temperature of the surroundings. Therefore, the amount of reflected radiation coming from the surface constitutes a significant error.

Determining the temperature of small cuts of meat with regular shapes is quite simple. Determining the temperature of irregular cuts of meat, and particularly large pieces, is more difficult. Possibly the most difficult problem is ascertaining deep leg temperature in beef carcasses. The Meat Research Corporation in Australia (1995) recommend that the temperature sensor should touch the *Trochanter Major* (aitch bone), which is the 'knob' of bone on the opposite side of the femur to the hip joint. To locate the sensor in this position it should be inserted through the 'Pope's Eye' at an angle about 15 to 20° below the horizontal. It should be aimed at an imaginary vertical line approximately one-third of the distance from the Achilles Tendon to the last tailbone. Because conduction occurs along the steel shaft of a probe it is important that the probe is inserted as far as possible into the meat. For example, to take the temperature of a cut of meat it is better practice to insert the probe to its full depth along the long axis of the cut rather than to insert the probe to half its length through the short axis (CSIRO, 1991).

15.7 Optimising the design and operation of meat refrigeration

In specifying refrigeration equipment the function of the equipment must be absolutely clear. Refrigeration equipment is always used to control temperature.

Either the meat passing through the process is to be maintained at its initial temperature, e.g., as in a refrigerated store or a packing operation, or the temperature of the meat is to be reduced, e.g., in a blast freezer. These two functions require very different equipment. If a room is to serve several functions then each function must be clearly identified. The optimum conditions needed for that function must be evaluated and a clear compromise between the conflicting uses made. The result will inevitably be a room that does not perform any function completely effectively. There are three stages in obtaining a refrigeration plant. The first is determining the process specification, the second is drawing up the engineering specification, i.e. turning processing conditions into terms which a refrigeration engineer can understand, independent of the food process and finally the procurement, the third and final stage being procurement of the plant.

Poor design in existing chillers/freezers is due to a mismatch between what the room was originally designed to do and how it is actually used. The first task in designing such plant is therefore the preparation of a clear specification by the user of how the room will be used. In preparing this specification the user would do well to consult with all parties concerned; these may be officials enforcing legislation, customers, other departments within the company and engineering consultants or contractors – but the ultimate decisions taken in forming this specification are the user's alone.

The aim of drawing up an engineering specification is to turn the processing conditions into a specification that any refrigeration engineer can then construct and deliver without knowledge of the meat process involved. If the first part of the process specification has been completed then the engineering specification will be largely in place. It consists of: the environmental conditions within the refrigerated enclosure, air temperature, air velocity and humidity; the way the air will move within the refrigerated enclosure; the size of the equipment; the refrigeration load profile; the ambient design conditions and the defrost requirements. The final phase of the engineering specification should be drawing up a schedule for testing the engineering specification prior to handing over the equipment. This test will be in engineering and not product terms.

The engineering specification should be sent out to tender. If tenders have been selected for the quality of their equipment and all accept the tender conditions and say they can meet the design and test conditions specified, the lowest tender would normally be chosen. The contractor is normally responsible for the detailed engineering design, construction and commissioning and the only need is to check that this work is carried out in a professional way. His first responsibility is in carrying out the acceptance tests. These test the performance of the refrigeration equipment in terms of the engineering specification and the plant should not be accepted until satisfactory tests have been carried out. The plant can then be handed over and training given to the plant operators in the correct use of the refrigeration equipment. The plant then needs to be commissioned by the factory personnel, systematically increasing throughput until the process tests can be carried out. These ensure that the original process specification in fact achieves the intended results in terms of temperatures, throughputs and yield.

15.8 Sources of further information and advice

Food Refrigeration and Process Engineering Research Centre, University of Bristol, Churchill Building, Langford, North Somerset, BS40 5DU, UK. http://www.frperc.bris.ac.uk

International Institute of Refrigeration, 177 Boulevard Malesherbes, 75017 Paris, France, www.iifiir.org

Processing and Preservation Technology, AgResearch Food Systems and Technology, Private Bag 3123, Hamilton, New Zealand.

15.9 References

ASHBY, B. J. and JAMES, G. N. (1974) Effects of freezing and packaging methods on shrinkage and freezer burn on pork bellies in frozen storage. *Journal of Food Science.* Vol. 39, pp. 1136–1139.

BENDALL, J. R. (1972) The influence of rate of chilling on the development of rigor and 'cold shortening'. In *Meat Chilling: Why and How? Meat Research Institute Symposium No. 2* (ed. C. L. Cutting). 3.1–3.6.

CORNFORTH, D. (1994) Colour – its basis and importance. Chapter 2, pp. 34–78. In *Quality Attributes and Their Measurement in Meat, Poultry and Fish Products* (eds. A. M. Pearson and T. R. Dutson). Advances in Meat Research Series, Volume 9, Blackie Academic & Professional, UK.

CSIRO (1991) Thermometers. *Meat Research News Letter.* 91/2. CSIRO Division of Food Processing, Meat Research Laboratory.

CUTTING, C. L. (1970) The influence of freezing practice on the quality of meat and fish. *Proceedings of the Institute of Refrigeration.* Vol. 66, p. 51.

DEATHERAGE, F. E. and HARSHEM, A. (1947) Relation of tenderness of beef to ageing time at 33–35°F. *Food Research.* Vol. 12, p. 164.

DOTY, D. M. and PIERCE, J. C. (1961) Beef muscle characteristics as related to carcass grade, carcass weight and degree of ageing. *Technical Bulletin 1231*, (Agricultural Marketing Service, United States Department of Agriculture).

DRANSFIELD, E. (1986) Conditioning of meat. Recent advances and developments in the refrigeration of meat chilling, Meeting of IIR Commission C2, Bristol (UK). Section 1, pp. 61–68.

EVANS J. A., RUSSELL S. L. and JAMES S. J. (1994) An evaluation of infrared non-contact thermometers for food use. *Developments in food science* 36, Elsevier Science, pp. 43–50.

FELDHUSEN, F., KIRSCHNER, T., KOCH, R., GRESE, W. and WENZEL, S. (1995a) Influence on meat colour of spray-chilling the surface of pig carcasses. *Meat science.* Vol. 40, pp. 245–251.

FELDHUSEN, F., WARNATZ, A., ERDMANN, R. and WENZEL, S. (1995b) Influence of storage time on parameters of colour stability of beef. *Meat Science.* Vol. 40, pp. 235–243.

GIGIEL, A. J., SWAIN, M. V. L. and JAMES, S. J. (1985) Effects of chilling hot boned meat with solid carbon dioxide. *Journal of Food Technology.* Vol. 20, pp. 615–622.

GRUJIC, R., PETROVIC, L., PIKULA, B. and AMIDZIC, L. (1993) Definition of the optimum freezing rate. 1. Investigations of structure and ultrastructure of beef M. Longissimus dorsi frozen at different rates. *Meat Science.* Vol. 33, 3, pp. 301–318.

GUENTHER, J. J. and HENRICKSON, R. L. (1962) Temperatures, methods used in freezing determine tenderness, colour of meat. *Quick Frozen Foods.* Vol. 25, p. 115.

HONIKEL, K. O. (1986) Influence of chilling on biochemical changes and quality of pork. Recent advances and developments in the refrigeration of meat chilling, Meeting of IIR Commission C2, Bristol (UK). Section 1, pp. 45–53.

JAKOBSSON, B. and BENGTSSON, N. E. (1969) The influence of high freezing rates on the quality of frozen ground beef and small cuts of beef. *Proceedings of the 15th European Meeting of Meat Research Workers.* p. 482.

JAKOBSSON, B. and BENGTSSON, N. E. (1973) Freezing of raw beef: influence of ageing, freezing rate and cooking method on quality and yield. *Journal of Food Science.* Vol. 38, p. 560.

JAMES S. J. and EVANS J. A. (1994) The accuracy of non contact temperature measurement of chilled and frozen food. IChemE Food Engineering Symposium, University of Bath 19–21/9/94.

LEE, B. H., SINARD, R. E., LALEYE, L. C. and HOLLEY, R. A. (1985) Effects of temperature and storage duration on the microflora, physiochemical and sensory changes of vacuum-packaged or nitrogen-packed pork. *Meat Science.* Vol. 13, pp. 99–112.

LOCKER, R. N. (1985) In *Advances in Meat Research*, (eds Pearson, A. M. and Dutson, T. R.) Vol. 1, pp. 1–44. AVI publishing Co., Westport, Conn.

LOCKER, R. N. and DAINES, G. J. (1973) The effect of repeated freeze-thaw cycles on tenderness and cooking loss in beef. *Journal of the Science of Food and Agriculture.* Vol. 24, pp. 1273–1275.

MACDOUGALL, D. B. (1972) The effect of time and storage temperature on consumer quality. In *Meat Chilling: Why and How? Meat Research Institute Symposium No. 2* (ed. C. L. Cutting). 8.1–8.11.

MALTON, R. (1984) *National Cold Storage Federation Handbook*, 17–25.

MALTON, R. and JAMES, S. J. (1983) Drip loss from wrapped meat on retail display. *Meat Industry*, May, pp. 39–41.

MEAT RESEARCH CORPORATION (1995) Measurement of temperatures in fresh and processed meats. *Meat Technology Update*, 95/1. Australian Meat Technology Meat Research Newsletter.

OFFER, G., MEAD, G. and DRANSFIELD, E. (1988) Setting the scene: the effects of chilling on microbial growth and the eating quality of meat. *Meat Chilling, IFR-BL: Subject Day.*

PEARSON, A. M. (1994) Introduction to quality attributes and their measurement in meat, poultry and fish products. Chapter 1, pp. 1–33. In *Quality Attributes and Their Measurement in Meat, Poultry and Fish Products* (eds. A. M. Pearson and T. R. Dutson). Advances in Meat Research Series, Volume 9, Blackie Academic & Professional, UK.

PETROVIC, L, GRUJIC, R. and PETROVIC, M. (1993) Definition of the optimal freezing rate – 2 Investigations of the physico-chemical properties of beef M. *longissimus dorsi* frozen at different freezing rates. *Meat Science*. Vol. 33, pp. 319–331.

SACKS, B., CASEY, N. H., BOSHOF, E. and VANZYL, H. (1993) Influence of freezing method on thaw drip and protein loss of low-voltage electrically stimulated and non-stimulated sheeps muscle. *Meat Science*. Vol. 34, 2, pp. 235–243.

TAYLOR, A. A. (1972) Influence of carcass chilling rate on drip in meat. In *Meat Chilling: Why and How?* (ed. C. L. Cutting) Meat Research Institute Symposium No. 2, 5.1–5.8.

TAYLOR, A. A. (1982) The measurement of drip loss from meat. *Meat Research Institute Internal Report.*

TAYLOR, A. A., PERRY, A. M. and WARKUP, C. C. (1995) Improving pork quality by electrical stimulation or pelvic suspension of carcasses. *Meat Science*. Vol. 39, pp. 327–337.

TAYLOR, H. F. (1930) Solving problems of rapid freezing. *Food Industries*. Vol. 2, p. 146.

TAYLOR, H. F. (1931) What happens during quick freezing. *Food Industries*. Vol. 3, p. 205.

TUMA, H. J., HENRICKSON, R. L., STEPHENS, D. F. and MOORE, R. (1962) Influence of marbling and animal age on factors associated with beef quality. *Journal of Animal Science*. Vol. 21, p. 848.

TUMA, H. J., HENDRICKSTON, R. L., CEDILLA, G. V. and STEPHENS, D. I. (1963) Variation in the physical and chemical characteristics of the eye-muscle from animals differing in age. *Journal of Animal Science*. Vol. 22, p. 354.

ZARITZKY, N. E., AÑON, M. C. and CALVELO, A. (1983) Rate of freezing effect on the colour of frozen beef liver. *Meat Science*. Vol. 7, pp. 299–312.

16

High pressure processing of meat

M. de Lamballerie-Anton, ENITIAA, Richard G. Taylor and Joseph Culioli, INRA

16.1 Introduction: high pressure treatment and meat quality

High pressure technology, defined as a pressure treatment between 100 and 1000 MPa, is derived from the ceramic industry. It is of increasing interest to food processing because of its potential to decrease the level of microbial contamination without any heat treatment. Industrial high pressure food products are mainly manufactured in:

- Japan (fruit jams and juices, sake, ham, fish and rice products)
- the USA (oysters and fruit juices)
- Mexico (fruit juices)
- Spain (ham and other meat products)

Research on the effect of high pressure on food began one hundred years ago (Hite, 1899) but only expanded significantly in the 1990s.

High pressure processing of meat has been an active topic of research because of its potential to extend shelf-life (Ledward, 2002). However, pressure treatment brings about changes in the constituent molecules of meat, and could affect functional properties of meat such as colour and gelation properties. This chapter deals with the effects of high pressure on the constituents and quality attributes of meat products, and with current and potential industrial applications of this technique. Meat and meat products are solid foods so high pressure treatment requires the use of a discontinuous system. The most common technique involves vacuum wrapping and putting meat in a pressurizing vessel. In the presence of the pressure medium (water), the pressure increases at a rate of 100 or 200 MPa/min (Tonello, 2001). When the appropriate pressure level is reached, valves are closed and the pressure remains constant (Fig. 16.1). The

314 Meat processing

Fig. 16.1 High pressure processing vessel (courtesy of ACB Pressure Systems, France).

food industry generally uses pressure between 200 and 800 MPa, with the duration of pressure kept as short as possible (about 15 min. max.).

16.2 General effect of high pressure on food components

Reactions of food components under pressure are governed by Le Chatelier's principle, according to which a process associated with a decrease in volume is favored by an increase in pressure and *vice versa*. Volume changes are due to molecular conformation modifications, intramolecular interactions, solvent variation, and chemical reactions. Weak energy bonds such as hydrogen, hydrophobic or ionic bonds are affected by high pressure treatment, whereas covalent bonds are not modified. The main effects of high pressure on meat components are on water, proteins (including enzymes) and lipids.

16.2.1 Water

At 600 MPa and 22°C, water is compressed by 15%. Meat and meat products usually contain significant amounts of water and tend to compress by a similar amount. Packaging must take this compression into account. Adiabatic compression of water also induces an increase of temperature about 2 or 3°C for each 100 MPa (Hayashi, 1991) which can be dissipated by heat transfer between the food and the pressurizing water (Cheftel and Dumay, 1997). High pressure also induces the reversible dissociation of water, decreasing pH by 0.73 when pressure is increased from 0.1 to 100 MPa (Cheftel, 1991). This change is significant because of its potential effect on proteins (see section 16.2.2). In addition, pressure decreases the melting point of ice below 220 MPa, allowing thawing under pressure at subzero temperature.

16.2.2 Proteins and enzymes

The effect of high pressure on proteins is highly dependent on the structure of the macromolecule and the composition of the medium around it (pH, ionic strength and temperature). Generally the primary structure is not modified by pressure, whereas secondary structure is only affected by very high pressure treatment (Mozhaev et al., 1994), and tertiary and quaternary structures are modified from 100 MPa (Galazka et al., 1996). These changes in structure can alter the activity of endogenous meat enzymes. However, studies of the effect of high pressure on enzyme activity are complicated because enzymes react differently when they have been extracted from food (Hendrickx et al., 1998). High pressure treatment induces activation or inactivation of enzymes depending on conditions of buffer and pressure (see section 16.4).

16.2.3 Carbohydrates and lipids

High pressure acts only on polysaccharides. Glycogen is the only polysaccharide present in muscle, but post-mortem meat generally does not contain any residual glycogen because of the post-mortem glycolysis (Lawrie, 1998). Meat products therefore, are not affected by pressure modification of carbohydrates.

High pressure treatment can induce some modification of lipids, but it is difficult to make general conclusions about lipid changes in meat with regards to oxidation and free fatty acids formation resulting from pressure. High pressure does increase the melting point of lipids. At room temperature triglycerides can crystallise under pressure, which may alter the structure and, as a consequence, the permeability of cell membranes. This mechanism could be involved in the inactivation of bacteria.

16.3 Structural changes due to high pressure treatment of muscle

The effects of high pressure treatment on muscle structure have been examined for beef (Jung et al., 2000a; Kennick et al., 1980; Locker and Wild 1984; Macfarlane et al., 1980; Ueno et al., 1999; Suzuki et al., 1992), sheep (Kennick et al., 1980; Macfarlane and Morton 1978), poultry (Yuste et al., 1998) and fish (Ashie et al., 1997). In addition some of these effects have been discussed in a previous review of the effects of pressure treatment on meat (Cheftel and Culioli, 1997). Before discussing these effects it is necessary to give a brief review of post-mortem meat changes at atmospheric pressure.

Normal muscle structure has been thoroughly reviewed (Bendall, 1973; Bloom and Fawcett, 1975; Goll et al., 1984) and, for the purpose of this chapter, can be simply described as a composite structure composed of myofibers organized in bundles by connective tissue. Most of the structural changes related to high pressure treatment of meat have examined the ultrastructure of muscle, especially the attachment of sarcomeres to each other and to the endomysium by the cytoskeleton (cytoskeletal structure has been reviewed by Thornell and Price, 1991; Squire, 1997). Interpreting the effects of high pressure is complicated by inherent variability in meat structure and the rapid post-mortem changes. There are ultrastructural differences between muscle fiber types which include Z line width and mitochondrial content. There are also differences in post-mortem changes related to fiber type (Abbot et al., 1977; Stromer et al., 1967). Normal post-mortem changes include:

- fiber-fiber detachment in fish and mammals (Taylor and Koohmaraie, 1998; Taylor et al., 2002; Will et al., 1980)
- breaks in I bands and intermediate filaments (Taylor et al., 1995; Ho et al., 1997)
- extensive disruption of the reticulum (Will et al., 1980)

In general structural changes are similar in fowl (MacNaughtan, 1978), pork (Abbot et al., 1977), beef (Davey and Gilbert, 1969; Gann and Merkel, 1978; Will et al., 1980) and sheep (Taylor and Koohmaraie, 1998). However, fish do not show *I* band breaks (Papa et al., 1997, Taylor et al., 2002), one of the major features of other meat animals. Connective tissue of mammals is stable post-mortem with few structural changes for several weeks (McCormick, 1994; Purslow, 1994).

Structural changes due to pressure treatment of meat are very dependent on time post-mortem, temperature, pressure and, to a lesser degree, on species or on muscle type. If pressure treatment is applied pre-rigor there is extensive contraction of four different sheep and beef muscles groups, with 103 MPa causing shrinkage of up to 48% (Kennick et al., 1980) and disruption of the sarcolemma. Similar results are reported for sheep (Macfarlane and Morton 1978) and beef (Bouton et al., 1980, Elgasim and Kennick, 1982), with pre-rigor treatment causing contraction of fibers, contraction bands and sarcolemma disruption. Detailed structural studies using SEM and TEM also reveal changes in sarcomeric structure with loss of *M* lines and some gaps in the Z line (Elgasim and Kennick, 1982).

Treatment of meat post-rigor does not result in extensive contraction (Jung et al., 2000a, Macfarlane and Morton, 1978, Suzuki et al., 1992) and can improve tenderness. In general, at ambient or lower temperatures and low pressures, there is little or no structural change, for example at:

- 130 MPa and 10°C for beef (Jung et al., 2000a)
- 100 MPa and 20°C for bluefish (Ashie et al., 1997)
- 100 MPa and 10°C for beef (Suzuki et al., 1992; Ueno et al., 1999)

At these respective temperatures, increasing the pressure causes M line loss, disruption of thin filament organization, thickening of Z lines (Jung et al., 2000a, Suzuki et al., 1992), some fiber breaks and also disruption of endomysium and perimysium organization (Ashie et al., 1997, Ueno et al., 1999). These effects have been produced, for example, by increasing the pressure to:

- 325 MPa and above at 10C for beef (Jung *et al.*, 2000a)
- 200 or 300 MPa for bluefish (Ashie *et al.*, 1997)
- 200 MPa and 10°C for beef (Suzuki *et al.*, 1992, Ueno *et al.*, 1999)

At 150 MPa and 0°C for 3h there is loss of M lines and I band disruption in beef (Macfarlane *et al.*, 1980), and similar changes at 150 MPa and 20°C for 5 min. (Suzuki *et al.*, 1990). Due to the great variability of parameters involved, it is difficult to say that 150 MPa is a critical pressure which induces ultrastructural change. The general conclusion is that, at ambient temperatures or less and pressures of less than 150 MPa, there is little to no structural change in meat.

Studies such as these show that I band disruption, M line loss and endomysium/sarcolemma detachment are hallmarks of high pressure-induced structural changes in meat. Other frequently reported fiber-related changes include increased space between fibers (Elgasim and Kennick, 1982, Jung *et al.*, 2000a, Yuste *et al.*, 1998). In addition both fibers (Yuste *et al.*, 1998) and sarcomeres (Jung *et al.*, 2000a) have increased diameters after pressure treatment. Suzuki *et al.*, (1991) have examined structural changes in pressure-treated preparations of purified beef myofibers and also found M line loss, Z line thickening and I band disruption as reported for whole meat. However, they also observed more extensive changes, notably A band breaks, not reported for whole meat.

Combining pressure with heat treatment increases effects on structure and also causes changes at lower pressures. Treatment of sheep at 100 MPa for 1h at 25°C, for example, causes M line loss and I band disorganization (Macfarlane and Morton, 1978). At temperatures closer to normal cooking temperatures, pressure-induced structural changes can be extensive. Locker and Wild found ultrastructural changes in beef treated at 60 MPa and 50°C to 65°C which included disruption of thin filament organization, M line loss and increased N lines in the I band indicating protein aggregation (Locker and Wild, 1984). At 60°C and higher pressures of 150 MPa (Bouton *et al.*, 1977, Macfarlane and MacKenzie, 1986), beef muscle shows thicker Z lines, thin filament disorganization, M line loss and also breaks within the A band. This muscle structure is normally very stable post-mortem and even at cooking temperatures (Jones *et al.*, 1977, Leander *et al.*, 1980).

The connective tissue ultrastructure has also been examined in high pressure treated meat. As mentioned above, the endomysium detaches normally in meat and this is enhanced by pressure treatment. Under certain conditions there can be breaks in the endomysium (Ueno *et al.*, 1999, Ashie *et al.*, 1997). These changes tend to be minor and other authors have reported no changes in connective tissue structure due to high pressure (Suzuki *et al.*, 1993, Yuste *et al.*, 1998).

There is one report of changes in lysosomal ultrastructure in beef treated at high pressure. Jung *et al.*, (2000b) examined lysosome ultrastructure in beef stored for 48 hours and then treated at pressures from 0 to 600 MPa for various times at 10°C. At pressures above 300 MPa the lysosomes were swollen and had disrupted membranes. Under these conditions the content of free lysosomal enzymes increases, and may contribute to some of the changes observed. These observations are similar to lysosomal changes after 14 days of meat ageing (Chambers *et al.*, 1994), suggesting that high pressure accelerates normal lysosomal changes.

16.4 Influence on enzyme release and activity

Enzymatic reactions are highly susceptible to high pressure treatment, and muscle enzymes are particularly significant in the tenderization of meat during ageing. As a result, the effect of high pressure on meat enzymes has been extensively studied.

16.4.1 Calpain system and proteasome
High pressure treatment increases proteolytic activity of muscular enzymes (Homma *et al.*, 1994). Homma *et al.* (1995) showed a decrease of calpain activity in meat when it was pressurized, and that barosensibility was different according to calpain type. Proteasome activity is enhanced at 150 MPa, but decreases at higher pressure values (Otsuka *et al.*, 1998).

16.4.2 Lysosomal enzymes
High pressure treatment of muscle induces a release of lysosomal enzymes (Homma *et al.*, 1994, Elgasim *et al.*, 1983) due to lysosome membrane breakdown (Jung *et al.*, 2000b). Generally, activity of cathepsin D and acid phosphatase in pressurized meat samples is higher than in control samples at two days post-mortem and during storage (Jung *et al.*, 2000c).

16.4.3 ATPase and glycolytic activity
Yamamoto *et al.* (1993) showed that myosin ATPase activity decreased as soon as 70 MPa was applied, and that activity fell to zero after 210 MPa. Horgan (1980) showed a decrease of Ca^{2+}-dependant ATPase activity in pre- and post-rigor meat. But glycolysis may increase with pressure because of phosphorylase activation associated with changes of pH and free Ca^{2+} (Elkhalifa *et al.*, 1984).

16.5 High pressure effects on the sensory and functional properties of meat

16.5.1 Texture
Results presented in Table 16.1 show that several groups have found that high pressure can tenderize meat when applied pre-rigor. However, high pressure treatment does not have these effects if combined with commercial conditions

Table 16.1 Effect of high pressure treatment on meat tenderness (from Jung, 2000d)

Time post-mortem	High pressure treatment	Cooking conditions	Effect on tenderness	Authors
pre-rigor	As far as 130 MPa 30–40°C, 1–8 min	90°C, 1h	T WE	Macfarlane (1973)
pre-rigor post-rigor	20 to 140 MPa 25 to 60°C, 2.5 min to 1h	75 and 80°C, 1.5 h	T WE	Bouton et al. (1977)
pre-rigor	103 MPa 35°C, 2 min	Microwave + grill (80°C)	T	Riffero and Holmes (1983)
pre-rigor	103.5 MPa 37°C, 2 min	80°C, 40 min	T	Elgasim and Kennick (1982)
pre-rigor	103 MPa 35°C, 2 min	80°C, 40 min	T	Kennick et al. (1980)
post-rigor	Pre-cooking 45°C, 45 min 150 MPa 60°C, 30 min	80°C, 90 min 80°C, 24 h	T	Bouton et al. (1980)
post-rigor	60 MPa 25–65°C, 5–20 min	80°C, 40 min raw	T WE	Locker and Wild (1984)
post-rigor	150 MPa 40 to 80°C, 1 to 4h	80°C, 1 h	T WE	Beilken et al. (1990)
post-rigor	Pre-cooking 45°C, 45 min 150 MPa 60°C, 30 min	80°C, 2 h	T	Ratcliff et al. (1977)
post-rigor	Pre-cooking 45°C, 45 min 150 MPa 60°C, 30 min	80°C, 1 h	T	Robertson et al. (1984)
post-rigor	150 MPa 30 or 60°C 1.6 or 16 h	80°C, 1 h	WE T	Macfarlane and McKenzie (1986)
post-rigor	150 MPa 0°C, 3h	25, 50 or 80°C, 1 h	WE H	Macfarlane et al. (1980)
post-rigor	500 MPa 2°C, 10 or 30 min		H	Yuste et al. (1998)
post-rigor	400 to 600 MPa 20 to 50°C, 15 min	80°C, 20 min	WE	Margey et al. (1997)
post-rigor	200 to 350 MPa 20°C, 5 to 100 min	raw	T	Mussa (1999)
post-rigor	520 MPa, 260 s, 10°C	65°C, 1 h	H	Jung et al. (2000c)

T: Tenderizing WE: Without Effect H: Hardening.

for meat ageing, i.e. post-rigor and at low temperature. High pressure treatment of post-rigor muscle causes a significant increase in lysosomal enzyme activiy, but does not improve meat tenderness or reduce the ageing time.

16.5.2 Color and lipid oxidation

Meat color characteristics in the CIELAB system, i.e. L* (lightness), a* (redness) and b* (yellowness) are modified by high pressure processing. L* increases from 250 MPa to 350 MPa and then becomes constant for higher pressure values. As a consequence, bovine meat appears lighter and less red (Carlez et al., 1995, Shigehisa et al., 1991). This modification of lightness is due both to myoglobin denaturation (Carlez, 1994) and myofibrillar proteins denaturation (Goutefongea et al., 1995). The a* index decreases when meat has been pressurized at higher pressures (400–500 MPa), because of the increase of the metmyoglobin content (Ledward, 1998), while the b* index remains constant (Riffero and Holmes, 1983) and the meat becomes brown. Metmyoglobin formation can be prevented by complete removal of oxygen through vacuum packaging with an oxygen scavenger, or by previous formation of nitrosylmyoglobin, as in processed brined products (Carlez et al., 1995; Goutefongea et al., 1995). Although high pressure treatment induces visible modifications of the color of raw meat, after cooking the color difference is greatly reduced (Jung et al., 2000c).

Cheah and Ledward (1996) showed that high pressure (800 MPa, 20 min) treated pork mince samples revealed faster oxidation than control samples, and that pressure treatment at greater than 300–400 MPa caused conversion of reduced myoglobin/oxymyoglobin to the denatured ferric form. Cheah and Ledward (1997) also demonstrated that iron released from metal complexes during pressure treatment catalysed lipid oxidation in meat. According to Orlien and Hansen, (2000), 500 MPa is a critical pressure for lipid oxidation and development of rancidity in chicken breast muscle. Lipid oxidation at higher pressures is not related to the release of non-hem Fe or catalytic activity of metmyoglobin, but could be linked to membrane damage.

16.5.3 Gelation and emulsifying properties

The meat products industry depends on exploiting the functional properties of myofibrillar proteins, including water binding, gelling and emulsifying properties. High pressure may induce gelation of myofibrillar proteins without heating (Hermansson et al., 1986). When myofibrillar proteins have been previously submitted to high pressure treatment, heated gels are stronger (Ikeuchi et al., 1992a). These modifications are mainly linked to the increase of hydrophobicity and sulfhydryl interactions of myofibrillar proteins (Ikeuchi et al., 1992b, Chapleau et al., 2002).

Crehan and Troy (2000) have shown that the emulsion stability of meat was increased in frankfurters made with 1.5% NaCl after exposure of the meat to 150 MPa. Generally speaking, high pressure improves meat binding properties, partially compensating for a reduction of the NaCl content of meat products.

16.6 Pressure assisted freezing and thawing

As Fig. 16.2 shows, the phase change temperature of water decreases from 0°C down to -22°C when pressure increases up to 220 MPa. The opposite effect is observed above this pressure. This phenomenon can be used to achieve rapid thawing or freezing of foods, such as meat, which contain a significant amount of water. Slow freezing results in larger ice crystals, which generally damage the

High pressure processing of meat 321

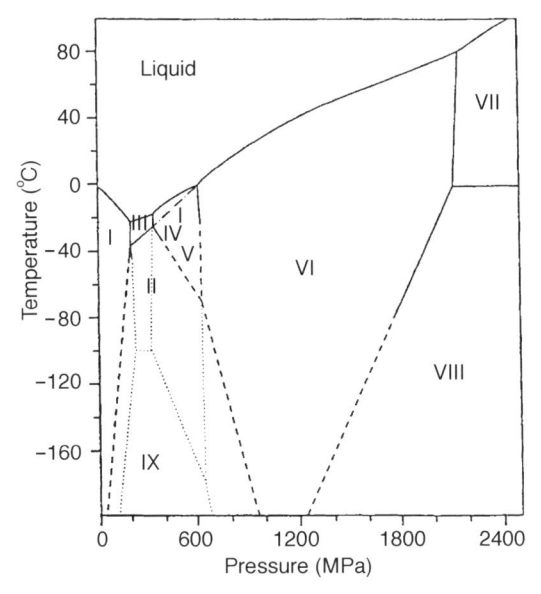

Fig. 16.2 Phase diagram of water under pressure (from Kalichevsky *et al.*, 1995).

texture of the food, whereas a rapid freezing rate usually preserves food texture (Sanz *et al.*, 1999). Rapid freezing using high pressure can be achieved by cooling at −20°C and 200 MPa. In these conditions water remains in the liquid state. Upon release of pressure, instantaneous and homogeneous crystalisation occurs with formation of very small crystals. This method has been shown to preserve the textural properties of pork meat as well as traditional methods (Martino *et al.*, 1998).

High pressure thawing also offers several advantages in comparison to thawing at atmospheric pressure, including the reduction of the thawing time by 2 to 5-fold in comparison with conventional processes, and partial destruction or growth limitation of pathogens (Haack and Heinz, 2001). Zhao *et al.*, (1998) have shown that high pressure thawing maintains the organoleptic properties of bovine meat. Additional studies are necessary to better understand water holding capacity and protein denaturation during high pressure thawing (Knorr *et al.*, 1998).

16.7 Effects on microflora

The use of high pressure for food preservation was proposed more than a century ago. First applied to milk (Hite, 1899), high pressure was then used to treat fruits and vegetables (Hite *et al.*, 1914), and other food products such as meat. The resistance of bacteria, enzymes and toxins to pressure treatments has been investigated for many years since the pioneering work published by Larson *et al.*, (1918) and Basset and Macheboeuf (1932). However, the first industrial application of high pressure preservation was not developed until the 1990s when large scale equipment were developed for continuous treatment of liquids such as fruit juice.

High pressure processing effectively inactivates spoilage micro-organisms as well as food borne pathogens (Cheftel, 1995). This inactivation is due to widespread damages of microorganisms through modification of morphology and of several vulnerable components such as cell membranes, ribosomes and enzymes, including those involved in the replication and transcription of DNA (Yuste *et al.*, 2001a). The effect of high pressure on bacterial survival is influenced by a number of interacting factors such as magnitude and duration of the treatment, temperature, environmental conditions, bacteria species and development phase (Patterson *et al.*, 1995).

At ambient temperature vegetative cells are inactivated between 400 and 600 MPa. In general, gram-positive bacteria (*Listeria monocytogenes*, *Staphylococcus aureus*) are more resistant to pressure than gram-negative (*Pseudomonas*, *Salmonella spp*, *Yersinia enterocolitica*, *Vibrio parahaemolyticus*), but large differences can exist between strains within the same species. Moreover, *cocci* are more resistant than rods because of fewer morphological changes under pressure. In addition, cultures in the exponential growth phase have been shown to be far more sensitive than cultures in the logarithmic growth or stationary phase (Hoover *et al.*, 1989). In contrast, spores at ambient temperature can resist pressures up to 1000 MPa, temperatures above 70°C being necessary to obtain a significant level of inactivation. However, it has also been shown that lower pressures (250 MPa) associated with mild temperatures (40°C) can inactivate spores in a two stage process, pressure first inducing germination and then inactivating the baro-sensitive germinated spores. In addition, pressurisation can inactivate some parasites such as *Trichinella spiralis* but its efficiency on inactivation of viruses is very limited (Cheftel, 1995).

The application of high pressure can produce a population of stressed cells which can be revived in certain environmental conditions. As a consequence, pressure inactivation of bacteria may not be effective. Attachment of bacteria to certain food constituents such as protein, carbohydrates, and lipids may also confer a baro-protection which limits the effectiveness of high pressure. Indeed, pressure resistance of bacteria has been shown to be dependent on culture medium. UHT milk, for example, exhibits a protective effect on *Staphylococcus aureus*, *Listeria monocytogenes* and *Escherichia coli* O157:H7 compared to pork slurry or poultry meat, whereas high pressure is the most effective when the bacteria are inoculated in pH 7.0 phosphate buffers (Patterson *et al.*, 1995). It has also been shown that the use of inoculated flora may lead to over-estimation of the effectiveness of high pressure. Carlez *et al.*, (1994) noted that minced meat endogenous flora was more resistant to pressure than inoculated collection strains.

A process cycle at both low and high pressures may be more effective in inactivating vegetative bacteria than continuous treatment (Yuste *et al.*, 2001a). However, the success of this approach depends on conditions such as intensity and duration of pressure together with other factors such as temperature (Yuste *et al.*, 2001b). In general, full sterilization of food products is not possible with high pressure at levels below 500 MPa. Most industrial applications operate at a ceiling of 400 to 500 MPa maximum, and products require chilled storage to maximize shelf-life. The combination of high pressure with other physical treatments (such as radiation, pulsed electric fields or ultrasound, for example) or chemical preservation methods (such as bacteriocins, chitosans or antioxidants) has been proposed in order to enhance its efficiency and/or reduce the severity of the other treatments (Yuste *et al.*, 2001a).

The fact that pressure-stressed cells may be less resistant to heat could explain the efficiency of combined high pressure and moderate temperature treatment on bacteria inactivation. High pressure is well adapted to 'pasteurization' of animal food products sensitive to heat. In particular, it can be applied to foie gras from fatty goose or duck, increasing its microbial safety and shelf-life. Conditions such as pressure at 400 MPa and temperature in the 50–60°C range applied for 5 to 15 min. have been shown to be efficient for preservation without the lipid loss associated with thermal pasteurisation (El Moueffak *et al.*, 1996). High pressure can also be used to increase the safety of meats cooked at low temperature over a long period. This technique increases the tenderness of meat as long as the temperature is kept no higher than 60–65°C to avoid contraction of collagen and resulting cooking losses. Finally, pressure can be used for preservation of already packaged meat products which may have been contaminated earlier during processing. In particular, packaged sliced ham and salami emulsion-type sausages are well adapted to pressure treatment, in part due to the stability of their pink or red color under pressure (Cheftel and Culioli, 1997).

16.8 Current applications and future prospects

Stabilization of fruit products (such as fruit juices, jam and avocado paste) is currently the main field of application for high pressure in the food industry. Japan was the first country to manufacture pressurized products such as fruit jam on a commercial scale during the 1990s. The main applications of high pressure treatment in the meat industry are in stabilizing meat products and texturing of meat paste in combination with thermal processing. As an example, high pressure treated, cooked and sliced ham has been produced since 1999 in Spain by the Espuña Company and has extended product shelf life to several weeks (Fig. 16.3). The same company has also expanded the use of high pressure to the manufacture of products such as meat 'tapas'. There remain a number of constraints on development, including cost. Early estimates suggested that high pressure processing could be as much as twenty times the cost of conventional thermal technologies (Manvell, 1996). In addition, more research is required to establish process parameters for microbial and enzyme inactivation. Such research is essential in the context of regulatory approval (for example, in meeting the EU's Novel Food Regulation) permitting a larger-scale commercial

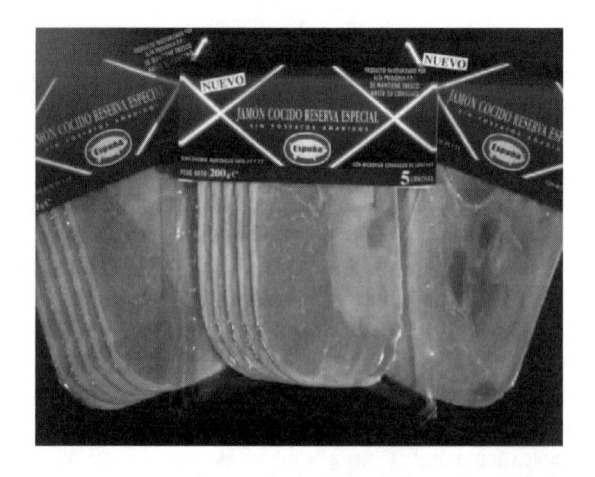

Fig. 16.3 Meat products produced using high pressure treatment – Espuña Company, Spain (courtesy of ACB Pressure Systems, France).

breakthrough for this technology (European Community, 1997). The development of kinetic data and appropriate models describing, for example, the combined effects of pressure and temperature on the inactivation of pathogenic and spoilage organisms, will provide a foundation for further development (Ludikhuyze *et al.*, 2001).

16.9 References

ABBOT M.T., PEARSON A.M., PRICE J.F. and HOOPER G.R. (1977) Ultrastructural changes during autolysis of red and white porcine muscle. *J. Food Sci.*, 42, 1185–1188.

ASHIE I.N.A., SIMPSON B.K. and RAMASWAMY H.S. (1997) Changes in texture and microstructure of pressure-treated fish muscle tissue during chilled storage. *J. Muscle Foods*, 8, 13–32.

BASSET J. and MACHEBOEUF M.A. (1932) Etude sur les effets biologiques des ultra-pressions: résistance des bactéries des diastases et des toxines aux pressions élevées. *Comptes rendus hebdomadaires des séances de l'Académie des Sciences*, 196, 1431–1439.

BEILKEN S.L., MACFARLANE J.J. and JONES P.N. (1990) Effect of high pressure during heat treatment on the Warner-Bratzler shear force values of selected beef muscles. *J. Food Sci.*, 55, 15–18.

BENDALL J.R. (1973) *Post-mortem* changes in muscle. In: *The Structure and Function of Muscle*. G. H. Bourne (ed.) Academic Press, NY, pp 243–309.

BLOOM W. and FAWCETT D.W. (1975) *A Textbook of Histology*. WB Saunders Co., London.

BOUTON P.E., FORD A.L., HARRIS P.V., MACFARLANE J.J. and O'SHEA J.M. (1977) Pressure-heat treatment of *post-rigor* muscle: effects on tenderness. *J. Food Sci.*, 42, 132–135.

BOUTON P.E., HARRIS P.V. and MACFARLANE J.J. (1980) Pressure-heat treatment of meat: effect of prior aging treatments on shear properties. *J. Food Sci.*, 45, 276–278.

CARLEZ A. (1994) Traitements par hautes pressions d'aliments d'origine musculaire: destruction microbienne, modifications de couleur, gélification protéique. PhD, University of Montpellier II, 252 pp.

CARLEZ A., ROSEC J.P., RICHARD N. and CHEFTEL J.C. (1994) Bacterial growth during chilled storage of pressure-treated minced meat. *Lebens.-Wiss. u.-Technol.*, 27, 48–54.

CARLEZ A., VECIANA-NOGUES T. and CHEFTEL J.-C. (1995) Changes in colour and myoglobin of minced beef meat due to high pressure processing. *Lebensm.-Wiss. u.-Technol.*, 28, 528–538.

CHAMBERS J.J., REVILLE W.J. and ZEECE M.G. (1994) Lysosomal integrity in *post-mortem* bovine skeletal muscle. *Sci. des Alim.*, 14, 441–457.

CHAPLEAU N., DELÉPINE S. and DE LAMBALLERIE-ANTON M. (2002) Effect of pressure treatment on hydrophobicity and sulfhydryl interactions of myofibrillar proteins. In: *Trends in High Pressure Bioscience and Biotechnology*, R. Hayashi (ed.) Elsevier Science 55–62.

CHEAH P.B. and LEDWARD D.A. (1996) High pressure effects on lipid oxidation in minced pork. *Meat Sci.*, 43, 123–134.

CHEAH P.B. and LEDWARD D.A. (1997) Catalytic mechanism of lipid oxidation following high pressure treatment in pork fat and meat. *J. Food Sci.* 62 (6) 1135–1138.

CHEFTEL J.-C. (1991) Applications des hautes pressions en technologie alimentaire. *IAA*, 141–153.

CHEFTEL J.-C. (1995) Review: High-pressure, microbial inactivation and food preservation. *Food Sci. Technol. Intern.*, 1, 75–90.

CHEFTEL J.-C. and CULIOLI J. (1997) Effects of high pressure on meat: a review.

Meat Sci., 46, 211–236.
CHEFTEL J.-C. and DUMAY E. (1997) Les hautes pressions: principes et potentialités. In: *La conservation des aliments*, Tec&Doc, Lavoisier (eds), 195–216.
CREHAN C.M. and TROY D.J. (2000) Effect of salt levels and high hydrostatic pressure processing on frankfurters formulated with 1.5 and 2.5% salt. *Meat Sci.*, 55 (1) 123–130.
DAVEY C.L. and GILBERT K.V. (1969) Studies of meat tenderness. 7. Changes in the fine structure of meat during aging. *J. Food Sci.*, 34, 69–74.
ELGASIM E.A. and KENNICK W.H. (1982) Effect of high pressure on meat microstructure. *Food Microstruct.*, 1, 75–82.
ELGASIM E.A., KENNICK W.H., ANGLEMIER A.F., KOOHMARAIE M. and ELKHALIFA E.A. (1983) Effect of *pre-rigor* pressurization on bovine lysosomal enzyme activity. *Food Microstruct.*, 2, 91–97.
ELKHALIFA E.A., ANGLEMIER A.F., KENNICK W.H. and ELGASIM E.A. (1984) Influence of *prerigor* pressurization on *post-mortem* beef muscle creatine phosphokinase activity and degradation of creatine phosphate and adenosine triphosphate. *J. Food Sci.*, 49, 595–597.
EL MOUEFFAK A., CRUZ C., ANTOINE M., MONTURY M., DEMAZEAU G., LARGETEAU A., ROY B. and ZUBER F. (1996) High pressure and pasteurization effect on duck *foie gras*. *Intern. J. Food Sci. Technol.*, 30, 737–743.
EUROPEAN COMMUNITY (1997) Règlement CE nE 258/97 du Parlement Européen et du Conseil du 27 janvier 1997 relatif aux nouveaux ingrédients alimentaires, in *Journal Officiel des Communautés Européennes*. 27 January. pp. L 43/1–L 43/7.
GALAZKA V.B., SUMNER I.G. and LEDWARD D.A. (1996) Changes in protein-protein and protein polysaccharide interactions induced by high pressure. *Food Chem.*, 57, 393–398.
GANN G.L. and MERKEL R.A. (1978) Ultrastructural changes in bovine *longissimus* muscle during *post-mortem* ageing. *Meat Sci.*, 2, 129–144.
GOLL D.E., ROBSON R.M. and STROMER M.H. (1984) Skeletal Muscle. In: *Duke's Physiology of Domestic Animals*, 10th edn. Swenson M.J. (ed.) Cornell University Press, Ithaca NY, pp. 548–580.
GOUTEFONGEA R., RAMPON V., NICOLAS N. and DUMONT J.-P. (1995) Meat color changes under high pressure treatment. 41st ICoMST, Am. Meat Sci. Assoc. (eds), II, 384–385.
HAACK E. and HEINZ V. (2001) Improvement of food safety by high pressure processing. II Studies on use in the meat industry. *Fleischwirtschaft*, 81 (6) 38–41.
HAYASHI R. (1991) High pressure in food processing and preservation: principle, application and development. *High Press. Res.*, 7, 15–21.
HENDRICKX M., LUDIKHUYZE L., VAN DEN BROECK I. and WEEMAES C. (1998) Effects of high pressure on enzymes related to food quality. *Trends Food Sci. Technol.*, 9, 197–203.
HERMANSSON A.M., HARBITZ, O. and LANGTON, M. (1986) Formation of two types

of gels from bovine myosin. *J. Sci. Food Agric.*, 37, 69–84.

HITE B.H. (1899) The effect of pressure in the preservation of milk. Bulletin No. 58, 15–35. West Virginia Agricultural Experimental Station, Morgantown, Virginia.

HITE B.H., GIDDINGS N. and WEAKLY C. (1914) The effects of pressure on certain microorganisms encountered in the preservation of fruits and vegetables. *Bull. West Virginia Agricultural Experiment Station of Morgantown W. VA.*, 146, 1–67.

HO C.-Y., STROMER M.H., ROUSE G. and ROBSON R.M. (1997) Effects of electrical stimulation and *post-mortem* storage on changes in titin, nebulin, desmin, troponin-T, and muscle ultrastructure in *Bos indicus* crossbred cattle. *J. Anim. Sci.*, 75, 366–376.

HOMMA N., IKEUCHI Y. and SUZUKI A. (1994) Effects of high pressure treatment on the proteolytic enzymes in meat. *Meat Sci.*, 38, 219–228.

HOMMA N., IKEUCHI Y. and SUZUKI A. (1995) Levels of calpain and calpastatin in meat subjected to high pressure. *Meat Sci.*, 41, 251–260.

HOOVER D.G., METRICK C., PAPINEAU A.M., FARKAS D.F. and KNORR D. (1989) Biological effects of high hydrostatic pressure on food micro-organisms. *Food Technol.*, 43, 99–107.

HORGAN D.J. (1980) Effect of pressure treatment on the sarcoplasmic reticulum of red and white muscles. *Meat Sci.*, 5, 297–305.

IKEUCHI Y., TANJI H., KIM K. and SUZUKI A. (1992a) Dynamic rheological measurements on heat-induced pressurized actomyosin gels. *J. Agric. Food Chem.*, 40, 1751–1755.

IKEUCHI Y., TANJI H., KIM K. and SUZUKI A. (1992b) Mechanism of heat-induced gelation of pressurized actomyosin: pressure-induced changes in actin and myosin in actomyosin. *J. Agric. Food Chem.*, 40, 1756–1761.

JONES S.B., CARROLL R.J. and CAVANAUGH J.R. (1977) Structural changes in heated bovine muscle: a scanning electron microscopy study. *J. Food Sci.*, 42, 125–131.

JUNG S., DE LAMBALLERIE-ANTON M. and GHOUL M. (2000a) Modifications of ultrastructure and myofibrillar proteins of *post-rigor* beef treated by high pressure. *Food Science and Technology* 33, 313–319.

JUNG S., DE LAMBALLERIE-ANTON M., TAYLOR R.G. and GHOUL M. (2000b) High pressure effects on lysosome integrity and lysosomal enzyme activity in bovine muscle. *J. Agric. Food Chem.* 48, 2467–2471.

JUNG S., GHOUL M. and DE LAMBALLERIE-ANTON M. (2000c) Changes in lysosomal enzyme activities and shear values of high pressure treated meat during ageing. *Meat Sci.* 56, 3 239–246.

JUNG S. (2000d) Etude de l'effet des hautes pressions sur la texture, l'ultrastructure et un système enzymatique de viande bovine. PHD., University of Nantes, 171 pp.

KALICHEVSKY M.T., KNORR D. and LILLFORD P.J. (1995) Potential food applications of high pressure effects on ice-water transitions. *Trends Food Sci. Technol.*, 6, 253–259.

KENNICK W.H., ELGASIM E.A., HOLMES Z.A. and MEYER P.F. (1980) The effects of pressurization of *pre-rigor* muscle on *post-rigor* meat characteristics. *Meat Sci* 4, 33–40.

KNORR D., SCHLUETER O. and HEINZ V. (1998) Impact of high hydrostatic pressure on phase transitions of foods. *Food Technol.*, 52, 42–45.

LARSON W.P., HARTZELL T.B. and DIEHL H.S. (1918) The effect of high pressures on bacteria. *J. Infectious Diseases*, 22, 271–281.

LAWRIE R.A. (1998) *Meat Science.* Woodhead Publishing Limited, Cambridge, 336 pp.

LEANDER R.C., HEDRICK H.B., BROWN M.F. and WHITE J.A. (1980) Comparison of structural changes in bovine *longissimus* and *semitendinosus* muscles during cooking. *J. Food Sci.*, 45, 1–6.

LEDWARD D.A. (1998) High pressure processing of meat and fish. In: *Fresh novel foods by high pressure*, Autio K., VTT Biotechnology and Food Research (eds), 165–175.

LEDWARD D.A. (2002) High Pressure Processing of Meat and Fish *High Pressure Resesarch in press.*

LOCKER R.H. and WILD D.J.C. (1984) Tenderization of meat by pressure-heat involves weakening of the gap filaments in the myofibril. *Meat Sci.*, 10, 207–233.

LUDIKHUYZE L., LOEY VAN A., INDRAWATI and HENDRICKX, M. (2001), Combined high pressure thermal treatment of foods. In: *Thermal technologies in food processing*, Richardson, P (ed.), Woodhead Publishing Ltd, Cambridge.

MACFARLANE J.J. (1973) *Pre-rigor* pressurization of muscle: effects on pH, shear value and taste panel assessment. *J. Food Sci.*, 38, 294–298.

MACFARLANE J.J. and MCKENZIE I.J. (1986) Pressure-accelerated changes in the proteins of muscle and their influence on Warner-Bratzler shear values. *J. Food Sci.*, 51, 516–517.

MACFARLANE J.J. and MORTON D.J. (1978) Effects of pressure treatment on the ultrastructure of striated muscle. *Meat Sci.*, 2, 281–288.

MACFARLANE J.J., MCKENZIE I.J. and TURNER R.H. (1980) Pressure induced pH and length changes in meat. *Meat Sci.*, 7, 169-181.

MACFARLANE J.J., MCKENZIE I.J. and TURNER R.H. (1980) Pressure treatment of meat: effects on thermal transitions and shear values. *Meat Sci.*, 5, 307–317.

MACNAUGHTAN A.F. (1978) A histological study of *post mortem* changes in the skeletal muscle of the fowl (*Gallus domesticus*). II. The cytoarchitecture. *J. Anat.*, 126, 7–20.

MCCORMICK R.J. (1994) The flexibility of the collagen compartment of muscle. *Meat Sci.*, 36, 79–91.

MANVELL, C. (1996), Opportunities and problems of minimal processing and minimal-processed foods. Paper presented at the EFFoST Conference on Minimal Processing of Foods. Cologne, November, 1996.

MARGEY D.M., PATTERSON M.F. and MOSS B.W. (1997) Effect of various temperature/pressure combinations on the microbiological and quality attributes of poultry meat. In: *High Pressure Research in the Biosciences and Biotechnology*, Heremans K. (eds), 307–310.

MARTINO M.N., OTERO L., SANZ P.D. and ZARITZKY N.E. (1998) Size and location of ice crystals in pork frozen by high-pressure-assisted freezing as compared to classical methods. *Meat Sci.*, 50, 303–313.

MOZHAEV V.V., HEREMANS K., FRANK J., MASSON P. and BALNY C. (1994) Exploiting the effects of high pressure in biotechnological applications. *Trends Biochem.*, 12, 493–501.

MUSSA D.M. (1999) High pressure processing of milk and muscle foods: evaluation of process kinetics, safety and quality changes. PhD, McGill University, 264 pp.

ORLIEN V. and HANSEN E. (2000) Lipid oxidation in high pressure processed chicken breast muscle during chill storage: critical working pressure in relation to axidation mechanism. *Eur. Food Res. & Technol.*, 211 (2) 99–104.

OTSUKA Y., HOMMA N., SHIGA K., USHIKI J., IKEUCHI Y. and SUZUKI A. (1998) Purification and properties of rabbit muscle proteasome, and its effect on myofibrillar structure. *Meat Sci.*, 49, 365–378.

PAPA I., TAYLOR R.G., VENTRE F., LEBART M.C., ROUSTAN C., OUALI A. and BENYAMIN Y. (1997) Cleavage and sarcolemma detachment are early *post-mortem* changes in bass (*Dicentrarchus labrax*) white muscle. *J Food Sci.*, 62, 917–921.

PATTERSON M.F., QUINN M., SIMPSON R. and GILMOUR A. (1995) Effects of high pressure on vegetative pathogens. In: *High Pressure Processing of Foods*. Ledward D.A., Johnston D.E., Earnshaw R.G. and Hasting A.P.M. (eds), Nottingham University Press, Nottingham, 47–63.

PURSLOW P.P. (1994) The structural basis of meat toughness: what role does the collagenous component play? *ICoMST Proc.*, 40, 27–34.

RATCLIFF D., BOUTON P.E., FORD A.L., HARRIS P.V., MACFARLANE J.J. and O'SHEA J.M. (1977) Pressure-heat treatment of *post-rigor* muscle: objective-subjective measurements. *J. Food Sci.*, 42, 857–859.

RIFFERO L.M. and HOLMES Z.A. (1983) Characteristics of *pre-rigor* pressurized versus conventionally processed beef cooked by microwaves and by broiling. *J. Food Sci.*, 48, 346–350.

ROBERTSON J., BOUTON P.E., HARRIS P.V., MACFARLANE J.J. and SHORTOSE W.R. (1984) Pressure-heat treatment of meat: a comparaison of beef and buffalo meat. *Meat Sci.*, 10, 285–292.

SANZ, P. D., DE ELVIRA, C., MARTINO, M., ZARITZKY, N., OTERO, L. and CARRASCO, J. A. (1999) Freezing rate simulation as an aid to reducing crystallization damage in foods. *Meat Sci.*, 52 (3) 275–278.

SHIGEHISA T., OHMORI, T., SAITO, A., TAJI, S. and HAYASHI, R. (1991) Effects of high hydrostatic pressure on characteristics of pork slurries and inactivation of microorganisms associated with meat and meat products. *Int. J. Food Microbiol.*, 12, 207–216.

SQUIRE J.M. (1997) Architecture and function in the muscle sarcomere. *Curr. Opinion Str. Biol.*, 7, 247–257.

STROMER M.H., GOLL D.E. and ROTH L.E. (1967) Morphology of rigor-shortened bovine muscle and the effect of trypsin on *pre-* and *post-rigor* myofibrils. *J. Cell. Biol.*, 34, 431–445.

SUZUKI A., SUZUKI N., IKEUCHI Y. and SAITO M. (1991) Effects of high pressure treatment on the ultrastructure and solubilization of isolated myofibrils. *Agric. Biol. Chem.*, 55, 2467–2473.

SUZUKI A., WATANABE M., IKEUCHI Y. and SAITO M. (1992) Pressure effects on the texture, ultrastructure and myofibrillar proteins of beef skeletal muscle. 38th ICoMST, 423–426.

SUZUKI A., WATANABE M., IKEUCHI Y., SAITO M. and TAKAHASHI K. (1993) Effects of high-pressure treatment on the ultrastructure and thermal behavior of beef intramuscular collagen. *Meat Sci.*, 35, 17–25.

SUZUKI A., WATANABE M., IWAMURA K., IKEUCHI Y. and SAITO M. (1990) Effect of high pressure treatment on the ultrastructure and myofibrillar protein of beef skeletal muscle. *Agric. Biol. Chem.*, 54, 3085–3091.

TAYLOR R.G. and KOOHMARAIE M. (1998) Effects of *post-mortem* storage on the ultrastructure of the endomysium and myofibrils in normal and callipyge longissimus. *J. Anim. Sci.*, 76, 2811–2817.

TAYLOR R.G., FJAERA S.O. and SKJERVOLD P.O. (2002) Salmon fillet texture is determined by myofiber-myofiber and myofiber-myocommata attachment. *J. Food Sci.*, in press.

TAYLOR R.G., GEESINK G.H., THOMPSON V.F., KOOHMARAIE M. and GOLL D.E. (1995) Is Z-disk degradation responsible for *post-mortem* tenderization? *J. Anim. Sci.*, 73, 1351–1367.

THORNELL L.-E. and PRICE M.G. (1991) The cytoskeleton in muscle cells in relation to function. *Biochem. Soc. Trans.*, 19, 1116–1119.

TONELLO C. (2001) Les équipements pour les traitements hautes pressions des aliments. In: *Traitements ionisants et hautes pressions des aliments*. M. Federighi and J.L. Tholozan (eds), Editions Polytechnica, Economica, Paris 151–161.

UENO Y., IKEUCHI Y. and SUZUKI A. (1999) Effect of high pressure treatment on intramuscular connective tissue. *Meat Sci.*, 52, 143–150.

WILL P.A., OWNBY C.L. and HENRICKSON R.L. (1980) Ultrastructural *post-mortem* changes in electrically stimulated bovine muscle. *J. Food Sci.*, 45, 21–34.

YAMAMOTO K., HAYASHI S. and YASUI T. (1993) Hydrostatic pressure-induced aggregation of myosin molecules in 0.5 M KCl at pH 6.0. *Biosci. Biotech. Biochem.*, 57, 383–389.

YUSTE J., RASZL S. and MOR-MUR M. (1998) Microscopic changes in poultry breast muscle treated with high hydrostatic pressure. 44th ICoMST, 550–551.

YUSTE J., CAPPELLAS M., PLA R., FUNG D.Y.C. and MOR-MUR M. (2001a). High pressure processing for food safety and preservation: a review. *J. Rapid Methods and Automation in Microbiology*, 9, 1–10.
YUSTE J., PLA R., CAPELLAS M., SENDRA E., BELTRAN E. and MOR-MUR M. (2001b). Oscillatory high pressure processing applied to mechanically recovered poultry meat for bacterial inactivation. *J. Food Sci.*, 66, 482–484.
ZHAO Y., FLORES R.A. and OLSON D.G. (1998) High hydrostatic pressure effects on rapid thawing of frozen beef. *J. Food Sci.*, 63, 272–275.

17

Processing and quality control of restructured meat

P. Sheard, University of Bristol

17.1 Introduction

A variety of restructured meats have been marketed successfully over the last 30 years in the UK, the USA and elsewhere. The success of these products has arisen from consumer demand for convenience, variety, consistent quality and, also, the economic desirability for the manufacturer to upgrade meat raw materials. Products are referred to, variously, as 'restructured', 'reformed', 'flaked and formed', 'chopped and shaped' and 'chunked and formed' determined to a large extent by the size of the constituent pieces (Franklin and Cross, 1982; Sheard and Jolley, 1988). The term 'intermediate value products' is also used (Breidenstein, 1982), suggesting that this type of product is perceived by the consumer, and marketed, as intermediate in value between traditional burgers and intact muscle steak.

In the UK, products are usually frozen, may be breaded and coated, and are usually rib or steak-shaped, with an appropriate coined name (e.g. joysteak, grillsteak or ribsteak). Restructured meats may also be used as an alternative to ordinary diced meat in canned meats or ready meals. Some products are sold in a bun or as the main component in a ready meal. The largest, and most easily identified, sector is that for grillsteaks. A UK survey of grillsteak-type products showed that most have a meat content of 93% or above, though some were as low as 55% (Jolley et al., 1988). Beef was the main ingredient in the products surveyed, though restructured ribs are usually pork and there are some lamb products. Poultry meat also features strongly. All the products included in the survey contained salt; phosphate, soya and caseinate were common in products with lower meat contents.

Meat restructuring 'involves the assembly of meat pieces into a cohesive product which aims to simulate or retain the texture of high quality muscle'

(Sheard and Jolley, 1988). This definition is useful in that it conveys something of the way in which the products are made: comminution followed by re-assembly or the binding together of the constituent pieces. It also prescribes the objective in terms of the desired texture and the way in which that is achieved: either by simulation (which is necessarily the case where the meats have been finely comminuted to a paste or emulsion) or by retaining the typical fibrous texture of good quality whole muscle, as is the case where the piece size is relatively large (e.g. chunked and formed products). Cohesion is developed during cooking by the gelation of meat proteins solubilised during processing by the action of salt.

Most of the published research has emanated from the United States (see for example, Franklin and Cross, 1982; Pearson and Dutson, 1987) where there appears to be a greater use of fresh meat than in the UK and more product is targeted at the food service (catering) sector (Field, 1982), including the armed forces (Shults, 1982). This review, for obvious reasons, focuses on the situation in the UK where the use of semi-frozen meat is more common. Section 17.2 describes the sequence of operations involved in the manufacture of UK-style grillsteaks (tempering, pre-breaking, flaking, mixing, forming and freezing). The same stages, with a few modifications, are employed in the manufacture of restructured roasts and diced meat. Factors affecting product quality are discussed in Sections 17.3–17.6 which highlight the need for accurate temperature control at the pre-break stage and explains the role of salt and phosphate in producing an adequate bind. Also included is some previously unpublished data from this laboratory. Some recent developments are highlighted in Section 17.7.

17.2 Product manufacture

The manufacture of UK-style grillsteaks involves a series of operations, summarised in Fig. 17.1 and described below.

17.2.1 Raw materials

Most products are made from frozen meat (at $-18°C$) and distributed frozen. Frozen meat is usually purchased as 27kg blocks specified with respect to animal (e.g. cow or steer and age), cut (forequarter cuts such as chuck and blade, flank and brisket, neck and shin) and amount of fat, usually on the basis of visual lean. Before the crisis surrounding bovine spongiform encephalopathy (BSE) it was common to use cow meat (5–8 years old) for grillsteak production. Arguably (see for example Cross *et al.*, 1976), the quality of grillsteaks made post-BSE has improved given the requirement to use animals of less than 30 months which yields more tender meat than that from older animals. The manufacturer is unlikely to have any control over factors, other than those specified, which might influence processing quality before the blocks arrive at

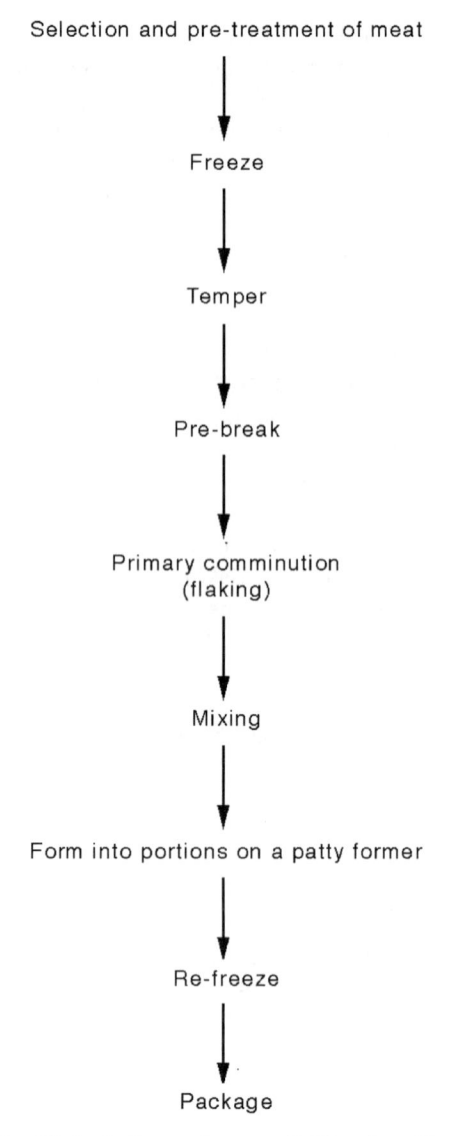

Selection and pre-treatment of meat
↓
Freeze
↓
Temper
↓
Pre-break
↓
Primary comminution (flaking)
↓
Mixing
↓
Form into portions on a patty former
↓
Re-freeze
↓
Package

Fig. 17.1 Method of manufacture of UK-style grillsteaks.

the factory. Other raw materials may include mechanically separated meat, salt, phosphate, onion, rusk, spices, monosodium glutamate, hydrolysed vegetable protein, sodium bisulphite and casein.

17.2.2 Tempering

Meat at 18°C is extremely hard and processing at frozen temperatures can lead to equipment damage (Koberna, 1986). It is usual, therefore, to temper frozen blocks, i.e., to increase the temperature of the block, usually to a target in the range −2 to −10°C, to facilitate further processing. There does not appear to be any ideal temperature and, in practice, the target temperature is usually based on pragmatic considerations and past practice ('we've always done it like this'). In one company, the target temperature is governed by the capability of the former at the end of the processing line which operates well only at low temperatures. Increasing meat temperature to the target, reduces the ice content but, more importantly, modifies the mechanical properties of the block (see Section 17.3).

Two methods of tempering are used: cold room or microwave, or sometimes a two-stage temper (microwave followed by cold room tempering), the latter giving better control over temperature. The main disadvantage with cold room tempering is that several days are required to reach the target temperature which, in turn, requires production to be forecast several days in advance. Cold room tempering often gives rise to temperature variation within (the surface being warmer than the interior) and between blocks (James and Crow, 1986). The rate at which meat tempers depends on the way in which blocks are stacked, air speed, temperature and position within the cold room. Poor stacking can markedly extend tempering times (James and Crow, 1986). The importance of accurate temperature control at this stage is not always recognised and, at worst, product may be tempered wherever there is space in the factory at ambient temperatures.

The main advantage of microwave tempering is speed; it also requires less space. Experiments carried out at Bristol demonstrate that frozen blocks, processed directly from frozen storage, can be tempered in about five minutes to −3°C (0 to −5°C), using a 30kW, 896MHz microwave (James and Crow, 1986). Because it is difficult to meet the target temperature consistently, there may be occasions when the meat is either under-tempered (i.e. too cold) or over tempered (i.e. too warm), leading to variability in particle size distribution and product quality (Koberna, 1986).

17.2.3 Pre-breaking

A variety of designs of pre-breaker exist such as chippers, flakers, guillotines and grinders. These differ markedly in the mode of action but all effect a reduction in size sufficient for further processing. In the UK pre-breaking by grinding is probably the most common, using a kidney-shaped pre-break plate and 'knife'. Most models require the meat to be tempered before pre-breaking. However, there is an attraction of pre-breaking blocks directly from the freezer store and the more powerful machines can do so, but at the expense of product quality. Product cohesiveness is adversely affected when meat is pre-broken by grinding at low temperatures, and accompanied by increased cooking losses (Ellery, 1985; Jolley et al., 1986), probably through damage to fat cells generating more liquid fat during cooking (Evans and Ranken, 1975), which could interfere with product cohesiveness (Jolley and Purslow, 1988).

17.2.4 Flaking

Flaking appears to be the method of choice for grillsteak manufacture, though ostensibly there is no reason why other methods of comminution could not be used. Comminution by flaking is relatively new, the equipment being developed in the 1960s by Urschel International, based on Comitrol machines used originally for cutting, slicing and dicing vegetables. The main purpose of comminution is to reduce the size of sinew and gristle, which would otherwise result in objectionable toughness. Fat is also comminuted and not easily discernible in the final product, especially using smaller flaking heads, even at high fat contents.

Cutting is effected by an impeller that rotates at 3000 rpm (50 revs a sec) and forces meat against a stationary cutting head (Fig. 17.2). The severity of comminution can be controlled by selecting the appropriate cutting head which can vary in aperture size (ranging from 60 to 1600 thousandths of an inch, i.e., 1.5 to 40.6mm) and the number of cutting stations (Anon., 1980). It is usual, where meats of different quality are used, to use a smaller aperture size (<120) for low-quality meats (with high fat and connective tissue levels) and a larger aperture size (180, 240 or 390) for the better quality meats (Anon., 1980).

17.2.5 Mixing

Salt, typically 0.5–1%, is added during mixing, together with any other added ingredients such as rusk, phosphate, onion powder, wheat gluten, caseinate, dextrose, sulphite (as preservative), soya flour, concentrate or isolate, MSG, pepper and spices. Water, added to cheapen formulation costs, may necessitate the addition of rusk or soya to reduce cooking losses. A small quantity of added

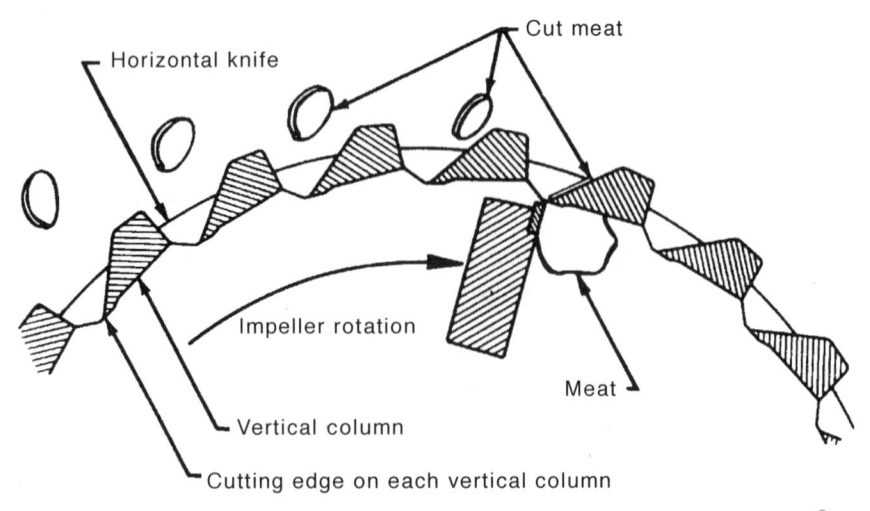

Fig. 17.2 Schematic diagram illustrating the cutting principle of the Comitrol® processor by Urschel Laboratories (Anon., 1980 with permission).

water may be used to compensate for evaporative losses during processing, with little effect on eating quality or cook losses. Salt and phosphate have a synergistic effect when used together, usually at a ratio of about 1:4. Flavour constraints limit the amount of salt that can be added to about 1.5%.

Initially the comminuted meat mass is relatively free flowing but as mixing proceeds becomes tacky and tends to 'ball together' due, presumably, to protein extraction. Mixing times vary but typically are around five minutes, depending on the product and type of mixer. This allows sufficient time to distribute any added ingredients uniformly through the meat mass and, more importantly, to solubilise myofibrillar proteins. Though it is difficult to prove, using analytical techniques, protein solubilisation can be inferred from the clear effects that added salt has on cooking losses and cohesive strength.

Short mixing times are usually associated with a loose, friable texture whilst excessive mixing times usually result in an objectionably rubbery texture. Optimal mixing times are usually determined by trial and error or a subjective appraisal of the appearance of the mix. Invariably, optimal mixing times vary from experiment to experiment (Booren *et al.*, 1981a, b; Coon *et al.*, 1983; Durland *et al.*, 1982; Noble et al., 1985; Pepper and Schmidt, 1975) indicating that the 'best' mixing time may depend upon the properties of the meat being mixed and also vary from machine to machine.

17.2.6 Forming
The use of a high-speed patty former is the most popular means of imparting shape to the finished product and achieving accurate portion control. A wide range of formers are available but all operate on roughly similar principles, employing a reciprocating mould plate and a means of transferring the meat from a hopper and to the mould. Mould plates (round, steak or rib-shaped, etc.) can be changed according to the product required. Most formers operate well over a relatively small temperature range, outside of which weight control becomes more variable and product can become ragged (Koberna, 1986). This presumably is related to the influence of temperature on the viscosity and flow behaviour of the meat mass.

An alternative method of forming is to extrude the meat into a log which is then tempered and sliced, as required. In the production of Bernard Matthews turkey roasts, lean turkey meat is co-extruded with a coating of fat in cylindrical form for subdivision into joints or steaks. Some products may be battered, coated and fried.

17.2.7 Freezing
Salt does not produce a good bind in the raw state and so most products are therefore frozen, usually cryogenically, using liquid nitrogen or liquid carbon dioxide, sprayed onto the food. Cryogenic freezing is rapid and has the advantage of reducing evaporative losses (Tomlins, 1995). However, it is

expensive and some companies now use conventional blast freezing and compensate for the higher evaporative losses by adding water.

Salt is a pro-oxidant increasing the rate of lipid oxidation (see, for example, Chen et al., 1984) and also discolouration. Even at freezer temperatures, salt-treated comminuted meat can discolour within a few days so most products are packaged in cardboard outers. Frozen product can usually be stored for up to three months, without any adverse effects on flavour due to lipid oxidation.

17.3 Factors affecting product quality: temperature, ice content, particle size and mechanical properties

Conceptually, a restructured meat product can be viewed as comprising meat pieces in a matrix of solubilised meat protein (Fig. 17.3). This model, though simplistic, suggests three key factors that might affect eating quality: (i) the nature of the particles (their size, shape, surface morphology and fibre direction) including their orientation and composition, (ii) the amount and composition of the exudate and (iii) the relative proportion of pieces to matrix. Thus, some authors have suggested 'optimal' size reduction and 'optimal' adhesion between meat pieces ('bind') to be key determinants of product quality (Jolley and Purslow, 1988). In order to control and manipulate particle size and adhesion, it

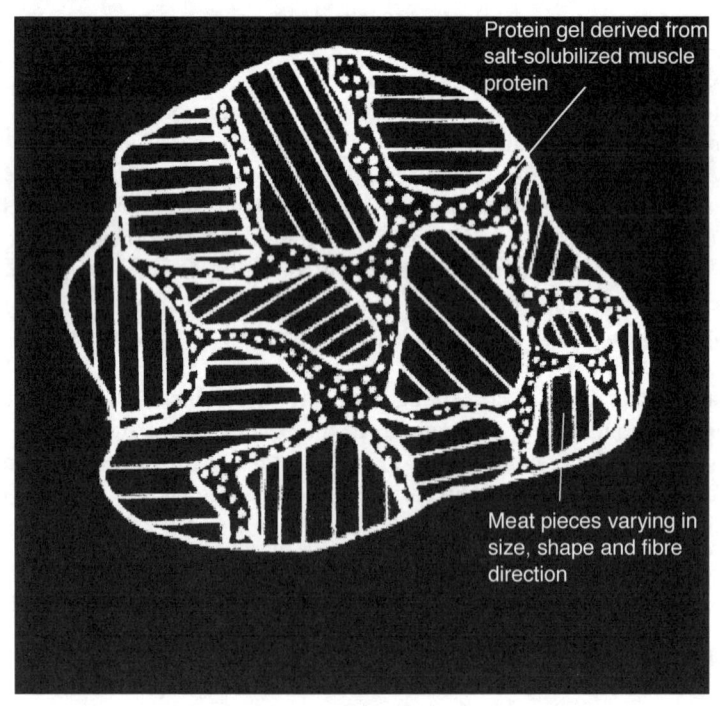

Fig. 17.3 Schematic diagram of a restructured meat product (Adapted from Jolley and Purslow, 1988).

is important to understand the factors affecting them. As well as having a direct effect on eating quality, particle size also affects the appearance of the product, and the available surface area for protein extraction. In grillsteak manufacture, the size reduction is a two-stage operation, involving pre-breaking (which achieves a relatively coarse comminution) and flaking (a fine comminution). The temperature of the meat is critical because of its effects on ice content and mechanical properties.

17.3.1 Ice content

Lean meat at slaughter contains about 75% water. The proportion which freezes depends on the temperature and can be expressed by the equation $I = 1 - ifp/T$ where I is the fraction of freezable water, ifp is the initial freezing point (about $-1°C$ in lean meat) and T is the temperature in °C (Fig. 17.4). The rationale for this behaviour is that solutes naturally present in the meat reduce the freezing point and these become progressively concentrated in the unfrozen liquor, resulting in an ionic strength of about 1M at $-5°C$ and 2M at $-15°C$ (Offer and Knight, 1988b). It is evident from Fig. 17.4 that some water remains unfrozen even at very low temperatures ($< -30°C$); also, the ice content changes most rapidly between -1 and $-10°C$, i.e., the temperature range in which processing takes place. Salt causes a depression of the initial freezing point (Table 17.1) which has a marked effect on ice content (Fig. 17.4).

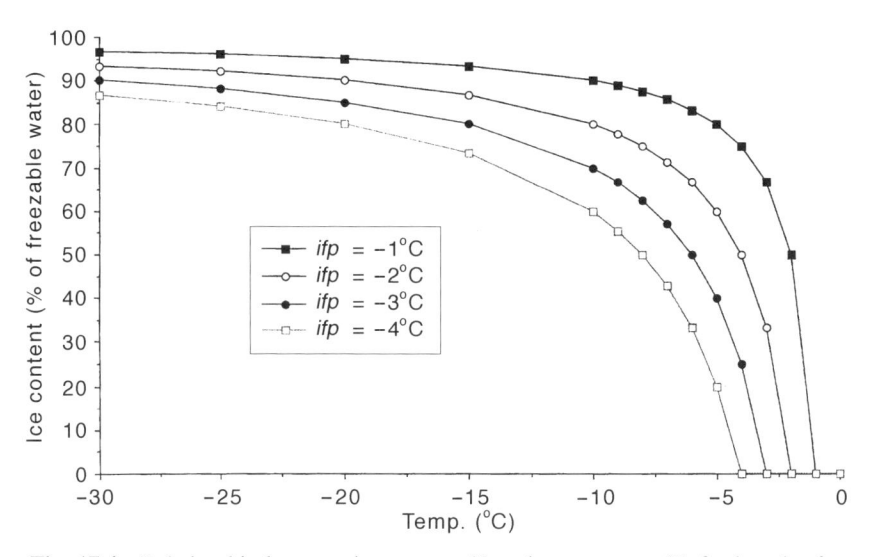

Fig. 17.4 Relationship between ice content (I) and temperature (T) for lean beef at initial freezing points (ifp) of -1, -2, -3 and $-4°C$ calculated using the expression $I = (1 - ifp/T) \times 100\%$.

340 Meat processing

Table 17.1 Effect of sodium chloride and sodium tripolyphosphate on the initial freezing point (°C) of meat (Sheard et al., 1990c)

Phosphate level (%)	Sodium chloride level (%)			
	0	1.0	4.0	means
0	−0.82	−1.53	−4.24	−2.20
0.25	−0.96	−1.57	−4.31	−2.28
0.50	−0.92	−1.82	−4.46	−2.40
means	−0.90	−1.64	−4.34	−2.29

17.3.2 Changes in ice content and temperature during processing

Temperatures usually rise during processing, whilst the ice content falls (Sheard et al., 1989, 1990b, 1991a, b). However, the *ifp* is lowered by the addition of salt during mixing – causing ice to melt – and it is not uncommon to see the temperature fall by 0.5 to 1°C during mixing with salt (Sheard et al., 1990c). Temperatures at the end of mixing and forming are typically around −2°C and unlikely to rise much above this unless product has been standing for long periods due to equipment failure. Differences in pre-breaking temperature can lead to relatively large differences in ice content at the end of the processing line, though differences in temperatures are only small and difficult to measure. Small temperature differences at the end of the processing line often reflect large variability in temperature at the beginning of the line. Unless temperature is well controlled at the tempering and pre-break stage, variation in product quality can be expected. Temperature measurements at the tempering stage are therefore essential for quality control purposes.

17.3.3 Size reduction

The factors influencing particle size have been reviewed elsewhere (Sheard et al., 1990a), are summarised in Table 17.2 and illustrated in Table 17.3. Particle size was measured by collecting particles on a pre-weighed perspex sheet, analysed using a video image analysis technique and presented in various ways (size distribution, number of particles per g, surface area per g or mean thickness). The technique is rapid and powerful providing particles are discrete; if not, as with mince, particle size may be measured using a wet sieving technique (Sheard et al., 1991a).

Meat pre-broken by grinding ranged in size from tiny fragments of less than 1mm in diameter to large, irregularly shaped pieces approximately 4–5 cm in diameter (Ellery, 1985; Sheard et al., 1989). Temperature had a major effect on particle size (Sheard et al., 1989, 1990b), even small differences (−5 and −3.5°C) leading to significant differences in particle size (Sheard et al., 1989); lower temperatures producing a larger proportion of smaller particles. There is a lack of information on the characteristics of particles produced by other methods of pre-breaking.

Table 17.2 Factors potentially affecting the particle size of pre-broken and flake-cut meat

1. Pre-breaking	
Temperature	Major effect. Cold meat is more brittle (Sheard *et al.*, 1989, 1990b)
Raw material	Small differences (Sheard *et al.*, 1989)
2. Flaking	
Aperture size	Major effect. Aperture size affects diameter and particle thickness (Sheard *et al.*, 1990a, 1991b)
Temperature	Major effect. Cold meat is more brittle. Temperatures >1°C produces mince-like strands (Sheard *et al.*, 1990a, b).
Pre-breaking	Major effect. Meat pre-broken by grinding breaks up far more readily than intact meat pieces (Sheard *et al.*, 1990b)
Meat species	Small differences between lean beef, turkey breast and turkey thigh meat (Sheard *et al.*, 1990a)
Number of cutting stations	Affects particle thickness (Sheard *et al.*, 1991a)
Impeller speed and design	Minor effects (Sheard *et al.*, 1991a)

Table 17.3 Particle size of meat pre-broken by grinding at −3 and −7°C, re-tempered and then flaked at −3 and −7°C (Adapted from Sheard *et al.*, 1990b)

Method of comminution	temp (°C)	aperture size (mm)	no./g	area/g (mm^2/g)	thickness (mm)
Pre-breaking	−7	—	2.7	410	—
	−3	—	0.8	221	—
Flaking	−7	6.1	368	2420	0.43
		19.0	108	1251	0.80
	−3	6.1	164	939	1.08
		19.0	30	651	1.55

Compared to meat pre-broken by grinding, flaking results in a huge increase in the number and surface area of particles (Table 17.3). Individual particles may be fractions of a mm thick (i.e. only a few fibres thick), ranging from <1mm to about 10 mm in the longest dimension, with an average particle weight of 0.01g (at 100 particles/g). One might expect such small dimensions to have implications for texture and eating quality. Temperature and aperture size appear to be the most important factors to influence particle size, though not the only ones (Sheard *et al.*, 1990a, b, 1991a, b) (Table 17.2).

Providing the temperature is low enough (usually less than −2°C), flaking appears to result in a true cutting action, leading to discrete flake-shaped (i.e. broad and thin) particles, resembling fine snow particles (Sheard *et al.*, 1990a). However, flaking at higher temperatures results in the production of mince-like strands, very different from those produced at lower temperatures (Sheard *et al.*,

1991a). The use of high-speed photography illustrates this difference in cutting action very clearly. At low temperatures, individual pieces of meat are cut very quickly (within about a third of a revolution or 0.007 s) but at higher temperatures individual pieces deform against the cutting head and are carried around the inside of the flaking head for many revolutions without being cut before emerging as mince-like strands (Sheard *et al.*, 1990a).

17.3.4 Mechanical properties

In order to understand and predict the effects of high-speed processing on the behaviour of meat during comminution, various authors have investigated the mechanical properties and fracture behaviour of meat under different conditions, to determine tensile strength, work of fracture and other properties using traditional ways employed by materials scientists (Munro, 1983; Dobraszczyk *et al.*, 1987). In common with many biological materials (Ashby, 1983; Atkins, 1987; Atkins and Mai, 1986), the mechanical properties of meat depend on temperature, water content (or ice content), the strain rate (i.e. the rate at which the sample is deformed) and the fibre direction (Dobraszczyk *et al.*, 1987; Munro, 1983; Purslow, 1985). At low temperatures ($< -15°C$) meat behaves in a brittle way (breaking suddenly in such a way that the broken ends may be refitted to regain essentially the original dimensions) but exhibits viscoelastic behaviour (in which samples exhibit extensive deformation before eventual fracture) at higher temperatures ($> -10°C$) (Dobraszczyk *et al.*, 1987; Munro, 1983). Water, it seems, acts as a plasticiser, allowing the material to deform to a greater extent before fracturing (Atkins, 1987). Munro (1983) showed that meat is highly anisotropic (i.e. the structure and properties of the material depend on fibre direction) above the *ifp* but only slightly so below the *ifp* where the ratio of tensile strengths falls from 5:1 (above the *ifp*) to 2:1. This observation is important because a highly anisotropic material is more likely to produce particles with a preferred fibre direction than a material that is isotropic or slightly anisotropic.

Brittle behaviour is favoured by high strain rates (Atkins and Mai, 1986), though most workers have used strain rates much lower than those encountered during meat processing (Dobraszczyk *et al.*, 1987; Munro, 1983). Recent results from this laboratory, using a Charpy impact test, and strain rates comparable to those used in high-speed comminuting equipment, are shown in Fig. 17.5. Below $-10°C$ fracture was complete and, from the surface appearance, specimens apparently failed in a brittle fashion. At intermediate temperatures (-10 to $-5°C$), fracture was complete but was accompanied by an increase in the amount of energy absorbed which can be attributed to an increase in the amount of energy absorbed plastically, resulting in greater deformation. At still higher temperatures ($> -5°C$), fracture was incomplete, and non-existent at -2 to $-3°C$ when specimens merely deflected on impact. There was an anisotropy of impact energy according to fibre orientation of about 2:1; this is far less than in thawed (Munro, 1983) or cooked meat (Purslow, 1985) and is probably

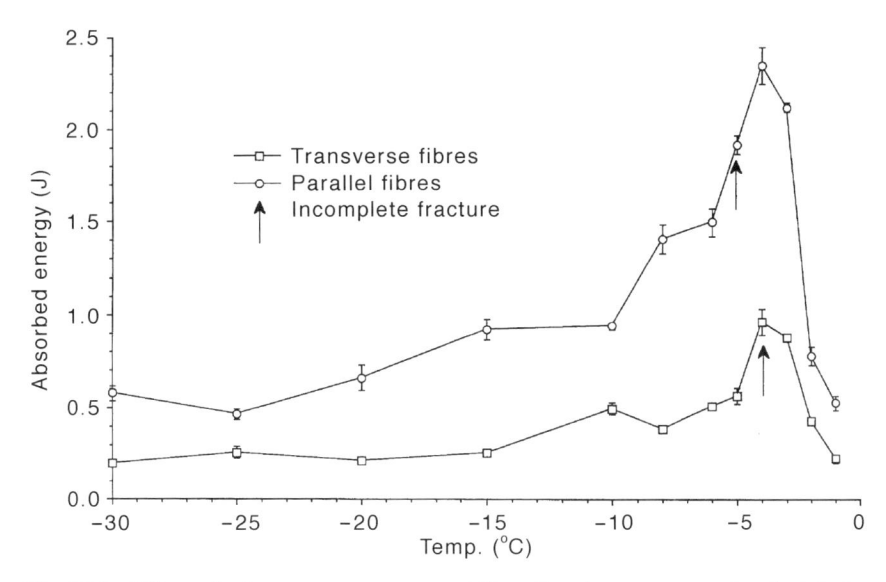

Fig. 17.5 Effect of sub-zero temperatures and fibre direction on energy absorbed during Charpy impact testing of frozen and semi-frozen meat.

insufficient to result in a preferred fibre direction during comminution. It can be inferred that at low temperatures (−10°C) fracture is easily induced but poorly controlled, leading to shattering; at high temperatures (close to the *ifp*) fracture is difficult to induce without causing extensive deformation, whilst at intermediate temperatures (−5 to −10°C) fracture can be induced but without causing extensive shattering and is therefore easier to control.

17.4 Factors affecting product quality: protein solubility and related effects

As mentioned earlier, the re-assembly of meat pieces once they have been comminuted is an essential requirement in restructuring. Good adhesion ('bind') between meat pieces is widely regarded as a key determinant of quality, usually achieved by adding salt, sometimes in conjunction with phosphate, added dry during mixing. Other means of achieving adhesion are discussed later. In addition to its effect on adhesion, salt also increases water retention, reduces cooking losses and may also increase tenderness slightly. The underlying mechanism involves depolymerisation of myosin and dissociation of actomyosin, at appropriate concentrations, which in turn (i) allows expansion of the myofibrillar lattice (Offer and Knight, 1988a), thus improving water retention characteristics and reducing weight loss on cooking, (ii) effects adhesion on cooking via gelation of the solubilised myosin which binds the constituent meat pieces together and (iii) effects a tenderisation partly due to the

Table 17.4 Factors affecting the extraction of myofibrillar protein from meat and adhesion between meat pieces

Properties of raw meat ingredients		
	Rigor state	Myofibrillar proteins more easily extracted in pre-rigor state
	Muscle type	Poor quality meats have low salt extractable protein contents (Saffle and Galbreath, 1964)
	Fresh or frozen	Conflicting results (e.g. compare Saffle and Galbreath, 1964 and Acton and Saffle, 1969)
Processing conditions		
	Particle size	Reducing meat particle size improves protein extraction (Acton, 1972)
	Extraction time	More protein extraction at longer times
	Temperature	Conflicting results (compare Bard, 1965 and Gillet et al., 1977)
Addition of salt		Causes depolymerisation of myosin and dissociation of actomyosin at appropriate concentrations
Addition of phosphate		Causes depolymerisation and dissociation, after enzymic hydrolysis to pyrophosphate, the active form. Hydrolysis at 0°C is slow
Adhesion		
	Protein type	Myosin is better than actomyosin. Sarcoplasmic proteins have little adhesive strength
	Fibre orientation	90/90 better than 0/0 or 90/0

increased water retention and partly as a result of the disassembly of the myofibrillar filaments.

In order to control and manipulate the beneficial action of salt and phosphate on the properties of restructured meats, it is important to understand the factors affecting them (Table 17.4). These have been reviewed in detail elsewhere (Jolley and Purslow, 1988) and need not be covered here, other than to highlight the following issues.

17.4.1 Effective salt concentration

Myosin is the most abundant muscle protein (Lawrie, 1998) and exists as discrete thick filaments in the myofibril. Its solubility with respect to ionic strength and pH have been investigated widely and it is well known that myosin is insoluble at physiological ionic strength (0.15–0.2M) and high ionic strengths but soluble at intermediate ionic strengths 0.3–0.6M at pH 5.5. This behaviour of myosin produces a curvilinear relationship with increasing salt concentration: with zero extraction at very low and high ionic strengths and maximum extraction at 5–10% NaCl (Bard, 1965; Callow, 1932; Grabowska and Hamm, 1979) i.e., much higher salt levels than those found in restructured meats. Similar relationships with increasing salt concentration have been found for water holding and cooking loss (Callow, 1932; Wierbicki et al., 1957). Other studies in model systems have shown that meat swelling is a highly co-operative phenomenom and no myosin extraction occurs below 0.6M (i.e. about 3% sodium chloride (Offer and Trinick, 1983). It seems reasonable, therefore, to question the assumption that myosin is solubilised in restructured meats where the salt concentration, 0.5–1% (0.17M), is much lower than the minimum required to evince any myosin extraction.

This apparent conflict has been resolved by arguing that the effective salt concentration during mixing is much higher than to be expected on a simple weight basis due to localised salt concentrations, particularly during the initial stages of mixing (Jolley and Purslow, 1988). Complete equilibration of salt within the time allowed for mixing, typically about five minutes, is unlikely given that salt diffusion through meat is slow (Sheard et al., 1990c). In addition, this concentrative effect would be exacerbated below the initial freezing point when large quantities of water are present as ice and, therefore, unavailable as a solvent. In other words, comparisons between model systems and the conditions pertaining during manufacture are quite different due to differences in the solvent:meat ratio. Given the observable effects of added salt on meat binding, water holding and cooking losses (see later), it seems highly likely, therefore, that some, though not necessarily maximal, protein extraction occurs during the manufacture of restructured meats.

17.4.2 Effect of temperature

The influence of temperature on the extractability of muscle protein has been a matter of some controversy. Gillett et al. (1977) found that maximum protein extraction at 7°C, was about a third higher than at −4°C or 20°C, suggesting that processing above the *ifp* may be beneficial as far as protein extraction is concerned. By contrast, Bard (1965) reported maximum protein extraction at −5°C, with a two- to three-fold reduction at 0°C. Recent data from this laboratory, investigating the influence of temperature (−7 to 15°C) across a range of salt concentrations (0 to 5M), is helpful in understanding these contradictory results. The amount of protein solubilised increased from about 1 mg/ml at 0M NaCl to a maximum at 2M NaCl which declined at higher salt concentrations (Fig 17.6). At 2M NaCl, salt-induced protein solubilisation increased with decreasing temperature but this relationship did not apply at most other salt concentrations.

17.4.3 Phosphate hydrolysis

Phosphate, like salt, is a mild structure breaker but its chemistry is more complicated. Though it can be used alone to improve water retention, juiciness and tenderness at levels of 0.25–5 g/100g (Sheard et al., 1999), in restructured meats it is usually used in conjunction with salt. Commercial mixtures may include pyrophosphate and tripolyphosphate as well as longer chain and cyclical

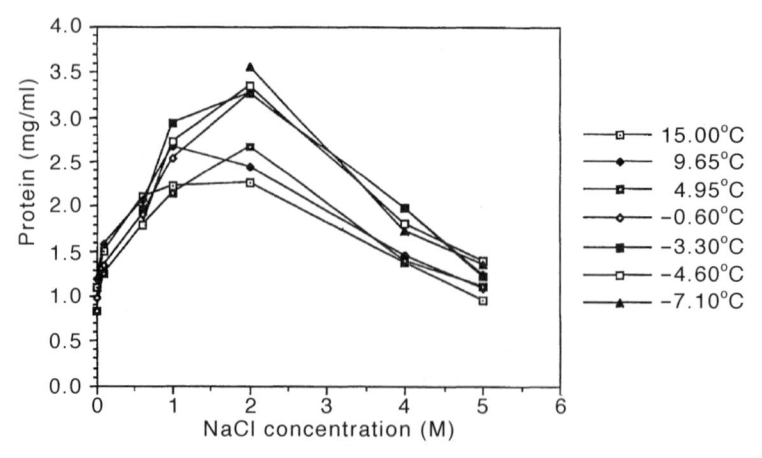

Fig. 17.6 Effect of temperature on the amount of protein solubilised from meat in different concentrations of sodium chloride (from Sheard and Savage, 1998).

phosphates (Iles, 1973). The active form is believed to be pyrophosphate (the diphosphate form, $(P_2O_7)^{4-}$) (Yasui et al., 1964) which is produced by hydrolysis from tripolyphosphate $((P_3O_{10})^{5-})$ and is itself hydrolysed to orthophosphate (the monophosphate form, PO_4^{3-}). The hydrolysis depends on pH, temperature and the ionic species. In beef, the tripolyphosphatase activity of myosin is about 15 times that of the diphosphatase activity (Neraal and Hamm, 1973), thus favouring the formation of the active form, pyrophosphate. The pH optima for tripolyphosphatase and diphosphatase, 5.6 and 6.7 respectively, likewise favours the formation of pyrophosphate at the normal ultimate meat pH (about 5.5), as does sodium chloride which increases the tripolyphosphatase activity but decreases the diphosphatase activity (Belton et al., 1987; Neraal and Hamm, 1973). The rate of hydrolysis doubles with every 10°C rise until above 40°C the enzyme is progressively denatured (Neraal and Hamm, 1973). At 20°C, 0.5% tripolyphosphate is broken down in about 8 to 20 minutes, whilst 0.5% diphosphate requires 2 to 15 hours (Neraal and Hamm, 1973). The hydrolysis appears to be much slower at 0°C (Neraal and Hamm, 1973; Sutton 1973). Given the above, one might not expect the hydrolysis to orthophosphate to be complete within normal processing times though it seems there is sufficient pyrophosphate, even at 0°C, to exert measurable effects on meat binding, cook loss and other properties.

17.4.4 Adhesion

Studies, using 'model' adhesive junctions, in which the adhesive is sandwiched between two meat pieces, have shown that myosin is a stronger adhesive than actomyosin (Macfarlane et al., 1977) and that the strength of the joint varies with the orientation of the muscle fibre (Purslow et al., 1987). It seems likely that the binding strength of restructured meat products will be determined by the weaker junctions which will be the first to fail during mastication. Sarcoplasmic

Table 17.5 Effect of sodium chloride and sodium tripolyphosphate (TPP) on cooking losses (%) from beef grillsteaks (Adapted from Sheard et al., 1990c)

TPP level (%)	NaCl level (%)					
	0	0.5	1.0	2.0	4.0	means
0	34.8	33.0	31.9	31.3	31.2	32.4
0.25	33.0	31.5	25.8	25.6	26.7	28.5
0.50	29.2	25.6	24.8	23.1	25.7	25.7
means	32.2	30.0	27.5	26.7	27.9	28.9

Table 17.6 Effect of salt on TBA values and cooking losses on cooked pork grillsteaks (Adapted from Schwartz and Mandigo, 1976)

	Salt (%)			
	0	0.75	1.50	2.25
TBA[a]	0.11	0.50	0.84	0.94
Cooking loss (%)	37.5	21.0	14.6	13.4

[a] mg malonaldehyde per kg meat

proteins are poor binders and probably play little part in meat binding except in situations where little or no myofibrillar protein is solubilised (Jolley and Purslow, 1988). However, acceptable products can be produced without the addition of salt or phosphate. In this case, adhesion is probably achieved partly by physical entanglement pieces and the aggregation of sarcoplasmic proteins.

17.4.5 Cooking losses
Cooking losses are important because cook loss not only affects the final weight and, therefore, portion size but also the perceived juiciness of the product and possibly texture as well. Losses can vary widely depending on product formulation, the conditions of processing and the cooking method. For a given cooking method, salt and phosphate levels are probably the most important factors (Mandigo, 1982; Moore et al., 1976) and, together, they have a marked effect on weight loss during cooking (Table 17.5), though salt has an adverse effect on the development of oxidative rancidity (Table 17.6). The type of meat is also important (Table 17.7), possibly because more myosin is extracted from white fibres than red fibres under equivalent conditions of ionic strength and pH (Xiong, 1994; Xiong and Brekke, 1989).

17.5 Factors affecting product quality: cooking distortion
Cooking distortion is a persistent problem with products – including burgers and grillsteaks – formed using a high-speed patty former (Jolley and Rangeley,

Table 17.7 Cooking losses (%) from UK-style grillsteaks made from beef forequarter, turkey breast and turkey thigh meat formulated with salt (0 or 0.75%), sodium tripolyphosphate (TPP) (0 and 0.25%) and water (0 and 2%) (University of Bristol, unpublished data)

Treatment	Beef	Breast	Thigh	Means
None	32.7	22.8	29.6	29.4
Water	37.4	23.0	32.2	31.6
Salt	26.6	20.2	22.1	24.0
TPP	29.7	18.0	23.7	24.3
Water + salt	28.6	20.0	21.8	24.1
Water +TPP	29.7	22.9	25.7	26.8
Salt + TPP	25.4	18.8	17.9	20.6
Water + salt +TPP	23.6	20.5	20.2	21.9
Means	30.2	20.8	25.0	25.3

1986). The most common type is most readily seen in burgers that, on cooking, shrink preferentially in one direction and, thus, adopt an oval shape. This is sometimes referred to as uneven lateral shrinkage. Using ultraviolet light (which causes fat and connective tissue to fluoresce), Mounsdon and Jolley (1987) demonstrated that lateral shrinkage is due to alignment of connective tissue on the surface of the product. This was attributed to friction generated between the surface of the product and the reciprocating plate during patty forming. On cooking, the connective tissue shrinks, generating tension and leading to preferential product shrinkage in the direction of alignment.

Shrinkage during cooking can also lead to an increase in height (doming) of the product and the development of an air-filled vacuole within the middle of the product. Sometimes fluid accumulates in the vacuole, a phenomenon sometimes referred to as 'welling'. Doming and welling are less common than lateral shrinkage, more variable in nature and difficult to control. Some types of flame-grilled grillsteaks seem especially vulnerable to distortion. The causes of doming and welling were investigated in two factory-based trials conducted by Jolley and Rangeley (1986). Selected results from the first trial, shown in Table 17.8, demonstrate that all patties shrank more along the base (27–35%) than the width (14–21%), indicating that the connective tissue was aligned preferentially along the length of the product. The height increase in the worst case was 41%, with fluid losses on puncturing up to 6.7g. The largest single influence on distortion was the design of patty former (of the two formers used, one was significantly worse than the other), though the causes were highly interactive (Table 17.9). This explains why the problems are not easy to control and suggests that the results obtained in one system may not necessarily apply elsewhere due to machine specificity. Perhaps more importantly, the results also indicate potential difficulties in scaling up from laboratory-based trials to factory conditions. Thus, optimal conditions pertaining under laboratory conditions may not be optimal in the factory situation.

Processing and quality control of restructured meat 349

Table 17.8 Effect of former type (1 and 2), delay between blending and forming (0 or 35 min.) and effectiveness of freezing ('fast', 'slow') on dimensional changes and weight of fluid collected on puncturing (welling) for UK-style grillsteaks (Jolley and Rangeley, 1986 with permission)

Former	1				2			
Delay (min.)	0		35		0		35	
Freezing rate	fast	slow	fast	slow	fast	slow	fast	slow
Shrinkage, base (%)	30	31	35	32	28	29	27	27
Shrinkage, width (%)	15	21	22	20	14	14	16	14
Doming (%)	23	31	41	33	12	14	8	9
Welling (g)	3.1	6.7	3.3	6.4	2.0	1.6	1.6	1.2

Table 17.9 Level of significance of main and primary effects (Jolley and Rangeley, 1986 with permission)

	Former	Delay	Freezing	Former × delay	Former × freezing	Delay × freezing
Shrinkage, base (%)	***					
Shrinkage, width (%)	***	X	(X)			**
Doming (%)	X	(X)	(X)	***		*
Welling (g)	X		X		***	

* $p < 0.05$, ** $p < 0.01$, *** $p < 0.001$
X main effect significant ($p < 0.05$ or better), but significant interaction with at least one other main effect
(X) main effect not significant on its own but involved in one or two significant interactions

In addition to the conditions pertaining during manufacture, cooking method, cook temperature and product thickness are also important factors (Campbell and Mandigo, 1978; Campbell *et al.*, 1977), presumably because of their influence on time-temperature regimes during cooking and, thus, the rate of shrinkage. Thicker products were more prone to height increase, especially at higher temperatures (Table 17.10).

17.6 Sensory and consumer testing

A wide range of sensory tests – ranking, category and profiling techniques – have been used to assess the tactile, appearance, texture, flavour, juiciness and hedonic (i.e. liking) attributes of restructured meats (e.g. Berry and Civille, 1986; Cardello *et al.*, 1983; Ford *et al.*, 1978). The texture profiles that have been developed (e.g. Berry and Civille, 1986; Cardello *et al.*, 1983; Savage *et al.*, 1990) use similar descriptors to those developed for burgers (e.g. Berry and Leddy, 1984; Dransfield *et al.*, 1985). Several authors, of which Cardello *et al.* (1983) and Berry *et al.* (1987) are the most convincing, have shown that texture,

Table 17.10 Influence of product thickness and cooking temperature on dimensional changes and weight losses in restructured pork patties cooked to 77°C on a rotary hearth oven (Adapted from Campbell *et al.*, 1977)

	Portion thickness (cm)		
	1.27	1.90	2.54
Thickness (%)	106.8	110.9	113.0
Cook loss (%)	40.5	35.6	31.9
	Cook temperature (°C)				
	149	177	205	233	261
Thickness (%)	97.5	107.6	113.4	115.0	117.5
Cook loss (%)	36.8	32.9	37.2	35.7	37.3

appearance and acceptability, as assessed by a trained sensory panel, depends upon aperture size. Fat becomes less detectable visibly with decreasing aperture size, as does the amount of connective tissue perceived during mastication. Other studies have shown that product cohesion, among other factors, depends upon the level of salt or added myosin (Ford *et al.*, 1978; Savage *et al.*, 1990), with perceived rubberiness increasing with level of myosin. As might be expected, juiciness and moisture release increase with increasing fat level (Berry *et al.*, 1985).

Consumer tests, involving large numbers (typically in excess of 100) of untrained assessors, are concerned primarily with preference or liking decisions (Nute, 1996). In one trial, the adhesion between meat pieces was systematically varied by adding different concentrations of crude myosin (0, 1.75, 3.5, 5.25 and 7% protein) (Savage *et al.*, 1990). Tensile adhesive strength measurements increased almost linearly with increasing levels of added myosin. However, consumer preference varied between individuals; some preferred weakly bound products whilst others preferred products that were strongly bound, and there was no overall preference for any one product. Clearly, these results challenge the assumption that an increase in adhesion necessarily results in a 'better' product.

In another trial (Nute *et al.*, 1988), eight formulations of restructured steaks were assessed for texture, saltiness, juiciness, taste, meatiness and overall liking by consumers in two regions of the UK (north and south). Steaks varied in fat level (12 and 20%), salt level (0.5 and 1%), temper (long and short) and mixing time (6 or 12 mins.). Analysis of variance revealed that consumers were able to perceive differences in saltiness and juiciness, although there was no significant difference in overall liking. However, internal preference mapping showed that consumers in the south could be segmented into four categories according to their preference for fat and salt level (high fat and high salt, high fat and low salt, low fat and high salt, low fat and low salt).

In another trial, consumers were asked to evaluate restructured beef steaks made using different aperture sizes, ranging from 1.5 mm to 40.6mm (Cardello *et al.*, 1983). All the flaked products were regarded as being different from ground beef patties and, also, from intact muscle (ribeye steak). However, there was no significant difference in overall acceptability, acceptability of texture, flavour or appearance between any of the flaked products, despite the large range of aperture sizes used, and it is tempting to speculate that this, too, is due to differences in consumer preference.

Taken together these trials challenge the idea of generating an 'optimal' product (in terms of separate characteristics such as bind, texture, or flavour or, indeed, in terms of the product's overall characteristics) since consumers vary markedly in their individual preferences. Of course, there is considerable scope for generating products with particular characteristics (e.g. a good bind or a firm bite) and marketing these in such a way to target certain segments of the population in order to match product characteristics with individual preferences.

17.7 Future trends

17.7.1 Fresh product

Salt is the ingredient of choice to bind restructured meats. There is, however, growing consumer pressure against the use of salt in foodstuffs, for medical reasons. Technically, also, salt has its drawbacks, being a pro-oxidant it accelerates the production of metmyoglobin, thus reducing colour stability and accelerates the rate of lipid oxidation. This, and the fact that product cohesion is poor in the raw state, effectively precludes the marketing of restructured meats in the chilled state. Among consumers there is a perception that fresh is better than frozen and consumers are willing to pay a premium for fresh product.

The potential to market grillsteak-type products in the fresh state is illustrated by the growth in sales of freshly chilled burgers whose manufacture is very similar to that of grillsteaks, and, like grillsteaks, traditionally have sold in a frozen form. In developing a market for fresh grillsteak type products, the main requirement is to maintain colour stability and minimise bacterial growth. The former can be achieved satisfactorily by using alternative binders to salt, that do not accelerate the formation of metmyoglobin, and using an appropriate packaging system to extend the colour shelf-life (e.g. a modified atmosphere pack containing 70% oxygen and 30% carbon dioxide). Bacterial growth can be minimised by using sodium bisulphite.

One innovative solution to overcoming these problems involves the use of the polysaccharide sodium alginate, whose use in meat for this application was first advocated in the patent of Schmidt and Means (1986). Cross-linking is achieved chemically, rather than thermally, between divalent cations, usually calcium, and the guluronic acid moities of alginate. Products made using alginate, packed under modified atmospheres retain a bright red, fresh colour for at least a week stored at 2°C (Richardson *et al.*, 1989). The process works best above the ifp, but once gel formation is complete, products can be frozen without impairing product cohesion. The rate of bind development, and its ultimate strength, in the raw and cooked state, can be controlled by the type of calcium salt, the amount of added alginate and alginate type (Trout, 1989; Richardson *et al.*, 1989). Weak acids such as citric acid or lactic acid can also be used to control the rate of gelation by accelerating the release of calcium. Although glucono-delta-lactone (gdl) is not itself acidic, it may also be used for this purpose as it is slowly hydrolysed in meat to gluconic acid; gdl also prevents the undesirable after-taste sometimes experienced with the alginate/$CaCO_3$ system described by Means and Schmidt (1986). The rate of gelation must be such that the gel is not broken down during mixing and forming (Richardson *et al.*, 1989). It is desirable to avoid the use of sodium chloride as this interferes with the development of bind, as does collagen (Richardson *et al.*, 1989).

Another system capable of producing cohesion in the raw state involves the use of extracted plasma thrombin and fibrinogen (Wijngaards and Paardekooper, 1988). This ingenious application of the blood clotting mechanism was developed at the Netherlands Centre for Meat Technology. Gel formation is the result of the conversion of fibrinogen into fibrin by the enzyme thrombin. Fibrin molecules, in turn, become covalently cross-linked by the action of transglutaminase, present in the partially purified fibrinogen, which also cross-links fibrin and collagen. An advantage of this system is that a well bound product can be made even where there is a high level of collagen. Gel formation is complete in about 12 hours depending on temperature and pH, the ultimate gel strength being determined by the fibrin concentration.

A third system employs purified transglutaminase (Kuraishi *et al.*, 1997; Nonaka *et al.*, 1989). Though relatively expensive, it is very effective binder. The enzyme may be added dry, using casein as a carrier, or added with water. The rate of bind development depends on enzyme concentration, temperature and time. At chill temperatures, bind develops in three to six hours.

17.7.2 Engineered texture
Most process development and product formulation has been largely empirical, on a trial and error basis, without any real understanding of the underlying principles that govern end product quality. However, there are exceptions which deliberately set to out to overcome some of the obstacles encountered in developing a comminuted product with a steak-like texture. These so-called 'second generation' products (Jolley *et al.*, 1988), apparently based on the patent application of Bradshaw and Hughes (1986), overtly claim to have steak-like characteristics. They comprise a 'texture-imparting' phase of relatively large pieces and a 'succulence-imparting' phase of finely comminuted meat, presumably a lower grade raw material having a higher fat content. Alignment of the pieces reputedly occurs during forming. Though examples like this are few, the very fact that such products have been developed is encouraging and augurs well for the future.

17.8 Sources of further information and advice

This chapter has considered the processing of restructured meats, a subject that has been reviewed extensively elsewhere (Franklin and Cross, 1982; Jolley and Purslow, 1988; Pearson and Dutson, 1987; Sheard and Savage, 1998). In conclusion it is worth emphasising four points.

1. Good control over temperature is vital when processing below the *ifp* to avoid variability in product control. Practically, this is achieved by good temperature control at the start of the processing line (i.e. at the temperature/pre-break stage) and keeping conditions constant at later stages of processing.
2. Some quality problems are highly interactive and machine specific and, therefore, difficult to control. The results obtained in one system need not necessarily apply under other conditions.
3. There are enormous obstacles to producing a steak-like texture and this goal, perhaps, is unattainable using existing methods of comminution, mixing and forming.
4. Consumers vary widely in their individual preferences – for degree of bind, optimal level of fat and salt, etc. – and there is scope for targeting segments of the population to match product characteristics with their particular preferences.

17.9 References

ACTON J C (1972), 'The effect of meat particle size on extractable protein, cooking loss and binding strength in chicken loaves', *J Food Sci*, 37, 240–243.

ANON. (1980), *Facts, flakes and fabricated meats*, Urschel Labs Inc.

ASHBY M F (1983), 'The mechanical properties of cellular solids', *Metallurgical Trans* A, 14A, 1755–1768.

ATKINS A G (1987), 'The basic principles of mechanical failure in biological systems', in Blanshard J M V and Lillford P, *Food Structure and Behaviour*, London, Academic Press, 149–176.

ATKINS A G and MAI Y-W (1986), 'Deformation transitions', *J Materials Sci*, 21, 1093–1110.

BARD J C (1965), 'Some factors influencing extractability of salt-soluble proteins', *Proc Meat Industry Research Conference*, 96–98.

BELTON P S, PACKER K J and SOUTHON T E (1987), 'P Nmr studies of the hydrolysis of added phosphates in chicken meat', *J Sci Food Agric*, 39, 283–291.

BERRY B W and CIVILLE G V (1986), 'Development of a texture profile panel for evaluating restructured beef steaks varying in meat particle size', *J Sensory Studies*, 1, 15–26.

BERRY B W and LEDDY K F (1984), 'Effects of fat level and cooking method on

sensory and textural properties of ground beef patties, *J Food Sci*, 49, 870–875.
BERRY B W, SMITH J J and SECRIST J L (1985), 'Effects of fat level on sensory, cooking and instron properties of restructured beef steaks', *J Animal Sci*, 60, 434–439.
BERRY B W, SMITH J J and SECRIST J L (1987), 'Effects of flake size on textural and cooking propertries of restructured beef and pork steaks', *J Food Sci*, 52, 558–563.
BOOREN A M, JONES K W, MANDIGO R W and OLSON D G (1981a), 'Effects of blade tenderisation, vacuum mixing, salt addition and mixing time on binding of meat pieces into sectioned and formed beef steaks', *J Food Sci*, 46, 1678–1680.
BOOREN A M, MANDIGO R W, OLSON D G and JONES K W (1981b), 'Effect of muscle type and mixing time on sectioned and formed beef steaks', *J Food Sci*, 46, 1665–1668, 1672.
BRADSHAW N J and HUGHES D (1986), 'Meat product', European Patent application 0 175 397.
BREIDENSTEIN B C (1982), *Intermediate Value Beef Products*, Chicago, National Live Stock and Meat Board.
CALLOW E H (1932), 'The theory of curing', *Report of the Food Investigation Board for 1931*, 144–147.
CAMPBELL, J F and MANDIGO R W (1978), 'Properties of restructured pork patties as affected by cooking method, frozen storage and reheating method', *J Food Sci*, 43, 1648–1651.
CAMPBELL J F, NEER K L and MANDIGO R W (1977), 'Effects of portion thickness and cooking temperature on the dimensional properties and composition of restructured pork', *J Food Sci*, 42, 179–181.
CARDELLO A V, SEGARS R A, SECRIST J, SMITH J, COHEN S H and ROSENKRANS R (1983), 'Sensory and instrumental texture properties of flaked and formed beef', *Food Microstructure*, 2, 119–133.
CHEN C-C, PEARSON A M, GRAY J I and MERKEL R A (1984), 'Effects of some salt and some antioxidants upon the TBA numbers of meat', *Food Chem*, 14, 167–172.
COON F P, CALKINS C R and MANDIGO R W (1983), 'Pre- and post-rigor sectioned and formed beef steaks manufactured with different salt levels, mixing times and tempering times', *J Food Sci*, 48, 1731–1734.
CROSS H R, CURTIS GREEN E, STANFIELD M S and FRANKS W J (1976), 'Effect of quality grade and cut formulation on the palatability of ground beef patties', *J Food Sci*, 41, 9–11.
DOBRASZCZYK B J, ATKINS A G, JERONIMIDIS G and PURLSOW P P (1987), 'Fracture toughness of frozen meat', *Meat Sci*, 21, 25–49.
DRANSFIELD E, JONES R C D and ROBINSON J M (1985), 'Development and application of a texture profile for UK beefburgers', *J Texture Studies*, 15, 337–356.
DURLAND P R, SEIDEMAN S C, COSTELLO W J and QUENZER N M (1982), 'Physical

and sensory properties of restructured beef steaks formulated with various flake sizes and mixing times', *J Food Protection*, 45, 127–131.

ELLERY A (1985), Changes in ice content during the manufacture of grill-steaks and their influence on product quality, MSc Thesis, University of Bristol.

EVANS G G and RANKEN M D (1975), 'Fat cooking losses from non-emulsified meat products', *J Food Tech*, 10, 63–71.

FIELD R A (1982), 'New restructured meat products – foodservice and retail', in Franklin K R and Cross H R, *Proceedings of Meat Science and Technology*, Chicago, National Live Stock and Meat Board, 285–298.

FORD A L, JONES P N, MACFARLANE J J, SCHMIDT G R and TURNER R H (1978), 'Binding of meat pieces: objective and subjective assessment of restructured steakettes containing added myosin and/or sarcoplasmic protein', *J Food Sci*, 43, 815–818.

FRANKLIN K R and CROSS H R (eds) (1982), *Proceedings of International Symposium on Meat Science and Technology*, Chicago, National Live Stock and Meat Board.

GILLETT T A, MEIBURG D E, BROWN C L and SIMON S (1977), 'Parameters affecting meat protein extraction and interpretation of model system data for meat emulsion formation', *J Food Sci*, 42, 1606–1610.

GRABOWSKA J and HAMM R (1979), 'Protein solubility and water binding under the conditions obtaining in Bruhwurst mixtures', *Fleischwirtschaft*, 59, 1166–1172.

ILES N A (1973), *Phosphates in meat and meat products – a survey*, BFMIRA Scientific and Technical Survey No 81.

JAMES S J and CROW N (1986) 'Thawing and tempering: industrial practice', Proceedings of subject day, Meat thawing/tempering and product quality, Bristol, IFR.

JOLLEY P D and PURSLOW P P (1988), 'Reformed meat products – fundamental concepts and new developments', in Mitchell J R and Blanshard J M V, *Food Structure – its Creation and Evaluation*, Surrey, Butterworths, 231–264.

JOLLEY P D and RANGELEY W R D (1986), 'Practical factors influencing the shape of a cooked, flaked and formed beef product', *Proc European Meeting Meat Research Workers*, 32, 353–356.

JOLLEY P D, ELLERY A, HALL L and SHEARD P R (1986), 'Keeping your temper – why it is important for the manufacturer', Paper presented at IFR subject day on Meat Thawing, Tempering and Product Quality.

JOLLEY P D, HALL L D, MACFIE H J H and NUTE G R (1988), 'Current UK restructuring techniques: unit operations, ingredients and interactions', in Krol B, van Roon P P S and Houben J H, *Trends in Modern Meat Processing* 2, 125–129.

KOBERNA F (1986), 'Tempering temperature requirements for cutting and processing equipment', Proceedings of subject day, Meat thawing/tempering and product quality, Bristol, IFR.

KURAISHI C, SAKAMOTO J, YAMAZAKI K, SUSA Y, KUHARA C and SOEDA T (1997),

'Production of restructured meat using microbial transglutaminase without salt or cooking', *J Food Sci*, 62, 488–490, 515

LAWRIE R A (1998), *Meat Science* (6th edition), Cambridge, Woodhead Publishing Ltd.

MACFARLANE J J, SCHMIDT G R and TURNER R H (1977), 'Binding of meat pieces: a comparison of myosin, actomyosin and sarcoplasmic proteins as binding agents', *J Food Sci*, 42, 1603–1605.

MANDIGO R W (1982), 'Processing systems – mixing, temperature control and raw materials' in Franklin K R and Cross H R, *Proceedings of Meat Science and Technology Conference*, Chicago, National Live Stock and Meat Board, 235–244.

MEANS W J and SCHMIDT G R (1986), 'Algin/calcium gel as a raw and cooked binder in structured beef steaks', *J Food Sci*, 51, 60–65.

MOORE S L, THENO D M, ANDERSON C R and SCHMIDT G R (1976), 'Effect of salt, phosphate and some nonmeat proteins on binding strength and cook yield of a beef roll', *J Food Sci*, 41, 424–426.

MOUNSDON R K and JOLLEY P D (1987), 'The changing shape of burgers', *British J Photography* (April), 414, 415, 433.

MUNRO P A (1983), 'The tensile properties of frozen and thawed lean beef', *Meat Sci*, 9, 43–61.

NERAAL R and HAMM R (1973), 'Enzymatic breakdown of added tripolyphosphate and diphosphate in meat', *Proc 19th Meeting European Meat Research Workers*, 4, 1419–1427.

NOBLE B J, SEIDEMAN S C, QUENZER N M and COSTELLO W J (1985), 'The effect of slice thickness and mixing time on the palatability and cooking characteristics of restructured beef steaks', *J Food Quality*, 7, 201–208.

NONAKA M, TANAKA H, OKIYAMA A, MOTOKI M, ANDO H, UMEDA K and MATSURA A (1989), 'Polymerization of several proteins by Ca independent transglutaminase derived from micro-organisms', *Agric Biological Chemistry*, 53, 2619–2623.

NUTE G R (1996), 'Assessment by sensory and consumer panelling', in Taylor A A, Raimundo A, Severini M and Smulders F J M, *Meat Quality and Meat Packaging*, Utrecht, ECCEAMST, 243–255.

NUTE G R, MACFIE J H and GREENHOFF K (1988),'Practical application of preference mapping', in Thomson D M H, *Food Acceptability*, London, Elsevier Applied Science, 377–386.

OFFER G and KNIGHT P (1988a), 'The structural basis of water holding in meat Part 1 General principles and water uptake in meat processing', in Lawrie R A, *Developments in Meat Science 4*, London, Elsevier Applied Science, 63–171.

OFFER G and KNIGHT P (1988b), 'The structural basis of water holding in meat Part 2 Drip losses', in Lawrie R A, *Developments in Meat Science 4*, London, Elsevier Applied Science, London, 173–244.

OFFER G and TRINICK J (1983), 'On the mechanism of water holding in meat: the swelling and shrinking of myofibrils', *Meat Sci*, 8, 245–281.

PEARSON A M and DUTSON T R (ed.) (1987), *Advances in Meat Research Vol 3 Restructured Meat and Poultry Products*, New York, Van Nostrand Reinhold.

PEPPER F H and SCHMIDT G (1975), 'Effect of blending time, salt, phosphate and hot-boned beef on binding strength and cook yield of beef rolls', *J Food Sci*, 40, 227–230.

PURSLOW P P (1985), 'The physical basis of meat texture: observations on the fracture behaviour of cooked bovine M.semitendinosus', *Meat Sci*, 12, 39–60.

PURSLOW P P, DONNELLY S M and SAVAGE A W J (1987), 'Variations in the tensile adhesive strength due to test configurations', *Meat Sci*, 19, 227–242.

RICHARDSON R I, NORTJE G L and TAYLOR A A (1989), 'Modified atmosphere storage of accelerated processed beef restructured with an algin/calcium system', *Proc 35th Int Congress Meat Sci and Tech*, 5.43, 918–923.

SAVAGE A W J, DONNELLY S M, JOLLEY P D, PURSLOW P P and NUTE G R (1990), 'The influence of varying degrees of adhesion as determined by mechanical tests on the sensory and consumer acceptance of a meat product', *Meat Sci*, 28, 141–158.

SCHMIDT G R and MEANS W J (1986), 'Process for preparing algin/calcium gel structured meat products', US patent 4,603,054.

SCHWARTZ W C and MANDIGO R W (1976), 'Effect of salt, sodium tripolyphosphate and storage on restructured pork', *J Food Sci*, 41, 1266–1269.

SHEARD P R and JOLLEY P D (1988), 'Restructured and reformed meat products', *Food Technology International Europe*, 129–132.

SHEARD P R and SAVAGE A W J (1998), 'Restructured meat products in the UK: basic principles and new developments', in, JD Buckley J D, Honikel K O and Smulders F J M, *New Meat Products Development*, Utrecht, ECCEAMST, 135–149.

SHEARD P R, JOLLEY P D, HALL L D and NEWMAN P B (1989), 'Technical note: the effect of temperature and raw material on the size distribution of meat particles pre-broken by grinding', *Int J Food Sci and Tech*, 24, 421–427.

SHEARD P R, COUSINS A, JOLLEY P D and VOYLE C A (1990a), 'Particle characteristics of flake-cut meat', *Food Structure*, 9, 45–56.

SHEARD P R, JOLLEY P D, MOUNSDON R K and HALL L D (1990b), 'Factors influencing the particle size distribution of flaked meat I Effect of temperature, aperture size and pre-breaking before flaking', *Int J Food Sci and Tech*, 25, 483–505.

SHEARD P R, JOLLEY P D, KATIB A M A, ROBINSON J M and MORLEY M J (1990c), 'Influence of sodium chloride and sodium tripolyphosphate on the quality of UK-style grillsteaks: relationship to freezing point depression', *Int J Food Sci and Tech*, 25, 643–656.

SHEARD P R, FOSTER-SMITH A and JOLLEY P D (1991a), 'Factors influencing the particle size distribution of flaked meat II Effect of aperture size, number of cutting stations, rotational speed and impeller design', *Int J Food Sci and Tech*, 26, 65–81.

SHEARD P R, JOLLEY P D and RUSH P A J (1991b), 'Effect of temperature on the particle size distribution of flake-cut meat', *Int J Food Sci and Tech*, 26, 199–205.

SHEARD P R, NUTE G R, RICHARDSON R I, PERRY A and TAYLOR A A (1999), 'Injection of water and polyphosphate into pork to improve juiciness and tenderness after cooking', *Meat Sci*, 51, 371–376.

SHULTS G W (1982), 'New restructured meat products for the military', in Franklin K R and Cross H R, *Proceedings of Meat Science and Technology*, Chicago, National Live Stock and Meat Board, 279–284.

SUTTON A H (1973), 'The hydrolysis of sodium triphosphate in cod and beef muscle', *J Food Tech*, 8, 185–195.

TOMLINS R (1995), 'Cryogenic freezing and chilling of food', *Food Technology International Europe*, Sterling Publications, London, 145–146 and 148–149.

TROUT G R (1989), 'The effect of calcium carbonate and sodium alginate on the color and bind strength of restructured beef steaks', *Meat Sci*, 25, 163–175.

WIERBICKI E, CAHILL V R and DEATHERAGE F E (1957), 'Effects of added sodium chloride, potassium chloride, calcium chloride, magnesium chloride and citric acid on meat shrinkage at 70°C and of added sodium chloride on drip losses after freezing and thawing', *Food Tech*, 11, 74–76.

WIJNGAARDS G and PAARDEKOOPER E J C (1988), 'Preparation of a composite meat product by means of an enzymatically formed protein gel', in by Krol B, Van Roon P P S and Houben J H, *Trends in Modern Meat Technology* 2, 125–129.

XIONG Y L (1994), 'Myofibrillar protein from different muscle fiber types: implications of biochemical and functional properties in meat processing', *Critical Reviews Food Sci and Nutrition*, 34, 293–320.

XIONG Y L and BREKKE, C J (1989), 'Changes in protein solubility and gelation properties of chicken myofibrils during storage', *J Food Sci*, 54, 1141–1144.

YASUI T, FUKAZAWA T, TAKAHASHI K, SAKANISHI M and HASHIMOTO Y (1964), 'Specific interaction of inorganic polyphosphates with myosin B', *Agric Food Chemistry*, 12, 399–404.

18

Quality control of fermented meat products

D. Demeyer, Ghent University and L. Stahnke, Chr. Hansen A/S, Hørsholm

18.1 Introduction: the product

This deals with 'fermented sausage', a product with an ancient origin in continental Europe and receiving increasing interest in other parts of the world like Australia. The basic concept of its preparation involves comminution of muscle and fat tissue with salt, nitrate and/or nitrite and spices often including sugar, starter cultures and other additives such as non-meat proteins. After stuffing the mixture into a casing, the resulting sausage is left to ferment and dry, often in two consecutive and separate stages. The presence of salt, the lowering of water activity (a_w) and the exclusion of O_2 selects for salt tolerant lactic acid bacteria, producing lactic acid from carbohydrates added and/or present. This lowers pH to final values between 4.5 and 5.5, inducing denaturation of salt solubilised protein to a gel structure that can be sliced. The adequate (fast) reduction of pH and the lowered a_w ensure both product stability and safety. Once these basic requirements are met, the production technology allows for many but imprecise variations, yielding a variety of different products, presenting a considerable challenge to standardisation and management of quality. The different technologies involved lead to two general types of products:

- *Northern type products* (NP) containing beef and pork and characterized by relatively short ripening periods, up to about three weeks, involving clearly separated fermentation (about three days) and drying periods. Fermentation temperatures do not normally exceed 30°C. Some products, particularly in the US, are not dried but pasteurised after fermentation (Pearson and Gillet, 1996). Rapid acidulation to final pH values just below 5 followed by product-dependent weight losses during drying ensures safety and shelf-life. Smoking is applied to add specific flavour (taste and aroma).

360 Meat processing

- *Mediterranean or Southern type products* (MP) are predominantly pure pork products and involve longer ripening periods, up to several months. Fermentation occurs at lower temperatures ($\leq 20°C$ versus $> 25°C$) and acidulation to final pH values above 5 is therefore slower and often not clearly separated from drying. Smoke is not applied, except for the typical Hungarian sausage, and shelf-life is mainly determined by drying and lowered water activity.

Variations within these basic technologies yield products varying in moisture content between 25 and 50% (Acton and Dick, 1976). In all types, very complex and interrelated physical, chemical and biochemical changes in the protein, fat and carbohydrate fraction, brought about by meat as well as microbial enzyme activity, determine both safety and sensory quality of the product (for a recent review, see Ordóñez *et al.*, 1999).

18.2 The quality concept

As with all meat and meat products, fermented sausages are subject to a quality concept that has been enlarged, essentially because of the definite condition of surplus meat production in the industrialised countries. In the complex set of variable interacting criteria that determine purchase (consumption?) of a meat product, indirect criteria, associated with the production of meat and rarely perceptible or measurable on the product, have become more important (Demeyer, 1997). Direct quality criteria are measurable on the product and mainly relate to safety and sensory evaluation. The safety of fermented sausages is mainly determined by the inactivation of pathogens through the development of desirable added (and/or present) microorganisms with the associated decrease in pH (Lücke, 2000). It is also covered, together with indirect quality characteristics, by the traceability of the raw materials used, as laid down in the requirements for integrated quality control (HACCP concept, brand certification; Jago *et al.*, 2000) and receiving increasing attention from the consumer (Gellynck and Verbeke, 2001).

This chapter will not deal with these aspects, but will focus on sensory quality. Indeed, whereas microbial and (bio)chemical changes as well as processing technology in relation to meat fermentation have been discussed extensively (see e.g. Montel *et al.*, 1998, Ordóñez *et al.*, 1999, Kottke *et al.*, 1996 respectively) little quantitative and comparative information is available on the factors affecting sensory quality and its control, except for the German literature (Rödel, 1985). It should also be clear that the importance of indirect quality characteristics may increase, subject to changes in the relative importance of the many conflicting interests associated with the production of meat and meat products (Demeyer, 1997). In this respect it is worthwhile to note that the energy cost of Northern type fermented sausage production amounts to $3.6.10^3$ MJ per 100 kg final product, a value close to tenfold higher than for other meat products and mainly due to the maintenance of ripening conditions (Stiebing et al., 1981).

Within sensory quality characteristics, flavour is very important. Whereas purchase and rejection of the product are initiated respectively by appearance (colour) and texture, flavour is the feature that convinces the consumer to buy the product again (Verplaetse, 1994a). The typical cured meat colour is associated with the formation of the nitric oxide heme pigment, stabilised by the denaturation of the globin component (Acton and Dick, 1977). In some types of Mediterranean products, such as Spanish chorizo sausage, its importance is shared with that of the colour of added chilli peppers (Fernández-Fernández et al., 1998). The importance of the cured meat colour for fermented sausage has of course also been diminished by the recent legalisation in the EU of colouring agents such as Monascus red (Angkak), cochenille and betanin, derived from yeast, a scale insect and red beet respectively. Methods for their determination have been perfected (Brockman, 1998) and although their use may facilitate technology, it also allows for the use of raw materials subject to less demanding quality characteristics. The aptitude of the sausage for slicing is brought about by the combination of gel formation because of acidulation of salt solubilised proteins, followed by drying. Again, the basic interaction of salt extracted muscle protein with pH decrease and water loss is affected by a number of additives, including, e.g., milk and soy proteins, as well as polysaccharides.

Flavour is a complex sensory reaction involving taste, smell (odour) and texture of a product. Odour or aroma is by far the most important component, because of the high sensitivity of the nasal receptors for the numerous volatile components released during chewing and ingestion. The number of aroma compounds derived from spices and smoking (Northern types) exceeds that of compounds derived from metabolism (Schmidt and Berger, 1998). The latter however are considered very important for the specific sausage flavour (Stahnke, 1995b). They are derived from changes in the lipid, protein and, to a less extent, the carbohydrate fraction of the sausage brought about by interaction between muscle and microbial metabolism as well as chemical reactions. Proteins and lipids are initially subject to hydrolysis catalysed by meat (muscle and adipose tissue) enzymes. This probably facilitates further microbial (bacterial) metabolism of peptides, amino acids and fatty acids formed (Demeyer, 2000). The relative importance of these processes is closely related to the ingredient composition, to the rate of pH decrease during fermentation, and thus to processing technology.

18.3 Sensory quality and its measurement

The sensory quality of a meat product can be measured in several ways by either sensory or instrumental methods or a combination of both. All methods have their advantages and disadvantages. Preferably, sensory quality of a dried sausage is measured by a sensory panel: a consumer panel, trained laboratory panel or an expert panel depending on the purpose in question. Sensory evaluations can be rather laborious and expensive and are therefore often replaced to various extents by instrumental methods, not only in development and research projects but also for regular quality control in the factory. One must realise however, that quality profiles obtained by merely instrumental means often give too simple results making it difficult or impossible to link the profile to the actual quality perceived by the consumer eating the product. During the eating process, flavour is released from the product in a complex manner depending on the matrix itself and by the anatomical and physiological characteristics of the person eating the food. Also, the perceived flavour is related to the previous experience and present expectations of the individual (Rothe, 1988). The following paragraph will briefly describe the sensory methods available for measuring sensory quality in general. The specific sensory and instrumental methods applied for dried sausage evaluation are described separately for appearance, texture and flavour together with the present knowledge on how these quality concepts develop in relation to processing technology.

Several sensory methods have been developed and are described in excellent textbooks (Piggott, 1984; Meilgaard et al., 1991). A brief survey only is presented here. Sensory evaluation methods can be divided into difference tests, scaling and ranking tests and descriptive tests. In general, difference tests can be accomplished with untrained panellists, whose numbers depend on the size of the difference, whereas descriptive tests need a carefully trained panel in order to get reasonable results (Meilgaard et al., 1991). Difference tests allow the investigator to determine if an ingredient or process change causes a significant difference in the sensory perception of the product. Comparison tests (triangle, paired comparison, duo-trio, etc.) only indicate if a difference exists or not, whereas ranking tests also give information about the direction of the difference.

For estimation of the magnitude of the difference as well, more elaborate scaling methods and a trained panel are necessary (Meilgaard et al., 1991). However, scaling methods using hedonic category rating (e.g. like/dislike scales) are generally used with untrained assessors since trained panellists are unlikely to give true affective responses (Land and Shepherd, 1984). Descriptive tests break down the overall sensory attributes into separate descriptors. The term 'rancid', for example, is often used as one of the descriptors for evaluating the flavour of fermented meat products. During training the panel is exposed to reference samples that represent the individual flavour or other descriptors both in order to learn what they stand for and to be able to quantify their magnitude on the same intensity scale (Bett and Grimm, 1994).

Trained sensory panels typically consist of five to twenty members selected on the basis of their ability to taste and smell and their availability and interest. After selection, the panellists are trained in the basic principles of sensory perception, descriptor development and flavour-intensity measurement. The training period may require many hours of practising to perfect evaluation skills depending on the purpose of the specific sensory analysis, a worthwhile effort, since the output from a well-performing sensory panel is as objective and reliable as data obtained from instrumental analysis. Over the years different formal and systemised descriptive procedures have been developed. Examples are The Flavour Profile Method, The Texture Profile Method and The Quantitative Descriptive Analysis (QDA) Method (Meilgaard *et al.*, 1991).

18.4 Appearance and colour: measurement and development

The major factor determining the appearance of a fermented meat product is the colour of the product. However, visual evaluation also involves other characteristics that may be covered by the term 'structure', a property of significant importance in evaluation of dry sausage slices. Examples of this are particle size, uniformity of particles, glistening of fat, stickiness and more (Meilgaard *et al.*, 1991). Evaluation of appearance involves considerable psychophysical elements, but can be rendered objective by image analysis technology (Roudot *et al.*, 1992). Such technology has shown that the surface occupied by fat particles in slices cut perpendicular to the sausage length, changes very little with drying and is about 45% of the total surface area (Colas and Simatos, 1976). Further discussion will be limited to colour.

18.4.1 Sensory measurement of colour

The appearance of a fermented sausage has most often been evaluated sensorially by hedonic scaling methods or by descriptive analysis either using a point scale, a ranking scale or a continuous line scale with two anchors (Stahnke *et al.*, 2002; Hagen *et al.*, 2000; Sanz *et al.*, 1997; Diaz *et al.*, 1997; Dellaglio *et al.*, 1996; Næs *et al.*, 1995). Typical colour attributes for fermented sausage are whiteness, hue, colour intensity and colour tone of fat and meat particles. In some cases, colour attributes have been evaluated following the Natural Colour System (NCS), a system based on the resemblance of the sample colour to the six elementary colours white, black, yellow, red, blue and green (Scandinavian Color Institute 2001; Stahnke *et al.*, 2002).

18.4.2 Instrumental measurement of colour

The method of choice for objective measurement of colour is the use of CIE tristimulus values (X,Y,Z), derived from the sausage surface reflectance of specified light sources under specific conditions and transformed into colour co-ordinates in order to obtain a uniform distribution. They reflect lightness, redness and yellowness as respectively L, a and b (Hunter, 1975) or L^*, a^* and b^* (CIE, 1976) and interconversion between both systems is provided for (http://www.colorpro.com/info/tools/convert.htm). Additional psychophysical parameters chromaticity, hue and redness index can be derived from the co-ordinates. Both redness co-ordinates are well correlated with each other and with sensory colour evaluation (Ansorena *et al.*, 1997) as well as with pigment nitrosation (Üren and Babayiğit, 1997). The same units can be used for the determination of colour changes after exposure to light, e.g., colour stability.

18.4.3 Colour development

The formation of nitrosomyoglobin is the net result of a series of complicated reactions involving the formation of nitrogen oxide (NO) and its reaction with myoglobin or metmyoglobin producing nitrosylated pigments with red and greyish colour, respectively. In the Mediterranian type of processing, the (added) substrate for NO production is often nitrate, whereas in the Northern type of processing, sodium nitrite, added as colouring salt, is used. Use of the former involves bacterial reduction to nitrite, a process generally considered to be inhibited by pH values below 5.2 (Ally *et al.*, 1992). The latter, however, acts upon addition as a very reactive oxidant, and is reduced to NO immediately after preparation of the sausage mix, associated with the oxidative formation of metmyoglobin and resulting in an immediate greyish discolouration of the mix, with lowering of L^* and a^* values (Pribis and Svrzic, 1995). Development of the stable red colour during fermentation and drying requires the subsequent reduction of the (nitrosylated) metmyoglobin back to (nitrosylated) myoglobin together with the denaturation of the pigment globin moiety.

The rates of both the initial oxidation and the subsequent reduction, as well as the stability of the colour formed to later oxidation, are determined by a complex set of factors linked to both processing and raw materials. These include, e.g., the amounts of nitrite used, the rate of pH drop during fermentation, the use of anti-oxidant additives and the anti-oxidant activities of the starter bacteria used. The reducing activity of the meat used and/or its pH are also important factors. In general, the susceptibilities of (cured) meat pigment and lipid to oxidation are tightly linked and increase with decreases of pH and of the reducing environment. Mechanisms determining the latter include, e.g., the use of oxidative muscle and thus higher pigment and iron concentrations and higher lipolytic activity as well as the presence of lower concentrations of a series of endogenous and/or added antioxidative compounds such as α-tocopherol (vit.E), numerous phenolic compounds of plant origin and, possibly, carnosine. The interactions of such compounds with membrane structures and enzymes, as well as with spices involve pro- as well as antioxidant activities of e.g. nitrite and free fatty acids (!).

As pointed out recently by Bertelsen *et al.*, (2000), much remains to be investigated in relation to the outstanding colour stability of Parma ham, for example where colour is formed without added nitrate/nitrite. It would seem that for the Northern ripening process the use of sodium ascorbate (e.g. 600 ppm) with minimal amounts of sodium nitrite (e.g. 150 ppm) is sufficient to obtain an acceptable colour stability also reflected in a low redox potential (Ally *et al.*, 1992), minimal lipid oxidation (Zanardi *et al.*, 2000) and efficient oxygen consumption (Torfs and Demeyer, in preparation). The latter may be related to the use of starter organisms with antioxidant activities (catalase, superoxide dismutase and/or nitrate reductase activities) also contributing to flavour development (Barrière *et al.*, 2001). pH values below 4.9 have been considered harmful for colour development (Stiebing and Rödel, 1989). It is evident that colour control can involve the introduction of limit values for colour co-ordinates, taking into account considerable variability: Dellaglio *et al.*, (1996) reported coefficients of variation between 5 and 11% for CIELAB colour co-ordinates for the interior of the same sausage brand, whereas Torfs and Demeyer (in preparation) could detect significant differences both within and between production units (cutters) in two companies.

18.5 Texture: measurement and development

Three senses – touch, sight and hearing – may be involved in sensory assessment of texture, but in the majority of cases the sense of touch plays the most important role (Brennan, 1984) The 'in-mouth' texture is the parameter that is normally measured when evaluating dried sausage texture by sensory means – either by descriptive analysis or by hedonic methods. Quantification has mostly been accomplished by a point scale or a continuous line scale (Stahnke *et al.*, 2002; Bruna *et al.*, 2001; Garcia *et al.*, 2001; Hagen *et al.*, 2000; Mendoza *et al.*, 2001; Patarata *et al.*, 1997; Diaz *et al.*, 1997). Typical texture attributes used for fermented sausage in descriptive analysis are: hardness, fattiness, juiciness, stickiness, tenderness, granularity, fibrousness and clamminess.

The rheology of dry sausage is that of a visco-elastic body but its structural complexity is better reflected in empirical units, rather than in physical laws. Objective but empirical determination of food texture has been elaborated extensively, also for meat and meat products (Honikel, 1998). The force (Newton) necessary to penetrate the sausage surface or interior (sausage slice) under standardised conditions and referred to as 'hardness' has been used mainly for texture evaluation of dry sausage (Touraille and Salé, 1976).

18.5.1 Development of texture
During comminution, the added salt solubilises muscle proteins, which coagulate and form a gel surrounding lard and meat particles upon the acidification brought about by fermentation. The pH necessary for coagulation increases with increasing salt concentration and is 5.3 at the often-used salt concentrations between 2 and 3% (Ten Cate, 1960). According to Rödel (1985), hardness increases sharply when sausage pH reaches 5.4 and further increases gradually until $pH = 4.9$ (Rödel, 1985). More detailed studies have shown that myosin is the major protein solubilised by salt and that myosin filaments swell and are progressively fragmented in a 'halolytic' process that involves increasing loss of the typical band pattern of the myofibrils, depending on intensity of chopping and NaCl concentration. At the periphery of the

myofibrils, swollen and partly dissolved proteins form a network that can be considered an 'adhesive substance' holding meat, connective tissue and fat particles together. This network consists of filamentous aggregates whose dimensions and formation depend on factors such as pH and NaCl concentration, determining, for example, the relative rates of filament formation and aggregation (Katsaras and Budras, 1992).

Coagulation by acidulation or heating involves the formation of more stable and more intensive aggregations, associated with the release of water. The gel formed by coagulation is further stabilised by the release of water occupying spaces between the aggregates and forms a matrix surrounding fat and connective tissue particles that determines sausage texture. It has been shown that proteolytic damage to the myosin molecule, brought about, e.g., by ageing or electrical stimulation of beef carcasses, lowers the strength of heat coagulated myosin gels (Demeyer and Samejima, 1991). Muscle cathepsin D-like activity degrades sausage myosin mainly during fermentation (Verplaetse et al., 1989, Verplaetse, 1994b), but significant changes require pH < 5.1 (Molly et al., 1997). It is therefore clear that acidulation during fermentation induces two opposing effects on texture development: coagulation of the myosin sol into a gel as well as accelerating proteolytic cleaving of myosin molecules, lowering their contribution to gel strength. A moderate negative correlation was indeed found between proteolytic activity and texture (Santamaria et al., 1994), and the clear tenderising effect upon addition of exogenous protease to a sausage has been related to myosin degradation (Melendo et al., 1996).

Different relative rates of acid induced coagulation and proteolysis may explain the positive relationship found between initial rates of acidulation and hardness development as illustrated in Table 18.1. Increased rates of acidulation increase rates of drying, a process also determining rate of texture development. In the Northern ripening process, drying during fermentation is very limited, as well as pH change during drying. The data in Table 18.1 show that texture development during fermentation is determined by the drop in pH, irrespective of small weight changes, whereas further texture development during drying is then determined by the loss of water only. Both hardness development and weight loss show exponential changes during ripening. However, the latter has been reported to decelerate with time whereas the former accelerates (Touraille and Salé, 1976). It is clear that numerous factors affect the interrelated rates of both acidulation and drying.

18.5.2 Sausage composition and size

The use of PSE pork (Townsend et al., 1980), the use of spices (Vandendriessche et al., 1980), starter organisms (Demeyer et al., 1986) and soy protein (Stiebing, 1998) are known to increase rates of acidulation and drying and thus of texture development. Substitution of KCl and $CaCl_2$ for NaCl at equivalent ionic strength significantly reduces hardness as well as sensory texture and colour intensity (Gimeno et al., 1999). A higher fat content of the

Table 18.1 Rates of acidification, drying and texture development

Experiment	pH		% DM		Hardness (Newton)	
	Rates of change (c)[d]					
Expt. 1: effect of starters[a]						
None	−0.23		0.03		0.01	
Starter sausage	−1.06		0.06		0.06	
Starter organisms (n = 4)	−0.58±0.06		0.04±0.01		0.03±0.01	
	Absolute changes after 2 or 3 and 21 d of ripening[e]					
Expt. 2: effect of comminution[b]						
Mean fat particle size (mm²)	2−3d	21d	2−3d	21d	2−3d	21d
24.4	−0.7	−0.9	−1.2	9.6	64	213
3.8−6.0	−0.6	−0.9	−0.9	8.8	49	203
1.5	−0.7	−1.0	0.6	7.3	61	145
Expt. 3: effect of sausage diameter[c]			% weight loss			
60 mm	−1.0	−1.0	9.0	30.5	83	184
140 mm	−1.0	−1.0	4.9	17.9	87	129

[a] Data relate to experiments reported by Demeyer *et al.*, (1984). [b] Data relate to experiments reported by Verplaetse *et al.*, (1990). [c] Demeyer and Claeys (unpublished). [d] Initial rates of change calculated from exponential kinetic models proposed by Demeyer *et al.*, (1986). [e] Values after ripening-start.

sausage mix decreases both rate of pH drop and rate of hardness development (Rödel, 1985; Touraille and Salé, 1976). The use of between 50 and 60% lard in the sausage mix (equivalent to 40 and 50% fat), can result in a spreadable product (Klettner, 1989). Increasing sausage diameter clearly decreases the rate of drying and thus rate of hardness development (Rödel, 1985). An increase in sausage diameter also decreases the rate of pH decline because of an increasing contribution of proteolytic processes to metabolism (Demeyer *et al.*, 1986) (Lois *et al.*, 1987).

18.5.3 Conditions of batter preparation

According to Rödel (1985) and Tourraille and Salé (1976), the degree of comminution does not affect texture development. Lowering the average size of fat particles to 1.5 mm, however was found to decrease initial as well as final values for hardness (Table 18.1); the findings probably related to the smearing of fat around protein particles. An increased degree of lipolysis for the fat used may intensify these effects (Touraille and Salé, 1976). In a detailed study under practical conditions, Van 't Hooft (1999) concluded that the mean processing factors that significantly determine binding and structure are (i) meat temperature (−2 better than −4°C), fat temperature (−18 better than −10°C), salt chopping time (60 better than 20 sec.) and vacuum treatment (90 better than

0 sec.). The latter factor reflects the presence of air pockets in the sausage batter, leading to a difference between weight and volume loss. Salt chopping time could be related to extracted myosin and a remarkable finding was that, contrary to GMP, the use of blunt rather than sharp cutter knives is to be preferred for good texture.

18.5.4 Fermentation and drying conditions

Air speed, temperature and humidity not only affect rate of drying, but also result in a gradient change of water activity (a_w), important for the establishment of product safety as well as for the development of texture, colour and flavour. Values for a_w can be estimated with acceptable accuracy from company-specific relationships with sausage composition and rates of drying (De Jaeger et al., 1984) or measured continuously during ripening (Stiebing and Rödel, 1992). Hardness values are subject to variability and coefficients of variation fluctuate between 5 and 15%, values being higher for slices than for the whole sausage (Touraille and Salé, 1976). Nevertheless, limits for hardness can be imposed for the final product and controlled by measurement or estimation from correlations with pH and drying losses.

18.6 Flavour: measurement and development

In general, the term flavour is defined as the overall impression perceived via the chemical senses from a product in the mouth. Defined in this manner, flavour includes the sensation of taste and aroma as well as trigeminal feelings, such as astringency, the pain from hot spices, metallic note from blood, etc. Texture, appearance and the sounds of the food during chewing have an influence on the perceived flavour as well, but they are not commonly included in the definition of flavour (Meilgaard et al., 1991). However, one must be aware that the temporal order of the sensation has a great influence on the total flavour impression, i.e., the order of stimulation is very important for how the food is perceived and liked. During eating the consumer is first of all confronted with the appearance and colour of the food and later on with its odour. This gives rise to certain expectations on how the food will taste. Finally, during the chewing process, the consumer is confronted with texture, taste and aroma, which together will create the final impression of the flavour (Rothe, 1988).

The sensation of taste is caused by primarily non-volatile compounds in the food interacting with the taste buds on the surface of the tongue as well as in the mucosa of the palate and areas of the throat. The sensation of aroma is caused by volatiles in the food evaporating from the food during the chewing process and travelling through the nasopharynx to the nasal cavity, where they react with the olfactory receptors producing an electrical signal, which is transmitted to the olfactory bulb in the front brain (Rothe, 1988). The discrepancy between taste and aroma should be kept in mind when analysing flavour either by sensory or instrumental means. However, it is also true that sub-threshold concentrations of non-volatile compounds may affect sensitivity to an aroma compound (Dalton *et al.*, 2000). Such 'taste-olfaction integration' of senses is apparent from the aroma enhancement due to the glutamate-umami taste and peptides in fermented sausages may have a similar effect.

18.6.1 Sensory measurement of aroma, taste and flavour
The aroma (or odour), taste and flavour of dried sausage is commonly measured by hedonic methods or descriptive analysis, either using a point scale, a ranking scale or a continuous line scale (Mendoza *et al.*, 2001; Hagen *et al.*, 2000; Sanz *et al.*, 1997; Diaz *et al.*, 1997; Dellaglio *et al.*, 1996). Typical taste attributes are: acidity, saltiness, sweetness, metallicness, bitterness, umami and acidic aftertaste. Preferably taste should be measured while the panellists have a clip on their nose in order to prevent air from travelling through the nasopharynx to the nasal cavity and confusing the flavour impression with the tasting sensation (Bingham *et al.*, 1990). Aroma (or odour) and flavour attributes that are frequently used are: overall intensity, meat type (pork, beef, etc.), fresh meat, sour-sweet, acid, vinegar, tanginess, sour socks, spices, pepper, flowery, nutty, garlic, maturity, cured, dry sausage, butter, cheese, sourdough, fatty, rancid, nauseous, burned, solvent, smoked (Stahnke *et al.*, 2002; Hagen *et al.*, 2000; Stahnke *et al.*, 1999; Zalacain *et al.*, 1997; Viallon *et al.*, 1996; Dellaglio *et al.*, 1996; Stahnke, 1995c; Berdagué *et al.*, 1993; Acton *et al.*, 1972).

During training of panellists for dry sausage evaluation, chemical standards may be included to exemplify the qualitative description of the various attributes. This was recently done by Erkkilä *et al.*, (2001) who used lactic acid, acetic acid, arginine, alanine and salt to describe the flavour of lactic and acetic acid, bitterness, sweetness and saltiness, respectively.

18.6.2 Instrumental measurement of aroma and taste compounds
A huge number of methods have been developed for analysing the flavour compounds of meat products and other foods, but one should realise that the composition of the final aroma sample is highly reflected by the choice of method. Also, the reproducibiliy of flavour analyses is in general lower than for other analytical methods, in particular on complex matrices such as meat products. Standard deviations within the same sausage range between 5 and more than 10% for some volatiles and between-batch variability may exceed 50% (Schmidt and Berger, 1998, Hinrichsen and Pedersen 1995; Mateo and Zumalacárregui 1996; Meynier *et al.*, 1999, Stahnke *et al.*, 2002). Flavour research has been primarily confined to the study of the volatile and the semi-volatile compounds since they are the most important contributors to the characteristic flavour of most foods (Cronin, 1982). The following paragraphs will therefore focus on the measurement of volatile compounds and only slightly on the non-volatile. The basic principles include four steps: Collection of the

flavour compounds, concentration, separation and detection. (Bett and Grimm, 1994).

Collection of compounds
Depending on the type of flavour compounds, their polarity, volatility, etc., and the kind of matrix in which they are embedded (raw meat/cooked meat, lipid content, structure, etc.) different collection methods are preferred and basically, three different methods exist; (i) direct extraction with a solvent (organic solvent, water, super-critical fluids), (ii) distillation combined with a transfer of the volatiles into a small amount of solvent (e.g., steam distillation, Likens-Nickerson, high vacuum distillation of extracts) and (iii) headspace collection, in which the volatiles in the air above the sample are sampled and collected in a cold-trap or on an adsorbent (dynamic headspace sampling), purge and trap sampling, solid phase micro-extraction, etc.). However, some methods are harsher than others giving rise to extensive artefact formation if not well suited to the product. For example, steam distillation at atmospheric pressure, although claimed to induce less variability (Schmidt and Berger, 1998), would not be the best choice for collecting aroma compounds from a raw meat product, since compounds not originally present in the product will be produced by thermal reactions during the distillation process (Parliament, 1997; Wampler, 1997). Although its use decreases analytical variability, it may increase the amounts of volatiles extracted from fermented sausage more than twentyfold compared to head space analysis (Demeyer *et al.*, 2000).

Concentration
Depending on the collection method and the concentration of compounds, the aroma sample may have to be concentrated to a smaller volume. For extracts the solvent is commonly evaporated on a Vigreux column in a thermostatted water bath or by blow down of an inert gas, usually nitrogen, on the surface of the extract. In headspace methods using adsorbent traps the aroma sample is located in the trap, which may be extracted with a minute amount of solvent or desorbed by heat (see below). During the concentration step there is a risk of highly volatile compounds being lost due to evaporation together with the solvent (Parliament, 1997; Hartman *et al.*, 1993; Burgard and Kuznicki, 1990).

Separation
The separation of the flavour sample is primarily performed by gas chromatography (GC) (for volatiles) or, for non-volatiles and compounds with low volatility, by high pressure liquid chromatography (HPLC). The presence of impurities or of numerous compounds may necessitate pre-separation (Cronin, 1982; Merritt and Robertson, 1982). Volatiles collected on adsorbent traps are often injected directly into the GC by a thermal desorber. Aroma extracts are preferably injected directly on-column at as low a temperature as possible since high temperatures may decompose labile compounds and produce artefacts (Wampler, 1997; Merritt and Robertson, 1982). Analysis of the peptide fraction

merits special mention: early work studying proteolysis during dry sausage fermentation has used semi-quantitative SDS-PAGE (Verplaetse *et al.*, 1989; Verplaetse, 1994b; Verplaetse *et al.*,1992). Such a method is limited however to the molecular weights (MW) > 5 kD. Size exclusion chromatography (gel filtration) by HPLC can be used to isolate smaller MW fractions for further analysis by reversed phase HPLC (Lambregts *et al.*, 1998).

Detection
Detection of flavour compounds in gas chromatographic analysis is mostly performed by mass spectrometry (MS) and by flame ionisation detection (FID), which responds to any compound that is combustible in a hydrogen flame. Element-selective detectors are available for organic compounds containing halogens, nitrogen, sulphur and phosphorus, but are rarely used in flavour research. The advantage of the mass spectrometer compared to the other detectors is based on its ability both to quantify and identify the compounds at the same time. However, in many applications the sensitivity and the linear range of the MS are much less than the other detectors (Rood, 1999).

A highly efficient, commonly used detection principle in instrumental flavour analysis method is olfactometry. In gas chromatography olfactometry (GCO) all or part of the effluent from the column is led to an outlet outside the GC-oven, where a human subject sniffs the compounds as they elute. The odours are rated qualitatively and sometimes also quantitatively and make it possible to identify the more important odorous compounds in the food sample. While instrumental detectors quantify the individual components of the food sample, the peak areas do not necessarily correspond to the flavour intensity. The human nose is much more sensitive than the instrumental detector to many flavour compounds (Bett and Grimm, 1994). Different protocols for analysing flavour samples by GCO have been developed (Blank, 1997) and modified versions are commonly used (Meynier *et al.*, 1999; Stahnke, 1995c; van Ruth and Roozen, 1995).

The newest principle for detection of volatile flavour compounds is the 'electronic nose', a device based on an array of sensors each having a partial specificity for each volatile compound in the gas phase thus producing an odour fingerprint that can be identified by a pattern recognition system without need for prior separation (Strike *et al.*, 1999). The main advantage of electronic noses is rapid analysis, enabling quick decisison making, e.g., in relation to quality control. Electronic nose sensing was shown to be sufficiently accurate as an approximation of human olfaction apparatus, but further development is necessary although its successful use in discriminating between sausage types has been reported (Vernat-Rossi *et al.*, 1996; Eklov *et al.*, 1998). For analysis of peptides by reversed phase HPLC, detection sensitivity may be a limiting factor, necessitating the use of mass spectrometry. Analogous to GCO, the liquid fractions from gel filtration or preparative HPLC on non-volatile flavour compounds may be tasted by a sensory panel and the taste of the eluted compounds evaluated (Henriksen and Stahnke, 1997) but this approach has not been much used by flavour researchers.

18.6.3 Chemical compounds related to the development of flavour

Even though the flavour impression is probably the most important component of the eating quality of fermented sausage, research on the matter has been rather scarce and until quite recently it was not directly focused on the analysis of flavour compounds (see section 18.6.2.). On the other hand much research has been aimed at understanding the chemistry of cooked meat flavour (Mottram, 1998). But one should be careful not to confuse results from meat flavour studies with fermented sausage flavour since flavour compounds in cooked meats primarily are derived through thermal processes, whereas compounds in raw dried meat products such as salami and dry ham mainly arise from both endogenous and microbial enzyme reactions taking place during the fermentation and drying steps.

18.7 Taste and aroma: measurement and development

The acid taste is an important component of the overall taste of fermented meat products, often sought after in the Northern process, whereas it may be rejected in the Mediterranean product. It is positively correlated with the D-lactate (Bucharles et al., 1984) and acetate (Demeyer, 1992) content. Only limited amounts of nitrogen-containing or protein-related compounds are found in the volatile aroma compounds (Berdagué et al., 1993), yet, flavour evaluation of fermented sausage has been mainly related to extent of proteolysis (Demeyer, 1992), specifically in relation to surface mould growth (Lücke, 1998). Indeed, the fatty acids present in sausage lipids are themselves too long to be of sensory relevance. Although such findings suggest that most of the significant aroma compounds are derived from the protein fraction of the sausage, it is known that intensity of proteolysis reflects release of peptides affecting taste, rather than aroma (Nishimura et al., 1988) as shown for cheese (Fox, 1989) and raw ham (Hansen-Møller et al., 1997). Also, the non-protein nitrogen fraction will affect sausage pH (Demeyer et al., 1979) and sausage pH may affect liberation of aroma determining acid compounds during chewing (Dainty and Blom, 1995; Dirinck, personal communication). Work on proteolysis in dry sausage has involved the initial degradation of myosin and actin (Verplaetse et al., 1992; Verplaetse, 1994b; Molly et al., 1997; Harnie et al., 2000) and preliminary work on the peptide and free amino acid fraction has been reported (Lambregts et al., 1998; Demeyer, 2000).

The use of antibiotics and paucibacterial meat incubations has clearly established that initial proteolytic changes mainly involve myosin and actin degradation through the action of cathepsin D-like enzymes. The contribution of bacteria in further endo- and, mainly, exoproteolytic changes increases down to ammonia production, the end of the proteolytic chain. Mediterranean, low temperature ripening, lowers rate of pH drop and thus, cathepsin D activity and initial protein degradation as well as the proportion of smaller peptides (1–10 kDa). The contribution to flavour of ATP metabolites such as IMP and hypoxanthine is also recognised and that of free higher fatty acids generally considered of less importance (Verplaetse, 1994a). Studies on the relative importance of volatiles and water solubles in flavour analysis should use multivariate statistics involving data sets for both volatile and water soluble compounds.

18.7.1 Aroma

The raw sausage mince does not contain any volatile compounds of sensory importance since it has little or no aroma. On the other hand, it contains a large number of aroma precursors, which during the fermentation, drying and maturation steps are converted by endogenous enzymes, microbial activity and chemical reactions into a large number of volatile compounds of both sensory and non-sensory importance. The first study on the aroma profile of a fermented sausage appeared in 1990 and since then different sausages from all over Europe have been analysed by various analytical methods (Berger *et al.*, 1990; Berdagué *et al.*, 1993; Johansson *et al.*, 1994; Stahnke, 1994, 1995b; Mateo and Zumalacárregui, 1996; Schmidt and Berger, 1998; Stahnke *et al.*, 1999; Viallon *et al.*, 1996; Meynier *et al.*, 1999; Sunesen *et al.*, 2001; Stahnke *et al.*, 2002). The volatiles present in fermented sausage consist of a wide variety of compounds from many different chemical classes depending on ingredient levels, spices, meat origin, smoke, starter cultures, processing conditions, packaging conditions etc., alkanes, alkenes, aldehydes, ketones, acids, alcohols, esters, sulphur compounds, furans, lactones, aromatics, terpenes, nitriles and more. Until now more than 200 compounds have been identified, but not all of them are of sensory relevance. In particular, compounds such as the alkanes and straight chain alcohols have sensory threshold values much too high for them to have any influence on fermented sausage flavour (Grosch, 1982).

By using gas chromatography olfactometry (GCO) or multivariate statistics combining sensory and instrumentally determined flavour profiles, it has been shown that the aroma compounds creating the basic cured sausage flavour are most likely to consist of compounds from microbial degradation of fatty acids and of the amino acids valine, leucine, isoleucine and methionine together with compounds from carbohydrate catabolism. More specifically, different branched aldehydes and acids, ketones, various sulfides, diacetyl, acetaldehyde, acetic acid and perhaps also certain ethyl esters (Stahnke *et al.*, 2002; Stahnke 2000; Meynier *et al.*, 1999; Stahnke *et al.*, 1999; Stahnke 1994, 1995c; Schmidt and Berger 1998; Montel *et al.*, 1996; Berdagué *et al.*, 1993). Compounds originating from chemical autoxidation of lipids such as hexanal, octanal, 1-octene-3-one, etc., are of great importance but may not be involved with the cured flavour attribute, but rather contribute to the rancid notes. Smoked sausages contain specific top notes attributed to 2-furfurylmercaptan and guaiacol, whereas Mediterranean moulded sausages contain popcorn notes due to 2-acetyl-1-pyrroline (Stahnke 2000). There have also been reports stating that degradation products of sulphur compounds in garlic may contribute positively

374   Meat processing

to maturity and cured flavour (Stahnke, 1998; Stahnke et al., 2002). However, flavour researchers still need to confirm these findings by reconstituting the proposed flavour compounds into a mixture holding the fermented sausage flavour.

18.7.2   The origin of the aroma compounds

Figure 18.1 gives an overview of the overall enzymatic and chemical reactions that may lead to sausage flavour compounds by degradation of carbohydrate, protein and fat in the sausage mince. Depending on sausage ingredients and processing conditions, some reactions will be more pronounced than others, but how these factors exactly determine the overall pattern is not clear and much more research is needed to understand in particular, the role of starter cultures and their interactions with both the background flora and the endogenous enzymes.

18.7.3   Aroma compounds from carbohydrate catabolism

During the fermentation period most of the added carbohydrate is converted into lactic acid and different amounts of side products depending on the applied lactic acid bacteria, the type and content of carbohydrate, temperature and other processing parameters. The additional starter cultures of, for example, staphylococci or yeast probably exert some effect in converting sugars to products other than lactic acid, but are of course in strong competition with the lactic acid bacteria. Volatile compounds in fermented sausage, generally

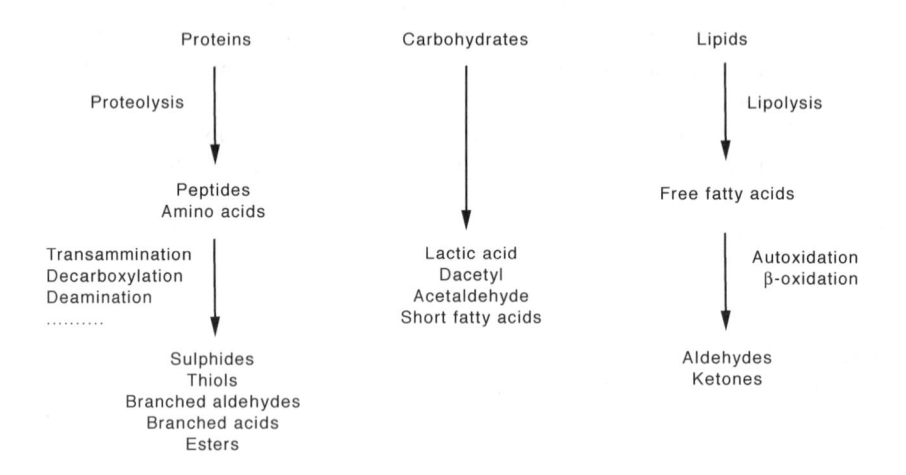

Proteins	Carbohydrates	Lipids
Proteolysis ↓		Lipolysis ↓
Peptides Amino acids	↓	Free fatty acids
Transammination Decarboxylation Deamination .... ↓	Lactic acid Dacetyl Acetaldehyde Short fatty acids	Autoxidation β-oxidation ↓
Sulphides Thiols Branched aldehydes Branched acids Esters		Aldehydes Ketones

Fig. 18.1   Simplified overview of the primary pathways leading to important flavour compounds in fermented sausage.

considered to be derived from carbohydrate catabolism, are acetic, propionic and butyric acids, acetaldehyde, diacetyl, acetoin, 2,3-butandiol, ethanol, acetone, 2-propanol and more (Gottschalk, 1986; Demeyer, 1982; Stahnke, 1999). However, the compounds are derived from pyruvate, which may originate from many sources other than carbohydrate during microbial metabolism (Demeyer *et al.*, 1986).

18.7.4 Aroma compounds from protein degradation

As mentioned earlier, extensive proteolysis takes place in fermented sausages creating peptides and free amino acids. During maturation amino acids and small peptides are taken up by the microorganisms and converted into numerous aroma compounds by different pathways (Fig. 18.1). Some of the more important are the biochemical conversions of the amino acids leucine, isoleucine, valine, methionine and phenylalanine into the sensory important branched aldehydes and corresponding secondary products, such as acids, alcohols and esters (Montel *et al.*, 1998; Stahnke *et al.*, 2002). The microorganisms responsible for those conversions are primarily species from the *Micrococcaceae* family. It has been shown, both in model experiments and in sausages, that different staphylococci and micrococci (kocuria) produce 2- and 3-methylbutanal, 2-methylpropanal, 2- and 3-methylbutanoic acid, 2-methylpropanoic acid, 2- and 3-methylbutanol, ethyl-2- and 3-methylbutanoate, methional, phenylacetaldehyde, phenylethanol and many more (Berdagué *et al.*, 1993; Stahnke, 1994, 1999; Montel et al., 1996; Masson *et al.*, 1999; Larrouture *et al.*, 2000).

The amount of the compounds is highly influenced by the processing conditions. In minimal media, model minces and in sausages it has been shown that parameters such as temperature, pH, glucose, salt, nitrite, nitrate and ascorbate all influence the amount of aroma compounds in one way or the other (Stahnke, 1995b, 1999; Masson *et al.*, 1999; Larrouture *et al.*, 2000). Results indicate that for *Staphylococcus* the reactions are negatively correlated with their growth, i.e., it seems as if the organisms produce more of the above-mentioned compounds when they are in the resting phase than when in active growth but this still needs to be studied further (Stahnke, 1999). It has been suggested that the branched-chained aldehydes could also arise from the non-enzymatic Strecker reactions between the corresponding amino acids and a diketone, such as diacetyl (Stahnke, 1995b; Barbieri *et al.*, 1992). This would explain the presence of different pyrazines in unspiced fermented sausages (Stahnke, 1995b; Johansson *et al.*, 1994; Berdagué *et al.*, 1993). Apart from aldehydes, the Strecker degradation results in various keto-amines that dimerize into different pyrazines (Hurrell, 1982). Indeed, studies have shown that the amount of 2- and 3-methylbutanal and 2-methylpropanal was of the same magnitude in sausages with or without microbial growth (Stahnke, 1994). However, the Strecker degradation is favoured by high temperature and very low water activity, i.e., conditions not prevailing in fermented sausage (Hurrell, 1982).

18.7.5 Aroma compounds from lipid degradation

During the fermentation and maturation periods the lipid fraction of the sausage mince is partly hydrolysed by lipolytic reactions in which triglycerides and phospholipids are liberating free fatty acids. Residual mono- and diglycerides have also been detected and it was shown that more unsaturated fatty acids were liberated preferentially, probably because of a preferential membrane phospholipid degradation as well as a positional and/or fatty acid specificity of the meat lipases (Demeyer et al., 1974; Molly et al., 1997). Lipolysis has been extensively studied over the years since free fatty acids are believed to be important precursors for oxidation products of relevance for flavour. Nevertheless, a direct correlation between lipolysis and maturity development has not been established (Montel et al., 1998). Recent results indicate that methyl ketones from microbial β-oxidation of free fatty acids may be important for maturity (Stahnke et al., 2002) but perhaps the amount of free fatty acids is so plentiful that increased amounts of this precursor do not influence the flavour profile. One should bear in mind that aroma compounds are present in the ppb to ppm levels whereas the level of free fatty acids are between 0.5 to 7% depending on sausage type (Nagy et al., 1989; Dominguez and Zumalacárregui,1991; Stahnke, 1994; Johansson et al., 1994; Navarro et al., 1997).

Lipolysis is caused both by microbial enzymes and endogenous enzymes in the meat and fat and there has been much debate about which mechanisms are the dominant. However, the most recent results from sterile model minces and sausages with added antibiotics show that the major part of the lipolytic breakdown is attributed to endogenous enzymes even if strongly lipolytic strains of *Staphylococcus* are used as a starter culture (Molly et al., 1997; Stahnke, 1994). The pH of the sausage mince may be decisive for the degree of lipolysis arising from microorgansims since pH is a major factor influencing the amount of *Micrococcaceae*, their production of lipases and their activity (Søndergaard and Stahnke, 2002; Hierro et al., 1997; Sørensen and Jakobsen, 1996). It has also been shown in sausages that the amount of free fatty acids is increased by high fermentation temperature and reduced salt levels (Stahnke, 1995a). The partial glycerides and/or the free fatty acids produced during lipolysis may oxidise via different pathways chemically or microbially. It is not clear whether free fatty acids are oxidised faster than intact glycerides. Although addition of lipases increased lipid oxidation during maturation (Ansorena et al., 1998), other work has shown that increased lipolysis was not associated with increased rancidity (Nagy et al., 1989) (Fernandez and Rodriguez, 1991).

Chemical autoxidation of unsaturated fatty acids produces a whole range of volatile aldehydes, ketones, alchols, etc., some of which are very potent aroma compounds. As mentioned above, gas chromatography olfactometry has shown that compounds such as hexanal, octanal and 1-octene-3-one are important for the overall flavour (Meynier et al., 1999; Schmidt and Berger, 1998; Stahnke, 1995c). In general, the influence of autoxidation processes will increase during maturation and storage depending on the sausage ingredients. It has been shown that ascorbate prevents autoxidation (Houben and Krol, 1986) and that nitrate may increase it (Stahnke, 1995b). Additionally, species belonging to the genus *Staphylococcus* have been reported to prevent autoxidation, possibly due to their capability of forming catalase and superoxide dismutase that degrade hydrogen peroxide (H_2O_2) and superoxide (O_2^-), respectively (Barriere *et al.*, 2001; Talon *et al.*, 2000).

As mentioned previously, methyl ketones (2-alkanones) can be formed during microbial metabolism, either directly by decarboxylation of free β-keto acids or by β-oxidation of free fatty acids. Their sensory threshold values are quite high though, compared to other lipid oxidation compounds (Grosch, 1982). The 2-alkanones may be further reduced to 2-alkanols by alcohol dehydrogenase in the microorganism. The level of methyl ketones increases steadily over time in Mediterranean fermented sausages (Sunesen *et al.*, 2001; Croizet *et al.*, 1992) and the Penicillium growing on the surface may be responsible. However, North European non-moulded sausages also contain methyl ketones, be it in slightly lower amounts (Stahnke *et al.*, 1999). Model experiments show that both *Staphylococcus* and *Penicillium* species are capable of producing methyl ketones (Stahnke, 1999; Montel *et al.*, 1996; Larsen, 1998).

18.8 The control and improvement of quality

It is evident that the methods discussed both for panel and instrumental analysis can be used at various stages of the production process. The results, associated with the information on quality development discussed above, can then be used to improve the progressive interactions between raw materials, microorganisms and processing.

18.8.1 Raw materials

Muscle lipase and protease activities determine to a large extent liberation of free fatty acids, peptides, and even amino acids during dry sausage processing. Although the effects of animal species on such changes (Demeyer *et al.*, 1992), as well as on flavour (Fournaud, 1978) have been reported for fermented sausage, their relation to the formation of flavour compounds from fatty acid oxidation and bacterial amino acid metabolism is not clear (Demeyer, 2000). Relationships of muscle protease and lipase activity with both carcass and meat quality have however been demonstrated (Toldrá and Flores, 2000; Claeys *et al.*, 2000) and may be used for the specification of raw materials, similar to a suggestion made for raw ham production (Toldrá and Flores, 1998) and in addition to the classic criteria related to pH (< 5.8), inner temperature ($< 4°C$) and fatty acid unsaturation ($< 12\%$ w/w of polyunsaturated fatty acids in total fatty acids) as described, e.g., by Stiebing (1994). Sugar (max. 2%), nitrite and nitrate addition may be optimised in relation to the findings discussed earlier.

18.8.2 Starter cultures

The application of starter cultures for sausage fermentation is today a natural part of industrial sausage production. The microorganisms used as starter cultures are commonly divided into two groups: lactic acid bacteria, primarily responsible for the acidification process, and flavouring microorganisms, often capable of nitrate reduction. The first group consists of *Lactobacillus* and *Pediococcus*, the second group of *Micrococcaceae* (*Staphylococcus, Kocuria* (formerly *Micrococcus*)), yeasts (*Debaryomyces*) and moulds (*Penicillium*) (Jessen, 1995). Lactic acid bacteria are used to ensure acidulation and avoid faulty production due to contamination with pathogens and other undesired microorganisms that may proliferate without the presence of a competitive starter inoculation. *Micrococcaceae* species and yeasts are added to speed up and intensify flavour development and to ensure that enough nitrate reducing microorganisms are present in sausage minces with added nitrate. Mould is primarily used to prevent growth of spontanous fungi capable of mycotoxin production and to shorten the onset of mould coverage. The choice of starters depends of course on the type of sausage to be produced. In the fermentation process bacteria, yeasts and fungi all contribute to the final sensory quality of the fermented sausage (Lücke 1998).

Lactic acid bacteria

The lactic acid bacteria primarily affect flavour formation by producing the taste component lactic acid of either the D(−) or L(+) configuration or a mixture of both. However, under certain processing conditions the bacteria may switch to other pathways producing end products such as acetic acid, ethanol, acetoin, formate and strong-smelling sulfides (Jessen, 1995). Lactic acid bacteria have been reported to be weakly proteolytic and lipolytic and thereby capable of influencing flavour formation, but generally these characteristics are not sought for in the selection of lactic acid bacteria for fermented sausage. Acidulation rate, lag phase, competitiveness towards endogenous flora, carbohydrate fermentation pattern and optimum growth temperatures are the typical control points. Production of hydrogen peroxide and biogenic amines are also of great concern as well as production of bacteriocins, phage resistance, stability during manufacturing and more (Jessen, 1995).

Micrococcaceae

Several studies have shown that it is possible to affect the sensory quality of the final product by changing the type or content of the starter culture. Studies have also shown that species from the *Micrococcaceae* family are the major flavour contributing microorganisms (Berdagué *et al.*, 1993; Montel *et al.*, 1996; Stahnke *et al.*, 2002). Model and sausage experiments have shown that different strains have very different volatile profiles, which are affected differently by processing parameters (Stahnke, 1995b; Sørensen and Jakobsen, 1996; Søndergaard and Stahnke, 2002; Stahnke, 1999; Larrouture *et al.*, 2000; Masson *et al.*, 1999). The *Micrococcaceae* species primarily sold as starter cultures are different strains of *S. xylosus, S. carnosus* and *Kocuria varians* (Jessen, 1995). However, at the low pH conditions prevailing during North European sausage production the amount of *Micrococcaceae* is drastically reduced. Even if added in high numbers these species begin to die out shortly after onset of fermentation (Johansson *et al.*, 1994). Selection of low pH tolerant *Micrococcaceae* species for sausage fermentation is thus highly relevant.

Other characteristics looked for are lipolytic and proteolytic capability, nitrate reductase and catalase production. The latter two activities are considered important for colour formation and colour stability and perhaps also for preventing lipid oxidation (Jessen, 1995; Talon *et al.*, 2000). However, based on the most recent knowledge as presented in the previous paragraphs, lipolytic and proteolytic activity of the starter cultures are probably not as relevant as previously suspected. In fact it may be more relevant for the culture manufacturers to look at the capability of degrading amino acids or forming methyl ketones from fatty acids. Nevertheless, there is still a long way to go before control of flavour formation in fermented sausage by control of growth or pre-history of *Micrococcaceae* is a practical reality. Most of the flavour developing pathways are not identified and their regulation on both gene and pathway level even poorer understood. Only very recently, the first staphylococcal gene for one of the enzymes in the catabolic pathway of leucine, isolucine and valine into the branched aldehydes and acids was sequenced (EMBL database, 2001).

Yeast
In commercial starter culture preparations yeast is offered for both exterior and interior use, but it is used in much smaller amounts than the *Micrococcaceae* (Jessen, 1995; Andersen, 2001). *Debaryomyces hansenii* is the dominant yeast species identified in naturally fermented products and is also the species primarily sold as starter culture (Encinas *et al.*, 2000; Metaxopoulos *et al.*, 1996; Jessen, 1995). The influence of yeast on the sensory quality of fermented sausages is not well documented. It seems to be highly dependent on the other microorganisms present (Gehlen *et al.*, 1991) and studies have shown that yeast may or may not have an influence on sausage flavour (Gehlen *et al.*, 1991; Miteva *et al.*, 1986; Olesen and Stahnke, 2000).

Due to its high oxygen demand *D. hansenii* is mainly observed in the periphery of the sausages where it stabilises colour, degrades lactic and acetic acid and produces ammonia (Geisen *et al.*, 1992; Gehlen *et al.*, 1991). The aroma compound production by *D. hansenii* has been studied in malt agar, minces and in sausages and it was shown that very small amounts of volatile compounds were produced compared to other yeast species (Westall, 1998; Olesen and Stahnke, 2000). In model minces and sausages *C. utilis* is a much stronger producer of volatiles, in particular acetates (Olesen and Stahnke, 2000). Both *C. utilis* and *D. hansenii* are lipolytic, but for *D. hansenii* the lipolytic activity is inhibited at low pH and temperatures (Miteva *et al.*, 1986; Sørensen, 1997). *D. hansenii* is a very salt tolerant species with a water activity minimum for growth of 0.84 in saline solution (Deak and Beuchat, 1996). However, garlic added to the sausage mince may prevent growth at a_w values as high as 0.95–0.96 (Olesen and Stahnke, 2000).

Fungi
Mediterranean air-dried sausages and certain Hungarian types are fermented with mould on the surface. The mould coverage gives the final product a characteristic flavour due to the fungal degradation of sausage ingredients acting together with the bacterial fermentation in the inside of the sausage. Unfortunately, research on the influence of mould on sausages is almost non-existent and it is not clear how fungi affect flavour. A few studies have shown, that application of different strains of *Penicillium nalgiovense, Penicillium camemberti* and *Penicillium chrysogenum* may give sensory differences in the final product but the cause is not known (Sunesen and Stahnke, 2001). A separate study has identified a specific popcorn-smelling compound (2-acetyl-1-pyrroline) as the most important volatile differentiating mould and non-moulded sausages (Stahnke, 2000). On the other hand it seems as if the other sensory differences are due to concentration changes in various compounds in the overall flavour pattern rather than to the existence of specific compounds from moulds. Other studies indicate that the mould may be responsible for much of the methyl ketones, 4-heptanone, 1-octene-3-ol and phenylacetaldehyde in moulded sausages (Sunesen *et al.*, 2001; Stahnke *et al.*, 1999; Croizet *et al.*, 1992; Kaminski *et al.*, 1974).

Moulds are lipolytic and proteolytic, perform β-oxidation of free fatty acids, produce ammonia and degrade lactic acid thereby increasing pH (Geisen, 1993; Trigueros *et al.*, 1995; Toledo *et al.*, 1997; Lücke, 1998; Larsen, 1998; Selgas *et al.*, 1999). All of these activities will affect flavour as described in the previous paragraphs. Also, it has been claimed that mould coverage reduces the risk of drying faults and delays rancidity by consuming oxygen (Lücke, 1998), but this has never actually been proven. However, the strongest argument for using fungal starter cultures is to prevent growth of mycotoxin-producing fungi and to produce sausages with an even, white surface without discolouring. This requires that the fungal starter has a short lag phase compared to the house-flora, high competitiveness, good adhesive properties and whitish colour while still creating a final product with the correct sensory quality (Jessen, 1995; Lücke, 1998). The mould primarily used as starter is *Penicillium nalgiovense*, since only certain strains of this species have been declared non-toxinogenic (Jessen, 1995). Additionally, many *Penicillia* are capable of producing the antibiotic penicillin (Andersen and Frisvad, 1994). Gene technology is presently being used in the construction of new well-performing and safe cultures (Geisen 1993).

18.8.3 Processing conditions
Customary rates of pH drop in Northern processing (4.8 after a three-day fermentation) are somewhat lower than those reported earlier by Stiebing and

Rödel (1987a, b) (about 5.3 after three days) who discussed the effect of air speed, temperature and relative humidity on the interrelated rates of drying (weight losses of 15 and 30% after 21 days for 60 and 140 mm diameter sausages respectively), acidulation and texture development. As discussed above, development of colour, appearance, and flavour are affected by rates of acidulation and drying, and thus indirectly by other processing variables such as addition of spices (Vandendriessche et al., 1980), decontamination of spices (Bolander et al., 1995) manganese addition, degrees of comminution and salting (Demeyer, 2001). It should be realised that the number of processing variables that may affect sensory quality is very large. As an example, it has been shown that the development of the mould-associated flavour was significantly improved by the use of natural, rather than artificial casing (Roncales et al., 1991).

18.9 Future trends in quality development

Product development in the production of fermented sausages follows the general trends of the food industry: acceleration of production with guaranteed sensory quality as well as safety and introduction of (new) functional properties.

18.9.1 Accelerated production

The use of low fermentation temperatures and extended drying periods leads to specific delicately flavoured Mediterranean products. An accelerated fermentation at higher temperatures and using specific starter cultures, similar to the Northern process, has been investigated for Mediterranean products, in order both to reduce the greater safety risks and production costs. However, the increased rate of pH drop (4.8 vs. 5.5 after three days) interferes with the typical flavour development, obviously related to the inhibition of Micrococcal metabolism (Martuscelli et al., 2000) and/or the stimulation of muscle protease activity (Molly et al., 1997). This finding has motivated extensive research into the acceleration of flavour development through addition of enzymes. As recently discussed, (Ordóñez et al., 1999) proteinases and lipases could be used to accelerate and increase proteolysis and free fatty acid formation. However, their activity is difficult to control and often leads to faulty (softer) texture development. These authors state correctly that addition of such enzymes may provide more substrate for flavour-producing microorganisms, but substrate concentration is obviously not a limiting factor in flavour development (Demeyer et al., 2001). Stimulatory effects of cell-free preparations on aroma development (Hagen et al., 1996; Bruna et al., 2001) may be related to the presence of non-enzymatic stimulatory agents such as manganese in the preparation used (Hagen, pers. comm.).

18.9.2 Fermented meat products as functional foods

Improvement of the nutritional value of meat products has been tried for years, e.g. by replacement or lowering of fat (Arganosa et al., 1988) and salt content (Gimeno et al., 1999). More recently, the successful enrichment with calcium (Gimeno et al., 2000) and inulin (Mendoza et al., 2001) was reported and the concept of 'functional foods', e.g., foods containing naturally, or by addition, ingredients with clearly identified beneficial effects on a target function of the human body and/or lowering the risk of disease (Diplock et al., 1999), were introduced into the meat industry (Jiménez-Colmenero et al., 2001). A major target function could be the lowering of the consumer body's antioxidant protection system through the production of meat and meat products containing natural or added antioxidants. A well investigated case is the addition of vitamin E for improvement of colour stability, a characteristic reflecting the antioxidant status of the meat. In fresh meat, improvement of colour stability by vitamin E was much better when supplied with the diet, than post-mortem (Mitsumoto et al., 1993). For dry sausage production, these findings should stimulate the use of raw materials with improved antioxidant status through selection of animal and muscle and/or dietary treatments reflecting, e.g., glutathione peroxidase activity and soluble selenium content (Daun et al., 2001), conjugated linoleic acid content (Raes et al., 2001) and animals fed diets enriched in polyphenols (Lopez-Bote et al., 2000; Tang et al., 2000). Other possibilities are reflected in the identification of angiotensin I-converting enzyme inhibitors in the peptide fraction of fermented sausages (Arihara et al., 1999) and the introduction of probiotic starter cultures (Lücke, 2000; Erkhilä et al., 2001) (Sameshima et al., 1998).

18.10 References

ACTON J C and DICK R L (1976), 'Composition of some commercial dry sausages', *J Food Sci*, 41, 971–972.

ACTON J C and DICK R L (1977), 'Cured color development during fermented sausage processing', *J Food Sci*, 42 (4), 895–897.

ACTON J C, WILLIAMS J G and JOHNSON M G (1972), 'Effect of fermentation temperature on changes in meat properties and flavour of summer sausage', *J Milk Food Technol*, 35, 264–268.

ALLY G, COURS D and DEMEYER D (1992), 'Effect of Nitrate, Nitrite and Ascorbate on Colour and Colour Stability of Dry, Fermented Sausage Prepared Using "Back Slopping"', *Meat Sci*, 32, 279–287.

ANDERSEN L (2001), Pers. comm. Chr. Hansen A/S, Hørsholm, Denmark.

ANDERSEN S J and FRISVAD J C (1994), 'Penicillin production by *Penicillium nalgiovense*', *Appl Microb*, 19 (6), 486–488.

ANSORENA D, DE PEÑA M P, ASTIASARÁN I and BELLO J (1997), 'Colour Evaluation of Chorizo de Pamplona, a Spanish Dry Fermented Sausage: Comparison Between the CIE L*a*b* and the Hunter Lab Systems with Illuminants D65 and C', *Meat Sci*, 46 (4), 313–318.

ANSORENA D, ZAPELENA M J, ASTIASARÁN I and BELLO J (1998), 'Addition of Palatase M (lipase from Rhizomucor miehei) to dry fermented sausages: consequences on lipid fraction and study of the further oxidation process by GC-MS', *J Agric Food Chem*, 46 (8), 3244–3248.

ARGANOSA G C, HENRICKSON R L and RAO B R (1988), 'Collagen as a lean or fat replacement in pork sausage', *J Food Quality*, 10 (5), 319–333.

ARIHARA K, MUKAI T and ITOH M (1999), 'Angiotensin I-converting enzyme inhibitors derived from muscle proteins, *45th ICoMST*, Yokohama, Japan, I, 676–677.

BARBIERI G, BOLZONI L, PAROLARI G, VIRGILI R, BUTTINI R, CARERI M and MANGIA A (1992), 'Flavor compounds of dry-cured ham', *J Agric Food Chem*, 40, 2389–2394.

BARRIERE C, CENTENO D, LEBERT A, LEROY-SETRIN S, BERDAGUE J L and TALON R (2001), 'Roles of superoxide dismutase and catalase of *Staphylococcus xylosus* in the inhibition of linoleic acid oxidation' *FEMS Microb Let*, 201 (2), 181–185.

BERDAGUÉ J L, MONTEIL P, MONTEL M C and TALON R (1993), 'Effects of starter cultures on the formation of flavour compounds in dry sausage', *Meat Sci*, 35, 275–287.

BERGER R G, MACKU C, GERMAN J B and SHIBAMOTO T (1990), 'Isolation and identification of dry salami volatiles', *J Food Sci*, 55, 1239–1242.

BERTELSEN G, JAKOBSEN M, JUNCHER D, MØLLER J, KRÖGER-OHLSEN M, WEBER C and SKIBSTED L H (2000), 'Oxidation, shelf-life and stability of meat and meat products', *46th ICoMST*, Buenos Aires, Argentina, 516–524.

BETT K L and GRIMM C C (1994), 'Flavor and aroma its measurement' in: Pearson A M and Dutson T R, *Advances in Meat Research Volume 9: Quality Attributes and their Measurement in Meat, Poultry and Fish Products*, London, Blackie Acad Prof, 202–221.

BINGHAM A F, BIRCH G G, GRAAF C, DE BEHAN J M and PERRING K D (1990), 'Sensory studies with sucrose-maltol mixtures', *Chemical Senses*, 15 (4), 447–451.

BLANK I (1997), 'Gas chromatography – Olfactometry in food aroma analysis', in: Marsili R, *Techniques for analysing food aroma*, New York, Marcel Dekker Inc, 293–329.

BOLANDER C R, TOMA R B, DAVIS R M and MEDORA N P (1995), 'Irradiated versus fumigated spices in sausage', *Int J Food Sci Nutr*, 46, 319–325.

BRENNAN J G (1984), 'Texture Perception and Measurements', in: J R Piggott, *Sensory analysis of Foods*, Essex, England, Elsevier Appl Sci Publish Ltd, 59–91.

BROCKMANN R (1998), 'Nachweis von Naturfarbstoffen in Rohwurst. Ergebnisse eines Ringversuches', *Fleischwirtsch*, 78 (2), 143–145.

BRUNA J M, HIERRO E M, DE LA HOZ L, MOTTRAM D S, FERNANDEZ M and ORDONEZ J A (2001), 'The contribution of *Penicillium aurantiogriseum* to the volatile composition and sensory quality of dry fermented sausages', *Meat Sci*, 59, 97–107.

BRUNA J M, ORDOÑEZ J A, FERNANDEZ M, HERRANZ B and DE LA HOZ L (2001), 'Microbial and physico-chemical changes during the ripening of dry fermented sausages superficially inoculated with or having added an intracellular cell-free extract of Penicillium aurantiogriseum', *Meat Sci*, 59 (1), 87–96.

BUCHARLES C, GIRARD J-P, SIRAMI J and PASCAL S (1984), cited in Montel *et al.*, 1998.

BURGARD D R and KUZNICKI J T (1990), *Chemometrics: Chemical and sensory data*, Boca Raton, CRC Press.

CIE (1976) *Colorimetry*, 2nd edn, Commission Internationale de l'Eclairage, Vienna.

CLAEYS E, DE SMET S, DEMEYER D, GEERS R and BUYS N (2000), 'Muscle enzyme activities in two pig lines', *Meat Sci*, 57, 257–263.

COLAS B and SIMATOS D (1976), 'Développement de méthodes pour l'étude de la structure de produits homogénéisés à base de viande (saucisson sec)', *L'Alimentation et la vie*, 64 (2), 214–227.

CROIZET F, DENOYER C, TRAN N and BERDAGUÉ J L (1992), 'Les composés volatils du saucisson sec. Evolution au cour de la maturation', *Viandes Prod Carnés*, 13, 167–170.

CRONIN D A (1982), 'Techniques of analysis of flavours', in: Morton I D and MacLeod A J, *Food Flavours. Part A. Introduction*, Amsterdam, Elsevier Sci Publish Company, 15–48.

DAINTY R and BLOM H (1995), 'Flavour chemistry of fermented sausages', in: Campbell-Platt G and Cook P E, *Fermented Meats*, London, Blackie Acad Prof, 176–193.

DALTON P, DOOLITTLE N, NAGATA H and BRESLIN P A S (2000), 'The merging of the senses: integration of subthreshold taste and smell', *Nature Neurosci*, 3 (5), 431-432.

DAUN C, JOHANSSON M, ÖNNING G and KESSON B (2001), 'Glutathione peroxidase activity, tissue and soluble selenium content in beef and pork in relation to meat ageing and pig RN phenotype', *Food Chem*, 73, 313–319.

DE JAEGER I, VAN STEENKISTE J P and DEMEYER D (1984), 'Calculation of water activity in industrial sausage production', *30th Europ Meet Meat Res Work*, Bristol, England.

DEAK T and BEUCHAT L R (1996), *Handbook of food spoilage yeasts*, New York, CRC Press.

DELLAGLIO S, CASIRAGHI E and POMPEI C (1996), 'Chemical, Physical and Sensory Attributes for the Characterization of an Italian Dry-cured Sausage', *Meat Sci*, 42 (1), 25–35.

DEMEYER D (1982), 'Stoichiometry of dry sausage fermentation', *Antonie van Leeuwenhoek*, 48, 414–416.

DEMEYER D I (1992), 'Meat fermentation as an integrated process', in: Smulders F J M, Toldrá F, Flores J and Prieto M, *New Technologies for Meat and Meat Products*, ECCEAMST, 21–36.

DEMEYER D (1997), 'An introduction to the OECD programme: meat quality and

the quality of animal production', *Food Chem*, 59 (4), 491–497.
DEMEYER D (2000), 'The relative contribution of protein vs lipid and meat vs microbes to flavour development in dry and fermented sausage', *46th ICoMST*, Buenos Aires, August, 526–527.
DEMEYER D (2001), 'Control of bioflavour and safety in Northern and Mediterranean fermented meat products (FMP) -FAIR CT 97-3227 : Extended executive summary' 53 pp. (available at : http://fltbwww.rug.ac.be/animalproduction).
DEMEYER D I and SAMEJIMA K (1991), 'Animal Biotechnology and Meat Processing', in: Fiems L, Cottyn B G and Demeyer D I, *Animal Biotechnology and the Quality of Meat Production. Developments in Animal and Veterinary Sci*, Amsterdam, Elsevier Sci Publ, 25, 127–143.
DEMEYER D, HOOZEE J and MESDOM H (1974) 'Specificity of lipolysis during dry sausage ripening', *J Food Sci*, 39, 293–296.
DEMEYER D I, VANDEKERCKHOVE P and MOERMANS R (1979), 'Compounds determining pH in dry sausage', *Meat Sci,* 3, 161–167.
DEMEYER D I, VERPLAETSE A and GISTELINCK M (1986), 'Fermentation of meat: an integrated process', *32nd Europ Meet Meat Res Work*, Ghent, Belgium, 241–246.
DEMEYER D, CLAEYS E, DENDOOVEN R and VOSS AKERØ B (1984), 'The effect of starter cultures on stoichiometry and kinetics of dry sausage metabolism', 30th *Europ Meet Meat Res Work*, Bristol, England, 282–283.
DEMEYER D, CLAEYS E, ÖTLES S, CARON L and VERPLAETSE A (1992), 'Effect of meat species on proteolysis during dry sausage fermentation', *38th ICoMST*, Clermont-Ferrand, France, 775–778.
DEMEYER D, RAEMAEKERS M, RIZZO A, HOLCK A, DE SMEDT A, TEN BRINK B, HAGEN N, MONTEL C, ZANARDI E, MURBREKK E, LEROY F, VANDENDRIESSCHE F, LORENTSEN K, VENEMA K, SUNESEN L, STAHNKE L, DE VUYST L, TALON R, CHIZZOLINI R and EEROLA S (2000), 'Control of bioflavour and safety in fermented sausages: first results of a European project', *Food Res Int*, 33, 171–180.
DIAZ O, FERNÁNDEZ M, GARCIA-DE F G, HOZ L DE LA and ORDÓNEZ J A (1997), 'Proteolysis in dry fermented sausages: the effect of selected exogenous proteases', *Meat Sci*, 46, 115–128.
DIPLOCK A T, AGGETT P J, ASHWELL M, BORNET F, FERN E B and ROBERFROID MB (1999), 'Scientific concepts of functional foods in Europe consensus document', *Brit J Nutr*, 81 (4), S1–S27.
DOMINGUEZ F M and ZUMALACÁRREGUI R J (1991), 'Lipolytic and oxidative changes in 'chorizo' during ripening', *Meat Sci*, 29, 99–107.
EKLOV T, JOHANSSON G, WINQUIST F and LUNDSTROM I (1998), 'Monitoring sausage fermentation using an electronic nose', *J Sci Food Agric*, 76 (4), 525–532.
EMBL DATABASE (2001), Nucleotide accession number AJ279090.
ENCINAS J-P, LOPEZ-DIAZ T-M, GARCIA-LOPEZ M-L, OTERO A and MORENO B (2000), 'Yeast populations on Spanish fermented sausages', *Meat Sci,* 54 (3), 203–208.

ERKKILÄ S, PETÄJÄ E, EEROLA S, LILLEBERG L, MATTILA-SANDHOLM T and SUIKHO M-L (2001), 'Flavour profiles of dry sausages fermented by selected novel meat starter cultures', *Meat Sci*, 58, 111–116.

FERNÁNDEZ M C D and RODRIGUEZ J M Z (1991), 'Lipolytic and oxidative changes in chorizo during ripening', *Meat Sci*, 29, 99–107.

FERNÁNDEZ-FERNÁNDEZ E, VÁZQUEZ-ODÉRIZ M L and ROMERO-RODRÍGUEZ M A (1998), 'Colour changes during manufacture of Galician chorizo sausage', *Z Lebensm Unters Forsch A*, 207, 18–21.

FOURNAUD J (1978), 'La microbiologie du saucisson sec', *L'Alimentation et la vie*, 64 (2–3), 82–92.

FOX P F (1989), 'Proteolysis during cheese manufacturing and ripening' *J Dairy Sci*, 72, 1379–1400.

GARCIA M L, CASAS C, TOLEDO V M and SELGAS M D (2001), 'Effect of selected mould strains on the sensory properties of dry fermented sausages', *Europ Food Res Technol*, 212, 287–291.

GEHLEN K H, MEISEL C, FISCHER A and HAMMES W P (1991), 'Influence of the yeast *Debaryomyces hansenii* on dry sausage fermentation', 37th *ICoMST*, Kulmbach, Germany, 2, 871–876.

GEISEN R (1993), 'Fungal starter cultures for fermented foods: molecular aspects', *Trends Food Sci & Technol*, 4, 251–256.

GEISEN R, LÜCKE F-K and KRÖKEL L (1992), 'Starter and protective cultures for meat and meat products', *Fleischwirtsch*, 72, 894–898.

GELLYNCK X and VERBEKE W (2001), 'Consumer perception of traceability in the meat chain', *Agrarwirtsch*, 50 (6), 368–374.

GIMENO O, ASTIASARÁN I and BELLO J (1999), 'Influence of Partial Replacement of NaCl with KCl and $CaCl_2$ on Texture and Color of Dry Fermented Sausages', *J Agric Food Chem*, 47, 873–877.

GIMENO O, ASTIASARÁN I and BELLO J (2000), 'Calcium ascorbate as a potential partial substitute for NaCl in dry fermented sausages: effect on colour, texture and hygienic quality at different concentrations', *Meat Sci*, 0, 1–7.

GOTTSCHALK G (1986), *Bacterial Metabolism, 2nd ed.*, New York, Springer-Verlag, 208–282.

GROSCH W (1982), 'Lipid degradation products and flavor', in: Morton I D, MacLeod A J, *Food Flavors, part A. Introduction*, Amsterdam, Elsevier, 325–398.

HAGEN B F, BERDAGUÉ J-L, HOLCK A L, NÆS H and BLOM H (1996), 'Bacterial Proteinase Reduces Maturation Time of Dry Fermented Sausages', *J Food Sci*, 61 (5), 1024–1029.

HAGEN B F, NÆS H and HOLCK A L (2000), 'Meat starters have individual requirements for Mn^{2+}', *Meat Sci*, 55, 161–168.

HANSEN-MØLLER J, HINRICHSEN L and JACOBSEN T (1997), 'Evaluation of peptides generated in Italian-style dry-cured ham during processing', *J Agric Food Chem*, 45, 3123–3128.

HARNIE E, CLAEYS E, RAEMAEKERS M and DEMEYER D (2000), 'Proteolysis and lipolysis in an aseptic meat model system for dry sausages', *46th ICoMST*,

Buenos Aires, Argentina, 238–239.
HARTMAN T G, LECH J, KARMAS K, SALINAS J, ROSEN R T and HO C-T (1993), 'Flavor characterization using adsorbent trapping-thermal desorption or direct thermal desorption and gas chromatography-mass spectrometry', in: HO C-T and Manley C H, *Flavor Measurement*, New York, Marcel Dekker, 37–60.
HENRIKSEN A P and STAHNKE L H (1997), 'Sensory and Chromatographic Evaluations of Water Soluble Fractions from Dried Sausages', *J Agric Food Chem*, 45, 2679–2684.
HIERRO E, HOZ L DE LA and ORDÓNEZ J A (1997), 'Contribution of microbial and meat endogenous enzymes to the lipolysis of dry fermented sausages', *J Agric Food Chem* 45, 2989–2995.
HINRICHSEN L L and PEDERSEN S B (1995), 'Relationship among flavor, volatile compounds, chemical changes, and microflora in Italian-type dry-cured ham during processing', *J Agric Food Chem*, 43, 2932–2940.
HONIKEL K O (1998), 'Reference Methods for the Assessment of Physical Characteristics of Meat', *Meat Sci*, 49 (4), 447–457.
HOUBEN J H and KROL B (1986), 'Effect of ascorbate and ascorbyl palmitate on lipid oxidation in semi-dry sausages manufactured from pork materials differing in stability towards oxidation', *Meat Sci*, 17 (3), 199–211.
HUNTER R S (1975) *The Measurement of Appearance*, New York, Wiley Interscience.
HURRELL R F (1982), 'Maillard reaction in flavour', in: Morton I D and MacLeod A J, *Food Flavours. Part A. Introduction*, Amsterdam, Elsevier Sci Publish, 399–437.
JAGO J, FISHER A and LE NEINDRE P (2000), 'Animal welfare and product quality', in: Balázs G E, Lynch J M, Schepers J S, Toutant J-P, Werner D and Werry P A Th, '*Biological Resource Management*', New York, Springer, 163–171.
JESSEN B (1995), 'Starter cultures for meat fermentations', in: Campbell-Platt G and Cook P E, *Fermented Meats*, London, Blackie Acad Prof, 130–159.
JIMÉNEZ-COLMENERO F, CARBALLO J and Cofrades S (2001), 'Healthier meat and meat products: their role as functional foods', *Meat Sci*, 59 5–13.
JOHANSSON G, BERDAGUÉ J L, LARSSON M, TRAN N and Borch E (1994), 'Lipolysis, proteolysis and formation of volatile components during ripening of a fermented sausage with *Pediococcus pentosaceus* and *Staphylococcus xylosus* as starter cultures', *Meat Sci*, 38, 203–218.
KAMINSKI E, STAWICKI S and WASOWICZ E (1974), 'Volatile flavor compounds produced by molds of *Aspergillus, Penicillium,* and fungi imperfecti', *Appl Microb,* 27, 1001–1004.
KATSARAS K and BUDRAS K-D (1992), 'Microstructure of Fermented Sausage', *Meat Sci*, 31, 121–134.
KLETTNER P-G (1989), 'Zur Herstellung feinzerkleinerter, streichfähiger Rohwurst', *Mitteilungsblatt der Bundesanstalt für Fleischforschung*, Kulmbach, 104, 214–220.

KOTTKE V, DAMM H, FISCHER A and LEUTZ U (1996), 'Engineering Aspects in Fermentation of Meat Products', *Meat Sci*, 46, S243–S255.

LAMBREGTS L, RAEMAEKERS M and DEMEYER D (1998), 'Peptides involved in flavour development of dry fermented sausage', *Med Fac Landbouww Univ Gent*, 63 (4b), 1537.

LAND D G and SHEPHERD R (1984), 'Scaling and Ranking Methods', in: Piggott J R, *Sensory analysis of Foods*, Essex, England, Elsevier Appl Sci Publish, 141–177.

LARROUTURE C, ARDAILLON V, PÉPIN M and MONTEL MC (2000), 'Ability of meat starter cultures to catabolize leucine and evaluation of the degradation products by using an HPLC method', *Food Microb*, 17 (5), 563–570.

LARSEN T O (1998), 'Volatile flavour production by *Penicillium caseifulvum*', *Int Dairy J*, 8, 883–887.

LOIS A L, GUTIÉRREZ L M, ZUMALACÁRREGUI J M and LÓPEZ A (1987), 'Changes in several constituents during the ripening of 'Chorizo' A Spanish dry sausage', *Meat Sci*, 19, 169–177.

LOPEZ-BOTE C J (2000) 'Dietary treatment and quality characteristics in Mediterranean meat products', in: Decker E, Faustman C and Lopez-Bote C J (eds) *Antioxidants in Muscle Foods: Nutritional strategies to improve quality*, New York, Wiley Interscience, pp. 345–366.

LÜCKE F K (1998), 'Fermented sausages', in: Wood B J B, *Microbiology of fermented foods, 2nd ed*, London, Blackie Acad Prof, 2, 441–483.

LÜCKE F K (2000), 'Utilization of microbes to process and preserve meat', *Meat Sci*, 56 (2), 105–115.

MARTUSCELLI M, CRUDELE M A, FIORE C, SUZZI G, AMATO D, DI GENNARO P and LAURITA C (2000), 'Effect of indigenous starter cultures on the quality of "salsiccia sotto sugna" salami', *Industrie Alimentari*, XXXIX, 817–823.

MASSON F, HINRICHSEN L, TALON R and MONTEL M C (1999), 'Factors influencing leucine catabolism by a strain of *Staphylococcus carnosus*', *Int J Food Microb*, 49, 173–178.

MATEO J and ZUMALACÁRREGUI J M (1996), 'Volatile compounds in chorizo and their changes during ripening', *Meat Sci*, 44, 255–273.

MEILGAARD M, CIVILLE G V and CARR B T (1991), *Sensory evaluation techniques, 2nd ed*, Boca Raton, CRC Press.

MELENDO JA, BELTRAN JA, JAIME I, SANCHO R and RONCALES P (1996), 'Limited proteolysis of myofibrillar proteins by bromelain decreases toughness of coarse dry sausage', *Food Chem*, 57 (3), 429–433.

MENDOZA E, GARCÍA M L, CASAS C and SELGAS M D (2001), 'Inulin as fat subsitute in low fat, dry fermented sausage', *Meat Sci*, 57, 387–393.

MERRITT JR C and ROBERTSON D H (1982), 'Techniques of analysis of flavours: Gas chromatography and mass spectrometry', in: Morton I D and MacLeod A J, *Food Flavours. Part A. Introduction*, Amsterdam, Elsevier Sci. Publish. Company, 49–78.

METAXOPOULOS J, STAVROPOULOS S, KAKOURI A and SAMELIS J (1996), 'Yeasts isolated from traditional Greek dry salami', *Ital J Food Sci*, 8 (1), 25–32.

MEYNIER A, NOVELLI E, CHIZZOLINI R, ZANARDI E and GANDEMER G (1999), 'Volatile compounds of commercial Milano salami', *Meat Sci*, 51, 175–183.
MITEVA E, KIROVA E, GADJEVA and RADEVA M (1986), 'Sensory aroma and taste profiles of raw-dried sausages manufactured with a lipolytically active yeast culture', *Nahrung*, 30, 829–832.
MITSUMOTO M, ARNOLD R N, SCHAEFER D M and CASSENS R G (1993), 'Dietary Versus Postmortem Supplementation of Vitamin E on Pigment and Lipid Stability in Ground Beef', *J Anim Sci*, 71, 1812–1816.
MOLLY K, DEMEYER D, JOHANSSON G, RAEMAEKERS M, GHISTELINCK M and GEENEN I (1997), 'The importance of meat enzymes in ripening and flavour generation in dry fermented sausages. First results of a European project', *Food Chem*, 59 (4), 539–545.
MONTEL M C, MASSON F and TALON R (1998), 'Bacterial Role in Flavour Development', *Meat Sci*, 49 (suppl 1), S111–S123.
MONTEL M C, REITZ J, TALON R, BERDAGUÉ J L and ROUSSET A S (1996), 'Biochemical activities of Micrococcaceae and their effects on the aromatic profiles and odours of a dry sausage model', *Food Microb*, 13, 489–499.
MOTTRAM D S (1998), 'Flavour formation in meat and meat products: A review', *Food Chem,* 62, 415–424.
NÆS H, HOLCK A L, AXELSSON L, ANDERSEN H J and BLOM H (1995), 'Accelerated ripening of dry fermented sausage by addition of a *Lactobacillus* proteinase', *Int J Food Sci Technol*, 29, 651–659.
NAGY A, MIHÁLYI V and INCZE K (1989), 'Ripening and storage of Hungarian salami. Chemical and organoleptic changes', *Fleischwirtsch*, 69, 587–588.
NAVARRO J L, NADAL M I, IZQUIERDO L and FLORES J (1997), 'Lipolysis in dry cured sausages as affected by processing conditions', *Meat Sci*, 45, 161–168.
NISHIMURA T, RHUE M R, OKITANI A and KATO H (1988), 'Components contributing to the improvement of meat taste during storage', *Agric Biol Chem*, 52, 2323–2330.
OLESEN P T and STAHNKE L H (2000), 'The influence of *Debaryomyces hansenii* and *Candida utilis* on the aroma formation in garlic spiced fermented sausages and model minces', *Meat Sci*, 56, 357–368.
ORDOÑEZ J A, HIERRO E M, BRUNA J M and HOZ L DE LA (1999), 'Changes in the components of dry-fermented sausages during ripening', *Crit Rev Food Sci Nutr*, 39, 329–367.
PARLIAMENT T H (1997), 'Solvent Extraction and Distillation Techniques', in: Marsili R, *Techniques for analysing food aroma*, New York, Marcel Dekker Inc, 1–26.
PATARATA L, JUDAS I, SILVA J A, ESTEVES A and MARTINS C (1997), 'Relationship between sensory, physical and chemical parameters of a Portuguese traditional sausage – "Alheira"', *43rd ICoMST*, Auckland, 446–447.
PEARSON A M and GILLETT T A (1996), *Processed Meats*, 3rd edn, New York,

Chapman & Hall, 242–290.
PIGGOTT J R (1984), *Sensory analysis of foods*. London, Elsevier.
PRIBIS V and SVRZIC G (1995), 'Farbbildung in Rohwürsten während der Herstellung. Einsatz moderner Farmesysteme bei der Untersuchung', *Fleischwirtsch*, 75 (6), 819–821.
RAES K, DE SMET S and DEMEYER D (2001), 'Effect of double-muscling in Belgian Blue young bulls on the intramuscular fatty acid composition with emphasis on conjugated linoleic acid and polyunsaturated fatty acids' *Anim Sci*, 73, 253–260
RÖDEL W (1985), 'Rohwurstreifung: Klima und andere Einflussgrössen' in: *Mikrobiologie und Qualität von Rohwurst und Rohschinken*, Bundesanstalt für Fleischforschung, Kulmbacher Reihe, 5, 60–84.
RONCALES P, AGUILERA M, BELTRAN J A, JAIME I and PEIRO J M (1991), 'The effect of natural or artificial casing on the ripening and sensory quality of a mould-covered dry sausage', *Int J Food Sci Technol*, 26, 83–89.
ROOD D (1999), *A practical guide to the care, maintenance, and troubleshooting of capillary gas chromatographic systems*, 3rd revised ed, Weinham, Wiley-VCH, 156–209.
ROTHE M (1988), *Introduction to aroma research*, Dordrecht, Kluwer Academic Publish, 1–7.
ROUDOT A-C, DUPRAT F, GROTTE M-G and O'LIDHA G (1992), 'Objective measurement of the visual aspect of dry sausage slices by image analysis', *Food Structure*, 11, 351–359.
SAMESHIMA T, MAGOME C, TAKESHITA K, ARIHARA K, ITOH M and KONDO Y (1998), 'Effect of intestinal *Lactobacillus* starter cultures on the behaviour of *Staphylococcis aureus* in fermented sausage', *Int J Food microb*, 41 (1), 1–7.
SANTAMARIA I, LIZARRAGA T, IRIATE J, ASTIASARAN I and BELLO J (1994), 'Die Korrelation zwischen sensorischen und den auf Stickstofffraktionen bezogenen Parametern bei spanischer Rohwurst', *Fleischwirtsch*, 74 (9), 999–1000.
SANZ Y, VILA R, TOLDRÁ F, NIETO P and FLORES J (1997), 'Effect of nitrate and nitrite curing salts on microbial changes and sensory quality of rapid ripened sausages', *Int J Food Microb*, 37, 225–229.
SCANDINAVIAN COLOR INSTITUTE (2001), www.ncscolor.com.
SCHMIDT S and BERGER R G (1998), 'Aroma compounds in fermented sausages of different origins' *Lebensm-Wiss u-Technol*, 31, 559–567.
SELGAS M D, CASAS C, TOLEDO V M and GARCÍA M L (1999), 'Effect of selected mould strains on lipolysis in dry fermented sausages', *Europ Food Res Technol*, 209 (5), 360–365.
SØNDERGAARD A and STAHNKE L H (2002), 'Growth and aroma production by *Staphylococcus xylosus, Staphylococcus carnosus* and *Staphylococcus equorum* – a comparative study in model systems', *Int J Food Microb*, in press.
SØRENSEN B B (1997), 'Lipolysis of pork fat by the meat starter culture

Debaryomyces hansenii at various environmental conditions', *Int J Food Microb*, 34, 187–193.

SØRENSEN B B and JAKOBSEN M (1996), 'The combined effects of environmental conditions related to meat fermentation on growth and lipase production by the starter culture *Staphylococcus xylosus*', *Food Microb*, 13, 265–274.

STAHNKE L H (1994), 'Aroma components from dried sausages fermented with *Staphylococcus xylosus*', *Meat Sci*, 38 (1), 39–53.

STAHNKE L H (1995a), 'Dried sausages fermented with *Staphylococcus xylosus* at different temperatures and with different ingredient levels – Part I. Chemical and Bacteriological Data', *Meat Sci*, 41 (2), 179–191.

STAHNKE L H (1995b), 'Dried sausages fermented with *Staphylococcus xylosus* at different temperatures and with different ingredient levels – Part II. Volatile components', *Meat Sci*, 41 (2), 193–209.

STAHNKE L H (1995c), 'Dried sausages fermented with *Staphylococcus xylosus* at different temperatures and with different ingredient levels – Part III. Sensory evaluation', *Meat Sci*, 41 (2), 211–223.

STAHNKE L H (1998), 'Character impact aroma compounds in fermented sausage', *44th ICoMST*, Barcelona, Spain, II 786–787.

STAHNKE L H (1999), 'Volatiles produced by *Staphylococcus xylosus* and *Staphylococcus carnosus* during growth in sausage minces. Part II. The Influence of Growth Parameters', *Lebensm-Wiss u-Technol*, 32, 365–371.

STAHNKE L H (2000), '2-acetyl-1-pyrroline – key aroma compound in Mediterranean dried sausages', in: Schieberle P and Engel K-H, *Frontiers of Flavour Science*, Garching, Deutsche Forschungsanstalt für Lebensmittelchemie, 361–365.

STAHNKE L H, SUNESEN L O and DE SMEDT A (1999), 'Sensory characteristics of European dried fermented sausages and the correlation to volatile profile', *Med Fac Landbouw Univ Gent*, 64 (5b), 559–566.

STAHNKE L H, HOLCK A, JENSEN A, NILSEN A and ZANARDI E (2002), 'Maturity acceleration by *Staphylococcus carnosus* in fermented sausage – relationship between maturity and flavor compounds', *J Food Sci*, in press.

STIEBING A (1994), 'Kritische Kontrollpunkte bei der Herstellung von Rohwurst', *Fleisch*, 48 (4), 297–301.

STIEBING A (1998), 'Einfluss von Proteinen auf den Reifungsverlauf von Rohwurst' *Fleischwirtsch*, 78 (11), 1140–1144.

STIEBING A, KLETTNER P-G and MÜLLER W-D (1981), 'Energieverbrauch bei der Fleischwarenherstellung', *Fleischwirtsch*, 61 (8), 1138–1145.

STIEBING A and RÖDEL W (1987a), 'Einfluß der Luftgeschwindigkeit auf den Reifungsverlauf bei Rohwurst', *Fleischwirtsch*, 67 (3), 236–240.

STIEBING A and RÖDEL W (1987b), 'Einfluß der relativen Luftfeuchtigkeit auf den Reifungsverlauf bei Rohwurst', *Fleischwirtsch*, 67 (9), 1020–1030.

STIEBING A and RÖDEL W (1989), 'Influence of pH on the drying behavior of dry sausage', *Fleischwirtsch*, 69 (10), 1530–1538.

STIEBING A and RÖDEL W (1992), 'Kontinuierliches Messen der Oberflächen-

Wasseraktivität von Rohwurst', *Fleischwirtsch*, 72 (4), 432–438.
STRIKE D J, MEIJERINK M G H and KOUDELKA-HEP M (1999), 'Electronic noses A mini-review', *Fresenius J Anal Chem*, 364, 499–505.
SUNESEN L O and STAHNKE L H (2001), 'Mould starter cultures for dry sausage – selection, application and effects', *Meat Sci*, submitted.
SUNESEN L O, DORIGONI V, ZANARDI E and STAHNKE L H (2001), 'Volatile compounds released during ripening in Italian dried sausage', *Meat Sci*, 58, 93–97.
TALON R, WALTER D and MONTEL M C (2000), 'Growth and effect of staphylococci and lactic acid bacteria on unsaturated free fatty acids', *Meat Sci*, 54, 41–47.
TANG S Z, KERRY J P, SHEEHAN D, BUCKLEY D J and MORRISSEY P A (2000), 'Dietary tea catechins and iron-induced lipid oxidation in chicken meat, liver and heart', *Meat Sci*, 56, 285–290.
TEN CATE L (1960), cited in Demeyer *et al.*, 1986.
TOLDRÁ F and FLORES M (1998), 'The Role of Muscle Proteases and Lipases in Flavor Development During the Processing of Dry-Cured Ham', *Crit Rev Food Sci*, 38 (4), 331–352.
TOLDRÁ F and FLORES M (2000), 'The use of muscle enzymes as predictors of pork meat quality', *Food Chem*, 69, 387–395.
TOLEDO V M, SELGAS M D, CASAS M C, ORDÓÑEZ J A and GARCIA M L (1997), 'Effect of selected mould strains on proteolysis in dry fermented sausages', *Z Lebensm Untersuch Forsch*, 204 (5), 385–390.
TOURAILLE C and SALÉ P (1976), 'Etude par des méthodes physiques et sensorielles de la consistance du saucisson sec', *L'Alimentation et la vie*, 64 (2), 192–213.
TOWNSEND W E, DAVIS C E, LYON C E and MESCHER S E (1980), 'Effect of pork quality on some chemical, physical, and processing properties of fermented dry sausage', *J Food Sci*, 45, 622–626.
TRIGUEROS G, GARCÍA M L, CASAS C, ORDÓÑEZ J A and SELGAS M D (1995), 'Proteolytic and lipolytic activities of mould strains isolated from Spanish dry fermented sausages', *Z Lebensm Untersuch Forsch*, 201 (3), 298–302.
ÜREN A and BABAYIĞIT D (1997), 'Colour Parameters of Turkish-type Fermented Sausage During Fermentation and Ripening', *Meat Sci*, 45 (4), 539–549.
VAN RUTH S M and ROOZEN J (1995), 'Volatile compounds of rehydrated French beans, bell peppers and leeks. Part 1. Flavour release in the mouth and in three mouth model systems', *Food Chem*, 53 (1), 15-22.
VAN 'T HOOFT B-J (1999), 'Development of binding and structure in semi-dry fermented sausages. A multifactorial approach', *Doctoral Thesis*, University of Utrecht, Faculty of Veterinary Medicine, 162 pp.
VANDENDRIESSCHE F, VANDEKERCKHOVE P and DEMEYER D (1980), 'The influence of spices on the fermentation of a Belgian dry sausage', *26th Europ Meet Meat Res Work*, Colorado Springs, USA, 128–133.
VERNAT-ROSSI V, GARCIA C, TALON R, DENOYER C and BERDAGUÉ J-L (1996), 'Rapid discrimination of meat products and bacterial strains using

semiconductor gas sensors', *Sensors and Actuators B*, 37, 43–48.
VERPLAETSE A (1994a), 'Influence of raw meat properties and processing technology on aroma quality of raw fermented meat products', *40th ICoMST*, The Hague, Netherlands, 45–65.
VERPLAETSE A (1994b), 'Muscle proteinases and lipases and quality of fermented meat products', *Med Fac Landbouww Univ Gent*, 59 (4b), 2229–2240.
VERPLAETSE A, DE BOSSCHERE M and DEMEYER D (1989), 'Proteolysis during dry sausage ripening', *35th ICoMST*, Copenhagen, Denmark, 3, 815–818.
VERPLAETSE A, VAN HOYE S and DEMEYER D (1990), 'The effect of chopping conditions on dry sausage metabolism', *36th ICoMST*, Havana, Cuba, III, 920–927.
VERPLAETSE A, DEMEYER D, GERARD S and BUYS E (1992), 'Endogenous and bacterial proteolysis in dry sausage fermentation', *38th ICoMST*, Clermont-Ferrand, France, 4, 851–854.
VIALLON C, BERDAGUÉ J L, MONTEL M C, TALON R, MARTIN J F, KONDJOYAN N and DENOYER C (1996), 'The effect of stage of ripening and packaging on volatile content and flavour of dry sausage', *Food Res Int*, 29, 667–674.
WAMPLER T P (1997), 'Analysis of food volatiles using headspace-gas chromatographic techniques', in: Marsili R, *Techniques for analysing food aroma*, New York, Marcel Dekker Inc, 27–58.
WESTALL S (1998), 'Characterisation of yeast species by their production of volatile metabolites', *J Food Myc*, 1 (4), 187–202.
ZALACAIN I, ZAPELENA M J, PAZ-DE P M, ASTIASARÁN I and BELLO J (1997), 'Use of lipase from *Rhizomucor miehei* in dry fermented sausages elaboration: microbial, chemical and sensory analysis', *Meat Sci* 45, 99–105.
ZANARDI E, NOVELLI E, GHIRETTI G P, DORIGONI V and CHIZZOLINI R (2000) 'Oxidative stability of lipids and cholesterol in salame Milano, coppa and Parma ham: dietary supplementation with vitamin E and oleic acid', meat sci 55, 169–175.

19
New techniques for analysing raw meat quality

A. M. Mullen, The National Food Centre, Dublin

19.1 Introduction

Analysis of raw meat quality encompasses a large variety of attributes and analytical methods. In the following section a broad definition of meat quality is presented and those attributes which are more pertinent to this discussion are highlighted. Laboratory-based methods tend to require a large input of resources. A brief overview, of some of the more popular laboratory methods is outlined in Section 19.2. There is a huge requirement for more rapid methods for the estimation of meat quality. Such methods need to be accurate, precise and suitable for use in a commercial situation. Currently, only a few early *postmortem* measurements are recorded in meat plants in attempts to obtain advance knowledge regarding quality (Section 19.3). However, a wide array of novel techniques has been tested for their usefulness as indicators of meat quality. Section 19.4 details the findings of some of these new methodologies. Animal genomics has benefited greatly from the advances that have recently been witnessed in the analysis of the human genome. Section 19.5 highlights some of the prospects this technology offers to deepen our understanding of meat quality and to the development of objective methods for analysing raw meat quality. While attempts have been made to present many of the recent advances in the area of meat quality analysis this chapter is not to be considered a definitive list of laboratory-based methods and relevant emerging technologies.

19.2 Defining meat quality

Meat quality can be defined in various ways from palatability to technological aspects to safety. A common definition of quality is that it is a 'measure of traits that are sought and valued by the consumer'. Hoffman (1990) described meat quality as the 'sum of all quality factors of meat in terms of the sensoric, nutritive, hygienic and toxicological and technological properties.' In the following chapter most emphasis will be on sensory, nutritive and technological aspects of meat quality. Sensory properties include tenderness, flavour and colour while nutritive factors include fat, protein and connective tissue content. Technological factors include such parameters as water holding capacity, pH, water distribution, etc. Most references will be regarding bovine, porcine and ovine muscle.

19.2.1 Difficulties for the meat industry
The variability of meat quality prevents the meat industry marketing its produce according to quality. Despite much work in understanding the scientific basis of quality attributes (tenderness, colour, water-holding capacity, juiciness) their evaluation, prediction and control remain most elusive within the meat processing plant (i.e. within 48 h post-slaughter). As a result meat produced today can not be guaranteed to possess the best quality attributes as its quality can only be truly assessed after purchase. Therefore the marketability of meat, the consistency of quality and the guaranteeing of set standards of product are made very difficult. The reasons for the variability of meat quality are numerous, but they emanate from the fact that these quality attributes are altered from post-slaughter conditions, right along the production chain, into the beef processing plant, in retail outlet and even in the purchaser's home.

19.2.2 Why the need for objective methods
Most of the laboratory-based methods (see below) require an expenditure of time, personnel and cost. Most procedures are generally not quick enough or adaptable enough to an 'on-line' or 'at-line' situation. Ideally, the ultimate eating quality of meat needs to be predicted in the early *postmortem* period (Mullen *et al.*, 1998a and Troy *et al.*, 1998). By this, we mean within 24–48 h post-slaughter, during which time the carcass is within the confines of the meat factory. There is also a need for a method of assessing meat quality at the point of sale. Presently, routine methods of measuring meat quality, within a typical European meat processing plant, revolve around a few measurements. These include measurements of both pH and temperature. Commercially available probes include hand held probes for the measurement of the electrical parameters of conductivity and impedance.

19.2.3 Laboratory-based methods
Many objective and subjective, laboratory-based, methods for characterising meat quality have been developed to aid the comprehensive assessment of quality attributes. The whole area of sensory analysis provides a complex array of tools for deciphering specific details relating to meat quality. The sensory

assessment depends on three principal considerations. First there are appearance characteristics including colour, form, size, shape, integrity, viscosity, etc. Second are textural characteristics, which may include tenderness, firmness, mouthfeel, bite and chewability. The third principal consideration includes flavour factors such as taste, odour, off flavours. In general the assessment of these sensory attributes requires trained panels of judges who can minimise subjectivity, (Singhal et al., 1997). In some circumstances, where untrained panelists are used, larger numbers of assessors may be required. A variety of sensory tests are available to the researcher each providing quite distinct types of information. Analytical tests include, for example, difference test such as triangular and paired comparisons, which determine if detectable differences exist between different samples. Descriptive analysis is used to quantitatively and qualitatively characterise sensory attributes. A scale used to measure the degree of liking or disliking is usually called a hedonic scale. Much use has been made in food research of this hedonic scale developed by Peryam and Pilgrim (1957) (see Love, 1994 for review on acceptability evaluation).

A wide range of objective tests are available for meat quality assessment (Chrystall, 1994). Some of the physical methods which have been developed to predict tenderness, as assessed by a sensory panel, include measuring the force required to shear, penetrate, bite, mince, compress or stretch meat (Szczesniak, 1963; Szczesniak and Torgeson, 1965). To date, no means with sufficient precision have been developed or identified to predict cooked meat tenderness, with sufficient accuracy, during the early *postmortem* period. However, the use of the Warner-Bratzler shear force (WBSF) blade (Warner, 1928; Bratzler and Smith, 1963) has become a standard method for estimating tenderness from cooked meat samples. Recently, in the US, a more rapid method for estimating tenderness has been developed (Shackelford et al., 1999). Many research groups have determined the relationship between subjective and objective measurements of tenderness and in many instances 'r' values of above 0.7 are observed.

Instrumental analysis of flavour volatiles requires precise GC methods following extraction and concentration of the compounds. A portion of the column effluent can be diverted from the end of the column, prior to the detector and routed through a port to be sniffed by a human subject for qualitative evaluation. Splitting the effluent in this manner with the use of a sniffer port allows for simultaneous human evaluation and instrumental qualification. The use of mass spectrometry greatly enhances this analysis. Other instruments such as high performance liquid chromatography (HPLC) are also used for flavour analysis.

The measurement of the water-holding capacity (WHC) of meat is carried out in many different ways (see Honikel and Hamm 1994, for review). All measure the inherent ability of the cellular and subcellular structures of meat to hold on to part of its own and/or added water. In spite of the variation in methods used, there are three main treatments that give rise to three basic procedures for measuring WHC, i.e. (i) applying no force, (ii) applying mechanical force and (iii) applying thermal force.

Colour is usually considered the most important sensory characteristic in the appearance of meat. Various systems exist for the objective measurement of colour with CIE L*a*b* or Hunter Lab being quite popular (see MacDougall, 1994 for review).

In the past meat was often described in a simplistic manner as a combination of muscle fat and moisture and small amounts of non-combustible material (ash). However, with new labeling laws, growing interest and focus on the nutritional value of food meat compositional analysis encompasses many other attributes (Ellis, 1994). In recent years, in particular, there has been a lot of interest in the analysis of fatty acids. Because of their reported health benefits the ratio of unsaturated to saturated fats, and the level of individual fatty acids such as conjugated linoleic acid have received a lot of attention. GC and HPLC are commonly used methods in the analysis of fatty acids.

19.3 Current state of art techniques

19.3.1 pH

A knowledge of pH and its importance in the quality of meat is an essential element in meat quality measurements. *Postmortem* glycolysis results in the accumulation of lactic acid and a decline in the pH of the muscle from about 7.2, at death, to roughly 5.5 after *rigor mortis* onset (Geesink, 1993). In pork, rapid pH decline can result in pale, soft, exudative (PSE) meat, which presents a large problem to pork producers. To date the most effective predictor of the occurrence of this condition is the measurement of the early *postmortem* pH decline. In pork pH measurements at 45 min *postmortem* (pH_{45}) are used to detect the presence of PSE conditions (Somers *et al.*, 1985). Recently Kircheim *et al.* (2001) showed pH_{45} displayed a high degree of reliability in correctly predicting PSE and RFN (reddish-pink, firm, non-exudative) meat while Eikelenboom *et al.* (1996) suggest ultimate pH is a good predictor of pork tenderness. In bovine muscle, measurements at 48 h *postmortem* (pH_{48}) are used to detect DFD meat. While Shackelford *et al.* (1999) reported that pH_3 was not an effective predictor of meat tenderness, O'Halloran *et al.* (1997) concluded that both the early *postmortem* and ultimate pH of muscle, has an important effect on the disruption of myofibrillar proteins and thus on meat tenderness. The relationship between pH and temperature up to 24 h *postmortem* is an important factor when considering ultimate meat quality. Recent research has focused on the manner in which even the temperature of the meat at the time of sampling will influence the pH (Jansen, 2001; Bruce *et al.*, 2001). In general glass or solid state electrodes are used to measure pH electrochemically. Investigations at The National Food Centre (NFC) have recently shown that significant variation in recorded pH values is introduced through the use of different types of meters and probes. For industrial situations a solid state electrode would be the most suitable, however, some difficulties have been noted in terms of drift within the probe, which may be associated with protein build up. Andersen *et al.* (1999) suggested that near infared (NIR) and visual range spectrometric methods are comparable to the precision of the standard glass electrode pH meter.

19.3.2 Electrical impedance and conductivity

The electrical conductivity of muscle changes when damage to differing degrees occurs to the membrane system of the muscle during *postmortem* glycolysis. Conductivity is a way of testing the intact cell membranes within a muscle tissue. Impedance, which is a combination of both resistance and capacitance, decreases when there is a disruption in membrane integrity (Kleibel *et al.*, 1983) and an increase in fluids within the muscle tissue (Pliquett *et al.*, 1995). Early *postmortem* measurements of these electrical parameters have been acquired from trials within the NFC and have been compared with sensory and technological attributes in pork. Pearson correlations revealed that impedance and conductivity measurements were significantly correlated to the colour, drip loss and tenderness values. Schoeberlein *et al.* (1999) ascertained that impedance measurements act as a good indicator of pork quality, while Lee *et al.* (2000) suggest that conductivity (24 h *postmortem*) may be a reliable predictor of water holding capacity in pork. The relationship between electrical impedance and beef tenderness has been explored by Lepetit *et al.* (2002). Using an arbitrary method of segregating bovine carcasses on the basis of colour, drip loss and pH, a stronger correlation ($r = 0.84$) between conductivity 24 h, and WBSF 14 days was observed (Mullen *et al.*, 2000) in one classification group. Some relationship exists between electrical measurements and tenderness, however, it would appear that this is not a simple, linear relationship. The usefulness of conductivity and impedance measurements as quality predictors has been addressed by many researchers and more details regarding impedance, can be read in Chapter 10 in this book and a paper by Pliquett and Pliquett (1998).

19.3.3 Colour

Colour measurements in the early *postmortem* period have been investigated for their relationship with meat tenderness. Jeremiah *et al.* (1991) conclude segregation (tenderness) based on both pH and colour appear to offer little in advantage over the use of pH alone. While more recent studies suggest a relationship between colour and tenderness (Wulf *et al.*, 1997), correlation coefficients were quite low ($r = -0.38$ (WBSF); $r = 0.37$ (sensory tenderness)). Wulf and Page (2000) suggested inclusion of muscle colour and pH would increase the accuracy and precision of the USDA quality grading standards for beef carcasses. A monochromatic fiber-optic probe was developed by MacDougall and Jones (1975) to measure the internal reflectance of meat. It has a peak sensitivity at 900 nm, where absorbance by red heme pigments is minimal (Swatland, 1995). The usefulness of the fiber-optic probe in segregating PSE and

DFD classes has been demonstrated (MacDougall and Jones, 1980; Garrido *et al.*, 1995). However, this probe has not been taken up by the industry to any great extent. Following analysis on a large number of topside pork muscles, we have shown both reflectance, using a hand held, non invasive probe, (Optostar) and the CIE L* value of colour, performed well as objective methods of segregating the visual quality of pork topsides prior to processing into hams (NFC results accepted for ICoMST 2002). Tan *et al.* (2000) concluded that a colour machine vision system was an accurate method of evaluating pork colour while van Oeckel *et al.*, (1999) found double density light transmission to be a useful method. Results from a multichannel fibre optic probe indicate that adipose and collagenous connective tissue have high reflectance values (Swatland, 2000).

19.4 Emerging technologies

In recent years, new technologies have been developed which show promise for exploitation and use in the meat plant. These include both physical and biochemical techniques, which are described in more detail below as well as in other chapters in this book. These show potential for use as indicators of meat quality and pave the way for further research into their predictive ability.

19.4.1 Ultrasound

Sound waves with frequencies above the audible range are called ultrasound waves. A number of variables characterising the propagation of an ultrasound signal, through a medium such as meat, can be measured. In addition the ultrasound measurements can also be made at a number of frequencies. Sound moves by compression waves, which are reflected and refracted when they move from one medium to the next. Sono-elastography is a method which uses ultrasonic pulses to track the internal displacements of small tissue elements in response to externally applied stress.

The mechanical properties of meat have been characterised by two sono-elastography parameters – the propagation velocity and the attenuation co-efficient of the mechanical wave. Optimisation of the measurement conditions have taken temperature, sample dimensions, probe dimensions and frequency range into account. Measurements taken from beef samples have indicated the usefulness of this technology as a non-intrusive method of visualising structural characteristics of beef (Ophir *et al.*, 1993; and Cross and Belk, 1994). These investigations indicated that measurements may describe muscle structure at the muscle bundle level. The evolution of meat during *rigor* onset and ageing was followed by pH measurements and mechanical resistance of myofibrils at 20% of deformation. Comparisons were made between these parameters and the variables from the sono-elastography analysis. Results indicate that sono-elastography can be used to follow *rigor* mortis onset and ageing in meat. Elastic deformation of intramuscular connective and adipose tissue caused by external stress is detected ultrasonically and has been proposed as a method of predicting beef quality (Swatland, 2001). Abouelkaram, et al., (2000) have analysed bovine muscle samples and investigated the influence of compositional and textural characteristics on ultrasonic measurements. The method used was based on the measurement of acoustic parameters (velocity, attenuation and backscattering). They conclude that the ultrasonic method used is robust enough to have real potential for muscle tissue characterisation.

A novel ultrasound technique has been developed with the potential to rapidly and accurately measure tenderness on small samples of meat (Allen et al., 2001). This technique has much higher resolution than previously applied methods and is, therefore, very sensitive to meat texture. The method is also sensitive to differences in meat composition, particularly fat and protein content. It is possible, therefore, that composition and texture could be measured simultaneously to give an overall picture of the likely eating quality of meat. This could be particularly useful in processed meats where it is important to know the composition and textural properties of intermediate emulsions in order to control the eating quality of the final product. Anisotropy of ultrasonic velocity is capable of tracking structural changes in ageing beef and has potential as an indicator of eating quality in beef (Dwyer et al., 2001a, b). However, further work on a broader spectrum of meat samples is needed to validate this approach for predicting beef eating quality.

19.4.2 Nuclear magnetic resonance (NMR)

NMR imaging measures energy differences between magnetic moments of naturally occurring, intrinsically magnetic atoms, when external fields are imposed. The relationship between water dynamics and muscle types has been explored using NMR. Various quantitative NMR parameters which are closely related to water dynamics, namely relaxation curves (T_{21} and T_{22}) and magnetic transfer parameters, have been acquired and analysed. According to Beauvallet and Renou (1992), NMR has proven to be a powerful technique for the study of transformations in muscle tissue and is of value in assessing meat quality. Further measurements, recorded at 3 h and 14 days *postmortem* in bovine muscle, detected a redistribution of water over the ageing process. This redistribution is being analysed to monitor its suitability in determining meat quality (water holding capacity). A pattern was seen to exist for some of the NMR measurements versus time *postmortem* for different pH falls. This was thought to be partly due to pH fall and partly due to muscle shortening. A study by Brondrum et al. (2000) concluded NMR to be a better predictor of water holding capacity, intramuscular fat and total water content, than the use of a fibre optic probe or visible or NIR spectroscopy. Laurent et al. (2000) published a good overview of the characterisation of muscle by NMR imaging and spectroscopic techniques. These techniques give non-invasive access to stuctural, metabolic and chemical information as well as water dynamics and fat distribution. However, the cost of this analysis may be prohibitive.

19.4.3 Image analysis

Image analysis of meat has been investigated with the view to identifying features which are characteristic of the meat sample. Textural attributes of photographic images were analysed using several feature extraction methods to evaluate the possibility of identifying bovine muscle (Basset *et al.*, 2000). This involved acquiring photographic images of meat samples under both ultraviolet and visible light. Classifications experiments were performed to identify the bovine muscle according to muscle type, age and breed. Classification of muscles from animals within one age category led to high identification rates, while classifications based on age of the animal and breeds proved more difficult.

19.4.4 Autofluorescence (AF) spectroscopy

AF was assessed for its potential to predict the fat and connective tissue content in meat (Wold, *et al.*, 1999b). Meat samples containing varying amounts of meat, fat and connective tissue were analysed. The results conclude that at different wavelengths along the spectra, AF may provide a rapid method for detection of connective tissue (380 nm) and fat (332 nm) content in meat. AF spectroscopy has also been investigated in conjunction with imaging to analyse *postmortem* changes in sensoric tenderness and WBSF measurements. An optimum excitation wavelength has been demonstrated for modeling the biochemical constituents. Good correlations have been obtained between the emission spectra and time *postmortem*. Combining spectral data with information from images can result in better prediction models for fat content (Wold, *et al.*, 1999a).

19.4.5 Near Infared (NIR) spectroscopy

Applications of NIR for the prediction of functional properties and quality variables in foods have emerged. NIR spectrometry is both rapid and non-destructive and can be calibrated to measure several components at one time. Infrared light is part of the broad spectrum of energy known as electromagnetic radiation. NIR spectrometry is concerned with a specific region of the infrared, adjacent to the red end of the visible spectrum. The frequency of wave oscillations in the IR region is of the same order as the natural mechanical vibrational frequencies of many chemical groups (Benson, 1993). The types of absorptions that dominate NIR are hydrogenic absorptions such as $-OH$, $-NH$ and $-CH$ vibrations. These types of absorptions are displayed by moisture and virtually all other major constituents of foodstuffs.

Many applications of NIR have dealt with the quantitative determination of compositional analysis of foods (Osborn and Fearn, 1986). NIR spectra reflect the organic constituents of food products and contain information about the conformation of components such as proteins and polysaccharides. NIR applications have also focused on direct estimation of quality parameters in foods. When using NIR spectra to predict the fat content it is advisable to mince or grind the meat into a homogenous mixture (Isaksson et al., 1992; Rodbotten et al., 2000). In this way a single NIR scan will probably detect the 'true' composition. Perhaps increasing the number of NIR scans substantially on intact meat samples may improve the prediction results for fat (Rodbotten et al., 2000).

The ability of NIR to detect changes in the state of water and hydrogen bond interactions in foods has been observed (Hildrum et al., 1995). Since such changes evidently occur in meat during tenderisation and ageing it was thought that a relationship may exist between NIR measurements and meat quality attributes. This was initially investigated by Mitsumota et al. (1991) who observed a strong relationship between NIR and both WBSF and compositional measurements of meat. Since then many studies concluded that NIR was a good predictor of meat tenderness as assessed by WBSF (Byrne, 1998; Park et al., 1998). However, these prediction models were not always easily replicated (NFC, unpublished work). Recently work at the NFC (Venel et al., 2001) focused on avoidance of a number of potential experimental errors which may have mitigated against the development of satisfactory models. While NIR (750–1100 nm) was unable to satisfactorily predict the selected organoleptic properties of bovine M. *semimembranosu*s (SM), better results were obtained for the WBSF values in the LD ($r = 0.51$). While the predictive performance was improved, when the sample set was segregated according to animal grade, sex, ultimate pH or day of bone out ($r = 0.54-0.72$), the accuracies were no better than previously reported (NFC, unpublished work). NIR spectra (1100–2500 nm) were collected *pre-rigor* in bovine M. *longissimus dorsi* (LD) (Rodbotten et al., 2000). While these spectra had higher overall absorbancies compared to *post-rigor* measurements the results did not support that early post mortem NIR spectra could be used as a predictor of beef tenderness.

19.4.6 Temperature

It is well known that temperature plays a major role in determining meat quality. Problems such as cold shortening, heat *rigor* and cold autolysis are associated with *postmortem* carcass temperature. May et al. (1992) concluded that an internal *longissimus dorsi* temperature (4.5 cm to 5.5 cm into the medial portion immediately anterior to the 9th rib) of 33°C at 2.5 h *postmortem* was related to the tenderness of the meat ($r = -0.63$ (WBSF); $r = 0.54$ (sensory tenderness)). Pork loins having low carcass temperature at 3 h *postmortem* have higher pH and water holding capacity and produced lower lightness and WBSF values (Park et al., 2001). Optimal beef tenderness after ageing for 14 days was associated with an intermediate pH decline of pH5.9–6.2 at 1.5 h *postmortem* or *rigor* temperature of 29–30°C at pH 6.0 (Hwang and Thompson, 2001). In lamb a temperature of approximately 15°C at onset of *rigor* appears to be optimal for tenderness (Geesink et al., 2000). Not all researchers however agree to the usefulness of temperature in the identification of differences in tenderness (Jones and Tatum, 1994). Perhaps strategic temperature measurements may provide useful information on the ultimate eating quality of meat.

19.4.7 Immunoassays

The application of immunoassay technology to the analysis of non-clinical samples in the Food and Agricultural sectors has been growing steadily for some time (Rittenburg and Grothaus, 1992). It is a powerful analytical tool that depends on the interaction between an antibody and the antigen being measured. It can provide a rapid, economical, highly sensitive and specific analysis and is relatively simple to perform. In comparison to other analytical methods such as electrophoresis and high performance liquid chromatography (HPLC), immunoassays allow a higher sample throughput with minimal sample processing. This analytical method has been used in the detection of food contaminants such as antibiotics, pesticides and microbiological organisms (Rittenburg, 1990). The potential of this method for investigating endogenous food components such as vitamins, enzymes and structural proteins has been realised (Finglas et al, 1992; Doumit et al, 1996).

Proteolytic degradation of key myofibrillar proteins has been shown to contribute to post mortem tenderisation (Troy et al., 1987; O'Halloran 1996; Boyer-Berri and Greaser, 1998). Degradation of many structurally important myofibrillar proteins, during the *postmortem* ageing of meat, has been observed by many research groups including NFC. Of these troponin T and its 30kDa myofibrillar proteolytic fragment have been related to meat tenderness (Buts et al., 1986; Troy et al., 1987). It has been suggested that the appearance of this 30kDa fragment could serve as an early *postmortem* indicator of meat tenderness. A soluble 1734.8Da fragment of the troponin T molecule has recently been isolated (Stoeva et al., 2000; Mullen et al., 1998b; Nakai et al., 1995) which appears to be related to meat tenderness (Mullen et al., 2000). Due to its solubility this fragment is more easily extracted from meat than the myofibrillar 30kDa and, therefore, it may be a more suitable candidate for routine factory analysis. An immuno-based assay was developed using polyclonal antibodies for the detection of this fragment. Initial results show a relationship between formation of the fragment and beef tenderness. The calpastatin/calpain proteolytic system which has been implicated in the tenderisation process has also been targeted for the development of an immunoassay test (Doumit et al., 1996; Koohmaraie, 1996b). Exoprotease activities can constitute a novel and adequate technique to predict early *postmortem* pork meat quality (Toldra and Flores, 2000). These enzymes remain stable up to 24 h *postmortem*. Development of immuno-based assays to these and other important enzymes and proteolytic fragments are underway with the ultimate aim of predicting meat quality traits. The analytical performance of these assays needs to be thoroughly validated and convenient sample preparation procedures designed for application of the assays to meat extracts. Evaluation of the efficacy of the assays in the prediction of meat tenderness is also necessary and the resulting data can then be used to select the assay or combination of assays which give the best correlation with alternative indices of tenderness. Data collected from the validation stages would also indicate whether a qualitative or quantitative assay format is needed for a final test system. There are some reservations about speed of immunoassays for use in a commercial situation (Koohmaraie, 1996b). However, ultimately, and more ideally, a successful assay could be developed into a more rapid test such as a 'dip-stick' type test, which would be more readily transferable to the meat industry.

19.4.8 Metabolites

Many metabolites exist in muscle, blood and urine which may contain valuable information regarding the variability of quality between different animals. Determination of some of these factors may be part of routine analysis in some laboratories. Results from these samples may provide an indicator of meat quality and may also provide a greater knowledge of the biochemical process occurring over this period (Troy *et al.*, 1998). Higher levels of metabolites such as creatinine phosphokinase, free fatty acids and β-hydroxybutyrate have been linked to the ultimate pH (pH_u). No relationship was observed between various blood biochemical parameters and pH_u (NFC, unpublished results). Nucleosides and nucleotides also play an important role in the biochemical changes which occur during the early *postmortem* period. There is a rapid decrease in high energy molecules such as tri- (ATP) and di- (ADP) nucleotides, early *postmortem*, with a simultaneous increase in mononucleotides (AMP, IMP, GMP, etc.), and nucleosides (adenosine, inosine, guanosine, etc.). Levels of hypoxanthine and inosine were different in exudative and normal pork meat up to 3 days *postmortem* (Battle, *et al.*, 2001). In the same study IMP and ATP only differed between 4 and 6 h *postmortem* with the IMP:ATP ratio differing up to 2 h *postmortem*. The authors suggest that 2 h is the optimum time for sampling when attempting to predict PSE. In another study the same authors suggest that nucleotide contents may serve as an index of some taste variations between pork quality classes but a direct link may be unlikely due to their rapid metabolism and the low degree of differences accounted for (Flores *et al.*, 1999). Other novel metabolites of interest include nitric oxide (NO) which is a gaseous intercellular messenger and has recently been proposed in ubiquitous roles, including those within a normal functioning skeletal muscle. WBSF values for striploins incubated with NO enhancers were seen to decrease while those of striploins incubated with NO inhibitors increased at days 3 and 6 of conditioning (Cook *et al.*, 1997). The mechanism of NO in tenderisation is not yet known, however NO can mediate its effects by free radicals and/or calcium changes which can in turn affect proteolytic enzymes. The non-enzymatic tenderisation of meat proposed by Takahashi (Takahashi 1996) and involving calcium ions could be partially explained by NO.

19.4.9 Tenderness probe

As mentioned above efforts have been made to develop physical methods to predict tenderness, as assessed by a sensory panel. As well as the Warner-Bratzler shear force method, other instruments have been developed and tested.

The MIRINZ (Meat Industry Research Institute of New Zealand) tenderness probe consists of two sets of pins on which meat samples are impaled (Jeremiah and Phillips, 2000). Tension is applied to the muscle fibres by one set of pins which rotate relative to a static set of pins. A torque signal is recorded against the angle of rotation. Results indicate it may be an alternative method to the Warner Bratzler. However, this method would still require cooking of the meat. Ideally the industry requires a method which would allow measurements to be taken on raw meat, either on the carcass or on primal or retail cuts.

19.5 The genetics of meat quality

In recent years great advances have been made which have enabled deeper understanding of the relationship between genomics and meat quality. It is anticipated that breeding programmes will be greatly improved through identification of polymorphisms in the DNA sequences which impact on quality traits. Functional genomics aims to provide further insight to the complex interplay of gene expression events involved in the development of meat quality. The importance of this expanding area of research is becoming more apparent in the understanding of meat quality (Mullen *et al.*, 2000 and many references in the following paragraph). The reader is also directed to Marshall (1999) and Solomon *et al.* (2000) for further information in this area. Complementary to the functional genomics approach is proteome analysis (Anderson *et al.*, 2000). The use of proteomics allows the characterisation of expressed protein within any given cell type. This is a very powerful technique as the phenotypic traits of an organism are ultimately manifested through the interaction of the environment and the various proteins expressed in its tissues (structural, enzymatic, metabolic and regulatory proteins) (Pandey and Mann, 2000).

DNA markers, (regions of chromosomes showing polymorphisms in the nucleic acid sequence) can be used to detect quantitative trait loci (QTL), or regions on a chromosome which have a substantial effect on quantitative traits such as tenderness, growth rate, intramuscular fat etc. This is referred to as QTL mapping. To successfully carry out QTL analysis, several hundred animals are required. These must be produced over several generations using pure bred animals and employing strict control of the breeding regimes. DNA markers linked to the QTL can be used to develop strategies for marker assisted selection (MAS) breeding programmes (for review read Dentine, 1999). Preliminary data from genome-wide scanning of DNA markers have revealed a number of putative QTL associated with meat traits, although relatively few results have been published to date. There have been several projects, world wide, designed to identify QTL affecting phenotypes such as meat quality, carcass attributes, growth and development of beef cattle and pigs. The following are some of the research groups that are currently undergoing programmes in this area: Jay Hetzel and Bill Barendse's group from the CSIRO in Australia (Hetzel and Davis, 1997 and Hetzel *et al.*, 1997), Yoshikazy Sugimoto from the Shirakawa

Institute of Animal Genetics in Japan, Steve Kappes' group at the USDA MARC in the US (Stone et al., 1999), animal genomics group at Texas A&M (Taylor and Davis, 1997, 1998), animal genetics group in Michigan State University, and Michel George's group from Belgium. Casas et al., (2000) has identified suggestive QTL for longissimus muscle area, hot carcass weight, marbling and WBSF.

QTL for carcass and growth traits have been identified (Stone et al., 1999). The chromosomal locations of at least five genes influencing beef tenderness and another four genes influencing marbling have been identified (Taylor and Davis, 1998 and Hetzel et al., 1997). A QTL influencing WBSF has also been recently identified (Keele et al., 1999). In pork QTL for IMF and backfat have been identified (for review see De Vries et al., 2000; Rothschild, 1997). In general, however, relatively little has been published concerning the localisation of genes influencing variations in beef cattle phenotypes.

19.5.1 Major genes and candidate genes

Although many quantitative traits seem to be controlled by multiple genes, individual genes can account for a relatively large amount of variation in some traits. These genes are referred to as major genes. Candidate genes are genes which are related physiologically or biochemically to the selected trait and are assumed to have an effect on trait performance. Using this approach several major genes have been detected. Recent evidence suggests that double muscling in some beef breeds is caused by a mutation of a gene located on bovine chromosome 2 that produces the protein myostatin (Grobet et al., 1997; Kambadur et al., 1997; Smith et al., 1997). Normally, myostatin serves to repress skeletal muscle growth, but the mutation appears to block this effect and permits extra muscle growth. Porcine stress syndrome (PSS) or malignant hypothermia (MH) is a genetic disorder that can result in sudden death or PSE meat in domestic pigs. Initially the genetic mutation responsible for this disorder was located to the MH locus but subsequent work demonstrated a mutation in a single (major) gene, the ryanodine receptor gene (see DeVries et al., 2000; Cunningham, 1999 for review). An example of a candidate gene for meat quality is provided by the gene for fatty acid binding protein (FABP). Polymorphisms in this gene (heart and adipocyte) have been found to be associated with variations in IMF in pigs (Gerbens et al., 1998a, b). In addition, the effect appears to be independent of back fat. Another relevant candidate approach is the research on calpain and calpastatin. There are many reports of the role of the calpain system, a set of calcium dependent proteases and their inhibitor calpastatin, both *in vivo* and *postmortem* in protein turnover and meat tenderness (Koohmaraie 1992, 1996a; Zamora et al., 1996). Three polymorphic sites have been identified in the pig calpastatin gene (Ernst et al., 1998). Significant associations between beef tenderness and calpastatin genotype were detected by Green et al. (1996a, b), but not by Lonergan et al. (1995).

19.5.2 Gene expression and microarray technology

Genomics is the study of an organism's whole genetic blueprint and the variations within that blueprint which make every individual in a population unique. Functional genomics seeks to relate differences in the genetic blueprint to physiology and phenotype. It is widely believed that thousands of genes and their products (proteins) in a given living organism, function in a complicated and orchestrated way to create the unique characteristics in individuals. Differences in gene expression or changes in the sequence of expressed genes contribute to the various phenotypes in any animal population. Traditional methods allow only one gene to be studied at a time. However, there is much excitement at present over the use of microarrays or biochips, and their potential contribution to the study of genomics. The technique, which involves laying down an ordered array of genetic elements onto a solid substrate, is useful because it enables genetic analysis of thousands of genes and markers in a single experiment (Zhao *et al.*, 1995; Duggan *et al.*, 1999; Sinclair, 1999). Combining microarray technology with methods of molecular biology allows comparison of gene expression to be made between individuals or between cell types etc. (Yang *et al.*, 1999; Liang and Pardee, 1992). Using this technology will allow patterns of gene expression to be compared between animals displaying extremes of quality traits of interest. In this manner a 'favourable' set of gene expression patterns may be defined for particular quality traits. These patterns could be monitored through various treatment regimes to determine an optimal set of conditions that should exist within the animal prior to processing. Many research groups world wide, including NFC, have dynamic programmes in place to determine the relationship between patterns of gene expression (and protein expression) and meat quality traits. This is an exciting and revolutionary area with much potential for real applications to improving the consistency of the quality of meat. While methodology requires investment in terms of time and development of specific tools, the benefits in the medium to long term are potentially great.

19.6 The future

Many innovative advances have been made in the area of measuring and predicting meat quality traits. It can be seen that in many cases the findings can be difficult to interpret and may be contradictory to others. Obviously factors such as experimental design, sampling methodology, sampling conditions, instrument type and data analysis are critical to the interpretation of results. The application of any of these techniques to an on-line situation will require large scale industrial based trials to verify and confirm their ultimate usefulness. Many advances have been made in the area of DNA and gene expression analysis. It is anticipated that this will contribute greatly to our understanding of meat quality traits. It is possible that early *postmortem* prediction of meat quality will require recording more than one 'on-line' measurement. Ultimately predicting meat quality attributes may require a more holistic approach which sets criteria at more that one point along the chain of animal production through to consumption.

19.7 Sources of further information and advice

A wide array of books are available for general reading on the science behind meat quality attributes and their measurements. A few of the more noteworthy include Singhal *et al.* (1997), Pearson and Dutson (1994), Kress-Rogers and Brimelow (2001), Taylor *et al.* (1996) and Lawrie (1998). Throughout this chapter many references have been cited, many of which are review articles or book chapters, rather than list all of these here again the reader should consult the relevant sections. In the area of meat quality prediction (especially using various probes) Chapter 10 in this book should be read, in conjunction with many of the published papers by this author. Some of the results discussed in this chapter are from research carried out under the EU FAIR programme (Fifth Framework), project number PL96-1107 and various nationally funded projects (Department of Agriculture, Food and Rural Development, Ireland).

19.8 References

ABOUELKARAM, S., SUCHORSKE, K., BUQUET, B., BERGE, P., CULIOLI, J., DELACHARTRE, P. and BASSET, O. (2000) Effects of bovine muscle composition and structure on ultrasonic measurements. *Food Chemistry*, 69 (4), 447–455.

ALLEN, P., DWYER, C., MULLEN, A.M., BUCKIN, V., SMYTH, C. and MORRISSEY. S. (2001) *Using ultrasound to measure beef tenderness and fat content. End of project report.*

ANDERSEN, J.R., BORGGAARD, C., RASMUSSEN, A.J. and HOUMOLLER, L.P. (1999) Optical measurements of pH in meat. *Meat Science*, 53, 135–141.

ANDERSON N.L., MATHESON A.M. and STEINER S. (2000) Proteomics: applications in basic and applied biology [Review] *Current Opinion in Biotechnology*, 11 (4), 408–412.

BASSET, O., BUQUET, B., ABOUELKARAM, S., DELACHARTRE, P. and CULIOLI, J. (2000) Application of tectural image analysis for the classification of bovine meat. *Food Chemistry*, 69 (4), 437–445.

BATTLE, M., ARISTOY, M.C. and TOLDRA, F. (2001) ATP metabolites during aging of exudative and non-exudative pork meats. *Journal of Food Science*, 66 (1), 68–71.

BEAUVALLET, C. and RENOU, J.P. (1992) Applications of NMR spectroscopy in meat research. *Trends in Food Science and Technology* 3, 241–246.

BENSON, I.B. (1993) Compositional analysis Using Near Infrared Absorbtion Spectroscopy, In: *Instruments and sensors for the food industry* (Ed.

Kress-Rogers, E.), 121–166.
BOYER-BERRI, C. and GREASER, M.L. (1998) Effect of portmortem storage on the Z-line region of titin bovine muscle. *Journal of Animal Science*, 76, 1034–1044.
BRATZLER, L.J. and SMITH, H.D. (1963) A comparison of the press method with taste panel and shear measurements of tenderness in beef and lamb muscles. *Journal of Food Science*, 28, 99.
BRONDRUM, J., MUNCK, L., HENCKEL, P., KARLSSON, A., TORNBERG, E. and ENGELSEN, S.B. (2000) Prediction of water holding capacity and composition of porcine meat by comparative spectroscopy. *Meat Science*, 55 (2), 177–185.
BRUCE, H.L., SCOTT, J.R. and THOMPSON, J.M. (2001) Application of an exponential model to early *postmortem* bovine muscle pH decline. *Meat Science*, 58, 39–44.
BUTS, B., CLAEYS, E. and DEYMEYER, D. (1986) Relation between concentration of troponin-T, 30,000 dalton and titin on SDS-PAGE and tenderness of bull *longissimus dorsi*. Proceedings of the 32nd European meeting of meat research workers, 175–178.
BYRNE, C. (1998) Near infrared reflectance spectroscopy and electrical measurements of beef muscle as indicators of meat quality, PhD Thesis, National University of Ireland, Cork.
CASAS, E., SHACKELFORD, S.D., KEELE, J.W., STONE, R.T., KAPPES, S.M. and KOOHMARAIE, M. (2000) Quantitative trait loci affecting growth and carcass composition of cattle segregating alternate forms of myostatin. *Journal of Animal Science*, 78, 560–569.
COOK, C.J., SCOTT, S.M. and DEVINE, C.E. (1997) The effect of endogenous nitric oxide on tenderness changes of meat. *ICOMST*, G1-5, 558–559.
CROSS, H.R. and BELK, K.E. (1994) Objective measurements of carcass and meat quality. *Meat Science*, 36, 191–202.
CHRYSTALL, B. (1994) Meat texture measurement. In: *Quality attributes and their measurement in meat, poultry and fish products.* Advances in meat research series, Volume 9. (Eds Pearson, A.M. and Dutson, T.R.), Blackie Academic & Professional, Chapman & Hall, UK.
CUNNINGHAM, E.P. (1999) The application of biotechnologies to enhance animal production in different farming systems. *Livestock Production Science*, 58, 1–24.
DE VRIES, A.G., FAUCITANO, L., SOSNICKI, A. and PLASTOW, G.S. (2000) The use of gene technology for optimal development of pork meat quality. *Food Chemistry*, 69, 397–405.
DENTINE, M.R. (1999) Marker assisted selection. In: *The Genetics of Cattle*, (Eds, Fries, R. and Ruvinsky, A.), CABI Publishing, Oxford, UK, 497–510.
DOUMIT M. E., LONERGAN S. M., ARBONA J.R., KILLEFER J. and KOOHMARAIE M. (1996) Development of an enzyme-linked immunosorbent assay (ELISA) for quantification of skeletal muscle calpastatin. *Journal of Animal Science*, 74, 2679–2686.

DUGGAN, D.J., BITTNER, M., CHEN, Y., MELTZER, P. and TRENT, J.M. (1999) Expression profiling using cDNA arrays. *Nat. Genet.*, Jan; 21 (1 Suppl.) 10–14

DWYER, C, MULLEN, A.M., ALLEN, P. and BUCKIN, V. (2001a). Anisotropy of ultrasonic velocity as a method of tracking *postmortem* ageing in beef. Proceedings 47th ICoMST, Poland. 4 (P4), 250.

DWYER, C, MULLEN, A.M., ALLEN, P. and BUCKIN, V. (2001b). Ultrasonic characterisation of dairy and meat systems. Proceedings, Food Science and Technology Research Conference, Cork. P 30.

EIKELENBOOM, G., HOVING-BOLINK, A.H. and VAN DER WAL, P.G. (1996) The eating quality of pork. I. The influence of ultimate pH. *Fleischwirtschaft*, (76), 4, 392–393.

ELLIS, R.L. (1994) *Food analysis and chemical residues in muscle foods.* Advances in meat research series, Volume 9. (Eds Pearson, A.M. and Dutson, T.R.), Blackie Academic & Profressional, Chapman & Hall, UK.

ERNST, C.W., ROBIC, C., YERLE, M., WANG, L. and ROTHSCHILD, M.F. (1998) Mapping of calpastatin and three microsatellites to porcine chromosome 2q2.1–q2.4. *Animal Genetics*, 29, 212–215.

FINGLAS P.M., ALCOCK S.A. and MORGAN M.R.A. (1992) The biospecific analysis of vitamins in food and biological materials. In: *Food safety and quality assurance.* (Eds. Morgan M.R.A., Smith C.J. and Williams P.A.), Elsevier Applied Science Publishers Ltd., London. 401–409.

FLORES, M., ARMERO, E., ARISTOY, M.C. and TOLDRA, F. (1999) Sensory characteristics of cooked pork loin as affected by nucleotide content and *postmortem* meat quality. *Meat Science*, 51 (1), 53–59

FREYWALD, K.H, PLIQUETT, F., SCHOBERLEIN, L. and PLIQUETT, U. (1995) Passive electrical properies of meat as characteristic of its quality. Poster/paper, 4th International Conference Elec. Bio. Imp., Heidelberg.

GARRIDO., M.D., PEDUAYE, J., BANON, S., LOPEZ, M.B. and LAENCINA, J. (1995) On-line methods for pork quality detection. *Food Control* 6, 111–113.

GEESINK, G.H. (1993) *Postmortem* muscle proteolysis and beef tenderness with special reference to the action of the calpain/calpastatin system. Ph.D. Thesis. The University of Utrecht, The Netherlands.

GEESINK, G.H., BEKHIT, A.D. and BICKERSTAFFE, R. (2000) *Rigor* temperature and meat quality characteristics of lamb longissimus muscle. *Journal of Animal Science*, 78 (11), 2842–2848.

GERBENS, F., JANSEN, A., VAN ERP, A.J.M., HARDERS, F., MEUWISSEN, T.H.E., RETTENBERGER, G., et al (1998a) The adipocyte fatty acid binding protein locus: characterisation and association with intramuscular fat content in pigs. *Mammalian Genome*, 9, 1022–1026.

GERBENS, F., VAN ERP, A.J.M., MEUWISSEN, T.H.E., VEERKAMP J.H. and TE PAS, M.F.W. (1998b) Heart fatty acid binding protein gene variants are associated with intramuscular fat content and back fat thickness in pigs. Proceedings 6th World Congress Genetics Applied to Livestock Production, 26, 201–204.

GREEN, H.U., COCKETT, N.E., TATUM, J.E., O'CONNOR, S.F., HANCOCK, D.L. and SMITH, G.C. (1996a) Association of a Taq1 calpastatin polymorphism with *postmortem* measures of beef tenderness in Bos taurus and Bos indicus-Bos taurus steers and heifers. *Journal of Animal Science*, 74 (Suppl. 1), 111.

GREEN, H.U., COCKETT, N.E., TATUM, J.E., O'CONNOR, S.F., HANCOCK, D.L. and SMITH, G.C. (1996b) Association of a Taq1 calpastatin polymorphism with *postmortem* measures of beef tenderness in Charolais- and Limousin-sired steers and heifers. *Journal of Animal Science*, 74 (Suppl. 1), 113.

GROBET, L., ROYO MARTIN, L.J., PONCELET, D., PIROTTIN, D., BROUWERS, B, RIQUET, J., SCHOEBERLEIN, A., DUNNER, S., MENISSIER, F., MASSABANDA, J., FRIES, R., HANSET, R. and GEORGES, M. (1997) A deletion in the bovine myostatin gene causes the double-muscled phenotype in cattle, *Nature Genetics*, 17, 71–74.

HETZEL, D.J.S. and DAVIS, G.P. (1997) Mapping quantitative trait loci: A new paradigm. Proceedings: Beef cattle genomics: Past, present and future, Texas A & M University, College Station, Texas, pp 24–25.

HETZEL, D.J.S., DAVIS, G.P., CORBET., N.J., SHORTHOSE, W.R., STARK, J., KUYPERS, R., SCACHERI, S., MAYNE, C., STEVENSON, R., MOORE, SS. and BYRNE, K. (1997) Detection of gene markers linked to carcass and meat quality traits in a tropical beef herd. Breeding. Responding to client needs, AAABG, Proc, of 12th conference, Australia, Part 1, 442–446.

HILDRUM, K.I., ISAKSSON, T., NAES, T., NILSEN, B.N., RODBOTTEN, M. and LEA, P. (1995) Near infared reflectance spectroscopy in the prediction of sensory properties of beef. Journal of *Near Infared Spectroscopy*, 3, 81–85.

HOFFMAN, K. (1990) Definition and measurement of meat quality. Proceedings of the 36th Ann. Int. Cong. of Meat Sci. and Tech., Cuba, p. 941.

HONIKEL, K.O. and HAMM, R. (1994) Measurement of water-holding capacity and juiciness. In: *Quality attributes and their measurement in meat, poultry and fish products*. Advances in meat research series, Volume 9. (Eds. Pearson, A.M. and Dutson, T.R.), Blackie Academic & Profressional, Chapman & Hall, UK.

HWANG, I.H. and THOMPSON, J.M. (2001) The interaction between pH and temperature decline early *postmortem* on the calpain system and objective tenderness in electrically stimulated beef longissimus dorsi muscle. *Meat Science*, 58 (2) 167–174.

ISAKSSON, T., MILLER, C.E. and NÆS, T. (1992) Nondestructive NIR and NIT determination of protein, fat and water in plastic wrapped, homogenised meat. *Applied Spectroscopy*, 42, 1685–1694.

JANSEN, M.L. (2001) Determination of meat pH temperature relationship using ISFET and glass electrode instruments. *Meat Science*, 58, 145–150.

JEREMIAH, L.E. and PHILLIPS, D.M. (2000) Evaluation of a probe for predicting beef tenderness. *Meat Science*, 55, 493–502.

JEREMIAH, L.E., TONG, A.K.W. and GIBSON, L.L. (1991) The usefulness of muscle colour and pH for segregating beef carcasses into tenderness groups. *Meat Science*, 30, 97–114.

JONES, B.K. and TATUM, J.D. (1994) Predictors of beef tenderness among carcasses produced under commercial conditions. *Journal of Animal Science*, 72, 1492–1501.

KAMBADUR, R., SHARMA, M., SMITH, T.P.L. and BASS, J.J. (1997) Mutations in myostatin (gdf8) in double muscled Belgian Blue and Piedmontese cattle. *Genome Research*, 7, 910–916.

KEELE, J.W., SHACKELFORD, S.D., STONE, R.T., KAPPES, S.M., KOOHMARAIE, M. and STONE R.T. (1999) A region on bovine chromosome 15 influences beef longissimus tenderness in steers. *Journal of Animal Science*, 77, 1364–1371.

KIRCHEIM, U., KINAST, C. and SCHONE, F. (2001) Early *postmortem* measurement as indicator of meat quality characteristics. *Fleischwirtschaft*, 81 (1) 98–90.

KLEIBEL, A. PSUTZNER, H. and KRAUSE, E. (1983) Measurement of dielectric loss factor routine method for recognising PSE muscle, *Fleischwirtschaft*, 63, 1183–1185.

KOOHMARAIE, M. (1992) The role of endogenous proteases in meat tenderness. Proc. 41st Annual Reciprocal Meat Conf., National Livestock and Meat Board, Chicago Ill, pp. 89–100.

KOOHMARAIE, M. (1996a) Biochemical factors regulating the toughening and tenderization processes of meat. *Meat Science*, 43, S193–S201.

KOOHMARAIE, M. (1996b) ELISA test for calpastatin. Reciprocal Meat Conference Proceedings, 49, 177–178.

KRESS-ROGERS, E. and BRIMELOW, C.J.B. (2001) *Instrumentation and sensors for the food industry*. Second edition. Woodhead Publishing Ltd, Cambridge.

LAURENT, W., BONNY, J.M. and RENOU, J.P. (2000) Muscle characterisation by NMR imaging and spectroscopic techniques. *Food Chemistry*, 69 (4), 419–426.

LAWRIE, R.A. (1998) *Meat Science*, 6th Edition, Woodhead Publishing Ltd, Cambridge.

LEE, S., NORMAN, J.M., GUNASEKARAN, S., VAN LAACK, R.L.J.M., KIM, B.C. and KAUFFMAN, R.G. (2000) Use of electrical conductivity to predict water-holding capacity in post-*rigor* pork. *Meat Science*, 55 (4) 385–389.

LEPETIT, J, SALE, P., FAVIER, R. and DALLE, R. (2002) Electrical impedance and tenderisation in bovine meat. *Meat Science*, 60 (1) 51–62.

LIANG P. and PARDEE, A. (1992) Differential display of eukaryotic messenger RNA by means of the polymerase chain reaction. *Science*, 257, 967–970.

LONERGAN, S.M., ERNST, C.W., BISHOP, M.D., CALKINS, C.R. and KOOHMARAIE, M. (1995) Relationship of restriction fragment length polymorphisms (RFLP) at the bovine calpastatin locus to calpastatin activity and meat tenderness. *Journal of Animal Science*, 73, 3608–3612.

LOVE, J. (1994) Product acceptability evaluation. In: *Quality attributes and their measurement in meat, poultry and fish products*. Advances in meat research series, Volume 9. (Eds Pearson, A.M. and Dutson, T.R.), Blackie Academic & Profressional, Chapman & Hall, UK.

MACDOUGALL, D.B. (1994) Colour of meat. In: *Quality attributes and their measurement in meat, poultry and fish products*. Advances in meat research series, Volume 9. (Eds Pearson, A.M. and Dutson, T.R.), Blackie Academic & Profressional, Chapman & Hall, UK.

MACDOUGALL, D.B. and JONES, S.T. (1975) The use of a fibre optic probe for the detection of pale pork. Proceedings of the 21st European meeting of meat research workers, Berne, August 31–September 5, research notes, 113–115.

MACDOUGALL, D.B. and JONES, S.T. (1980) Use of fibre optic probe for segregating pale, soft, exudative and dark, firm, dry carcasses. *Journal of the Science of Food & Agriculture* 31, 1371–1376.

MARSHALL, D.M. (1999) Genetics of meat quality. In: *Genetics of Cattle* (Eds Fries, R. and Ruvinsky, A.) CABI Publishing, Oxford, UK, 605–636.

MAY, S.G., DOLEZAL, H., G., GILL, D.R., RAY, F.K. and BUCHANAN, D.S. (1992) Effects of days fed, carcass grade traits, and subcutaneous fat removal on *postmortem* muscle characteristics and beef palatability. *Journal of Animal Science*, 70, 444–453.

MITSUMOTO, M., MAEDA, S., MITSUHASHI, T. and OZAWA, S. (1991) Near Infared Spectroscopy determination of physical and chemical characteristics in beef cuts. *Journal of Food Science*, 56, 1493–1496.

MULLEN, A.M., CASSERLY, U. and TROY, D.J. (1998a) Predicting Beef Quality at the early *postmortem* period. Proceedings, International Conference, Challenges for the Meat Industry in the Next Millennium, May 1998, Malahide Castle, Ireland.

MULLEN, A.M., STOEVA, S., VOELTER, W. and TROY, D.J. (1998b) Identification of bovine protein fragements produced during the ageing process. ICOMST, Barcelona, Spain, 1998.

MULLEN, A.M., MURRAY, B. and TROY, D.J. (2000) *Predicting the eating quality of meat. End of project report.* Teagasc.

NAKAI, Y., NISHIMURA, T., SHIMIZU, M. and ARAI, S. (1995) Effects of freezing on the proteolysis of beef during storage at 4°C. *Biosci. Biotech. Biochem.*, 2255–2258.

O'HALLORAN, G.R. (1996) Influence of pH on the endogenous proteolytic enzyme systems involved in tenderisation of beef. PhD thesis, The National University of Ireland.

O'HALLORAN, G.R., TROY, D.J. and BUCKLEY, D.J. (1997) The relationship between early *postmortem* pH measurements and the tenderisation process of beef muscle. *Meat Science*, 45, 239–251.

OPHIR, J., MILLER, R.K., PONNEKANTI, H., CESPEDES, I. and WHITTAKER, A.D. (1993) Elastography of beef muscle. Meat Science, 36, (1 and 2), 239–250.

OSBORN, B.G. and FEARN, T. (1986) *Near Infared spectroscopy in food analysis.* Longman Sci. & Tech. Harlow, Essex, UK.

PANDEY, A. and MANN, M. (2000) Proteomics study genes and genomes. *Nature*, 405 837–846.

PARK, B., CHEN, Y.R., HRUSCHKA, W.R., SHACKELFORD, S.D. and KOOHMARAIE, M. (1998) Near-infared reflectance analysis for predicting beef longissimus tenderness. *Journal of Animal Science*, 76, 2115–2120.

PARK, B.Y., CHO, S.H., YOO, Y.M., YOO, Y.M., KO,J.J., KIM, J.H, CHAE, H.S., AHN, J.N., LEE, J.M., KIM, Y.K. and YOON, S.K. (2001) Effect of carcass temperature at 3 hours *postmortem* on pork quality. *Journal of Animal Science and Technology*, 43 (6) 949–954.

PEARSON, A.M. and DUTSON, T.R. (1994) *Quality attributes and their measurement in meat, poultry and fish products.* Advances in meat research series, Volume 9, Blackie Academic & Professional, Chapman & Hall, UK.

PERYAM, D.R. and PILGRIM, F.J. (1957) Hedonic scale method of measuring food preferences. *Food Technology*, 11, 9.

PLIQUETT, U. and PLIQUETT, F. (1998) Conductivity of meat as a quality parameter: critical remarks. *Fleischwirtschaft*, 78, (9), 1010–1012.

PLIQUETT, F., PLIQUETT, U., SCHOBERLEU, L. and FREYWALD, K. (1995) Impedance measurements to characterise meat quality. *Fleischwirtschaft*, 75, 496.

RITTENBURG J.H. (ed.) (1990) *Development and application of immunoassay for food analysis.* Elsevier Applied Food Science Series. Elsevier Science Publishers.

RITTENBURG J.H. and GROTHAUS G.D. (1992) Immunoassays: Formats and applications. In: *Food safety and quality assurance.* Morgan M.R.A., Smith C.J. and Williams P.A., Eds. Elsevier Applied Science Publishers Ltd., London. 3–12.

RODBOTTEN, R., NILSEN, B.N. and HILDRUM, K.I. (2000) Prediction of beef quality attributes from post mortem near infared reflectance spectra. *Food Chemistry*, 69 (4), 427–436.

ROTHSCHILD, M.F. (1997) Identification of major genes and quantitative trait loci in swine. Proceedings National Swine Improvement Federation, Iowa.

SCHOEBERLEIN, L., SCHARNER, E., HONIKEL, K.O., ALTMAN, M. and PLIQUETT, F. (1999) The Py value as a characteristic feature of meat quality. *Fleischwirtschaft*, 79 (1), 116–120.

SHACKELFORD, S.D., WHEELER, T.L. and KOOHMARAIE, M. (1999) Tenderness classification of beef. II. Design and analysis of a system to measure beef longissimus shear force under commercial processing conditions. *Journal of Animal Science*, 77 (6), 1474–1481.

SINCLAIR, B. (1999) Everything's Great When It Sits on a Chip, *The Scientist*, 13 (11), 18.

SINGHAL, R.S., KULKARNI, P.R. and REGE, D.V. (1997) The development of the concept of food quality, safety and authenticity. In: *Handbook of indices of food quality and authenticity.* Woodhouse Publishing Limited, Cambridge, 9–34.

SMITH, T.P.L., LOPEZ-CORRALES, N.L., KAPPES, S.M. and SONSTEGARD, T.S. (1997) Myostatin maps to the interval containing the bovine *mhl* locus. *Mammalian Genome* 8, 742–744.

SOLOMON, M.B., EASTRIDGE, J.S., PURSEL, V.G., BEE, G. and MITCHELL, A.D. (2000) Implications of biotechnology: meat quality and value, 53rd Annual Reciprocal Meat Conference.

SOMERS, C., TARRANT, P.V. and SHERRINGTON, J. (1985) Evaluation of some objective method for measuring pork quality. *Meat Science*, 15, 63–76.

STOEVA, S., BYRNE, C., MULLEN, A.M., TROY, D.J. and VOELTER, W. (2000) Isolation and identification of proteolytic fragments from TCA soluble extracts of bovine M. *longissimus dorsi*. *Food Chemistry*, 69, 365–370.

STONE, R.T., KEELE, J.W., SHACKELFORD, S.D., KAPPES, S.M. and KOOHMARAIE, M. (1999) A primary screen of the bovine genome for quantitative trait loci affecting carcass and growth traits. *J. Anim. Sci.*, 77, 1370–1384.

SWATLAND, H.J. (1995) Objective assessment of meat yield and quality. *Trends in Food Science and Technology*, 6, 117–119.

SWATLAND, H.J. (2000) Measurements with an on-line probe. *Food Research International*, 33(9) 749–757.

SWATLAND, H.J. (2001) Elastic deformation in probe measurements on beef carcasses. *Journal of Muscle Foods*, 12 (2) 97–105.

SZCZESNIAK, A.S. (1963) Objective measurements of food textrue. *Journal of Food Science*, 28, 410.

SZCZESNIAK, A.S. and TORGESON, K.W. (1965) Methods of meat texture measurement viewed from a background of factors affecting tenderness. *Advances in Food Research*, 14, 134–165

TAKAHASHI, K. (1996) Structural weakening of skeletal muscle tissue during *postmortem* ageing of meat: the non-enzymatic mechanism of meat tendrisation. *Meat Science*, 43, S67–S80.

TAN, F.J., MORGAN, M.T., LUDAS, L.I., FORREST, J.C. and GERRARD, D.E. (2000) Assessment of fresh pork colour with machine vision. *Journal of Animal Science*, 78 (12) 3078–3085.

TAYLOR, J.F. and DAVIS, S.K. (1997) Mapping quantitative trait loci: A new paradigm. Proceedings: Beef cattle genomics: Past, present and future, Texas A & M University, College Station, Texas.

TAYLOR, J.F. and DAVIS, S.K. (1998) Detection of quantitative trait loci influencing growth and carcass quality traits. Proceedings: Challenges for the meat industry in the next millennium, Dublin, Ireland, Organised by The National Food Centre, Teagasc.

TAYLOR, S.A., RAIMUNDO, A., SEVERINI, M. and SMULDERS, F.J.M. (1996) *Meat quality and meat packaging*. Utrecht, ECCEAMST. I11.

TOLDRA, F. and FLORES, M. (2000) The use of muscle enzymes as predictors of pork meat quality. *Food Chemistry*, 69 (4), 371–377.

TROY, D.J., TARRANT P.V. and HARRINGTON, M.G. (1987) Changes in myofibrillar proteins from electrially stimulated beef. *Biochemical Society Transactions* 15, 297–298.

TROY, D.J., MULLEN, A.M. and CASSERLY, U. (1998) Predicting meat quality at the early *postmortem* period. Hygiene, quality and safety in the cold chain and air-conditioning. Conference Proceedings, Nantes.

VAN OECKEL, M.J., WARNANTS, N. and BOUCQUE, C.V. (1999) Measurement and prediction of pork colour. *Meat Science*, 52 (4) 347–354.

VENEL, C., MULLEN, A.M., DOWNEY, G. and TROY, D.J. (2001) Prediction of tenderness and other quality attributes of beef by near infared reflectance spectroscopy between 750 and 1100 nm; further studies. *Journal of Near Infared Spectroscopy*, 9, 185–198.

WARNER, K.F. (1928) *Proceedings of the American Society of Animal Production.* 21, 114.

WOLD, J.P., KVAAL, K. and EGELANDSDAL, B. (1999a) Quantification of intramuscular fat content in beef by combining autofluoresence spectra and autofluorescence images. *Applied Spectroscopy*, 53 (4), 448–456.

WOLD, J.P., KVAAL, K. and EGELANDSDAL, B. (1999b) Quantification of connective tissue (hydroxyproline) in ground beef by autofluorescence spectroscopy. *Journal of Food Science*, 64 (3), 377–383.

WULF, D.M. and PAGE, J.K. (2000) Using measurements of muscle colour, pH and electrical impedance to augment the current USDA beef quality standards and improve the accuracy and precision of sorting carcasses into palatability groups. *Journal of Animal Science*, 78, 2595–2607.

WULF, D. M., S. F. O'CONNOR, J. D. TATUM and G. C. SMITH. 1997. Using objective measurements of muscle color to predict beef longissimus tenderness. *Journal of Animal Science*, 75, 684–692.

YANG, G.P., ROSS, D.T., KUANG, W.W., BROWN, P.O. and WEIGEL, R.J. (1999) Combining SSH and cDNA microarrays for rapid identification of differentially expressed genes. *Nucleic Acids Research*, 27, 1517–23.

ZAMORA, F., DEBITON, E., LEPETIT, J., LEBERT, A., DRANSFIELD, E. AND Ouali, A. (1996) Predicting variability of ageing and toughness in beef M.*Longissimus dorsi et thoracis*. *Meat Sci.* 43, 321–333.

ZHAO, N., HASHIDA, H., TAKAHASHI, N., MISUMI, Y AND Sakaki, Y. (1995) High-density cDNA filter analysis: a novel approach for large-scale quantitative analysis of gene expression. *Gene*, 156, 207–213.

20
Meat packaging

H. M. Walsh and J. P. Kerry, University College Cork

20.1 Introduction

Meat packaging technology has evolved rapidly over the past two decades yet despite major developments in packaging materials and systems the fundamental principles of packaging muscle foods remain the same. Packaging fresh meat is carried out to delay spoilage, permit some enzymatic activity to improve tenderness, reduce weight loss, and, where applicable, to ensure an oxymyoglobin or cherry red colour in red meats at retail or customer level (Brody, 1997). When considering processed meat products, factors such as dehydration, lipid oxidation, discolouration and loss of aroma must be taken into account (Mondry, 1996).

Many meat packaging systems currently exist, each with different attributes and applications. These systems range from overwrap packaging for short-term chilled storage and retail display to 100% carbon dioxide atmosphere packaging for long-term chilled storage (Payne *et al.*, 1998). Considerable interest has developed recently within the retail sector in merchandising centrally processed, display-ready, meat cuts. Packs are prepared at the meat plant and from here, placed straight onto the supermarket shelves in this form. The notion of centralised fresh meat packaging began almost immediately after the corner butcher had moved to the backroom of the newly emerged supermarket in the 1950s (Brody, 2000).

The meat and retailing industry's movement towards centralised packaging and distribution has led to an increased interest in other types of packaging systems which may have potential use for the future. The only criterion customers have at the point of purchase of fresh meat cuts is visual appearance (Kropf, 1980). Consequently, desirable colour must be maintained during chilled storage, distribution and subsequent retail display if preservative packaging systems are to be effective (Jeremiah and Gibson, 1997). A longer shelf life is also required for long- and short-distance distribution. The use of modified atmosphere packaging (MAP) as well as vacuum packaging systems are just some of the options available to the processor/retailer. The shift from point-of-sale processing to central processing has also led to the introduction of mother or master packaging systems. These systems consist of a number of gas-permeable overwrapped trays, sealed in a master pack which contains a modified atmosphere. These may be superior to MAP, provided that the meat maintains an acceptable appearance for a sufficient period after bulk packs are opened and individual meat trays are displayed.

In conclusion, selecting a preservative packaging that is appropriate for a particular fresh/processed meat product for a particular niche market requires knowledge of the principal deteriorative processes to which the product is subjected, the general hygienic condition of the product when it is presented for packaging and the temperature history that the product can be expected to experience during its storage and distribution (Gill, 1991)

The dominant preference among consumers in their daily food consumption patterns (after taste and safety) is for convenience. Consumers want their food fast and easy (Kinsey, 1997). The family unit has decreased in size over recent years and more people are now living alone. There are now more women in the work-force, who do not desire complicated food preparation each evening. Longer commuting to and from work and increasingly busy life-styles also contribute to the need for convenience. Therefore, smaller servings of a wide choice of meat cuts, well presented in packs in retail display units are the obvious choice for consumers.

Centralised meat packaging offers a number of significant advantages to the industry and the consumer. For the industry, the overall costs of production are reduced. Transportation costs are lower and trimming, evaporation and drip losses are reduced also. There is also improved inventory and product control with the use of computerised product movement systems. This leads to accurate forecasting of required stock and can all be carried out from one central location (www.packstrat.com). Central processing also offers some economy of labour for the industry. Personnel work on a high-speed, automated line, each with their own specified job, and greater safety is achieved due to close supervision. At the very least, retailers must be knowledgeable about the differences between packaging treatments to make cost effective buying decisions, which in turn, will benefit the consumer (Schulter et al., 1994).

The advantages for the consumer include better sanitation and enhanced palatability due to controlled ageing of the product (www.ansci.uiuc.edu/meatscience/Library/packaging.htm). The choice of product available to the consumer has also improved. Fifteen to twenty years ago, only fresh poultry meat was seen in shelf-ready packaging. However, today a wide range of fresh and processed pork, lamb and beef products, are also available. Therefore, the main contributors to fresh meat deterioration leading to loss of profit, are, colour, microbiology and drip loss. Processed meats on the other hand, need protection from bacteriological contamination, oxidation, dehydration, discolouration and loss of aroma to ensure a good quality product is available to the consumer (Mondry, 1996). Before considering the form of packaging which will be used to pack muscle food products, it is important that factors which will determine shelf-life quality of a particular meat product are considered, measured and quantified where possible. For the purposes of this review, meat quality factors will be divided into those for fresh meat and those for processed meat products.

20.2 Factors influencing the quality of fresh and processed meat products

The principal factors that must be addressed in the preservation of fresh chilled meat are the retention of an attractive, fresh appearance for the product which is displayed and a delay in bacterial spoilage (Gill, 1996). Minimising drip losses is also of economic concern because such losses can increase the cost of the product, which is often traded on narrow profit margins. During storage, processed meats deteriorate in the first instance because of discolouration, secondly because of oxidative rancidity of fat and thirdly on account of microbial changes (Pearson and Tauber, 1984).

20.2.1 Meat colour

In the mind of the average consumer about to purchase meat, colour becomes synonymous with fresh red meat quality (Renerre and Labas, 1987). The colour of fresh red meat is of the utmost importance in marketing since it is the first quality attribute seen by the consumer who uses it as an indication of freshness and wholesomeness. The colour of fresh meat is not well correlated with the eating quality however the consumer still demands that beef has a bright cherry red colour (Taylor, 1996), lamb a brick red colour and pork and chicken a uniform pink colour.

Meat colour depends on myoglobin, a pigment with several forms (Renerre and Labas, 1987) (see Fig. 20.1). Myoglobin is concerned with the storage and transfer of oxygen within the muscle. Myoglobin concentrations are variable between species and between muscles. Light chicken meat, pork and beef contain 0.01, 1–3 and 3–6 mg/g myoglobin, respectively (Warriss, 1996). Deoxymyoglobin or reduced myoglobin is purple in colour and is the predominant muscle colour in the absence of oxygen. It is the expected muscle colour of vacuum packaged meat. When oxygen is introduced or when meat is exposed to air, deoxymyoglobin becomes oxygenated to oxymyoglobin which is responsible for the bright cherry red colour consumers expect to see on the supermarket shelves. If oxygen is removed, oxymyoglobin reverts to the basic dark purple myoglobin pigment (Taylor, 1996). Over a period of time and under atmospheric

420 Meat processing

```
                    Carboxymyoglobin
                    Cherry red
          +CO₂
                              +O₂
 Deoxymyoglobin                              Oxymyoglobin
 Purple                                      Bright red

        Oxidation        Metmyoglobin         Oxidation
                         Green/brown/grey
```

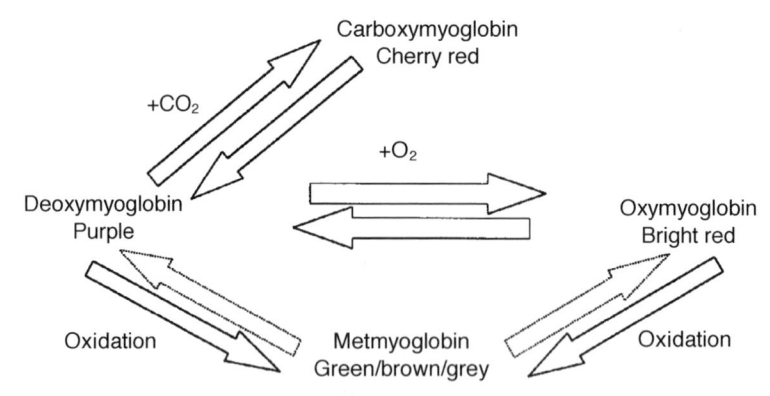

Fig. 20.1 The colour of fresh red meat related to the forms of the pigment myoglobin (Sorheim and Nissen, 2000).

conditions, which may vary from several hours to several days, oxymyoglobin is further oxidised to metmyoglobin. This is the pigment responsible for the brown discolouration associated with non-saleable meat (Kerry et al., 2000). Deoxymyoglobin and oxymyoglobin are heme proteins in which the iron is present in the ferrous state (Fe^{2+}) while metmyoglobin possesses the ferric (Fe^{3+}) form. The conversion of the ferrous to the ferric form is brought about by oxygenation.

Myoglobin can rapidly combine with or lose oxygen in response to the partial pressure of oxygen to which it is exposed (Livingston and Brown, 1981). Ledward (1970) found that the formation of metmyoglobin on bovine *M. semitendinosus* was maximal at a partial O_2 pressure of 6mm Hg at 0°C. Meat can be maintained in much higher or extremely low O_2 concentrations in order to avoid these colour-destructive O_2 tensions. Sorheim et al. (1995) reported that pork loin sections stored in CO_2 atmospheres with 1.0, 2.8 and 4.0% O_2 for 12 days were all discoloured but that sections stored in O_2 free CO_2 were not. Greene et al. (1971) reported that 40% metmyoglobin caused meat rejection by consumers.

The pigment of cured meats, nitrosylmyoglobin, is stable in the absence of oxygen or under vacuum but oxidation to metmyoglobin is rapid when oxygen is present. The rate of oxidation increases directly with increasing oxygen tension. Low humidity and high storage temperature may result in a brown discolouration on the surface due to chemical alteration and dehydration. Nitrosylmyoglobin is much more susceptible to light than myoglobin; cured meats can fade in one hour under retail display lighting conditions. Since light accelerates oxidative changes in the presence of oxygen, vacuum or inert gas packaging will help reduce the effect (Robertson, 1993). Lundquist (1987) discovered that holding vacuum packaged meats in the dark for 1–2 days before exposing them to display lights allowed residual surface oxygen to be depleted by microorganisms and tissue activity, thus reducing colour deterioration. He also reported that to inhibit colour changes in cured meat products, a lower level of available oxygen than that required to shift the microbial population from aerobic to anaerobic was required.

The degree of photooxidation of the nitrosyl meat pigment is highly affected by the oxygen pressure above the cured meat product but it can be minimised by the use of low oxygen transmission rate films in combination with vacuum packaging or packaging in modified atmospheres without oxygen or interactive packaging with oxygen absorbers (Andersen *et al.*, 1988, 1990).

20.2.2 Lipid oxidation

Lipid oxidation is a leading cause of quality deterioration in muscle foods (Rhee *et al.*, 1996), resulting in rancidity, off-flavours and off-odours as well as colour and texture deterioration (Kanner *et al.*, 1992). Lipid oxidation *in vivo* and in muscle foods, is believed to be initiated, in the highly unsaturated phospholipid fraction in subcellular membranes (Morrissey *et al.*, 2000). Immediately preslaughter, and during the early post slaughter phase, oxidation in the highly unsaturated phospholipid fraction is no longer highly controlled and the balance between the proxidative factors and antioxidative capacity is likely to favour oxidation. At this stage it is unlikely that the defensive mechanisms that exist in the live animal still function (Morrissey *et al.*, 1998).

Mechanical boning, mincing, restructuring and cooking can cause significant disruption of the cellular compartmentalisation structure which facilitates the meeting of proxidants with unsaturated fatty acids resulting in the generation of free radicals and propagation of the oxidative reaction (Buckley *et al.*, 1995). The major products of lipid oxidation are the hydroperoxides, which break down into secondary products such as aldehydes, alcohols, hydrocarbons and ketones. These secondary products formed during oxidation contribute to the off-flavours generated during storage (St Angelo and Spanier, 1993). A distinctive off-flavour develops rapidly in meat that has been precooked, chilled-stored and reheated. The term Warmed-Over Flavour (WOF) has been adopted to identify this flavour deterioration. Autoxidation of membrane phospholipids is largely accepted as causal in the formation of WOF. It is thought that the polyunsaturated fatty acids from polar phospholipids rather than triglycerides are responsible for the initial development of off-flavours and off-odours in raw and cooked meats (Renerre and Ladabie, 1993).

Lipid oxidation is not usually a limiting factor in conventional over-wrapped trays as air-permeable films allow odour volatiles to escape through the package. However, with modified atmosphere packages, the volatile products are retained within the package and can be clearly detected by consumers when opened (Zhao *et al.*, 1994). These odours can be termed as compartmentalised packaging odours and may arise through a complex interaction of product volatiles with packaging volatiles in a manipulated gaseous environment. Storage of meat in high O_2 atmospheres leads to a limited shelf life due to lipid oxidation. Taylor *et al.* (1990) reported that malonaldehyde in beef samples increased in modified atmosphere packs more rapidly than in vacuum packs but not until one week after retail packaging. Jackson *et al.* (1992) found that beef striploins packaged in 80% O_2 and 20% CO_2 at 3°C developed strong off-flavours as a result of lipid oxidation.

Cooked or pressure treated meats are more vulnerable to oxidation than fresh meats. Factors which influence the rate and course of oxidation of lipids are well known and include light, local oxygen concentration, high temperatures, presence of catalysts and water activity. Control of these factors can significantly reduce the extent of lipid oxidation in foods. St Angelo and Spanier (1993) discovered that heme-containing catalysts such as myoglobin, hemoglobin and cytochromes may also cause oxidation. Nitrites, which are added to meat during curing, are thought to prevent lipid oxidation or the formation of meat flavour deterioration. It has been well documented that vacuum packaging or modified atmosphere packaging in an anoxic gas has been shown to reduce lipid oxidation of precooked meats effectively (Nolan *et al.*, 1989; Spanier *et al.*, 1992).

Even though precooked meats are placed in an anoxic environment, lipid oxidation still proceeds due to residual oxygen in the product and the package. The residual concentration of oxygen in MAP packages is normally 0.5–2.0% (Smiddy *et al.*, 2002, Randall *et al.*, 1995). Bertelsen *et al.* (2000) stated that the oxidative stability may be further improved if a raw material with an optimum intrinsic oxidative stability, achieved by feeding supranutritional levels of vitamin E, is used for production of these products. The combined synergistic effects of supplementing animals with dietary α-tocopheryl acetate and the subsequent manipulation of packaging systems to improve meat quality and extend shelf-life of muscle food products have been shown by a number of researchers. Cannon *et al.* (1995) showed that packaging at low oxygen levels combined with dietary vitamin E was highly effective in decreasing TBARS in chill-stored precooked pork chops and roasts. Kerry *et al.* (1996) reported that α-tocopheryl actetate supplemented modified atmosphere packaged beef increased colour stability in steak cores when compared to aerobically packaged cores. O'Grady *et al.* (2001) found that lipid and oxymyoglobin oxidation in minced beef stored in 80% O_2 : 20% CO_2 were lower in muscle from vitamin E supplemented animals compared with unsupplemented animals.

20.2.3 Meat microbiology

The microbial population of fresh meat is affected by a number of factors such as species, health and handling of the live animal, slaughtering practices, chilling of the carcass, sanitation during fabrication, type of packaging used and handling through distribution and storage (Young *et al.*, 1988). Most microbial contamination of muscle tissue occurs after the animal has been slaughtered through two main sources, those derived from the slaughter environment and organisms from the intestinal tract. The predominant organisms on the surface of freshly prepared carcasses are gram-negative bacteria such as *Acinetobacter*, *Aeromonas*, *Pseudomonas* and *Moraxella*. *Enterobacter* and *Escherichia* are also found. Gram-positive organisms are less abundant but commonly include *Brochothrix*, other lactic acid bacteria and *Micrococcaceae*.

The wrapping materials for fresh over-wrapped meat are only slightly permeable to water vapour but highly permeable to oxygen and carbon dioxide.

Therefore, conditions for microbial growth are favourable. The predominant organisms in this case are the aerobic psychrotrophs, such as *Pseudomonas*, *Acinetobacter* and *Psychrobacter* spp. Pseudomonads in particular cause a putrid spoilage of the meat. Vacuum packaging and using high oxygen impermeable films can inhibit the growth of *Pseudomonas* but it does not affect the growth of *Lactobacillus* organisms, which are facultative anaerobes and tend to thrive in vacuum-packaged meats. Lactic acid bacteria have two particular advantages. Firstly, they develop at a slower rate than aerobic gram-negative flora, resulting in a longer shelf life. Secondly, the sour 'off' odour, which can be detected on opening of packs, is far less offensive than putrid odours.

There are inconsistencies in the literature relating to the growth of different bacterial strains in vacuum packaged meat. Newton and Rigg (1979) reported that these contradictions could be explained if films of different permeabilities were used in each study. High oxygen modified atmosphere packaged meat spoils aerobically with the spoilage flora being dominated by *Pseudomonas*. Their growth rate is only half that attained under non-preservative aerobic conditions. Low oxygen modified atmosphere packaged meat is normally spoiled by lactic acid bacteria if the atmosphere used is 100% CO_2. The growth rate of the lactic acid bacteria is nearly halved due to the bacteriostatic effect of CO_2, resulting in a longer shelf life than obtained for equivalent vacuum packaged products.

Many factors influence the nature of the microflora that develop in processed meat products during chill storage. The main factors are nitrite concentration, salt concentration (which affects the a_w), presence of oxygen and permeability of the packaging film. The overall pH of the product may also play a part (Zeuthen and Mead, 1996). The most frequently observed types of bacterial spoilage in vacuum packed, sliced, cooked, cured meats are a sweet/sour odour caused by lactobacilli, leuconostocs and streptococci (Mol *et al.*, 1971); a cheesy odour caused by *Brochothrix thermosphacta* (Egan *et al.*, 1980); a sulphide odour caused by *Enterobacteriaceae* and greening caused by hydrogen peroxide producing lactobacilli. Boerema *et al.* (1993) reported that the advantage of shelf-life extension afforded fresh meats by carbon dioxide controlled atmosphere packaging, over that attainable in vacuum packaging, does not apply for sliced cooked ham. The presence of yeasts and moulds can be a problem for some processed meat products, especially on dry meat surfaces. The most common species of yeasts encountered on meat products are *Candida* and *Rhodotorula*. The presence of mould growth on packaged meat is usually an indication of a defective packaging system. Yeasts can cause spoilage when counts are as low as 10^5/g meat, whereas, bacterial spoilage does not usually occur below populations of 10^7/g–10^8/g meat. McDaniel *et al.* (1984) reported that vacuum packed cooked beef steaks were organoleptically acceptable after 21 days storage at 4°C, while modified atmosphere packaged steaks were not fit for consumption after 14 days of storage. Hintlian and Hotchkiss (1987) found that high CO_2 MAP was effective in inhibiting the growth of *Pseudomonas fragi*, *Salmonella typhimurium*, *Staphylococcus aureus* and *Clostridium perfringes* on cooked, sliced roast beef.

20.2.4 Drip loss

One of the main quality attributes of fresh meat is its water-holding capacity because it influences consumer acceptance and the final weight of the product (Den Hertog-Meischke, 1997). The loss of exudates from muscle tissue is unavoidable. Any system prolonging the shelf life of packed chilled meat will be subject to accumulation of exudates or drip. The drip is thought to originate from the spaces between the fibre bundles and the perimysial network and the spaces between the fibres and the endomysial network (Offer and Cousins, 1992). These spaces appear during rigor development. Factors that may affect drip losses include rigor temperature and membrane integrity (Honikel, 1988; Honikel *et al.* 1986), preslaughter stress, processing factors, and packaging (Payne *et al.*, 1997). Exudate losses are exacerbated by cutting of meat into smaller portions. Losses of approximately 5% of the primal cut weight at the packing plant can be expected. The amount of drip in cut meat is also largely dependent on sample thickness, surface to volume ratio, orientation of cut surface with respect to muscle fibre axis and prevalence of large blood vessels (Farouk *et al.*, 1990).

Taylor *et al.* (1990) found that drip losses were lower for vacuum skin packaged samples than modified atmosphere packaged samples. Payne *et al.* (1998) investigated drip loss in beef under conventional vacuum packaging systems and non-vacuum packaging systems and found that drip loss can be reduced by using a packaging system that avoids applying a vacuum. The accumulation of juice from processed meat products is also a cause for concern in vacuum packs. Vacuum skin packs again have the advantage over vacuum packs for processed meats in that there is no excess of film around the product, leaving virtually no space for product juice to collect (Mondry, 1996). Condensation of moisture on the surface of the meat and the package may be prevented by carrying out packaging operations in an environment having a dew point temperature below the subject temperature (Rizvi, 1981). Antifog properties in fresh meat packaging film are provided by lowering the surface tension of the film through incorporating a wetting agent into the film formulation or by coating the surface with a wetting agent.

20.3 Vacuum packaging

Today the most widely used method employed to extend the storage life of fresh meat is vacuum packaging (Bell and Garout, 1994). In the USA, 97% of all beef is estimated to have been fabricated and transported as a vacuum packaged product (Humphreys, 1996). Vacuum packaging extends the storage life of chilled meats by maintaining an oxygen deficient environment within the pack. The air within the package must be evacuated effectively to nominal anoxic levels (less than 500ppm) to prevent irreversible browning due to low levels of residual oxygen. The exclusion of oxygen from the meat surface as soon as possible after the breaking of the carcass into primals preserves the meat's potential to reoxygenate following retail pack display.

The concept of 'boxed beef' was developed by French Scientists as early as 1932, to prolong the shelf-life of frozen meat stored as military provisions. It was not until 1966, however, that beef was centrally processed and distributed in this form. Today different packaging materials are used but the concept remains the same. The carcasses are divided into parts, deboned and trimmed, vacuum packed in heavy plastic bags, placed in corrugated paperboard boxes and shipped to retailers. Boxed beef is a preferred method of packaging by distributors and retailers because weight loss and drying can be prevented, it is hygienic, meat colour and quality are maintained as films of low O_2 permeability are used, operations can be centralised, rational distribution is possible and there is ease of inventory control (Tomioka, 1990).

The preservative effect of vacuum packaging is achieved by maintaining an oxygen depleted atmosphere since potent spoilage bacteria are inhibited in normal pH meat under optimum vacuum packaged conditions (Gill, 1991). The microenvironment within the pack will determine the type of microflora which develops (Devore and Solberg, 1974). When meat is first vacuum packaged any residual oxygen remaining in the pack is consumed by meat and muscle pigments (Hood and Mead, 1993) and CO_2 is produced as the end product of tissue and microbial respiration. Under a good vacuum the package headspace consists of $< 1\% O_2$ and 10–20% CO_2 (Lambert et al., 1991). This type of package severely restricts the growth of aerobic microorganisms such as Pseudomonas and favours facultative anaerobic organisms such as Lactobacillus (desirable) and *Brochothrix thermosphacta* (undesirable). These are slow growing bacteria, they cause less offensive types of spoilage at higher bacterial numbers than other aerobic organisms and will eventually cause spoilage but only after many weeks of storage (Muller, 1990). Under aerobic conditions the growth of *Brochothrix thermosphacta* at chill temperatures is inhibited at pH values below 5.8. However, the growth of this organism is often associated with the early spoilage of vacuum packaged high pH meat.

If vacuum packaging procedures are followed correctly, the storage life of meat can be extended (Muller, 1990). Care must be taken to ensure the initial bacterial load has been kept as low as possible by adherence to good hygiene procedures. The temperature must be maintained as near as 0°C as possible (Brody, 1989). High pH and DFD meat must also be avoided as early spoilage may occur (Humphreys, 1996). The success of packaging is thought also to depend largely upon the degree of vacuumisation achieved. High vacuum levels result in the most desirable muscle colour and fat appearance. Vacuum packaged meat of normal pH (< 5.8) can be stored for 12 (Eustace, 1989) to 14 weeks (Hood and Mead, 1993) at 0°C.

Vacuum packaging is a simple easily controlled process with any failure clearly identified by visible air pockets remaining in the pack. There is increased hygiene and simplified handling during distribution and storage as well as more efficient use of refrigerated storage. There is more efficient process control through the use of flow line techniques with subsequent increased productivity. There is also a significant reduction in skilled labour and waste at retail level associated with vacuum packaging (Humphreys, 1996).

However, vacuum packaging is considered unsuitable for red meats for retail display purposes since the oxygen depleted atmosphere causes the meat in these packages to be the purplish colour of deoxymyoglobin and therefore not acceptable to consumers (Gill, 1991). Seman *et al.* (1988) observed that vacuum packaged meat had greater colour stability than meat stored in carbon dioxide. Pork chops stored in vacuum were also more desirable in appearance and had more acceptable colour than aerobically stored chops (Doherty and Allen, 1998; Doherty *et al.*, 1996). The range of products to which straight vacuum packaging can be applied is limited however; vacuum packaging is ineffective for whole carcasses or cuts of shapes which prevent the packaging film being closely applied to all surfaces (Gill, 1996).

The formation of drip in meat packaging is unsightly, and represents a portion of the product that the consumer cannot use. Many studies reported that vacuum packaged meat produced higher levels of drip than meat packaged in modified atmospheres (Doherty *et al.*, 1996, Schulter *et al.*, 1994, O'Keefe and Hood, 1982). However, a very low level of drip is observed when meat is shrink-wrap vacuum packaged. This lower drip loss could be due to the fact that there is less space for drip to form, or due to softer packaging or a combination both (Payne *et al.*, 1998). Vacuum packaging is considered unsuitable for meat products that are sensitive to pressure, for example very thin slices of ham that when packed under vacuum are difficult to separate without damage. The introduction of CO_2 or N_2 flush at this stage could prevent slice adhesion and provide better colour retention due to the removal of residual O_2.

Vacuum packages for meat are formed in four basic ways. The first method is through heat shrinking a flexible packaging material around the primal cuts (Zagory, 1997). Shrink bags are supplied pre-made with a seal provided at either end and along the sides. When they are briefly exposed to heat, a built in tension, called 'locked in tension' is released and they shrink in both directions. This process increases film thickness, improving mechanical resistance and oxygen barrier properties. The level of drip is also reduced and handling of the product is improved. Most of these shrink bags are made from multi-ply formulations based on polyolefin resins, with either polyvinylidene chloride (PVDC) or ethyl vinyl alcohol (EVOH) as the gas barrier component (Humphreys, 1996).

The second method is by using a preformed plastic bag, also known as a pouch, in an evacuation chamber (Zagory, 1997). The basic pouch structure uses polyamide (PA) as the outer layer which provides barrier and physical strength properties with an inner core and sealing layer of polyethylene (PE) or linear low density polyethylene (LLDPE). These materials cannot be shrunk which may give rise to the accumulation of drip in the creases and folds (Humphreys, 1996).

The third method is the use of thermoforming trays in line from a base web (Zagory, 1997). After the product has been placed into the newly-formed tray, a second film web, coming from a second reel of film is placed on the top of the tray. The resulting pack is evacuated in this case and the top and bottom films are sealed in the machine's vacuum-sealing station. After cutting the sealed web across and longitudinally, finished single packs leave the machine (Mondry, 1996).

The final method is through vacuum skin packaging, in which the product acts as a forming mould (Zagory, 1997). In vacuum skin packaging the meat is placed in a rigid pre-formed tray or on the flat surface of a flexible base material. The top web is first softened by heat and air is evacuated. The top web then forms closely around the product, shrinking as it does, and forms a seal everywhere it comes into contact with the base (Humphreys, 1996).

A number of packaging systems exist for processed meats also. Vacuum packaging is used for thickly sliced meats and whole pieces of processed meats. Vacuum packaging with a steam shrink operation can also be used. This method removes loose film and wrinkles are reduced. The product must be able to withstand the heat treatment during shrinking. Vacuum packs from rigid trays are used for stacked slices of processed meat. The shape of the product is designed to fit tightly to the package dimensions. Modified atmosphere packs from flexible film are used for salamis, bulk packs of sliced meats and bulk packs of frankfurter sausages. These bulk packs are delivered to supermarkets where they are opened in the preparation area before displaying the product for sale as a fresh product. Modified atmosphere packs in rigid film or trays are used for thinly sliced products and for products which can damage the film when packed under vacuum. The gas or gas mixture in these packs is very much dependent on the type of product, the way in which the product is consumed and the type of film used. Skin packs are used for high quality expensive products. The film is softened before being skin-draped onto the product and therefore does not damage it. Difficult opening procedures and the high cost of the package are the main disadvantages of this system (Mondry, 1996).

The *sous vide* method of food preparation was developed in the mid-1970s in France (Creed, 1998) and has found widespread acceptance there. Acceptance of this processing method has been slower in the United States and most other European countries but continues to grow, particularly in the food service sector. *Sous vide* involves vacuum packaging of foods, usually in multilayer laminate plastic pouches, cooking the vacuum packaged product in a water bath, moist steam or pressure cooker, cooling rapidly in cold water and then storing under refrigeration (Zagory, 1997). It has been widely claimed that *sous vide* food is sensorially and nutritionally superior to that produced using cook-chill methods. Firstly, the low oxygen tension inside the pack inhibits both chemical oxidation and microbial activity. Secondly, the packaging prevents evaporative losses of water and flavour volatiles during heat treatment (Church, 1998). The majority of studies undertaken on the microbiological safety of *sous vide* products concern *Clostridium botulinum*. The public health risk posed by *sous vide* is exacerbated by its anaerobic nature which inhibits both chemical oxidation and aerobic spoilage organisms which usually provide the sensory indication of spoilage prior to it becoming unfit for consumption. Thus, a potential risk exists for *sous vide* products to be palatable yet microbiologically unsafe (Conner *et al.*, 1989). However, there is little evidence to date to indicate a significant risk

428   Meat processing

except in cases of extreme product abuse resulting from either a lack of uniformity of heat treatment or temperature/time abuse during chilled storage.

Canning is also a form of vacuum packaging which is often accompanied by thermal processing. Vacuum in the cans may be obtained by filling completely with the product at a high temperature using atmospheric pressure. The vacuum may also be obtained by machine vacuum, by 'steam-vac' closure or by thermal exhausting. The major reason for canning meat is to provide safe products that have desirable flavour, texture and appearance. Successful production of commercially sterile canned meat products requires that all viable microorganisms be either destroyed or rendered dormant. The process must also inactivate raw material enzyme systems. Commercially sterile canned meat products generally reach an internal temperature of at least 107°C, but this temperature may be as low as 101°C, depending on the salt and nitrite content. Some meat products merchandised in cans receive only a pasteurisation process and are referred to as 'perishable', which means that they must be kept refrigerated. (Pearson and Gillett, 1996). Five principal types of cans are used in the meat industry; square and pullman base, pear shaped, round sanitary, drawn aluminium, and oblong. The cans used are generally tin or chrome plated low carbon steel or aluminium. The most recent development in this area has been in the replacement of such materials with PET-aluminium based laminates. The resulting product has an improved flavour, greater nutritional value and energy costs are reduced. There are a wide range of meat-based canned products available on the supermarket shelves. The consumer is probably most familiar with beef stew, chilli con carne, meat balls in gravy, canned hams, spam and tongue and luncheon meat (Pearson and Gillett, 1996).

20.4   Modified atmosphere packaging

Modified atmosphere packaging (MAP) is the enclosure of food products in high gas-barrier materials in which the gaseous environment has been changed to slow respiration rates, reduce microbiological growth and retard enzymatic spoilage with the intent of extending shelf-life (Young et al., 1988). Distribution distance is now probably the main factor determining the form of packaging used for fresh meats. As discussed previously, the central preparation of retail packs is now commonplace within the trading sector. These central cutting systems can be operated with product in conventional over-wrapped trays, but only if times are short between meat cutting and display. For wider trading of retail-ready meat, a modified atmosphere must be used to extend the storage life of the product (Young et al., 1988). The advantages and disadvantages of modified atmospheres are shown in Table 20.1.

MAP systems most frequently use mixtures of CO_2, O_2 and/or N_2, in which each gas has a specific role to play in extending the shelf-life and maintaining the appearance of packaged meat (Young et al., 1988). Over a hundred years ago, a patent was granted for applying a gas mixture of CO_2 and CO for storage

Table 20.1 Advantages and disadvantages of modified atmosphere packaging (Wolfe, 1980)

Advantages	Disadvantages
Extended transit time	Visible added cost
Higher quality maintenance	Variable product requirements
Active inhibition of bacteria and moulds	Not universally effective
Reduced economic loss	Colour changes with red meats
	Atmosphere maintenance
	Temperature regulation

of meat. In the 1970s, MAP was introduced for retail meat in France and the UK. The market share of MAP in Western European countries as a percentage of the total retail meat market is 10–40% (Sorheim and Nissen, 2000). However, this method of meat packaging is not as popular in the USA. The demand for meat packaging in the USA is led by meat packing companies rather than by retailers. In addition, the distribution chain is neither as quick or as controlled as it is in Europe, and distances from processor to market are greater (Taylor, 1996).

20.4.1 Gases used in fresh and processed meat packaging

The most commonly used gases for the packaging of meat are CO_2, N_2 and O_2 although other gases including CO, nitrous oxide, argon, sulphur dioxide and ozone have been tried to a limited extent (Church, 1994). The EU classifies packaging gases as additives and has given each of them an E-number. Also according to EU legislation, foods packaged in modified atmospheres must be labelled with a phrase like 'Packaged in a protective atmosphere' (Sorheim and Nissen, 2000). Gases for the packaging of meat are seldom used alone but in mixtures, which vary according to the application (Sorheim *et al.*, 1997).

Oxygen
One of the major functions of oxygen is to maintain the red pigment, myoglobin in the oxymyoglobin state that is responsible for the bright red colour associated with freshness. Oxygen pressure levels over 240mm are thought to greatly increase and extend the fresh appearance of meats (Seideman and Durland, 1984). Removal of oxygen is particularly important for processed muscle foods held in modified atmosphere systems. In most cases, deterioration of meats is caused by oxidation of meat components or spoilage by aerobic microorganisms, both of which are accelerated in the presence of oxygen. The level of residual oxygen in processed meat modified atmosphere packs is therefore an important factor to consider. It may be attributed to a number of factors, such as the oxygen permeability of the packaging material, the ability of the food to trap air, poor sealing of the pack which may cause air to leak in, or inappropriate evacuation and/or gas flushing procedures (Smith *et al.*, 1986). Smiddy *et al.* (2001) reported that measurement of residual oxygen in packs in conjunction with oxygen

Carbon dioxide

Carbon dioxide is a known inhibitor of microbial growth including meat-borne microorganisms and these preservative properties were reported as early as 1882 (Finne, 1982). Gram-negative spoilage flora of refrigerated meat are especially sensitive to CO_2 while lactic acid bacteria are less affected (Enfors and Molin, 1984). The inhibitory effects of CO_2 have been attributed to alteration of the bacterial cells permeability, pH changes and enzymatic inhibition (King and Nagel, 1967). There appears to be an increase in the lag phase and generation time, which delays the overall increase of bacterial populations. Factors such as initial bacterial load, time of application, storage temperature and gas concentration will affect the desired end result.

Gill and Tan (1980) demonstrated that the level of CO_2 that gave maximum inhibition for the common spoilage organisms was approximately 200mm Hg, equivalent to 26% CO_2 in air. That is, for most of the organisms used in their study, 26% CO_2 gave the same inhibition as higher levels of CO_2. The inhibitory efficiency of CO_2 is increased at lower temperatures. This is thought to be due to the fact that the solubility of gases is much higher at lower temperatures; the CO_2 concentration in the medium will increase as the temperature is lowered.

A packaging system using an atmosphere of carbon dioxide alone is now in commercial use for chilled red meats that are transported to distant markets (Gill and Harrison, 1989). The first practical use of modified atmospheres containing elevated levels of carbon dioxide as a preservative in the handling of fresh meat was in the shipment of whole beef carcasses from Australia and New Zealand to Great Britain in the 1930s (Silliker and Wolfe, 1980).

Carbon dioxide reacts with water to form carbonic acid and can actually dissolve in fresh meat and also in the fat. As the gas dissolves in the water of fresh meat, the quantity of gas within the package diminishes and a partial vacuum is generated. This may bring about the collapse of the pack (Hirsch, 1991). Bruce et al. (1992) reported fissures in beef stored in 100% carbon dioxide controlled atmospheres, which they attributed to evolution of absorbed carbon dioxide from the meat during cooking. It is thought that a carbon dioxide level of 5 l/kg was used in this study while the optimum level to use to extend chilled storage life of red meats and pork is 1–2 l/kg (Gill and Penny, 1988) and 2 l/kg respectively (Jeremiah et al., 1996). Sorheim et al. (1996) were of the opinion that increasing atmospheric pressures concentrations of CO_2 had a negative effect on the ability of meat to hold water. Ledward (1970) reported that more than 30% CO_2 in red meat accelerated discolouration.

Experience has shown that as far as processed meats are concerned, about 95% of applications can be covered by a standard mixture of 70% N_2 and 30% CO_2. For boiled and cooked meat products the main danger comes from CO_2 dissolving in the product juice. Too much CO_2 will change the product's aroma and causes the pack to collapse. Too little CO_2 will mean that after a few days, no CO_2 will remain in the package. Close monitoring of the gas mixture composition is required. Pre-fried products are packed at higher levels of CO_2 (up to 50%). These packages are generally made from rigid film, as collapsed packs are a common feature due to dissolution of CO_2 (Mondry, 1996).

Nitrogen
Nitrogen is an inert gas and is abundantly available at relatively low cost, has neither colour nor odour and is chemically unreactive (Hirsch, 1991). It has a low solubility in both water and fat. In modified atmospheres nitrogen is used to displace oxygen in order to delay aerobic spoilage and oxidative deterioration. Another role of nitrogen is to act as a filler gas so as to prevent pack collapse (Day, 1992).

Carbon monoxide
The positive effect of carbon monoxide (CO) on meat colour was known and patented over 100 years ago (Church, 1994) but as yet CO has been applied commercially only to a limited extent in the MAP of meat. The Norwegian meat industry has been using a gas mixture of 60–70% CO_2, 30–40% N_2 and 0.3–0.4% CO for the packaging of beef, pork and lamb. Based on the literature, the presence of 0.4–1.0% CO in modified atmospheres used for the packaging of meat seems sufficient to produce a stable cherry red colour (Sorheim *et al.*, 1997). However, the undesirable pink colour, which sometimes arises in cooked white meat, can sometimes be linked to exposure to CO also. Sorheim *et al.* (2001) reported that persistent redness in cooked beef burgers was influenced by CO. The burgers, containing carboxymyoglobin, were cooked to an end point temperature of more than 80°C while still having traces of pink colour and uncooked appearance. However carbon monoxide is a toxic gas and therefore its use for food packaging is not allowed in most countries. It is not approved for meat packaging in the US or in the EU (Luno *et al.*, 1998). Very little information exists in the literature on the exposure to CO following the consumption of meat that has been treated with CO gas but it is considered highly improbable that CO exposure from meat packaged in an atmosphere containing up to 0.5% will represent a toxic threat to consumers through the formation of carboxyhaemoglobin (COHb) (Sorheim *et al.* 1997).

The safety of workers who come in contact with carbon monoxide in meat packaging factories is also a cause for concern. If pure CO or high concentrations of CO were used for mixing of gases in the plant, they would pose a clear risk. The practice of Norwegian gas suppliers is either to deliver CO as a 1% CO/99% N_2 mixture and then blend this mixture with CO_2 on site, or as a complete 0.3% CO/70% CO_2/30% N_2 mixture. This practice is recognised by Norwegian health authorities to be safe (Sorheim *et al.*, 2001).

In a study carried out by Luno *et al.* (2000) it was shown that an atmosphere containing 50% CO_2 and 0.50–0.75% CO in the presence of a low concentration of O_2 (24%) is able to extend the shelf life of fresh beef steaks by 5–10 days when compared with the storage life in an atmosphere of 70% O_2, 20% CO_2 and

10% N_2. The presence of CO and 50% CO_2 extends product shelf-life by inhibition of spoilage bacteria growth, delayed metmyoglobin formation, stabilisation of red colour measured by instrumental and sensory techniques, maintenance of fresh meat odour and slowing down of oxidative reactions (Luno *et al.* 2000). Jayasingh *et al.* (2001) recommended the pretreatment of beef steaks with 5% CO for 24 hours to ensure a high colour stability before continuing storage in anaerobic conditions. It is thought that the colour of cooked meat products can also benefit from exposure to CO, as 1% CO in a N_2 atmosphere stabilised the colour of bologna (Aasgaard, 1993). The reason for the colour improvement is unknown but CO may bind to partly undenatured mylglobin.

Sulphur dioxide and argon
Sulphur dioxide is very chemically reactive in aqueous solution and forms sulphite compounds, which are inhibitory to bacteria in acid conditions (pH < 4). It has found use in the control of microbial growth in some processed meat products, such as sausages. Some people display hypersensitivity to sulphite compounds in foods and their use has come under scrutiny in recent years.

Argon is a noble gas and is not known to have any chemical or biological activity. However, it is reported to have some antimicrobial effects. Argon is present in the atmosphere (0.90%) and is therefore relatively abundant (Zagory, 1997).

20.4.2 Modified atmosphere equipment

A variety of machines are now available, using mixtures of between 60–80% O_2 and 40–20% CO_2 for prolonging the useful retail life of fresh meat to approximately a week. The most common of these is the thermoforming machine which produces trays from a bottom web of plastic, flushes them with a gas mixture and then seals them with a top web of film. The use of modified gas atmospheres for the bulk gas flush process was developed in the mid-1970s. The basic bulk gas packing technique employs the use of a preformed pouch or rigid tray along with a high barrier flexible lidding material. Air surrounding the product inside the package is removed and replaced with a specific gas mixture. The modified atmosphere gas mixes used for bulk gas packing, as well as the performance requirements of the package, differ greatly from those for consumer retail packs. Oxygen is not normally used in bulk modified atmosphere packing, since extension of case life of most products depends on the absence of oxygen The most common gas mixtures selected for different meat products for bulk gas flushing can be seen in Table 20.2.

Thermoforming machines have been designed for gas flushing of retail sized packs but can also be used for bulk packing by removing the smaller die inserts. Packs of up to 5 kg can be produced at high speeds but they can be difficult to handle and tend to be expensive. Snorkel-type machines are specifically designed for bulk gas packing of meat and are now the most widely used. The air

Table 20.2 Typical MAP combinations showing bulk storage life (Down, 1996)

Bulk MAP product	Packaging	Gas mix	Temp (°C)	Life (max. days)
Whole chicken	5L PA/PE	100% CO_2	−2 to 0	20–25
Whole turkey	5L PA/PE	100% CO_2	0 to 2	14–20
Poultry portions	5LPA/PE	100% CO_2	−2 to 0	10–14
Pork primals	5LPA/PE	40%CO_2/60%N_2	0 to 2	10–14
Pork primals	PA/EVOH/PE	100%CO_2	0 to 2	20–25
Beef primals	PA/EVOH/PE	80%CO_2/20%N_2	−2 to 0	42
Lamb primals	PA/EVOH/PE	80%CO_2/20%N_2	0 to 2	21
Cured meats	PA/EVOH/PE	80%CO_2/20%N_2	0 to 2	21
Whole salmon/trout	5L PA/PE	80%CO_2/20%N_2	0 to 2	14–20

is evacuated from preformed pouches and replaced with gas through retractable snorkels, which are inserted into the mouths of the pouches until just before heat sealing (Down, 1996).

Modified atmosphere trays are designed to minimise contact with the underside of the meat and also be deep enough to avoid contact with the lid. The excessive size of the tray can add considerably to the packaging costs. Tray depth can be considerably reduced by vacuum skin-packing the meat to the base of the tray using an O_2 permeable plastic. The space above the skin-packed meat is flushed with a modified atmosphere, before being sealed with a barrier plastic lid. The meat is fixed to the base of the tray and, therefore, there is less likelihood of accidental contact with the lid or tray sides (Taylor, 1996).

20.4.3 The effectiveness of MAP on fresh and processed meats

The composition of the atmosphere within a modified atmosphere package determines to a large degree the extent and type of spoilage that develops during storage. Reports differ concerning the optimum gas mixture required to maintain satisfactory meat colour and to extend the microbiological shelf life of the product. The most commonly used gas mixture for fresh red meat is high O_2, which has a minimum of 60–70% O_2 and 30–40% CO_2. Packaging in high O_2 extends the time for occurrence of microbiological spoilage and discolouration of meat (Sorheim and Nissen, 2000). Gill (1991) stated that modified atmospheres containing high oxygen concentrations essentially double the time to spoilage (storage life) and improve colour stability. According to Young *et al.* (1988), oxygen partial pressures over 240 mm greatly enhance and extend retention of the fresh appearance of meat. Low O_2 MAP is sometimes used for bulk packaging product; the inhibitory effects of CO_2 are exploited without any particular regard for the preservation of meat colour (Gill, 1995).

Shay and Egan (1990) found that the increase in display life of retail cuts by modified atmosphere storage decreased as the time of storage in the vacuum pack increased. However, they concluded that modified atmosphere storage of beef and lamb in a mixture of 80% O_2 and 20% CO_2 can give up to a threefold extension of retail display life. Gill and Jones (1996) reported that pork chops stored in 67% oxygen and 33% carbon dioxide for up to 12 days remained acceptable in appearance during 48 hours of display. Marriott et al. (1977) found that atmospheres containing 60% carbon dioxide, 25% oxygen and 15% nitrogen improved the overall retail appearance and desirability of beef cuts. However, Okayama et al. (1987) showed that beef steaks, stored in 20% carbon dioxide and 80% oxygen for 13 days at 4°C underwent a change in surface pH.

Most chilled poultry products are sold pre-wrapped in O_2 permeable film, which prevents moisture loss and the spread of contaminating micooorganisms. Relatively little use is made of MAP due to high cost and the lack of any marked advantage in preservation. However, where O_2 impermeable films are used, mainly for turkey and duck, there are clear benefits in extending shelf-life (Hood and Mead, 1993). There have also been reports that high CO_2 atmosphere packaging of poultry extends the storage life up to three times that of storage in air (Baker et al., 1985).

Zeuthen and Mead (1996) suggested that 50% of the shelf life of modified atmosphere meats could be attributed to effective chilling (+2°C during processing, storage and display), 33% to high standard of hygiene during processing while the remaining 17% is affected by the quality of material used. Temperature is probably the most important single environmental factor influencing the growth of bacteria on MAP meat (Lambert et al., 1991).

The optimum storage temperature for chilled meat is the minimum that can be maintained indefinitely without overt freezing of the product. Gill (1991) stated that the optimum temperature for packaged meat is -1.5 ± 0.5°C. Even a small increase above optimum temperature will lead to large losses of meat storage life, irrespective of the packaging used. At temperatures of 0, 2 or 5°C, the storage life is about 70, 50, or 30%, respectively, of the storage life obtained at the optimum temperature (Gill, 1991). The importance of storage temperature was further emphasised by Sorheim and Nissen (2000) who reported that when MAP meat is stored at 8°C, *Salmonella spp.* and *Escherichia coli*, may pose a health risk to the consumer. These pathogens can tolerate high carbon dioxide concentrations and therefore strict temperature control is required. Temperature is also important for maintaining the colour of meat. High temperatures are known to decrease colour stability. At temperatures above 3°C, myoglobin is more readily oxidised to metmyoglobin (Faustman and Cassens, 1990). At low temperatures, oxygen diffusion into the meat is greater and so a deeper layer of oxymyoglobin is formed (Winstanley, 1979). O'Keeffe and Hood (1982) suggested that the rate of discolouration of beef is temperature dependent and under aerobic conditions, the rate of discolouration is two to five times higher at 10°C than at 0°C.

Under aerobic conditions, the dominant spoilage organisms are the strictly aerobic *Pseudomonads*. Glucose is abundant in most muscle tissue which allows *Pseudomonads* to grow to numbers of about $10^8/cm^2$ before the glucose becomes growth limiting. The bacteria then attack amino acids as sources of growth substrates. While the bacteria are consuming glucose, no offensive by-products are produced but when they commence utilising the amino acids as food sources, a variety of by-products are produced which are detected organoleptically as putrid odours and flavours (Gill, 1996). High oxygen atmospheres in conjunction with CO_2 inhibits growth of *Pseudomonads* and allows the slower growing organisms, like lactic acid bacteria to dominate. The creation of conditions where lactic acid bacteria predominate are preferred because unlike *Pseudomonads*, lactobacilli do not precipitate spoilage when they are increasing in numbers (Gill, 1991).

Ahmad and Marchello (1989) reported that a gas mixture of 10% CO_2:5% O_2:85% N_2 was the most effective modified atmosphere in reducing psychotrophic growth on beef steaks. Buys *et al.* (1994) found that higher mean pseudomonad counts were recorded for 25% CO_2:50% N_2:25% O_2 and 80% O_2:20% CO_2 bulk packaged samples than for 100% CO_2 and 75%CO_2:25% N_2 gas mixtures, respectively.

Cooked meats packed under modified atmospheres, containing CO_2 as the antimicrobial component, are more prevalent than ever in the supermarket (Devlieghere *et al.*, 1999). However, there are conflicting reports published on the shelf-life extending effect of modified atmospheres for cooked products. Reported gas compositions of modified atmosphere packaged processed meat products are shown in Table 20.3. The widespread use of modified atmosphere packaging for cured meat products has generated problems regarding colour stability of cured meat products stored under illumination during retail display. Light exposure (Andersen *et al.*, 1988) including intensity of light or illuminance, oxygen transmission rate of the packaging material (Yen *et al.*, 1988) and residual level of oxygen (Moller *et al.*, in press) have been found to affect the colour stability of cured meat. Results of a study carried out by Moller *et al.* (in press) concluded that it is important to consider all factors simultaneously, when optimising the colour stability of cured ham.

20.5 Bulk, master or mother packaging

Jeyamkondan *et al.* (2000) considered master packaging to be the most economical of all centralised packaging techniques. However, it must be integrated with strict temperature control in a narrow range just above freezing, good processing hygiene and maintenance of a completely anoxic atmosphere in the package headspace throughout the distribution period to maximise storage life. If properly applied, the storage life of retail ready meat can be extended for up to ten weeks in the master package followed by three days of retail display. While processed meats are not master packaged at present, this method of packaging may be considered safe and reliable for such products where requirements or applications may arise.

It is thought that a ratio of 1:3 in meat to gas volume is required to maintain an adequately preservative composition of the atmosphere in a retail pack. Packs

Table 20.3 Reported gas compositions of processed meat products (Church, 1993)

	Gas (%)		
	O_2	CO_2	N_2
Bacon, cured	<0.5	CO_2/N_2	Flushed
Bacon, sliced	–	20–35	65–80
Barbecue ribs	–	20–40	65–80
Beef, sliced cooked	10	75	15
Chicken, cooked	<0.2	30	70
Chicken thighs, breaded, baked	–	30	70
Chicken, breaded, flash fried	–	–	100
Cooked meat	–	20–25	75–80
Cooked meat	–	20–25	70–75
Cooked meat	–	20–40	60–80
Cooked meat, sliced	–	80	20
Cooked minced meat products	–	20	80
Corned beef	<0.3	60	40
Cured meat	–	50	50
Cured meat	–	20	80
Cured meat	–	40	60
Cured meat, bulk	–	35	65
Cured meat, retail	–	20	80
Frankfurters	–	–	100
Frankfurters	–	100	–
Ham	–	20–35	65–80
Ham, Italian, sliced	–	20	80
Ham, sliced <0.3	60	40	
Lasagne	–	70	30
Luncheon meat	–	100	–
Meat pie	–	50	50
Meat pies	–	25–50	50–75
Pasta stuffed with meat (<30% moisture)	–	50	50
Pasta stuffed with meat	–	80	20
Pizza, depending on topping	–	30–60	40–70
Pizza, ham	–	60	40
Poultry products	–	25	75
Ravioli	–	20	80
Roast beef, sliced, cooked	10	75	15
Roast pork, sliced	<0.3	60	40
Salami	–	20	80
Salami	–	20–35	65–80
Sausage, British fresh (raw, uncured)	<0.5	CO_2/N_2	Flushed
Sausage, British fresh (raw, uncured)	50	50	–
Sausage, sliced	–	20–30	70–80
Sausage, smoked	–	30	70
Sausage, Summer	–	–	100
Sausage, uncured	40	60	–
Sausage, Vienna	–	20	80
Sausage in pastry	–	80	20
Turkey, cooked	<0.2	30	70
Wieners, natural casings	<0.4	30	70

that are oversized for the product they contain increase the distribution and display costs per unit and they are also considered unfavourably by the customer (Gill and Jones, 1996). Therefore, modified atmosphere display packs in commercial use are often of a size sub-optimal for meat preservation and so confer only a modest extension of product storage life (Gill and Jones, 1996). Master packaging would alleviate the problems associated with oversized display packs. Minced beef and beef steaks that are master packaged under N_2 and CO_2 atmospheres can be stored for three weeks or longer than product freshly prepared for display from vacuum packaged meat of the same age (Gill and Jones, 1994a, b). Off-flavour development constituted the limiting factor in extending the chilled storage life of display-ready pork in controlled atmosphere masterpacks (Jeremiah and Gibson, 1997).

It has been reported that the most successful gas mixture in the pouch for master packaging is 25% CO_2, 50% N_2 and 25% O_2 although the odour scores indicated that this mixture could achieve a storage period of only 14 days at 0°C and subsequent shelf-life of two days. A study was carried out that investigated the influence on the shelf life of fresh pork of different centralised prepackaging techniques – PVC overwrapping, MAP using 25% CO_2 and 75% O_2, vacuum skin packaging and bulk gas flushing of the master pack with 100% CO_2. All of the packaging systems were equally efficient for the first four days of retail display but in the extended shelf-life study the master pack system demonstrated the most promising shelf-life results and was also judged superior on odour scores (Buys, 1996).

20.6 Controlled atmosphere packaging and active packaging systems

Modified atmosphere packaging is a packaging system where the pack atmosphere is altered initially and then allowed to change over time during storage. With controlled atmosphere packaging, the package atmosphere is altered initially and then maintained during the life of the package (Jeremiah, 2001). Discolouration in controlled atmosphere packaging can be prevented only by exclusion of essentially all oxygen from the package, which requires the use of special evacuation equipment and totally gas impermeable packaging materials (Gill, 1991). Gill and Jones (1994a, b) reported that discolouration induced by any residual oxygen in controlled atmosphere packs will resolve after 2–4 days. If CO_2 is a major or sole component of the input atmosphere, then the quantity of added gas must be adjusted to assure that the intended atmosphere persists after dissolution of the gas into the product (Gill, 1995).

For chilled meat, the most effective technology to date is the high CO_2 controlled atmosphere packaging system. This regime limits microbial deterioration through optimal storage temperatures (-1.5°C), high levels of CO_2, low residual oxygen ($\leq 0.05\%$), and use of a gas impermeable film (Jeremiah *et al.*, 1995).

Active packaging is an innovative concept that can be defined as a type of packaging that changes the condition of the packaging to extend shelf-life or improve safety or sensory properties while maintaining the quality of the food. Major active packaging techniques are concerned with substances that absorb oxygen, ethylene, moisture, carbon dioxide, flavours/odours and those which release carbon dioxide, antimicrobial agents, antioxidants and flavours (Vermeiren et al., 1999).

The most prevalent form of active packaging in the meat industry is based on oxygen scavenging. The majority of commercial O_2 scavengers used in the meat industry are based on iron powders, which are mixed with acids and/or salts and a humectant, to promote oxidation of the iron. The humectant may be dry or pre-wetted (Gill and McGinnis, 1995). Oxygen scavengers are generally formulated to reduce the O_2 concentration in a volume of air that is five times the rated O_2 absorbing capacity of the scavenger to < 100ppm within about a day, but the time taken to reach that value may vary from 0.5–4 days (Smith et al., 1986). A potential risk associated with O_2 scavengers is accidental ingestion of a large amount of iron, in spite of the label 'Do Not Eat' on the front of the pack. A welcome alternative to sachets is the incorporation of the O_2 scavenger into the packaging structure itself. Low molecular weight ingredients may be dissolved or dispersed in a plastic or the plastic may be made from a polymeric scavenger. Other recent developments include inserts in the form of flat packets, cards or sheets upon which the meat product may sit, as well as O_2 scavenging adhesive labels.

Some oxygen scavengers use an enzyme reactor surface that reacts with some substrate to scavenge incoming O_2. Another technique involves sealing of a small coil of an ethyl cellulose film containing a dissolved photosensitive dye and single O_2 acceptor in the headspace of a transparent package. Illumination of the film excites dye molecules, which sensitise any O_2 molecules to the singlet state. These molecules in turn then react with acceptor molecules and are consumed (Rooney, 1995).

In a recent study a natural antimicrobial agent, grapefruit seed extract, was incorporated on the food contact surface of multilayered polyethylene (PE) film by a co-extrusion or solution coating process and applied to minced beef. Results showed that both types of grapefruit seed extract-incorporated multilayer PE films contributed to a reduction of the growth rates of aerobic and coliform bacteria on minced beef when compared to plain PE film (Ha et al., 2001). Ouatter et al. (2000) undertook a study to evaluate the feasibility of using antimicrobial films, designed to slowly release bacterial inhibitors, to improve the preservation of vacuum-packaged processed meats during refrigerated storage. Results confirmed that the lactic acid bacteria, were not affected by the antimicrobial films under study but the growth of *Enterobacteriaceae* was delayed or completely inhibited as a result of film application.

Research in biosensor technology for the detection of pathogens is very prevalent at present, although to date, there has been no commercial success. Belcher (2000) proposed that biosensors will play a big role in the future of the packaging of meat products. Biosensors are defined as indicators of biological compounds that can be as simple as temperature sensitive paints or as complex as DNA-RNA probes. Infectious dosages of pathogens such as *Salmonella* or *E coli 0157-H7* are as low as ten cells and until biosensors have a detection limit as low as a single organism per ml, with rapid detection and at a low cost, they will not be considered viable. Two systems, which have been recently developed to the semi-commercial state in North America are the SIRA 'Food Sentinel' system and the toxin alert 'Toxin guard' system. The SIRA 'Food Sentinel' system uses a bar code monitoring system for the detection of specific food contaminants. The second system is for manufacturing flexible packaging materials that can detect and identify microbial materials in the package.

Another recent development available now on the market is the Cryovac Lid 550P/Barrier Foam Tray packaging system. This uses a high barrier polystyrene foam tray and a multi-ply lidstock that has a peelable interface and is the active part of the process. The product is held at its reduced/deoxymyoglobin state throughout distribution but once it reaches the retail store the barrier film is broken at the interface and is peeled off. A high oxygen permeable layer remains, which lets oxygen into the package and allows the meat to return to the red oxymyoglobin state in 15–30 minutes (Belcher, 2000). A new and innovative product, known as Fresh-R-Pax moisture absorbing trays, suitable for fresh cut meats and other high purge items (manufactured by Maxwell Chase Technologies and supplied by Balitmore Chemicals) is now available. Fresh-R-Pax moisture absorbing technology is marketed in pad, pouch or tray format (Rowan, 2001). Smiddy *et al.* (2002) had success detecting oxygen levels in packaged meat products using disposable phosphorescent oxygen sensors, placed in each pack and a fibre-optic phase detector.

20.7 Packaging materials used for meat products

A number of simple criteria must be adhered to when selecting particular packaging materials for use with meat products. The package is the primary means of displaying the contained meat product and providing product information and point of sale advertising. The package must also be cost effective relative to the contained food. The main requirements of any chilled muscle food package are listed in Table 20.4.

Plastic films are the materials of choice for the majority of meat products. A plastic film derives its basic properties from the monomer unit of the polymer from which it is made. Monomers composed of carbon and hydrogen produce polymers such as polyethylene and polypropylene, which are good barriers to moisture but relatively permeable to gases. The inclusion of chlorine in the monomer unit greatly reduces gas permeability but may make the polymer film brittle. This problem can be overcome by adding small quantities of other monomers such as acetals and acrylates which upset the regular polymer structure and make the film more flexible. Other compounds may also be added

Table 20.4 Functions of a packaging system for chilled muscle food products

• Contain the product	• Seal integrity
• Be compatible with the food	• Prevent microbial contamination
• Non toxic	• Protect from taints and odours
• Handle distribution stresses	• Be cost effective
• Prevent physical damage	• Have sales appeal
• Have appropriate gas permeability	• Communicate product information
• Control moisture loss or gain	• Easily openable
• Protect against light where necessary	• Be tolerant to storage temperatures
• Possess antifog properties	• Prevent dirt contamination
• Be tamper evident	• Conform to legal legislation
• Conform to environmental legislation	

to polymers to improve their handling on machines or enhance particular properties. The method by which the film is produced also affects its properties. Most meat packaging films are thermoplastic and are extruded from the molten stage through dies. They can be stretched to thinner gauges (thickness) before cooling and this may impart the ability to shrink when reheated (Taylor, 1996).

If the properties required of a packaging material cannot be satisfied by a single film, several films with individual desirable properties may be combined to give a satisfactory composite. These may be combined by laminating two or more films together. This is carried out by joining together previously extruded plastic films either by adhesives or by extruding an adhesion polymer melt between the layers. Polymers may be co-extruded together to form a single material by delivering individual molten resins by separate extruders to a combined round or flat plate die which maintains their separation in discrete but welded layers. A composite can also be produced by coating a film with another polymer. In this case a layer of molten plastic resin or a dissolved or dispersed polymer is applied onto a preformed film (Humphreys, 1996). Polyvinylidene chloride and ethyl vinyl alcohol are commonly used in this way to produce materials with very good gas and moisture properties (Taylor, 1996).

The choice of films for packaging meat is largely determined by their moisture and gas permeabilities. Most of the films used are moisture barriers in order to avoid weight loss from the meat. Gas permeability is much more variable and is specific to individual polymers. For retail cuts of fresh meat where retention of bright red colour is desired, packages with high oxygen transmission rates are used. However, for cuts of meat and processed products where extended storage life is the main concern, packages with low gas transmission rates are used (Newton and Rigg, 1979). The moisture and gas permeabilities of some commonly used plastic films for meat packaging are shown in Table 20.5.

A vacuum packaging film must have mechanical toughness, a high resistance to puncture (especially from bone-in meats) and abrasion, gas barrier properties (particularly to oxygen) be adequate for the application, have suitable optical properties and the ability to form a seal even in the presence of fat or meat juice

Table 20.5 Barrier properties of plastic materials typically used in packaging of meat (Roberts, 1990)

Material (25 micron thickness)	O_2 transmission rate (cc.m^2.day^{-1}.atm.O_2^1) 23°C; dry	Water vapour transmission rate (g.m^{-2}.day^{-1}) 38°C; 90% RH
EVA (ethyl vinyl acetate)	12,000	110–160
LPDE (low density polyethylene)	7,100	16–24
PC (polycarbonate)	4,300	180
PP (polypropylene)	3,000	10
PS (polystyrene)	2,500–5,000	110–160
HDPE (high density polyethylene)	2,100	6–8
Nylon 11 (polyamide)	350	60
UPVC (unplasticised polyvinyl chloride)	120–160	22–35
Nylon 6 (polyamide)	80	200
PET (polyester terephthalate)	50–100	20–30
Amorphous nylon	40	20
Aromatic nylon	2.4	25
PVDC extrusion (polyvinylidene chloride)	1.2–9.2	0.8–3.2
PVDC emulsion (polyvinylidene chloride)	0.8–3.4	0.3–1.0
EVOH (ethyl vinyl alcohol)	0.16–1.6	24–120

and film overlap. The films that best meet all these requirements are composites, which utilise the properties of two or more individual film materials to provide a good package (Eustace, 1981), each providing their own contribution to the structure. Most film products currently used in Australia for vacuum packaging fresh meat have either nylon or PVDC included as the primary barrier to oxygen. The individual polymer materials most often used for meat vacuum packs are listed in Table 20.6 with their main contributions to the overall structure.

During manufacture, shrink bags are stretched both longitudionally and transversely at a controlled rate and temperature which reorganises the polymeric chains and retains a built-in tension as the film is chilled. Later, when the material is exposed to heat, this tension is released and the film shrinks in both directions. Most shrink bags available on the market use multi-ply formulations based on polyolefin resins, with either PVDC or EVOH as the gas barrier component. Electronic cross-linking is sometimes used to improve the mechanical resistance of the basic material (Humphreys, 1996). Pouches can be produced by co-extrusion, adhesive or extrusion lamination. One of the basic pouch structures uses PA as the outer layer with an inner core and sealing layer of PE or LLDPE. The properties of pouches can be improved by incorporating oriented PA which reduces the susceptibility of its oxygen barrier properties to moisture or by adding aluminium foil layers which can offer an exceptionally high barrier to oxygen. Heat sealability can be improved by using more expensive EVA or ionomer resins (Humphreys, 1996). For vacuum skin-packaging, the highly formable top webs are based on ionomer resins with barrier layers of EVOH and the rigid bottom webs on polyvinyl chloride (PVC), polystyrene or PET.

Table 20.6 Individual polymer materials, common abbreviation and associated properties

Polymer materials	Abbreviations	Associated properties
Low density polyethylene	LDPE	Sealability, formability, moisture barrier, low cost
Linear low density polyethylene	LLDPE	Sealability, abuse resistance, moisture barrier, formability
Polypropylene	PP	Moisture barrier, thermal resistance, dimensional stability
Ethylene vinyl acetate copolymer	EVOH	Sealability, improved abuse resistance over LDPE, clarity
Polyesters	PET	Mechanical resistance, heat resistance, medium O_2 barrier
Ethylene vinyl alcohol	EVA	High O_2 barrier, good co-extrusion, processability, clarity
Polyamides	PA	Mechanical strength, O_2 barrier (moisture sensitive), formability
Polyvinylidene chloride	PVDC	High O_2 barrier (moisture stable), grease and fat barrier
High density polyethylene	HDPE	More gas impermeable than LDPE, low cost, strong, reduced clarity
Polyvinyl chloride	PVC	Versatile, shrink properties, sparkling clear, low cost
Polystyrene	PS	Excellent clarity, low cost, readily themoformed and injection moulded
Ionomer		Heat sealability, produce films of unusual toughness and clarity
Polycarbonate	PC	High clarity, strong, impact resistance, dimensional rigidity

Thermoforming is now the most common method of modified atmosphere packaging of meat and meat products. Trays are produced from a bottom web of plastic, evacuated and then flushed with the gas mixture before they are sealed with a top web of film. The trays are usually made from unplasticised polyvinyl chloride (UPVC) or PS and the lidding materials from PET/PS combinations, which may also include a PVDC or EVOH component to improve gas barrier properties. Trays are often designed with patterned bases to disperse the drip which may accumulate on storage but in most cases extra absorbent pads are also included. Condensation on the inside of the container lid can be avoided by minimising temperature fluctuations in the display cabinet and by incorporating an anti-fog coating on the inner surface of the lidding material.

The properties of the material from which processed meat packages are made are also of great importance. Optical properties, barrier properties, neutral behaviour regarding taste and smell, resistance to fats and oils, sealability and

Table 20.7 Materials used for packaging processed meat (Mondry, 1996)

Pack type	Bottom web materials	Top film materials (where applicable)
Flexible vacuum pack	PA/PE, co-extruded as 5 layer-films	
Flexible MAP pack	PA/PE PA/EVOH/PE PA/EVOH/PA/PE PP/EVOH/PE PE/EVOH/PE	OPA/PE (O = oriented = prestretched) PET/PVDC/PE
Rigid vacuum pack	A-PET (amorphous polyester) PVC or PVC/PE PS/EVOH/PE	OPA/PE PET/PVDC/PE OPA/PE/EVOH/PE PET/PE/EVOH/PE
Rigid MAP pack	PVC PVC/PE or PVC/EVOH/PE A-PET A-PET/PE or A-PET/EVOH/PE PS/EVOH/PE	OPA/PE PET/PVDC/PE OPA/PE/EVOH/PE PET/PE/EVOH/PE
Skin packs	PVC/PE PS/EVOH/PE A-PET A-PET/PE	Several combinations of up to seven or more layers but incorporating EVOH as gas barrier

tightness, thickness and machinability must be considered. In most cases multilayer films are used. The most common materials used for packaging of processed meat are shown in Table 20.7.

20.8 Future trends

Packaging is a prerequisite today demanding distribution of food products globally, nationally regionally. Product development in meat foods will continue, leading to a wider range of products with increasing complexity. Sales of ready-made meals are increasing and ready-made products containing meat puts high demands on creating a suitable packaging environment which will guarantee quality and maintain or extend shelf-life. There is a health and fitness trend sweeping across many countries. Hygienic and ecological demands on meat producers will also increase and these demands will affect the choice or selection of materials used and ultimately, the packaging properties obtained.

Research and development in the area of active packaging systems for meat products has received much attention recently and will continue to do so in the near future. The areas of interest are bio and chemical sensor technologies and anti-microbial agents. Johansson (2001) concluded in a recent report that food packaging, including the total packaging system, is set to undergo major development changes over the next few years. The way to face the challenge is the development of barrier materials, a packaging systems review and to see packaging as a means of reducing environmental load in the supply chain.

20.9 References

AASGAARD, J. (1993). Colour stability of packed meat products. *Fleischwirtschaft*, **73**, 428.

AHMAD, H.A. and MARCHELLO, J.A. (1989). Effect of gas atmosphere packaging on psychotrophic growth and succession on steak surfaces. *J. Food Sci.*, **54**, 274.

ANDERSEN, H.J., BERTELSEN, G., OHLEN, A. and SKIBSTED, L.H. (1990). Modified Packaging as protection against photodegradation of the colour of pasteurised, sliced ham. *Meat Sci.*, **22**, 283.

ANDERSEN, H.J., BERTELSEN, G., BOEGH-SOERENSEN, L., SHEK, C.K. and SKIBSTED, L.H. (1988). Effect of light and packaging conditions on the colour stability of sliced ham. *Meat Sci.*, **22**, 283.

BAKER, R.C., HOTCHKISS, J.H. and QURESHI, R.A. (1985). Elevated carbon dioxide atmospheres for packaging poultry I. Effects on ground chicken. *Poult. Sci.*, **64**, 328.

BELCHER, J. (2000). Commercial applications of active packaging and biosensor technology in meat packaging-A packaging manufacturer's perspective. Proceedings of the 47th Int. Cong. of Meat Sci. and Tech., Krakow, Poland, p. 744.

BELL R.G. and GAROUT, A.M. (1994). The effective product life of vacuum-packaged beef imported into Saudi Arabia by sea as assessed by chemical, microbiological and organoleptic criteria. *Meat Sci.*, **36**, 381.

BERTELSEN, G., JENSEN, C. and SKIBSTED, L.H. (2000). Alteration of cooked and processed meat properties via dietary supplementation of vitamin E. In: *Antioxidants in Muscle Foods*. (Decker, E.A., Faustman, C. and Lopez-Bote, C.J. eds). John Wiley & Sons, Inc., New York, pp. 367.

BOEREMA, J.A., PENNY, N., CUMMINGS, T.L. and BELL, G. (1993). Carbon doxide controlled atmosphere packaging of sliced ham. *Int. J. Food Sci. & Tech.*, **28**, 435.

BRODY, A.L. (2000). A retrospective on the technologies of centralised meat packaging, 1963 through 2000 and beyond. 2000 IFT Annual Meeting. 56-1, 1.

BRODY, A. (1997). Packaging of food. In *The Wiley Encyclopedia of Packaging*, 2nd Edn. (Brody, A.L. and Marsh, K.S. eds) John Wiley & Sons, 605 Third Avenue, New York, pp. 699.

BRODY, A.L. (1989). Modified atmosphere/vacuum packaging of meat. In *Controlled/Modified Atmosphere Packaging of Foods*. (Brody, A.L. ed.) Food & Nutrition Press, Trumbull, CT., pp. 17.

BRUCE, H. L., ARGANOSA, G.C., SZPACENKO, A., HAWRYSH, Z., PRICE, M.A. and WOLFE, F.H. (1992). Chemical and physical changes during the storage of

Cap-Tech beef. *Proc. Can. Meat Sci. Assoc. Symp.*, **7**, 14.
BUCKLEY, D.J., MORRISSEY, P.A. and GRAY, J.I. (1995). Influence of dietary vitamin E on the oxidative stability and quality of pig meat. *J. Anim. Sci.*, **73**, 3122.
BUYS, E. (1996). Packing fresh meat. *Food review*, **23**, 27.
BUYS, E.M., KRUGER, J. and NORTJE, G.L. (1994) Centralised bulk pre-packaging of fresh pork retail cuts in various gas atmospheres. *Meat Sci.*, **36**, 293.
CANNON, J.E., MORGAN, J.B., SCHMIDT, G.R., DELMORE, R.J. SOFOS, J.N., SMITH, G.C. and WILLIAMS, S.N. (1995). Vacuum-packaged precooked pork from hogs fed supplemental vitamin E: Chemical, shelf-life and sensory properties. *J. Food Sci.*, **60**, 1179.
CHURCH, I. (1998). The sensory quality, microbiological safety and shelf life of packaged foods. In: *Sous Vide and Cook Chill Processing for the Food Industry*. (Ghazala, S. ed.) Aspen publishers, Inc., Gaithersburg, Maryland, pp. 190.
CHURCH, N. (1994). Developments in MAP and related technologies. *Trends in Food Sci. and Tech.*, **5**, 345.
CHURCH, P.N. (1993). Meat Products. In *Principles and Applications of Modified Atmosphere Packaging of Foods.* (Parry, R.T. ed.) Blackie Academic and Professional, Wester Cleddens Road, Bishopsbriggs, Glasgow, UK, pp. 229.
CONNER, D., SCOTT, V., BERNARD, D. and KAUTTER, D. (1989). Potential *Clostridium botulinum* hazards associated with extended shelf life refrigerated foods: a review. *J. Food Safety*, **10**, 131.
CREED, P.G. (1998). Sensory and nutritional aspects of *sous vide* processed foods. In: *Sous Vide and cook-chill processing for the food industry.* (Ghazala, S. ed.) Aspen Publishers, Inc., Gaithersburg, Maryland, pp 57.
DAY, P.F. (1992). Chilled Food Packaging. In: *Chilled Foods, A Comprehensive Guide*. (Dennis, C. and Stringer, M. eds) Ellis Horwood, London, pp. 147.
DEN HERTOG-MEISCHKE, M.J.A., VAN LAACK, R.J.L.M. and SMULDERS, F.J.M. (1997). The water holding capacity of fresh meat. *Veterinary Quarterly*, **19**, 175.
DEVLIEGHERE, F., VAN BELLE, B. and DEBEVERE, J. (1999). Shelf life of modified atmosphere packed cooked meat products: a predictive model. *Int. J. of Food Micro.*, **46**, 57.
DEVORE, D.P. and SOLBERG, M. (1974). Oxygen uptake in postrigor bovine muscle. *J. Food Sci.*, **39**, 22.
DOHERTY, A.M. and ALLEN, P. (1998). The effects of oxygen scavengers on the colour stability and shelf life of CO_2 packaged pork. *J. Muscle Foods*, **9**, 351.
DOHERTY, A.M., SHERIDAN, J.J., ALLEN, P., MCDOWELL, D.A. and BLAIR, I.S. (1996). Physical characteristics of lamb primals packaged under vacuum or modified atmospheres. *Meat Sci.*, **42**, 315.
DOWN, N.F. (1996). Bulk gas packing of fresh meat. In *Meat Quality and Meat Packaging.* (Taylor, S.A., Raimundo, A., Severini, M. and Smulders, F.J.M. eds) Utrecht: ECCEAMST.III, 295.
EGAN, A.F., FORD, A.L. and SHAY, B.J. (1980). A comparison of *Microbacterium*

thermosphactum and lactobacilli as spoilage organisms of vacuum packed sliced luncheon meats. *J. Food Sci.*, **45**, 1745.

ENFORS, S.O. and MOLIN, G. (1984). Carbon dioxide evolution of refrigerated meat. *Meat Sci.*, **10**, 197.

EUSTACE, I.J. (1981). Some factors affecting oxygen transmission rates of plastic films for vacuum packaging of meat. *J. Food Technol.*, **16**, 73.

EUSTACE, I.J. (1989). Food packaging-selection of materials and systems. *Food Australia*, **41**, 884.

FAROUK, M.M., PRICE, J.F. and SALIH, A.M. (1990). Effect of an edible collagen film overwrap on exudation and lipid oxidation in beef round steak. *J. Food Sci.*, **55**, 1510.

FAUSTMAN, C. and CASSENS, R.G. (1990). The biochemical basis for discolouration in fresh meat. A review. *J. Muscle Foods*, **1**, 217.

FINNE, G. (1982). Modified and Controlled Atmosphere Storage of Foods. *Food Technology*, **February**, 128.

GILL, C.O. (1991). Extending the storage life of raw meat. I. Preservative atmospheres. Western Canada Research group on Extended storage of meat and meat products. Technical Bulletin 1, Dept of Applied Micro. and Food Sci., University of Saskatchewan, Saskatoon, Sk.

GILL, C.O. (1995). MAP and CAP of fresh red meats poultry and offals. *Principles of modified-atmosphere and sous vide product packaging.* (Farber J.M. and Dodds, K.L eds) Technomic Publishing Co., Lancaster, Basel, p. 105.

GILL, C.O. (1996). Extending the storage life of raw chilled meats. *Meat Sci.*, **43**, S99.

GILL, C.O. and HARRISON, C.L. (1989). The storage life of chilled pork packaged under carbon dioxide. *Meat Sci.*, **26**, 313.

GILL, C.O. and JONES, T. (1994a). The display of retail packs of ground beef after their storage in masterpackages under various atmospheres. *Meat Sci.*, **37**, 281.

GILL, C.O. and JONES, T. (1994b). The display life of retail packed beef steaks after their storage in master packs under various atmospheres. *Meat Sci.*, **38**, 385.

GILL, C.O. and JONES, T. (1996). The display life of retail packaged pork chops after their storage in master packs under atmospheres on N_2, CO_2 or O_2 + CO_2. *Meat Sci.*, **42**, 203.

GILL, C.O. and MCGINNIS, J.C. (1995). The use of oxygen scavengers to prevent the transient discolouration of ground beef packaged under controlled oxygen depleted atmospheres. *Meat Sci.*, **19**, 27.

GILL, C.O. and PENNY, N. (1988). The effect of the initial gas volume to meat ratio on the storage life of chilled beef packaged under carbon dioxide. *Meat Sci.*, **22**, 53.

GILL, C.O. and TAN, W.D. (1980). Effect of carbon dioxide on growth of meat spoilage bacteria. *Appl. Environ. Microbiol.*, **39**, 317.

GREENE, B.E., HSIN, I.M. and ZIPSER, M.W. (1971). Retardation of oxidative colour changes in raw ground beef. *J. Food Sci.*, 36, 940.

HA, J.U., KIM, Y.M. and LEE, D.S. (2001). Multilayered antimicrobial polyethylene films applied to the packaging of ground beef. *Packaging Technology and Science*, **14**, 55.

HIRSCH, A. (1991). Why controlled or modified atmosphere packaging? In: *Flexible Food Packaging, Questions and Answers.* Van Nostrand Reinhold, New York, p. 7.

HINTLIAN, C.B. and HOTCHKISS, J.H. (1987). Comparative growth of spoilage and pathogenic organisms on modified atmosphere packaged cooked beef. *J. Food Protection*, 50, 218.

HOOD, D.E. and MEAD, G.C. (1993). Modified atmosphere storage of fresh meat and poultry. In: *Principles and applications of modified atmosphere packaging of food.* (Parry, T. ed.) T. Blackie, Glasgow, UK, pp. 269.

HUMPHREYS, P. (1996). Vacuum packaging of fresh meat – An overview. In: *Meat Quality and Meat Packaging.* (Taylor, S.A., Raimundo, A., Severini, M. and Smulders, F.J.M. eds) Utrecht: ECCEAMST.III, 285.

JACKSON, T.C., ACUFF, G.R., VANDERZANT, C., SHARP, T.R. and SAVELL, J.W. (1992). Identification and evaluation of volatile compounds of vacuum and modified atmosphere packaged beef striploins. *Meat Sci.*, **31**, 175.

JAYASINGH, P., CORNFORTH, D.P., CARPENTER, C.E. and WHITTIER, D. (2001). Evaluation of carbon monoxide treatment in modified atmosphere packaging or vacuum packaging to increase colour stability of fresh beef. *Meat Sci.*, **59**, 317.

JEREMIAH, L.E. and GIBSON, L.L. (1997). The influence of controlled atmosphere storage on the flavour and texture profiles of display-ready pork cuts. *Food Res. Int.*, **30**, 117.

JEREMIAH, L.E., GIBSON, L.L. and ARGANOSA, G.C. (1996). The influence of CO_2 level on the storage life of chilled pork stored at 1.5°C. *J. Muscle Foods*, **7**, 139.

JEYAMKONDAN, S., JAYAS, D.S. and HOLLEY, R.A. (2000). Review of centralised packaging systems for distribution of retail-ready meat. *J. Food Protection.* **63**, 796.

JOHANSSON K. (2001). Future challenges for food packaging. *Food Technology International*, European Federation of Food Sci. and Tech., 45.

KANNER, J., HAREL, S. and GRANIT, R. (1992). Oxidative processes in meat and meat products: quality implications. In: Proceedings of the 38th Int. Cong. of Meat Sci. and Tech., Clermont-Ferrand, France, 111.

KERRY, J.P., BUCKLEY, D.J. and MORRISSEY, P.A. (2000). Improvement of oxidative stability of beef and lamb with Vitamin E. In: *Antioxidants in Muscle Foods.* (Decker, E.A., Faustman, C. and Lopez-Bote, C.J. eds.). John Wiley & Sons, Inc., New York, pp. 229.

KERRY, J.P., BURKLEY, E., O'SULLIVAN, M.G., LYNCH, A., BUCKLEY, D.J. and MORRISSEY, P.A. (1996). The effect on colour, oxidation and microbiological status of beef from vitamin E supplemented cattle. *Proceedings of the 42nd Int. Cong. of Meat Sci. and Tech., Lillehammer*, p. 94.

KING, A.D. and NAGEL, C.W. (1967). Growth inhibition of a *Pseudomonas* by carbon dioxide. *J Food Sci.*, **40**, 1229.

KINSEY, J.D. (1997). What's new – What's not, *Cereal Foods World*, November/December, **42**, 11, 919.

KROPF, D.H. (1980). Effect of retail display conditions on meat colour. *Recip. Meat Conf. Proc.*, Nat. Livestock and Meat Board, Chicago. **33**, 15.

LAMBERT, A.D., SMITH, J.P. and DODDS, K.L. (1991). Shelf life extension and microbiological safely of fresh meat-a review. *Food Micro.*, **8**, 267

LEDWARD, D. A. (1970). Metmyoglobin formation in beef stored in carbon dioxide enriched and oxygen depleted atmospheres. *J. Food Sci.*, **35**, 33.

LIVINGSTON, D.J. and BROWN, W.D. (1981). The chemistry of myoglobin and its reactions. *Food Tech.*, **May**, 244.

LUNDQUIST, B.R. (1987). Protective Packaging of Meat and Meat Products. In: *The Science of Meat and Meat Products*. (Price, J.F and Schweigert, B.S, eds) Food and Nutrition Press, Inc., Westport, Connecticut, USA, pp. 487

LUNO, M., BELTRAN, J.A. and RONCALES, P. (1998). Shelf-life extension and colour stabilisation of beef packaged in a low O_2 atmosphere containing CO: Loin steaks and ground meat. *Meat Sci.*, **48**, 75.

LUNO, M., RONCALES, D., DJENANE, D. and BELTRAN, J. A. (2000). Beef shelf life in low O_2 and high CO_2 atmospheres containing different low CO concentrations. *Meat Sci.*, **55**, 413.

MCDANIEL, M.C., MARCHELLO, J.A. and TINSLEY, A.M. (1984). Effect of different packaging treatments on microbial and sensory evaluation of precooked beef roasts. *J. Food Protection*, **47**, 23.

MARRIOTT, N.G., SMITH, G.C., HOKE, K.E. and CARPENTER, Z.L. (1977). Short term transoceanic shipments of fresh beef. *J of Food Sci.*, **42**, 321.

MOL, J.H.H., HIETBRINK, J.E.A., MOLLEN, H.W.M. and VAN TINTEREN, J. (1971). Observations on the microflora of vacuum packed sliced cooked meat products. *J. Applied Bacteriology*, **34**, 377.

MOLLER, J.K., JAKOBSEN, M., WEBER, C.J. and MARTINUSSEN, Y. (in press). Optimisation of colour stability of cured ham during packaging and retail display by a multifactorial design. *Meat Sci.* Available online 19 March 2002.

MOLLER, J.K., JENSEN, J.S., OLSEN, M.B., SKIBSTED, L.H. and BERTELSEN, G. (2000). Effect of residual oxygen on colour stability during chill storage of sliced, pasteurised ham packaged in modified atmospheres. *Meat Sci.*, **54**, 399.

MONDRY, H. (1996). Packaging systems for processed meat. In *Meat Quality and Meat Packaging*. (Taylor, S.A., Raimundo, A., Severini, M. and Smulders, F.J.M. eds) Utrecht: ECCEAMST.III, 323.

MORRISSEY, P.A., BUCKLEY, D.J. and GALVIN, K. (2000). In *Antioxidants in Muscle Foods*. (Decker, E.A., Faustman, C. and Lopez-Bote, C.J. eds) John Wiley & Sons, Inc., New York, pp. 263.

MORRISSEY, P.A., SHEEHY, P.J.A., GALVIN, K., KERRY, J.P. and BUCKLEY, D.J. (1998). Lipid stability in meat and meat products. *Meat Sci.*, **49**, S73.

MULLER, S.A. (1990). Packaging and meat quality. *J. Inst. Can. Sci. Tech.*, **23**, 22.

NOLAN, N.L., BOWERS, J.A. and KROPF, D.H. (1989). Lipid oxidation and sensory analysis of cooked pork and turkey stored under modified atmospheres. *J. Food Sci.*, **54**, 846.

NEWTON, K.G. and RIGG, W.J. (1979). The effect of film permeability on the storage life and microbiology of vacuum packed meat. *J. Appl. Bacteriol.* **47**, 433.

OFFER, G. and COUSINS, T. (1992). The mechanism of drip production: formation of two compartments of extracellular space in muscle *postmortem*. *J. Food Sci. and Agric.*, **58**, 107.

O'GRADY, M.N., MONAHAN, F.J., FALLON, R.J. and ALLEN, P. (2001). Effects of dietary supplementation with vitamin E and organic selenium on the oxidative stability of beef. *J. Animal Sci.*, **79**, 2827.

OKAYAMA, T., IMAI, T. and YAMANOUE, M. (1987). Effects of ascorbic acid and alpha-tocopherol on storage stability of beef steaks. *Meat Sci.*, **21**, 267.

O'KEEFE, M. and HOOD, D.E. (1982). Biochemical factors influencing metmyoglobin formation on beef from muscles of differing colour stability. *Meat Sci*, **5**, 267.

OUATTAR, B., SIMARD, R.E., PIETT, G., BEGIN, A. and HOLLEY, R.A. (2000). Inhibition of surface spoilage bacteria in processed meats by application of antimicrobial films prepared with chitosan. *Int. J. of Food Micro.*, **62**, 139.

PAYNE, S.R., DURHAM, C.J., SCOTT, S.M., PENNEY, N., BELL, R.G. and DEVINE, C.E. (1997). The effects of rigor temperature, electrical stimulation, storage duration and packaging systems on drip loss in beef. In Proceedings of the 43rd Int. Cong. of Meat Sci. and Tech., Auckland, pp. 592.

PAYNE, S.R., DURHAM, C.J., SCOTT, S.M. and DEVINE, C.E. (1998). The effects of non-vacuum packaging systems on drip loss from chilled beef. *Meat Sci.*, **49**, 277.

PEARSON, A.M. and GILLETT, T.A. (1996). The canning process; Canned meat formulations. In: *Processed Meats*. Chapman and Hall, New York, pp. 372, 390.

PEARSON, A.M. and TAUBER, F.W. (1984). Curing. In *Processed Meats*, 2nd Edn. AVI Publishing Company INC, Westport, Connecticut, pp. 47.

RANDALL, K., AHVENAINEN, R., LATVAKALA, K., HURME, E., MATTILASANDHOLM, T. and HYVONEN, L. (1995). Modified atmosphere-packed marinated chicken breast and rainbow trout quality as affected by package leakage. *J. Food Sci.*, **60**, 667.

RENERRE, M. and LABADIE, J. (1993). Fresh meat packaging and meat quality. Proceedings of the 39th Int. Cong. of Meat Sci. and Tech., Calgary, Canada, pp. 361.

RENERRE, M. and LABAS, R. (1987). Biochemical factors influencing metmyoglobin formation in beef muscles. *Meat Sci.*, **19**, 151.

RHEE, K.S., ANDERSON, L.M. and SAMS, A.R. (1996). Lipid oxidation potential of beef, chicken and pork. *J. Food Sci.*, **61**, 8.

RIZVI, S.S.H. (1981). Requirements for foods packaged in polymeric films. *CRC Crit. Rev. Food Sci. Nut.* **14**, 111.

ROBERTS, R.A. (1990). An overview of packaging materials for MAP. *Proc. Int. Conf. on Modified atmosphere Packaging, Stratford-upon Avon, UK.*

ROBERTSON, G.L. (1993). *Food Packaging: Principles and Practice.* (Hughes, H.H. ed.) Marcel Dekker Inc., New York.

ROONEY, M.L. (1995). Overview of active food packaging. In *Active Food Packaging.* (Rooney, M.L. ed.) Blackie Academic and Professional, Glasgow, pp. 1.

ROWAN, C. (2001). The shelf life challenge. *Food Engineering and Ingredients,* **December**, 20.

ST. ANGELO, A.J. and SPANIER, A.M. (1993). Lipid oxidation in meat. In: *Shelf life Studies of Foods and Beverages: Chemical, Biological, Physical and Nutritional Aspects.* (Charalambous, G. ed.) Elsevier Science Publishers, Amsterdam, The Netherlands. pp. 35.

SCHULTER, A.R., MILLER, M.F., JONES, D.K., MEADE, M.K., RAMSEY, C.B. and PATTERSON, L.L (1994). Effects of distribution packaging method and storage time on the physical properties and retail display characteristics of pork. *Meat Sci.*, **37**, 257.

SEIDEMAN, S.C. and DURLAND, P.R. (1984). The utilisation of modified gas atmosphere packaging for fresh meat: A review. *J. of Food Quality*, **6**, 239.

SEMAN, D.L., DREW, K.R., CLARKEN, P.A. and LITTLEJOHN, R.P. (1988). Influence of packaging method and length of chilled storage on microflora, tenderness and colour stability of venison loins. *Meat Sci.*, **22**, 267.

SHAY, B.J. and EGAN, A.F. (1990). Extending retail storage life of beef and lamb by modified atmosphere packaging. *Food Australia*, **42**, 399.

SILLIKER, J.H. and WOLFE, S.K. (1980). Microbiological safety considerations in controlled atmosphere storage of meats. *Food Tech.*, **March**, 59.

SMIDDY, M., FITZGERALD, M., KERRY, J.P., PAPKOVSKY, D.B. O'SULLIVAN, C.K. and GUILBAULT, G.G. (2002). Use of oxygen sensors to non-destructively measure the oxygen content in modified atmosphere and vacuum packed beef: impact of oxygen content on lipid oxidation. *Meat Sci.*, **61**, 285.

SMIDDY, M.A., KERRY, J.P. and PAPKOVSKY, D.B. (2001). Evaluation of oxygen content in commercial modified atmosphere packs (MAP) of processed cooked meats, taken at the point of manufacture. *Proceedings of the 47th Int. Cong. of Meat Sci. and Tech., Krakow, Poland*, p. 78.

SMITH, J.P., OORAIKUL, B., KOERSON, W.J. and JACKSON, E.D. (1986). Novel approach to oxygen control in modified atmosphere packaging of bakery products. *Food Micro.*, **3**, 315.

SORHEIM, O. and NISSEN, H. (2000). Current technology for MAP Meat. *Food Marketing and Technology*, **August**, 39.

SORHEIM, O., AUNE, T. and NESBAKKEN, T. (1997). Technological hygienic and toxicological aspects of CO used in MA packaging of meat. *Trends in Food Sci. and Tech.*, **8**, 307.

SORHEIM, O., LEA, P., NISSEN, H. and NESBAKKEN, T. (2001). Effects of a high CO_2/ low CO atmosphere on colour and yield of cooked ground beef patties.

Proceedings of the 47th Int. Cong. of Meat Sci. and Tech., Krakow, Poland.
SORHEIM, O., GRINI, J.A., NISSEN, H., ANDERSEN, H.J. and LEA, P. (1995). *Fleischwirtschaft*, **75**, 679.
SORHEIM, O., KROPF, D.H., HUNT, M.C. KARWOSKI, M.T. and WARREN, K.E. (1996). Effects of modified gas atmosphere packaging on pork loin colour, display life and drip loss. *Meat Sci.*, **43**, 203.
SPANIER, A.M., VERCELLOTTI, J.R. and JAMES, C., JR. (1992). Correlation of sensory, instrumental and chemical attributes of beef as influenced by meat structure and oxygen exclusion. *J. Food Sci.*, **57**, 10.
TAYLOR, S.A. (1996). Modified atmosphere packing of meat. In: *Meat Quality and Meat Packaging.* (Taylor, S.A., Raimundo, A., Severini, M. and Smulders, F.J.M. eds) Utrecht: ECCEAMST.III, 301.
TAYLOR, A.A., DOWN, N.F. and SHAW, B.G. (1990). A comparison of modified atmosphere and vacuum skin packaging for the storage of red meats. *Int. J. Fd. Sci. and Tech.*, **25**, 98.
TOMIOKA, Y. (1990). Fresh meat. In: *Food Packaging.* (Kadoya, T. ed) Academic Press, London, pp. 309.
VERMEIREN, L., DEVLIEGHERE, F., VAN BEEST, M., DE KRUIJF, N. and DEBEVERE, J. (1999). Developments in the active packaging of foods. *Trends in Food Sci. and Tech.*, **10**, 77.
WARRISS, P.D. (1996). Instrumental measurement of colour. In *Meat Quality and Meat Packaging.* (Taylor, S.A., Raimundo, A., Severini, M. and Smulders, F.J.M. eds) Utrecht: ECCEAMST.III, 221.
WINSTANLEY, M.A. (1979). The colour of meat. *Nutrition and Food Sci.*, **November**, 5.
WOLFE, S.K. (1980). Use of CO and CO_2 enriched atmospheres for meats, fish and produce. *Food Technol*, **March**, 55.
YEN, J.R., BROWN, R.B., DICK, R.L. and ACTON, J.C. (1988). Oxygen transmission rate of packaging films and light exposure effects on the colour stability of vacuum-packaged dry salami. *J. Food Sci.*, **53**, 1043.
YOUNG, L.L., REVIERE, R.D. and COLE, A.B. (1988). Fresh red meats: a place to apply modified atmospheres. *Food Technol.*, **49**, 65.
ZAGORY, D. (1997). Modified atmosphere packaging. In: *The Wiley encyclopedia of packaging technology.* (Brody, A.L. and Marsh, K.S. eds) John Wiley and Sons Inc., New York, pp. 650.
ZEUTHEN, P. and MEAD, G.C. (1996). Microbial spoilage of packaged meat and poultry. In *Meat Quality and Meat Packaging.* (Taylor, S.A., Raimundo, A., Severini, M. and Smulders, F.J.M. eds) Utrecht: ECCEAMST.III, 273.
ZHAO, Y., WELLS, J.H. and MCMILLIN, K.W. (1994). Applications of dynamic modified atmosphere packaging systems for fresh red meats: Review. *J. Muscle Foods*, **5**, 299.

Index

'A'-'not A' test 181–3
accelerated production 381
acetic acid 265, 266
acid taste 372
acidulation 366–7, 380–1
  see also pH
actin 34, 35, 372
active packaging systems 437–9
actomyosin 35
adenosine triphosphate (ATP) 198
adhesion 343, 344, 346–7
adjusted temperature coefficient 209, 210
adolescents 84, 85
adult equivalent (AE) 243–4
AFLP 233
aldehydes 114–15, 374, 376–7
  branched 374, 375
alginate 351–2
α-linolenic acid 73, 74
alternative forced choice (3-AFC) tests 181
amines
  biogenic 90–1
  heterocyclic 67
amino acids 78–9, 374, 375
amplification 232
anaemia 82–3
analytical methods 394–416
  AF spectroscopy 204–6, 401
  colour 201–3, 363–4, 398–9
  current techniques 397–9

electrical impedance 195–9, 207–8, 398
emerging technologies 399–405
future trends 407–8
genetics 405–7
image analysis 194, 401
immunoassays 232, 403–4
laboratory based methods 395–7
metabolites 404
microbiological hazards 231–3
need for objective methods 395
NIR 201, 204, 401–2
NMR 400
pH 199–201, 397–8
temperature 402
tenderness probe 404–5
ultrasound 399–400
androstenone 181, 207
animal subsystem 241, 242–4, 246, 247–9
animals
  campylobacteriosis 225
  E. coli infections 221–2
  salmonellosis 219–20
  welfare standards 20–1
  see also cattle; pigs; poultry; sheep
anisotropy 196
antibiotics 233
antimicrobial films 438
appearance 363–5
  see also colour
arachidonic acid 69

Index   453

Arcobacters 224, 226
argon 432
aroma
　aroma compounds *see* volatile flavour/
　　aroma compounds
　fermented meat products 361, 368–71,
　　372–7
asymmetric information 8–9
ATPase 318
attributes, quality *see* quality attributes
autofluorescence (AF) spectroscopy
　204–6, 401
automated slaughter line layout 294, 295
automated washing systems 261, 263
automatic evisceration system 288–90
automation 2, 283–96
　current developments in robotics 284–5
　evisceration process 287–90
　future trends 294–6
　and hygiene 294
　and management 294–6
　pig slaughtering 285–7
　secondary processes 290–4
　　boning of the fore–end 291–3
　　boning and trimming of belly 293
　　boning and trimming of loin 293–4
　　carcass cutting 290–1
　　cutting of the middle 291, 292
　　hind leg boning 293
autoxidation, lipid 108–10, 376–7

B vitamins 86–7
Bacillus anthracis 229
Bacillus cereus 229
bacteria *see* microbiological hazards
batter preparation 367–8
beef 90, 433–4
　breed and genetic effects on quality
　　38–44
　CLA 147
　consumption 11–12, 13
　fat content 70, 139–40, 142–3, 143–4,
　　145–6
　fatty acids 75, 76
　high concentrate diets 49–51
　modelling colour stability 124–8, 129,
　　133–4
　robotic equipment 284
　sampling 166–7, 177–8
　sensory comparisons between countries
　　189
　US customer satisfaction study 29–30,
　　31
beef cattle production 239–58

　challenges for modellers 244–51
　elements of 240–4
　future developments 254–5
　simple model of herd structure 251–4
belly: boning and trimming 293
binders 351–2
bioavailability of iron 83
biochips 233, 407
bioelectrical impedance 198–9
biogenic amines 90–1
biohydrogenation 75
biosensors 438–9
boar taint 181, 207
bone health 89–90
boning
　belly 293
　fore-end 291–3
　hind leg 293
　loin 293–4
Bos indicus cattle 39–43
Bos taurus cattle 39, 40–3
bovine mastitis 222
bovine spongiform encephalopathy (BSE)
　11, 230–1, 333
branched aldehydes 374, 375
breed
　fatness 141–2
　and genetic effects on meat quality
　　37–49
brittle behaviour 342
brucellae 229
bulk density theory 29
bulk packaging 435–7

Caliciviridae 231
calpain proteolytic system 34, 164, 318,
　403, 406
calpastatin 34, 164, 406
Campylobacter jejuni 224–6
cancer 64, 65, 66–9
candidate genes 406
canning 427–8
capacitance 196
CAPER system 263
carbohydrates 81, 315
　aroma compounds from carbohydrate
　　catabolism 374–5
carbon dioxide 437
　absorption 124
　MAP 430–1
carbon dioxide stunning 54–5, 286
carbon monoxide 431–2
carcass
　compulsory classification 21

carcass (*continued*)
  cutting 290–1
  decontamination 2, 259–82
    problems caused by irregular surfaces 261
    measurements of connective tissue 204–5
    weight and fatness 141–2
carnitine 91
carnosine 90
carrying capacity, farm 251–4
category scales 185–6
cathepsins 164
cattle
  breed and genetic effects on beef quality 38–44
  effect of high concentrate diets on beef quality 49–51
  pathogenic diseases 219, 220, 221–2, 225
  see also beef; beef cattle production
centralised meat packaging 417–19
characteristics, quality 8, 17, 18–19
chemical decontamination 263–7, 273, 274
chi-squared test 183
chicken 269
  see also poultry
children 80–1, 84
chilling 2, 297–312
  impact on colour 300–2
  impact on drip loss 302–3
  impact on evaporative weight loss 303–4
  impact on texture 299–300
  primary 304
  secondary 304
  temperature monitoring 306–8
chlorine 263, 264, 265, 274
chlorine dioxide 264, 265, 274
cholesterol 69, 77–8
choline 91
chromaticity coordinates 209–10
CIE colour system 209–10, 363–4
climate 240–1, 246, 246–7, 249–50
Clostridium botulinum 228–9
Clostridium perfringens 228
coagulation 365–6
cobalt 110
cold chain 304–6
cold room tempering 335
cold-shortening (cold-induced toughening) 34, 299
cold water washing 262
collagen 37, 162–3, 204–6

colorectal cancer (CRC) 65, 66–9
colour 35–7, 397, 419–21
  fermented meat products 361, 363–5
    colour development 364–5
  high pressure processing 319–20
  impact of chilling and freezing 300–2
  instrumental measurement 201–3, 363–4, 398–9
  on-line monitoring 201–3
    colour changes during cooking 209–10
  quality indicator 159, 165
  sensory measurement 363
  spectrophotometry 194, 202–3
colour stability 122–36
  external factors affecting 123–31
    cured ham 128–31
    fresh beef 124–8
  modelling dynamic changes in headspace composition 123–4
  future trends 134–5
  internal factors affecting 131–3
  meat packaging and 426, 431, 432, 434, 435
  validation of models 133–4
Comitrol processor 336
comminution 336
Committee on Medical Aspects of Food and Nutrition (COMA) 65, 66, 138
communication 17, 18–19
comparison tests 179–81, 362
compartmentalised packaging odours 421
competition 14
complexity 249
composite polymer films 440–1
composition
  analysis of 397, 401–2
  quality, structure and 27–37
computer modelling *see* modelling
concentration 370
concussive stunning 54–5
conditioning 299–300
conductivity, electrical 398
conjugated linoleic acid (CLA) 75–7, 89, 137, 146–7, 148–9
connective tissue 37, 162–3, 317, 401
  on-line quality monitoring 204–6
consistency 56
consumers 4, 5
  and fat 138
  perceptions of quality 6–14, 17–18
  testing restructured meat 350–1
consumption, meat 11–12, 13, 65, 147
  demand 91–2

energy intake 138
   recommendations and cancer avoidance 66–7
contamination 260
   see also decontamination
controlled atmosphere packaging 437–9
convenience 12, 418
cooked cured-meat pigment (CCMP) 114, 115, 116
cooked meats 435, 436
cooking 259
   decontamination processes and 267, 268, 271
   measuring changes during 208–11
   method and cancer risk 68
   restructured meat
      distortion 347–9, 350
      losses 347, 348
copper 110
coronary heart disease (CHD) 65, 138
Corynebacterium pseudotuberculosis 229
countries, comparisons between 189
credence quality attributes (CQA) 9–11, 16, 18, 21, 22
Creutzfeldt-Jakob disease, new variant (vCJD) 230–1
Cryovac Foam Tray packaging system 439
Cryptosporidium parvum 229–30
cues 7, 17–19
   consumers and cue processing 9–11
culling rates 252, 253, 254
cured meats
   colour stability 134
      external factors 128–31
      internal factors 132–3
   flavour quality 113–16
cutting
   carcass 290–1
   middle 291, 292
Cyclospora spp. 230

D vitamins 87–8
dark, firm and dry (DFD) meat 53
Debaryomyces hansenii 379–80
decision frames 8–9
decision support systems 245, 247, 248, 249, 254
decontamination 2, 259–82
   current techniques and their limitations 260–2
   difference between methods and treatments 261–2
   future trends 273–6
   novel methods apart from steam 272–3
   problems 261
   steam 262, 267–72, 273–4
   use of chemicals 263–7, 273, 274
   washing 262–3
degradation enzymes 34–5
'Deluge' system 263
demand 91–2
   see also consumption
deoxymyoglobin 419–20
descriptive tests 362–3
design, refrigeration 308–10
DFD meat 53
diet
   animal
      effects on fat content and composition of meat 144–7
      influences on raw meat quality 49–52
   human
      vegan diet 74, 80
      vegetarian diet 74, 79–81
difference tests 178, 362
diffusion 268
direct extraction 370
directional preference tests 179
disease
   cancer 64, 65, 66–9
   caused by microbiological hazards 217–31 *passim*, 259, 297
   CHD 65, 138
   fat content and 69–71
   vCJD 230–1
disinfection of equipment 294
disodium ethylenediaminentetraacetic acid 111
display 305
distillation 370
distortion, cooking 347–9, 350
'DNA chip' technology 233, 407
DNA markers 405
docosahexaenoic acid 73, 74, 141
domestic transport and storage 305–6
doming 348, 349
drip loss 167, 298, 398, 419
   impact of chilling and freezing 302–3
   packaging and 424, 428
drying 366–7, 368, 381
duo-trio test 181, 182

E vitamins 51, 382, 422
eating quality 6, 157
   determining 166
   experience quality attributes (EQA) 8, 9–11, 16, 21–2

eating quality (*continued*)
  quality indicators 160–5
  standards for 21
Echinococcus granulosus 230
economic subsystem 241, 244, 246, 248–9
eicosapentaenoic acid 73, 74, 141
elastin 204–6
electrical conductivity 398
electrical impedance 195–6, 207–8, 398
electrical stimulation 55–6
electrical stunning 54–5
electrode penetration depth 196–7
electromagnetic scanning 198
electronic nose 371
empirical models 244–5
emulsifying properties 320
emulsions 207–8
energy intake 138
engineering specification 309–10
Envirobot 285
enzymes 34–5
  eating quality and enzymatic activity 164
  effect of high pressure 315, 318
  lysosomal 317–18, 318
  proteolytic 34, 164, 318, 403, 406
Escherichia coli (E. coli) 221–4, 231, 434
  disease in animals 221–2
  disease in man 222–4
  steam decontamination 270–1
ethics 12–14, 20–1
ethyl vinyl acetate copolymer (EVA) 442
ethyl vinyl alcohol (EVA) 440, 442
European Union (EU)
  decontamination 274, 276
  packaging 429
  standards 19–21
evaporative weight loss 298, 303–4
evisceration 287–90
expenditure 65
experience quality attributes (EQA) 8, 9–11, 16, 21–2
external sensory panels 176

F–line robot series 287
farm carrying capacity 251–4
fat
  component and quality of raw meat 27, 28–32, 33
  intramuscular fat 28–32, 33, 140–1, 159–60, 161–2, 166–7, 206
  lipid hypothesis 65, 148
  sources in the diet 71, 72

technological quality 159–60
fat content 89, 137–53, 401
  animal effects on fat content and composition 141–4
  CLA 146–7
  dietary effects 144–7
  and disease 69–71
  fat and the consumer 138
  fatness 138–41, 141–3, 144, 148
  fatty acids *see* fatty acids
  future trends 147–9
  on-line monitoring 206
fatness 138–41, 141–3, 144, 148
fatty acid binding protein (FABP) gene 406
fatty acids 71–8, 89, 397
  cholesterol 69, 77–8
  CLA 75–7, 89, 137, 146–7, 148–9
  dietary effects in pork 51–2
  eating quality 161–2
  fat content 139, 141, 143–4, 144–6, 148
  flavour quality 110
  free fatty acids in fermented meat products 374, 376
  monounsaturated 73, 139, 141, 143, 148
  polyunsaturated 73–5, 89, 139, 141, 143, 144–6, 148
  saturated 72–3, 139, 141, 143, 148
  technological quality 160
  trans fatty acids 77
FEEDMAN 245, 247, 248, 249
fermented meat products 2, 359–93
  appearance and colour 361, 363–5
  control and improvement of quality 377–81
  flavour 368–72
  as functional foods 382
  future trends in quality development 381–2
  products 359–60
  quality 360–1
  sensory quality and its measurement 361–3
  taste and aroma 372–7
  texture 365–8
fibre-optic probe 194, 398–9
fibrinogen 352
fixed-choice profiling 186
flaking 336, 341–2
flame ionisation detection (FID) 371
flavour 105–21, 166, 361
  dietary impacts 50–1
  effect of ingredients on 110–16

Index 457

evaluation of aroma compounds and flavour quality 116–17
fat and 30–2, 33, 162
fermented meat products 368–72
lipid oxidation and meat flavour deterioration 108–10
on-line testing 207
role of lipids 106–8
see also aroma; taste; texture; volatile flavour/aroma compounds
flexible MAP packs 427, 443
flexible vacuum packs 443
flow (gases) 268
flow cytometry 232
fluorescence 204–6, 401
food-borne illnesses 217–31 *passim*, 259, 297
food hygiene *see* hygiene
food preparation 12
food supply chain 3–5, 17–18
forage 49, 241, 242
fore-ends, boning 291–3
forming 337
fracture behaviour 342–3
free-choice profiling 186
freezer burn 301, 304
freezing 197, 297–312
  impact on colour 300–2
  impact on drip loss 302–3
  impact on evaporative weight loss 303–4
  impact on texture 299–300
  pressure assisted 320–1
  primary 304
  restructured meat 337–8
  secondary 304
  temperature monitoring 306–8
fresh grillsteak-type products 351–2
Fresh-R-Pax moisture absorbing technology 439
Friedman value 184
functional foods 79–82, 148–9
  components in meat 90–1
  fermented meat products as 382
functional genomics 405, 407
fungi/moulds 378, 380, 423

gas chromatography (GC) 370–1
gas chromatography–mass spectrometry (GC–MS) 166
gas chromatography olfactometry (GCO) 166, 371
gases
  absorption in meat 124
  modelling changes in headspace gas composition 123–31
  permeabilities of packaging films 440, 441
  used in MAP 429–32
  see also under individual gases
gelatinisation 210–11
gelation 320
GEMANOVA model 125–8
gene expression 407
Generalised Procrustes Analysis (GPA) 187
genetic markers 43–4
genetics 37–49, 405–7
genotype 141–2, 143–4
Germ Plasm Evaluation (GPE) program 38–9
Giardia duodenalis 229
Giardia lamblia 229
glucono-delta-lactone (gdl) 352
glutathione 83, 90, 91
glycogen 315
glycolysis 318
grain-based diets 50–1
gram-negative bacteria 322, 422
gram-positive bacteria 322, 422
grapefruit seed extract 438
grass-fed beef 145–6
GRAZE 245
grillsteaks 332, 333–8
  fresh 351–2
  see also restructured meat

HACCP 19–20, 217
haem iron 83, 84
haem proteins 111, 114
haemoglobin 37, 55, 202
Halothane gene 45–7
ham 128–31
hardness 365, 366–8
headspace collection 370
headspace gas composition 123–31
health 149
  disease *see* disease
  functional foods *see* functional foods
  nutrition *see* nutrition
heat decontamination treatments 262
heat-insoluble collagen bonds 37
hedonic scales 396
  see also sensory analysis
herd structure model 251–4
heterocyclic amines 67
heterogeneous assays 232
hexanal 116–17

high concentrate diets 49–51
high pressure liquid chromatography (HPLC) 370–1
  reversed phase HPLC 371
high pressure processing 272, 273, 313–31
  current applications and future prospects 323–4
  effect on food components 314–15
  effects on microflora 321–3
  effects on sensory and functional properties 318–20
  enzyme release and activity 318
  and meat quality 313–14
  pressure assisted freezing and thawing 320–1
  structural changes 315–18
hind leg, boning 293
homocysteine 87
homogeneous assays 232
hot water washing 263
hunter-gatherer societies 81–2
hydroperoxide-dependent lipid peroxidation 110
hydroperoxides 421
hygiene 260
  automation and 294
  standards 19–20

ice
  content in restructured meat 339–40
  crystal formation 300, 301
image analysis 194, 401
immersion 264
immunoassays 232, 403–4
impedance, electrical 195–9, 207–8, 398
impedimetry 232
indole 207
infra-red thermometry 307–8
inoculation microbiology 275–6
inspection quality attributes (IQA) 8, 16
instrumental analysis
  colour 363–4, 397, 398–9
  comparison of sensory analysis with 186–8
  eating quality 166
  emerging technologies 399–405
  fermented meat products 363–4, 369–71
  flavour volatiles 369–71, 396
  impedance 195–9, 207–8, 398
  on-line monitoring 1–2, 193–216
  pH 199–201, 397–8
insurance theory 29
interactive prototyping 249

interface problem 246–9
intermediate value products *see* restructured meat
intramuscular fat (IMF) (marbling) 28–32, 33, 140–1, 159–60, 161–2, 166–7, 206
iron 82–4, 90, 438
  prooxidant effect 110–11
iron deficiency 82–3
isoelectric point 35

Japan: pork customer satisfaction study 30, 33
juiciness 28–9, 162

ketones 374, 376–7
KUKA robot 285

laboratory-based quality methods 395–7
  see also analytical methods
lactic acid 53, 374
  decontamination with 265, 266
lactic acid bacteria 378, 423
lamb 90, 433–4
  automated slaughtering 384
  dietary influences 51
  fat content 70, 139, 140
  fatty acids 75, 76, 139
  Spanish and UK sensory analysis 189
land 241–2, 246, 246–7
laser diffractometer 203–4
laser scanning 200–1
lateral shrinkage 348, 349
lean tissue *see* muscles
leg temperature measurement 308
light scattering 200, 208
linear low density polyethylene (LLDPE) 442
linoleic acid 69, 73
lipid hypothesis 65, 148
lipids 105–21
  aroma compounds from lipid degration 105–10, 374, 376–7
  colour and lipid oxidation 319–20
  effect of ingredients on flavour 110–16
  evaluation of aroma compounds and flavour quality 116–17
  high pressure processing and 315
  oxidation and quality deterioration in packaging 421–2
lipoic acid 90
lipolysis 376
Listeria monocytogenes 227–8
loin, boning and trimming 293–4

Index   459

long-term pre-slaughter stress 52–3
longissimus dorsi (LD) muscle
  colour stability 131, 132
  fat content 142–3, 161
  sampling 166–7, 167–8
low density polyethylene (LDPE) 442
low-temperature, long-time cooking
    technique 323
lubrication effect 29
lysosomal enzymes 317–18, 318

Maillard reaction 106–7, 108, 165
major genes 406
malignant hypothermia (MH) 45–7, 406
malonaldehyde 421
management, automation and 294–6
management models 245, 254
marbling (intramuscular fat) 28–32, 33,
    140–1, 159–60, 161–2, 166–7, 206
market subsystem 241, 244, 246, 248–9
mass spectrometry (MS) 371
  GC-MS 166
  MS-tester 197
master packaging systems 418, 435–7
mastitis, bovine 222
'meat factor' 83
meat flavour deterioration (MFD)
    108–10, 115–16
meat quality schemes 15–16
mechanical properties 342–3
mechanistic models 244–5
Mediterranean (Southern) type fermented
    products 360, 361, 364, 372, 377,
    380, 381
  see also fermented meat products
metabolites 404
metal ions 110–11
methyl ketones 377
metmyoglobin 165, 201, 320, 419–20
microarray technology 233, 407
microbiological hazards 2, 217–36, 259–60
  analytical methods 231–3
  Campylobacter jejuni 224–6
  Clostridium botulinum 228–9
  Clostridium perfringens 228
  E. coli 221–4, 231, 270–1, 434
  effects of high pressure processing
      321–3
  future trends 233–4
  Listeria monocytogenes 227–8
  main hazards 218–31
  other agents 230–1
  other bacteria 229
  packaging and 422–3, 425, 430, 434–5
  parasites 229–30
  Salmonellae 218–21, 270, 297, 434
  Staphylococcus aureus 227
  temperature and microbiological
      growth 297–8
  Yersinia enterocolitica 226, 270
  see also decontamination
Micrococcaceae 375, 376, 378, 378–9
micronutrients 82–8
  see also vitamins
microwave decontamination 272–3, 274
microwave tempering 335
middle, cutting of the 291, 292
minimum growth temperature 298
MIRINZ tenderness probe 405
mixing 336–7
modelling
  beef cattle production 2, 239–58
  colour stability 122–36
  coping with natural variability 249–50
  interface problem 246–9
  matching purpose and structure 244–6
  validation of models 133–4, 251
  verification of models 250
modified atmosphere packaging (MAP)
    421, 423, 428–35, 437
  advantages and disadvantages 429
  effectiveness 433–5
  equipment 432–3
  flexible and rigid packs 427, 443
  gas compositions 436
  gases 429–32
  modelling and colour stability 122–36
moisture permeability 440, 441
molecular typing 233
monounsaturated fatty acids (MUFA) 73,
    139, 141, 143, 148
mother packaging systems 418, 435–7
moulds 378, 380, 423
MS-tester 197
multi-dimensional scenario 249
muscles
  classification 401
  colour stability and muscle types 131,
      132
  growth 142–3
  muscle fibre component 32–7
  structural changes due to high pressure
      treatment 315–18
Mycobacterium paratuberculosis 229
myofibres 195
myofibrillar proteins 35, 200
myoglobin 35–7, 131, 159, 165, 201–2,
    419–20

myosin 34, 35, 187–8, 344–5, 346, 365–6, 372
myostatin 406
myristic acid 73

Napole gene 47–9
National Cholesterol Education program (NCEP) 77
Natural Colour System (NCS) 363
natural variability 249–50
near infrared (NIR)
  polarised 204
  spectrophotometry 201
  spectroscopy 401–2
niacin 86
nitric oxide 364, 404
nitrite-free curing systems 116
nitrites 364, 422
  and colour stability 132–3
  effects on flavour quality 113–16
nitrogen 431
nitrosomyoglobin 364
nitrosylhaemochrome 201
nitrosylmyoglobin 420–1
non-directional preference tests 179
non-haem iron 83, 84
Northern type fermented products 359, 360–1, 364, 366, 377, 380–1
  see also fermented meat products
Norwalk-like viruses (NLVs) 231
nuclear magnetic resonance (NMR) 400
nucleic acid based assays 232–3
nucleotides 91, 404
nutrition 1, 64–104, 149
  fatty acids in meat 71–8
  future trends 88–92
  meat and cancer 64, 65, 66–9
  meat, fat content and disease 69–71
  meat as a functional food 79–82
  micronutrients 82–8
  protein in meat 78–9

obesity 82, 89
objective approach to quality 5–6, 7–8
objective analytical methods 395, 396–7
off-flavours 109, 421
olfactometry 371
omnivores 81
on-line monitoring 193–216
  apparatus 193–4
  boar taint 207
  changes during cooking 208–11
  connective tissue 204–6
  electrical impedance 195–9

emulsions 207–8
marbling and fat content 206
meat colour and other properties 201–3
meat flavour 207
NIR spectrophotometry 201
pH 199–201
sarcomere length 203–4
water-holding capacity 203
optical fibres 194, 398–9
organic acids 264, 265–6, 269–70, 274
oscillating magnetic field (OMF) pulses 272, 273
osteomalacia 87, 88
oxygen 419–21
  MAP 429–30, 433, 435, 436
  modelling colour stability in meat 127, 128, 129, 130–1
oxygen scavengers 438
oxymyoglobin 165, 201, 419–20, 429
ozone 264, 267, 274

pack collapse 430–1
packaging 2, 417–49
  bulk, master or mother packaging 435–7
  centralised 417–19
  controlled atmosphere packaging and active packaging systems 437–9
  factors influencing quality of fresh and processed meat products 419–24
  future trends 442–3
  MAP see modified atmosphere packaging
  materials 439–42, 443
  permeability of packaging film 123–4
  vacuum packaging 423, 424, 424–8
paired comparison tests 179
palatability 28–9, 49–50
pale, soft and exudative (PSE) meat 45–7, 53–4, 197, 199, 397
paleolithic diets 81–2
palmitic acid 73
pantothenic acid 86
parasites 229–30
particle size 340–2
Pasteurella spp. 229
pasteurisation 323
pasture subsystem 246, 246–8
pathogens see microbiological hazards
Penicillia 380
peptides 370–1, 372, 374, 375
perceived quality 5–6, 7
permeability of packaging film 123–4
pH 53

acidulation and fermented meat
  products 366–7, 380–1
analytical techniques 199–201, 397–8
breed type and 45, 46
eating quality 164–5
isoelectric point 35
technological quality 158–9
phosphate hydrolysis 345–6
phospholipids (PL) 106–7, 110, 421
phosphorus 85
pigs
  pathogenic diseases 219, 221–2
  slaughtering and automation 284–96
  see also pork
plastic films 439–42, 443
polarisation 196
policy models 245–6, 254–5
polyamides (PA) 442
polyesters (PET) 442
polymerase chain reaction 232
polymers 439–42, 443
polyphosphates 264, 266
polypropylene (PP) 442
polyunsaturated fatty acids (PUFA) 73–5,
  89, 139, 141, 143, 144–6, 148
polyvinylidene chloride (PVDC) 440, 442
pork 167, 177, 434
  breed and genetic effects on quality
    44–9
    Halothane gene effects 45–7
    Napole gene effect 47–9
  customer satisfaction studies 30, 32, 33
  dietary effects 51–2, 144, 144–5
  fat content 70–1, 139, 140, 144
  fatty acids 75, 76, 139, 143, 144–5
  PSE 45–7, 199, 397
Pork Stress Syndrome (PSS) 45–7, 406
potential liveweight gain 248
pouches 426, 441
poultry 269
  automated slaughtering 284
  packaging 434
  pathogenic diseases 219–20, 221, 222,
    225
pre-breaking 335, 340, 341
price premium 3
process specification 309–10
processing industries 4–5
production 17, 18–19
propanal 117
proteasome 318
proteins 67
  degradation and aroma compounds 374,
    375

effect of high pressure 315
haem proteins 111, 114
historical development of diet 81–2
myofibrillar 35, 200
nutritional quality 78–9
sarcoplasmic 347
solubility 343–7
proteolysis 56, 366, 372
proteolytic enzymes 34, 164, 318, 403,
  406
proteome analysis 405
PSE meat 45–7, 53–4, 197, 199, 397
Pseudomonads 423, 434–5
pseudofluorescence 205–6
pulsed electric fields (PEF) 272, 273
pulsed field gel electrophoresis (PFGE)
  233
pulsed light 272, 274
pyrazines 375
pyrophosphate 346

quality 1, 3–24
  circle 16–19
  consumer perceptions 6–14, 17–18
  defining meat quality 394–7
  eating see eating quality
  improving meat and meat product
    quality 21–2
  objective and subjective views 5–6, 6–
    8
  raw meat see raw meat quality
  regulatory definitions 19–21
  sensory quality 360–1, 361–3
  supplier perceptions 14–16
  supply chain and 3–5
  technological quality 157–60
quality attributes (QA) 7, 8–12, 17,
  18–19, 21–2
  credence quality attributes 9–11, 16,
    18, 21, 22
  experience quality attributes 8, 9–11,
    16, 21–2
  inspection quality attributes 8, 16
quality characteristics (QC) 8, 17, 18–19
quality circle 16–19
quality cues see cues
quality function deployment 14–15
quality guidance approach 15
quality indicators 157–74
  determining eating quality 166
  eating quality 157, 160–5
  future trends 168
  sampling procedure 166–8
  technological quality 157–60

quality schemes 15–16
quantitative trait loci (QTL) analysis
     405–6

random amplified polymorphic DNA
     (RAPD) 233
ranking tests 183–4, 362
rating scales 185
raw materials
   fermented meat products 377
   restructured meat 333–4
raw meat quality 27–63
   breed and genetic effects 37–49
   dietary influences 49–52
   ensuring consistency 56
   future trends 57
   quality, meat composition and structure
      27–37
   rearing and 52
   slaughtering and 52–6
   storage 56
rearing environment 52
red, soft and exudative (RSE) meat 47–9,
     53–4
reduced-fat products 140
reflectance spectra 194, 209
reflected radiation 308
refrigeration 2, 297–312
   cold chain 304–6
   drip loss 302–3
   evaporative weight loss 303–4
   impact on colour 300–2
   impact on texture 299–300
   optimising design and operation
      308–10
   temperature monitoring 306–8
regulatory definitions of quality 19–21
Rendement Napole (RN) gene 47–9
research models 245
resistance 196
response surface plot 125, 126
restriction fragment length polymorphism
     (RFLP) 233
restructured meat 2, 332–58
   factors affecting product quality
      338–49
      cooking distortion 347–9, 350
      ice content 339–40
      mechanical properties 342–3
      particle size 340–2
      protein solubility 343–7
      temperature 340
   future trends 351–2
   product manufacture 333–8

flaking 336
forming 337
freezing 337–8
mixing 336–7
pre-breaking 335
raw materials 333–4
tempering 334–5
sensory and consumer testing 349–51
retail display 305
retailers 4, 5
reversed phase HPLC 371
riboflavin 86
rickets 87, 88
rigid MAP packs 427, 443
rigid vacuum packs 443
RNA amplification 232
robotics 284–5, 287
   see also automation
rotation/reflection 187
RSE meat 47–9, 53–4

safety 19–20
   see also decontamination;
      microbiological hazards
Salmonellae 218–21, 270, 297, 434
   salmonellosis in animals 219–20
   salmonellosis in man 220–1
salt
   effect on flavour quality 111–13
   restructured meat 339–40, 345, 346,
      347, 351
   effective salt concentration 344–5
samples
   sampling procedure 166–8
   using in sensory analysis training
      177–8
sarcomere length 34, 163–4, 203–4
sarcoplasmic proteins 347
satiety 82, 89
saturated fatty acids (SFA) 72–3, 139,
     141, 143, 148
sausage
   composition and size 366–7
   see also fermented meat products
scaling 187, 362
scoring 185
scrapie 230, 231
screening 176–7
SDS-PAGE 371
'second generation' restructured meat
     products 352
selenium 85, 90
sensitivity analysis 250
sensory analysis 1, 166, 175–92, 395–6

category scales 185–6
comparisons between countries 189
fermented meat products 361–2, 362–3
  aroma, taste and flavour 369
  colour 363
  restructured meat 349–50
  sensory panel *see* sensory panel
  sensory profile methods and comparisons with instrumental measurements 186–8
  sensory tests 178–85
sensory panel 176–8, 361–2, 362–3
  general conditions for assessment of samples 178
  screening criteria and training 176–7
  using samples in training 177–8
sensory quality 360–1
  measurement 361–3
separation 370–1
SFK-Danfotech robot series 287
shear force analysis 166
  Warner-Bratzler shear force 39, 40–3, 44, 396
sheep
  pathogenic disease 219, 221–2
  see also lamb
short-term preslaughter stress 53–4
shrink bags 426, 441
shrinkage during cooking 348, 349
signals 17, 18–19
SIRA 'Food Sentinel' system 439
skatole 181, 207
skin packs 424, 426, 427, 443
slaughtering
  automation 284–96
  and meat quality 52–6
snorkel packaging machines 432–3
sodium alginate 351–2
sodium chloride 111–12, 339, 340
  see also salt
sodium halides 112–13
sodium nitrite 364
sodium tripolyphosphate (STPP) 111
soft fat 206
sono-elastography 399–400
sous vide 427
spectrophotometry 194, 202–3
  NIR 201
spray decontamination systems 261–2, 263, 264
standards 19–21
Staphylococcus 375, 376, 378–9
  aureus 227
starter cultures 366–7, 378–80

steam decontamination 262, 267–72, 273–4
Steam Pasteurisation System (SPS) 271–2
stearic acid 73, 141
sterilisation of equipment 294
storage 56
  refrigeration 305
    domestic 305–6
strain theory 29
Strecker degradation 375
stress 52–4
  long-term 52–3
  short-term 53–4
structure 363
  changes due to high pressure treatment 315–18
  quality, meat composition and 27–37
  ultrasound analysis 399–400
stunning method 54–5, 286
subjective approach to quality 5–6, 6–7
sulphur dioxide 432
suppliers
  perceptions of quality 14–16
  problems of variability of meat quality 395
supply chain 3–5, 17–18
surfactants 261
survival rates 252

taste 368–71, 372–7
  see also flavour
taurine 79
TBARS test 116
technological quality 157–60
temperature 297–9, 402, 434
  adjusted temperature coefficient 209, 210
  measuring changes during cooking 208
  and microbiological growth 297–8
  monitoring 306–8
  restructured meat 339–40, 345, 346
  see also refrigeration
tempering 334–5
tenderness
  analytical methods 396, 397, 398, 400, 402, 403
  beef and breed effect 39–43
  chilling, freezing and 299–300
  connective tissue and 162–3
  degradation of muscle proteins 56
  effect of high pressure 318–19
  intramuscular fat and 29, 30, 161–2
  muscle fibre and 32–5
tenderness probe 404–5

464 Index

texture 166
  effects of high pressure 318–19
  fermented meat products 365–8
  impact of chilling and freezing
    299–300
  restructured meat and engineered
    texture 352
thaw-shortening 299
thawing 320–1
thermal contact 307
thermocouples 208
thermoforming 426, 432, 442
thermometers 208
thiamin 86
thickness 349, 350
thrombin 352
'Toxin guard' system 439
Toxoplasma gondii 230
training sensory panels 176–8
trans fatty acids 77
transglutaminase 352
transition metal ions 110–11
translation 187
transmissible spongiform
  encephalopathies (TSEs) 230–1
transport 305, 305–6
trays, packaging 426, 432, 433, 442
triacylglycerols (TAG) 106, 107, 110
triangular tests 180
Trichinella spiralis 230
trimming
  belly 293
  loin 293–4
trisodium phosphate (TSP) 263, 264, 266,
  274
Trochanter Major 308
troponin T molecule 403
trust 9, 10
'Tween 80' 261
two from five test 184–5

'Ulixes Sortierer' system 285
ultrasound 194, 272, 273, 274, 399–400
ultraviolet light (UV) 272, 274
uneven lateral shrinkage 348, 349
uniformity 168
United States (US) 274–5
  meat satisfaction studies 29–30, 31, 32

vacuum packaging 423, 424, 424–8
vacuum skin packaging 424, 426, 427,
  443

validation of models 133–4, 251
values 12–14, 20–1
variability, natural 249–50
variety 168
vegan diet 74, 80
vegetarian diet 74, 79–81
verification of models 250
verotoxigenic *E. coli* (VTEC) 221, 222,
  223–4
video image analysis (VIA) 194
viruses 231
vitamins
  B 86–7
  D 87–8
  E 51, 382, 422
volatile flavour/aroma compounds
    105–10, 373–7
  from carbohydrate catabolism 374–5
  collection 370
  concentration 370
  detection 371
  evaluation 116–17
  instrumental measurement 369–71, 396
  from lipid degradation 105–10, 374,
    376–7
  origin 374
  from protein degradation 374, 375
  separation 370–1
volume ratio 130

warmed-over flavour (WOF) 109, 421
Warner-Bratzler shear force 39, 40–3, 44,
  396
washing 262–3
  automated washing systems 261, 263
water
  effect of high pressure 314, 320–1
  washing with 261–2, 262–3, 274
water-holding capacity (WHC) 35, 158–9,
  164–5, 203, 396
weaning rates 252, 253, 254
weight loss, evaporative 298, 303–4
welling 348, 349
window of acceptibility 28
World Cancer Research Fund (WCRF) 66

yeasts 378, 379–80, 423
yellow fat 206
Yersinia enterocolitica 226, 270

Z lines 34
zinc 84–5